# Guanidino Compounds in Biology & Medicine: 2

John Libbey

JL

LONDON · PARIS · ROME · SYDNEY

# Guanidino Compounds
# in Biology & Medicine: 2

## Proceedings of the 4th
## International Symposium

*Edited by*
*P.P. De Deyn, B. Marescau, I.A. Qureshi and*
*A. Mori*

1997

**British Library Cataloguing in Publication Data**

Guanidino Compounds in Biology & Medicine: 2

I. De Deyn, Peter
574'.04

ISBN: 0 86196 5434

Published by
**John Libbey & Company Ltd,** 13 Smiths Yard, Summerley Street, London SW18 4HR, England
**Telephone: +44 (0) 181-947 2664**
**John Libbey & Company Pty Ltd,** Level 10, 15/17 Young Street, Sydney, NSW 2000, Australia
**John Libbey Eurotext Ltd,** 127 rue de la République 92120 Montrouge, France
**John Libbey - C.I.C. s.r.l.,** via Lazzaro Spallanazani 11, 00161 Rome, Italy

Printed and bound by Biddles Ltd, Guildford, UK

# CONTENTS

# Foreword

This book contains a series of selected papers presented at the 4th International Symposium on Guanidino Compounds in Biology and Medicine that was held in September 1994 in Montreal, Canada under the presidency of Professor Ijaz Aslam Qureshi. The first International Symposium on Guanidino Compounds in Biology and Medicine that was held in 1983 in Tokyo, Japan, coincided with the sixth Annual Guanidino Compound Research Meeting of the Japanese Guanidino Compound Research Association. The second International Symposium on Guanidino Compounds in Biology and Medicine took place in Susono City, Shizuoko, Japan in 1987. These meetings were realized under the inspiring and powerful impulse of Professor Akitane Mori and the respective presidents of the Japanese Guanidino Compound Research Association. The third International Symposium on Guanidino Compounds in Biology and Medicine was organized by Professors Peter Paul De Deyn and Bart Marescau in Antwerp, Belgium in 1991.

This book reflects the world-wide distribution of guanidino compound research. The book offers a comprehensive updated 'state-of-the-art' information on natural guanidino compounds in microorganisms, plants, invertebrates and mammals, including man. It illustrates the multidisciplinary approach to research on these natural substances.

The selection of papers covers biochemical, metabolic, (patho)physiological, clinical, therapeutic, diagnostic, electrophysiological, pharmacological, toxicological and analytical aspects of natural guanidino compounds.

Guanidino compounds and/or their synthetic or catalytic enzymes with their molecular genetics are covered. The diagnostic, therapeutic and prognostic value of guanidino compound studies in a variety of human diseases such as inborn errors of metabolism, epilepsy, metabolic encephalopathy, renal failure and cardiovascular diseases is presented as well.

We believe that this volume will further contribute to our aim of promoting a concerted action among all scientists involved with guanidino compound studies.

Finally, we would like to acknowledge the following institutions and companies that financially and logistically contributed to the realization of the symposium and this volume: Japanese Guanidino Compound Research Association, Department of Paediatrics (University of Montreal), Centre de Recherche Hôpital Sainte-Justine, Born-Bunge Foundation, University of Antwerp, Antwerp OCMW Medical Research Foundation and Schering Canada Inc.

*Prof. Dr Bart Marescau*
University of Antwerp
Laboratory of Neurochemistry and Behaviour
Department of Medicine-UIA
Born-Bunge Foundation
Universiteitsplein 1
2610 Wilrijk-Antwerp, Belgium

*Prof. Dr Peter De Deyn*
University of Antwerp
Laboratory of Neurochemistry and Behaviour
Department of Medicine-UIA
Born-Bunge Foundation
Universiteitsplein 1
2610 Wilrijk-Antwerp, Belgium;
Department of Neurology
General Hospital Middelheim
Lindendreef 1, 2020 Antwerp, Belgium

# Section I
## Physiological importance of arginine – nitric oxide

*Guanidino Compounds : 2*, eds. by P.P. De Deyn, B. Marescau, I.A. Qureshi and A. Mori.
©1997 John Libbey & Company Ltd., pp. 3–7.

# Chapter 1

# Structure–activity relationships of guanidino compounds on brain nitric oxide synthase activity

Isao YOKOI, Hideaki KABUTO, Hitoshi HABU, Kazunori INADA, Junji TOMA,
[1]Hiroshi ASAHARA and Akitane MORI

*Department of Neuroscience, Institute of Molecular and Cellular Medicine, and [1]Department of Orthopaedic Surgery, Okayama University Medical School, 2–5–1 Shikata-Cho, Okayama 700, Japan*

**Summary**

As nitric oxide (NO) is synthesized by nitric oxide synthase (NOS) from L-arginine (Arg), we examined the effect of 23 kinds of Arg analogues on NOS activity in rat brain homogenate. α-Guanidinoglutaric acid (2-GGA) inhibited NOS activity equally to $N^G$-monomethyl-L-arginine (MeArg) which is a known NOS inhibitor, and arcaine also acted as weak NOS inhibitor. Though MeArg inhibits nitric oxide synthase competitively, 2-GGA acts as a linear mixed-type inhibitor. Though almost all nitric oxide synthase inhibitors reported previously were synthesized by substituting one of the guanidino nitrogens of Arg, guanidino nitrogens of arcaine and 2-GGA were not substituted. Furthermore, 2-GGA and arcaine are both natural substances. They may be useful tools in elucidating the chemical nature and the physiologic function of NO and NOS.

## Introduction

In the brain, nitric oxide (NO), which is synthesized from the terminal amidino group of L-arginine (Arg) by nitric oxide synthase (NOS)[8,10], is thought to act both as an intra- and inter-cellular second messenger[5,7]. To understand the role of NO in brain function, it is important to clarify the physical and chemical properties of the active site as well as the mechanism of NOS. To accomplish this, potent NOS inhibitors are needed which have different active sites. Unfortunately, all known potent inhibitors have been synthesized by substituting one of the guanidino nitrogens of Arg, and they act as competitive inhibitors at the Arg binding site of NOS[1,3,4].

In this study, we examined the effect of Arg and 23 of its analogues (Fig. 1) (22 of them had a non-substituted guanidino group), on NOS activity in rat brain in order to investigate structure–activity relationships on the Arg binding site in NOS.

## Materials and methods

The brain of a male Sprague-Dawley rat was homogenized with ice-cold buffer (50 mM Tris-HCl and 2 mM EDTA, pH 7.4). The homogenate was centrifuged for 20 min at $21,000 \times g$, and the supernatant was passed over a 3.0 ml column of Dowex 50W-X8 (Na$^+$ form) to remove endogenous Arg. Incubations were initiated by addition of 60 µl of homogenate to 190 µl of incubation buffer containing (final concentration) 1 mM NADPH, 750 µM CaCl$_2$, 100 µM Arg analogue, and 100 µM

| | | | |
|---|---|---|---|
| **MeArg** | $H_3C-$ $HN-C-NH-CH_2-CH_2-CH_2-CH$ $-COOH$ <br> $\quad\quad\parallel\quad\quad\quad\quad\quad\quad\quad\quad\quad\mid$ <br> $\quad\quad NH\quad\quad\quad\quad\quad\quad\quad\quad NH_2$ | Canava | $H_2N-C-NH-O-CH_2-CH_2-CH-COOH$ <br> $\quad\quad\parallel\quad\quad\quad\quad\quad\quad\quad\quad\mid$ <br> $\quad\quad NH\quad\quad\quad\quad\quad\quad NH_2$ |
| HArg | $H_2N-C-NH-CH_2-CH_2-CH_2-CH_2-CH-COOH$ <br> $\quad\quad\parallel\quad\quad\quad\quad\quad\quad\quad\quad\quad\mid$ <br> $\quad\quad NH\quad\quad\quad\quad\quad\quad\quad\quad NH_2$ | OH−GVA | $H_2N-C-NH-CH_2-CH_2-CH_2-CH-COOH$ <br> $\quad\quad\parallel\quad\quad\quad\quad\quad\quad\quad\quad\mid$ <br> $\quad\quad NH\quad\quad\quad\quad\quad\quad OH$ |
| **Arg** | $H_2N-C-NH-CH_2-CH_2-CH_2-CH$ $-COOH$ <br> $\quad\quad\parallel\quad\quad\quad\quad\quad\quad\quad\quad\mid$ <br> $\quad\quad NH\quad\quad\quad\quad\quad\quad NH_2$ | GButyram | $H_2N-C-NH-CH_2-CH_2-CH_2-C=O$ <br> $\quad\quad\parallel\quad\quad\quad\quad\quad\quad\quad\quad\mid$ <br> $\quad\quad NH\quad\quad\quad\quad\quad\quad NH_2$ |
| A−GBA | $H_2N-C-NH-CH_2-CH_2-CH$ $-COOH$ <br> $\quad\quad\parallel\quad\quad\quad\quad\quad\quad\mid$ <br> $\quad\quad NH\quad\quad\quad\quad NH_2$ | Agmatine | $H_2N-C-NH-CH_2-CH_2-CH_2-CH_2$ <br> $\quad\quad\parallel\quad\quad\quad\quad\quad\quad\quad\quad\mid$ <br> $\quad\quad NH\quad\quad\quad\quad\quad\quad NH_2$ |
| A−GPA | $H_2N-C-NH-CH_2-CH$ $-COOH$ <br> $\quad\quad\parallel\quad\quad\quad\quad\mid$ <br> $\quad\quad NH\quad\quad NH_2$ | OH−Arg | $H_2N-C-NH-CH_2-CH$ $-CH_2-CH-COOH$ <br> $\quad\quad\parallel\quad\quad\quad\quad\mid\quad\quad\quad\quad\mid$ <br> $\quad\quad NH\quad\quad OH\quad\quad\quad NH_2$ |
| 6−GCA | $H_2N-C-NH-CH_2-CH_2-CH_2-CH_2-CH_2-COOH$ <br> $\quad\quad\parallel$ <br> $\quad\quad NH$ | OH−AA | $H_2N-C-NH-CH_2-CH$ $-CH_2-CH-COOH$ <br> $\quad\quad\parallel\quad\quad\quad\quad\mid\quad\quad\quad\mid$ <br> $\quad\quad NH\quad\quad OH\quad\quad NH-CO-CH_3$ |
| 5−GVA | $H_2N-C-NH-CH_2-CH_2-CH_2-CH_2-COOH$ <br> $\quad\quad\parallel$ <br> $\quad\quad NH$ | NA−Arg | $H_2N-C-NH-CH_2-CH_2-CH_2-CH-COOH$ <br> $\quad\quad\parallel\quad\quad\quad\quad\quad\quad\quad\quad\quad\mid$ <br> $\quad\quad NH\quad\quad\quad\quad\quad\quad NH-CO-CH_3$ |
| 4−GBA | $H_2N-C-NH-CH_2-CH_2-CH_2-COOH$ <br> $\quad\quad\parallel$ <br> $\quad\quad NH$ | Arcaine | $H_2N-C-NH-CH_2-CH_2-CH_2-CH_2$ <br> $\quad\quad\parallel\quad\quad\quad\quad\quad\quad\quad\quad\mid$ <br> $\quad\quad NH\quad\quad\quad\quad\quad NH-C(NH)-NH_2$ |
| 3−GPA | $H_2N-C-NH-CH_2-CH_2-COOH$ <br> $\quad\quad\parallel$ <br> $\quad\quad NH$ | **2−GGA** | $H_2N-C-NH-CH$ $-CH_2-CH_2-COOH$ <br> $\quad\quad\parallel\quad\quad\quad\mid$ <br> $\quad\quad NH\quad\quad COOH$ |
| 2−GAA | $H_2N-C-NH-CH_2-COOH$ <br> $\quad\quad\parallel$ <br> $\quad\quad NH$ | 2−GSA | $H_2N-C-NH-CH$ $-CH_2-COOH$ <br> $\quad\quad\parallel\quad\quad\quad\mid$ <br> $\quad\quad NH\quad\quad COOH$ |
| CTN | $H_2N-C-N$ $-CH_2-COOH$ <br> $\quad\quad\parallel\quad\mid$ <br> $\quad\quad NH\quad CH_3$ | 2−GEt | $H_2N-C-NH-CH_2-CH_2-OH$ <br> $\quad\quad\parallel$ <br> $\quad\quad NH$ |
| MGua | $H_2N-C-NH-CH_3$ <br> $\quad\quad\parallel$ <br> $\quad\quad NH$ | GES | $H_2N-C-NH-CH_2-CH_2-SO_3H$ <br> $\quad\quad\parallel$ <br> $\quad\quad NH$ |

*Fig. 1. Structures of L-arginine (Arg) and its analogues used. MeArg; $N^G$-methyl-L-arginine, HArg; homoarginine, Arg; L-arginine, A-GBA; α-amino-γ-guanidinobutyric acid, A-GPA; α-amino-β-guanidinopro-prionic acid, 6-GCA; ε-guanidinocaproic acid, 5-GVA; δ-guanidinovaleric acid, 4-GBA; γ-guanidinobutyric acid, 3-GPA, β-guanidinopropionic acid, 2-GAA; guanidinoacetic acid, CTN; creatine, MGua; methylgua-nidine, Canava; canavanine, OH-GVA; α-hydroxy-δ-guanidinovaleric acid (argininic acid), GButyram; 4-guanidino butyramide, OH-Arg; γ-hydroxyarginine, OH-AA; γ-hydroxy-N-acetyl-D, L-arginine, NA-Arg; N-acetyl-L-arginine, 2-GGA; α-guanidinoglutaric acid, 2-GSA; guanidinosuccinic acid, 2-GEt; 2-guanidinoe-thanol and GES; guanidinoethanesulfonic acid.*

Arg. After 15 min of incubation at 37 °C, the reaction was stopped by the addition of 180 μl of 0.3 N NaOH, and thereafter 180 μl of 5 per cent $ZnSO_4$ was supplemented. After 20 min of centrifugation at $10,000 \times g$, the content of nitrite and nitrate (NOx), which are stable end products of NO was determined, in the supernatant using a NOx analyzer TCI-NOX1000 (Tokyo Kasei Kogyo Co., Ltd., Tokyo) utilizing the flow injection method[6].

In other experiments, Arg concentrations varied from 2.5 μM to 40 μM with or without inhibitor in order to observe the enzyme kinetics.

All the values reported have been corrected for the background content of NOx after incubation with NADPH–free incubation medium[16]. Data were analyzed using paired student's *t*-test. For determination of inhibition constant ($K_i$), data were analyzed by non-linear least-squares regression[11] using the equations according to Segel[14]. As the lower sum of square residual (SSR) better fits the model, the inhibition model that gave lowest SSR was chosen as the type of inhibition[15].

## Results and discussion

NOx production without Arg analogues was $164.0 \pm 6.4$ pmol NOx formation/mg protein/min (mean $\pm$ SEM, n = 19). Among 23 kinds of guanidino compounds, $N^G$-monomethyl-L-arginine (MeArg), α-guanidinoglutaric acid (2-GGA) and arcaine inhibited NOS activity significantly (Fig. 2).

As NO is derived from the guanidino group of Arg[8,10], it is quite logical that strong inhibitors for NOS are produced either by adding a functional group on to, or by substituting one of the guanidino

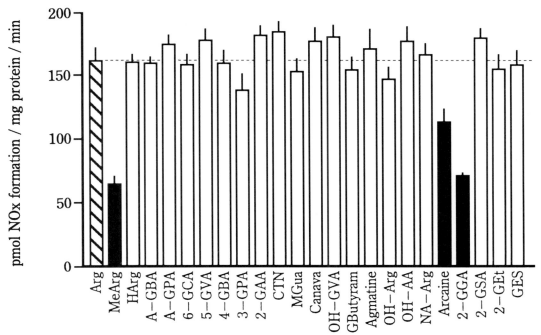

*Fig. 2. Effect of arginine analogues as inhibitors on nitric oxide synthase (NOS) activity. Brain homogenates were incubated with 100 μM of each arginine (Arg) analogue in the presence of 100 μM of Arg. Ordinate: NO formation (pmol NOx formation/mg protein/min). Open column; NOS was not inhibited, and closed column; NOS was significantly inhibited (mean ± SEM, n = 6~7). Abbreviations used are the same as for Fig. 1.*

nitrogens of Arg, i.e. MeArg[1], $N^G$-nitro-L-arginine[3], $N^G$-amino-L-arginine[4], or L-iminoethyl-ornithine ($N^G$-deamino-Arg)[9]. These amidino nitrogen-substituted analogues inhibit NOS activity competitively[9]. In contrast with those authentic inhibitors, 2-GGA and arcaine have an intact amidino group in their structures as shown in Fig. 1. In this study, we found that 2-GGA was as potent an inhibitor of NOS as MeArg, and that arcaine acts as a weak inhibitor. Therefore we investigated the inhibitory kinetics of 2-GGA in further studies.

The Michaelis constant ($K_m$) value and maximum velocity ($V_{max}$) of NOS in the brain homogenate were 5.64 ± 0.87 μM and 210 ± 11 pmol NOx formation/mg protein/min (mean ± SEM, n = 12), respectively. This $K_m$ of the rat brain NOS is similar to that of previous observations (8.4 μM[9]). In present study, MeArg inhibited NOS competitively, and the inhibition constant ($K_i$) value was calculated to be 3.51 ± 0.27 μM (mean ± SEM, n = 6) as shown in Fig. 3. On the other hand, 2-GGA inhibited NOS activity in a linear mixed manner. The calculated $K_i$ value and the α factor[14] were 2.74 ± 0.55 μM and 9.44 ± 2.54 (mean ± SEM, n = 8), respectively (Fig. 4). The $K_i$ values of MeArg and 2-GGA for Arg were almost equal. These data clearly show that 2-GGA is a strong NOS inhibitor.

2-GGA and arcaine are known as amidino-glutamic acid and as diamidino-putrescine, respectively, and they are naturally occurring[12,13]. 2-GGA is found in cerebral cortex even in small quantities, but it is increased in the cobalt-induced epileptic focus of the cat cerebral cortex[12]. Arcaine occurs naturally also in the marine mollusc and annelids[13]. Though MeArg residues are found in the myelin basic protein of mammalian brain[2], 2-GGA and arcaine also exist freely. These structural characteristics of 2-GGA and arcaine make them useful tools to study the enzyme–substrate relationship of NOS and NO functions in animals.

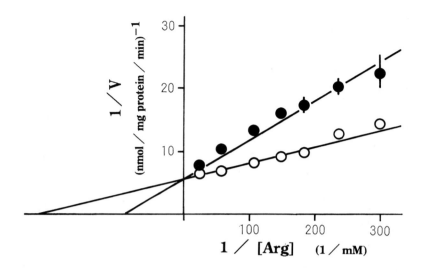

*Fig. 3. Effect of N$^G$-methyl-L-arginine (MeArg) on nitric oxide synthase (NOS) activity. Incubations were carried out without MeArg (◯) and with 5 μM of MeArg (●). MeArg inhibited NOS competitively. In the present study, the $K_i$ value was calculated to be 3.51 ± 0.27 μM (mean ± SEM, n = 6).*

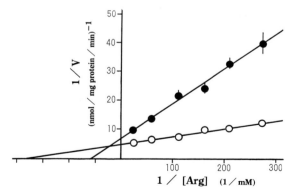

*Fig. 4. Effect of α-guanidinoglutaric acid (2–GGA) on nitric oxide synthase (NOS) activity. Incubations of brain homogenate were carried out without 2-GGA (●) or with 10 μM of 2-GGA (◯). The figure shows a Lineweaver–Burk plot, and vertical bars represent SEM. 2-GGA inhibited NOS activity in the linear mixed manner, and the apparent $K_i$ value and the α factor[14] was 2.74 ± 0.55 μM and 9.44 ± 2.54 (mean± SEM, n = 8), respectively.*

# References

1. Bredt, D.S. & Snyder, S.H. (1989): Nitric oxide mediates glutamate-linked enhancement of cGMP levels in the cerebellum. *Proc. Natl. Acad. Sci.* USA **86**, 9030–9033.

2. Carnegie, P.R. (1971): Amino acid sequence of the encephalitogenic basic protein from human myelin. *Biochem. J.* **123**, 57–67.

3. East, S.J. & Garthwaite, J. (1990): Nanomolar $N^G$-nitroarginine inhibits NMDA-induced cyclic GMP formation in rat cerebellum. *Eur. J. Pharmacol.* **184**, 311–313.

4. Fukuto, J.M., Wood, K.S., Byrns, R.E. & Ignarro, L. (1990): $N^G$-Amino-L-arginine: A new potent antagonist of L-arginine-mediated endothelium-dependent relaxation. *J. Biochem. Biophys. Res. Commun.* **168**, 458–465.

5. Garthwaite, J., Charles, S.L. & Chess-Williams, R. (1988): Endothelium-derived relaxing factor release on activation of NMDA receptors suggests role as intercellular messenger in the brain. *Nature* **336**, 385–388.

6. Habu, H., Yokoi, I., Kabuto, H. & Mori, A. (1994): Application of automated flow injection analysis to determine nitrite and nitrate in mouse brain. *NeuroReport* **5**, 1571–1573.

7. Hu, J. & El-Fakahany, E. (1993): Role of intercellular and intracellular communication by nitric oxide in coupling of muscarinic receptors to activation of guanylate cyclase in neuronal cells. *J. Neurochem.* **61**, 578–585.

8. Klatt, P., Schmidt, K., Uray, G. & Mayer, B. (1993): Multiple catalytic functions of brain nitric oxide synthase: Biochemical characterization, cofactor-requirement, and the role of $N^\omega$-hydroxyl-L-arginine as an intermediate. *J. Biol. Chem.* **268**, 14781–14787.

9. Knowles, R.G., Palacios, M., Palmer, R.M.J. & Moncada, S. (1990): Kinetic characteristics of nitric oxide synthase from rat brain. *Biochem. J.* **269**, 207–210.

10. Marletta, M.A., Yoon, P.S., Iyengar, R., Leaf, C.D. & Wishnok, J.S. (1988): Macrophage oxidation of L-arginine to nitrite and nitrate: nitric oxide is an intermediate. *Biochemistry* **27**, 8706–8711.

11. Marquardt, D.W. (1963): An algorithm for least-squares estimation of nonlinear parameters. *J. Soc. Indust. Appli. Math.* **11**, 431–441.

12. Mori, A., Akagi, M., Katayama, Y. & Watanabe, Y. (1980): Guanidinoglutaric acid in cobalt-induced epileptogenic cerebral cortex of cats. *J. Neurochem.* **35**, 603–605.

13. Robin, Y., Thoai, N.-V. & Roche, J. (1957): Sur la présence d'arcaine chez la Sansue, Hirudo medicinalis L. *Compt. Rend. Soc. Biol.* **157**, 2015–2017.

14. Segel I.H. (1976): *Biochemical Calculations,* 2nd edn., pp. 246–266. New York: John Wiley & Sons.

15. Watanabe, Y., Yokoi, I., Watanabe, S., Sugi, H. & Mori, A. (1988): Formation of 2-guanidinoethanol by a transamidination reaction from arginine and ethanolamine by the rat kidney and pancreas. *Life Sci.* **43**, 295–302.

16. Yokoi, I., Kabuto, H., Habu, H. & Mori, A. (1994): Guanidinoglutaric acid, an endogenous convulsant, as a novel nitric oxide synthase inhibitor. *J. Neurochem.* **63**, 1565–1567.

*Guanidino Compounds : 2*, eds. by P.P. De Deyn, B. Marescau, I.A. Qureshi and A. Mori.
©1997 John Libbey & Company Ltd., pp. 9–15.

# Chapter 2

# Cerebral nitric oxide synthase and NMDA subtype of glutamate receptors in *spf* mice with congenital ornithine transcarbamylase deficiency

L. RATNAKUMARI[1,2], I.A. QURESHI[1], B. MARESCAU[3] P.P. DE DEYN[3] and R.F. BUTTERWORTH[2]

[1]*Division of Medical Genetics, Sainte-Justine Hospital, 3175 Côté Sainte-Catherine, Montreal, Québec, Canada H2X 3J4;* [2]*Neuroscience Research Unit, Saint-Luc Hospital, Montréal, Canada H27 3J;* [3]*Laboratory of Neurochemistry & Behaviour, University of Antewrp, Born-Bunge Foundation, U.I.A., 2610, Antwerp, Belgium*

## Summary

The sparse-fur (*spf*) mutant mouse has an X-linked deficiency of hepatic ornithine transcarbamylase (OTC) and develops hyperammonemia in the post-natal period similar to congenital hyperammonemia type II syndrome seen in children. Significantly lower levels of serum arginine, citrulline and ornithine were previously reported in *spf* mutant mice. Since L-arginine is the precursor for nitric oxide (NO), an important messenger molecule in central nervous system (CNS), we studied the cerebral levels of arginine and the activities of nitric oxide synthase (NOS) in the brains of *spf* mutant mice and compared them with normal controls. The activity levels of NOS were significantly decreased in cerebral cortex, cerebellum, striatum and hippocampus (up to –42 per cent) of OTC deficient *spf* mice. As the NOS gets activated by the $Ca^{2+}$ influx subsequent to the NMDA receptor activation in the brain, we also measured the densities of NMDA binding sites in the *spf* mouse brain by autoradiography using [$^3$H] MK801 as a ligand. A generalized decrease (of up to –53 per cent) in the densities of NMDA binding sites was observed in the *spf* mutant mouse brain as compared to controls. Since NO has an important role in regulation of cell function and cellular communications, this observed decrease in NOS may have a role in the neuronal abnormalities seen in congenital OTC deficiency.

## Introduction

The deficiency of the second enzyme of the urea cycle, i.e. ornithine transcarbamylase, is the most common among the congenital hyperammonemias of primary origin[8]. Children with OTC deficiency often present with seizures and impaired cognitive function[20]. Neuropathology of the patients indicates the presence of Alzheimer type II astrocytes and a moderate to severe neuronal loss[1,20]. The chronically elevated levels of plasma ammonia are suggested to play an important role in the neurological deficit or abnormalities observed in this disorder.

The sparse-fur (*spf*) mouse with an X-linked deficiency of hepatic OTC serves as an appropriate animal model of congenital hyperammonemia[13,26]. With less than 10 per cent of normal liver OTC activity and a significantly increased urinary orotate excretion, the sparse-fur mutant mouse very closely resembles the congenital hyperammonemia type II syndrome seen in children[21]. In this volume, we reported significantly decreased levels of guanidino compounds including arginine, guanidinoacetic acid, α-N-acetyl arginine, argininic acid, creatine, γ-guanidinobutyric acid and guanindinosuccinic acid in the brains of the *spf* mice as compared to normal controls[28]. The cerebral citrulline levels were also significantly decreased in these mutants as compared to normal mice[28]. This evidence of reduced arginine and citrulline levels is consistent with the previously demonstrated reduced levels of these amino acids in serum of *spf* mutant mice[18,27,33].

Decreased levels of cerebral arginine in OTC deficiency might result in decreased levels of nitric oxide (NO). NO is produced by nitric oxide synthase (NOS) from arginine through oxidation of its guanidino nitrogen resulting in the formation of L-citrulline as a by-product[19]. NO appears to be a unique and simple molecule with diverse functions in signal transduction[16,35]. NO offers a new perspective on cell-to-cell communication with its ability to diffuse freely across cell membranes and initiate a cascade of events in different cell types. NOS activity has been demonstrated throughout the brain with predominant localization in neurons[4,15]. The neuronal NOS is constitutive, which mainly accounts for the role of NO in mediating rapid events like neurotransmission and blood vessel dilatation[12].

It is now known that the neuronal NOS is a calcium/calmodulin requiring enzyme[19]. In the brain, the activation of NMDA receptors triggers $Ca^{2+}$ influx, which in turn can activate a variety of $Ca^{2+}$-dependent enzymes including NOS[7]. The NMDA receptors are involved in certain forms of synaptic plasticity and through this subtype of receptors glutamate exerts trophic effects on neurons[10].The excitotoxic action of glutamate is suggested to be mediated primarily by NMDA subtype of receptors and in particular the high $Ca^{2+}$ conductance of this receptor ion channel[32]. Glutamatergic synaptic dysfunction has been implicated in the pathophysiology of the hyperammonemic disorders[5].

The aim of this study was to assess the role of chronic hyperammonemia and sustained low levels of arginine on the cerebral NO production. For this purpose we have measured the activity levels of NOS in four different brain regions of congenitally hyperammonemic *spf* mice and compared them with those of normal controls. The densities of NMDA type of glutamate receptors in the *spf* mutant mice brains were also measured by quantitative receptor autoradiography using [³H] MK801. The possible role of these factors on the neuronal integrity in congenital OTC deficiency is discussed.

## Materials and methods

### Materials

L-[³H]arginine (specific activity, 58 Ci/μmol) was purchased from Amersham International. (6R)-5,6,7,8-tetrahydro-L-biopterin dihydrochloride, NADPH, Calmodulin, Nω-nitro-L-arginine were obtained from Sigma Chemical Co., MO, USA. Dowex AG 50W-X8 (hydrogen form) was obtained from Sigma Chemical Co., and later converted to the Na⁺ form. Blood ammonia test kits were obtained from Technicon Instruments Corporation, Tarrytown, NY.

### Animals

The parent stock for the colony of sparse-fur mice was originally obtained from Dr. L. B. Russel of the Oak Ridge National Laboratories, Oak Ridge, TN, USA[13]. Male *spf* mice were the progeny of matings of homozygous affected *spf/spf* females with *spf* males[26]. All male progeny of these matings, being *spf*/Y, were separated by simple sexing; however, at the end of each experimental study, liver ornithine transcarbamylase activity was determined to check the mutant status of the animals. CD-1 strain mice obtained from Canadian Breeding Farms, St. Constant, Quebec, were used as normal controls. All animals were kept in a controlled environment (12 h dark/12 h light) with free access to water and food (Purina mouse chow, Ralston Purina, St. Louis, MO) at the animal house of the Saint-Justine Hospital. The animals were kept and experimented upon according to the guidelines of the Canadian Council on Animal Care (*Guide to the Care And Use of Experimental Animals*, Vol. 2, 1984).

## Blood ammonia estimation

The mice were decapitated and the blood was collected from the neck wound. Serum was separated by centrifuging at 3000 r.p.m for 10 min. Ammonia levels were estimated by employing a commercial blood ammonia test kit, which uses an ion exchange method followed by colorimetric measurement of isolated ammonia nitrogen with the Berthelot Phenate–Hypochlorite reaction[14].

## Hepatic OTC activity

Liver OTC activity was measured as described previously to verify the mutant status of the *spf* animals[26].

## Nitric oxide synthase assay

The brain regions were homogenized in 0.32 M sucrose/20 mM Hepes (pH 7.2) containing 0.5 mM EDTA, 1 mM DTT and centrifuged at 40,000 $\times g$ for 15 min. The supernatant was passed over a 1 ml column of Dowex AG 50W–X8 (Na[+] form) to remove endogenous arginine. NOS activity was determined as formation of L-[³H]citrulline from L-[³H]arginine as described previously[3]. The reaction mixture consists of 50 mM Hepes (pH 7.4), 0.2 mM NADPH, 1 mM EDTA, 1.25 mM CaCl$_2$, 1 mM dithiothreitol, 10 µM tetrahydrobiopterin, 10 µg/ml of calmodulin and 100 µM L-[³H]arginine. Blank values were determined in the absence of enzyme. After incubation for 10 min at −30 °C, assays were terminated with 2 ml of 20 mM Hepes, pH 5.5, consisting of 2 mM EDTA which were then applied to 1 ml columns of Dowex AG 50W–X8 (Na[+] form), and eluted with 2 ml of water. The amount of [³H]citrulline formed was estimated by a Beckman liquid scintillation counter.

## Measurement of amino acids

Amino acids were separated over a cation exchange column using lithium citrate buffers and were detected with the colorimetric ninhydrin method using a Biotronik LC 6001 (Biotronik, Maintal, Germany) amino acid analyser. Chromatographic conditions and characteristics have been described in detail by Pei[24].

## Quantitative receptor autoradiography

[³H]MK801 was obtained from New England Nuclear (Boston, MA, USA). Tritium-sensitive film was purchased from Amersham (Arlington heights, IL, USA). Mice were decapitated and the brains removed quickly and frozen in isopentane chilled on dry ice. Brain sections (20 µm thick) were cut using a microtome onto gelatin-coated glass microscopic slides and were kept at −70 °C till further use. The brain sections were thawed at room temperature for 30 min and incubated in 30 µM Hepes (pH 7.5) containing 25 nM [³H]MK801, 100 µM glutamate, 100 µM glycine and 1 mM EDTA[38]. At the end of the incubation, slides were rinsed with cold buffer and finally dipped in ice-cold distilled water and rapidly dried under a stream of cold air. Non-specific binding was determined in the presence of 200 µM ketamine.

Autoradiograms were prepared by apposing sections together with tissue-calibrated standards of known [³H]-concentrations (Amersham microscales) to [³H]-sensitive Hyperfilm (Amersham) for 3 weeks. Films were developed and tissue concentrations of [³H]MK801 were measured by quantitative densitometry analysis using a MCID computer-based densitometer and image analysis system (Imaging Research, Ontario, Canada). The amount of [³H]MK801 bound to various brain regions was calculated from the specific activity of the ligand. The brain regions of interest were defined according to the mouse brain atlas[34]. Protein concentrations were determined by the method of Lowry *et al.* [22]. Statistical analysis was done by the Student's *t*-test. All the data are expressed as mean ± SE.

# Results

**Table 1. Levels of cerebral amino acids, ammonia and hepatic ornithine transcarbamylase in control and hyperammonemic *spf* mutant mouse**

| Compound | Control | *spf* mutant |
|---|---|---|
| Arginine[1] | 128 ± 3.1 | 78 ± 7.5** |
| Citrulline[1] | 12 ± 0.3 | 8.0 ± 0.1** |
| Ornithine[1] | 11.7 ± 3.1 | 6.9 ±0.7* |
| Ammonia[2] | 1.41 ± 0.05 | 2.85 ± 0.16** |
| Hepatic OTC[3] | 64.2 ± 4.0 | 6.23 ± 0.8** |

Units: [1]nmol/g wet wt; [2]μmol/g wet wt; [3]μmol citrulline/min/mg protein. Each value is mean ± SE of 5 animals.
*$P < 0.01$ and **$P < 0.001$ compared to control by Student's $t$-test.

**Table 2. Activities of nitric oxide synthase in different brain regions of the ornithine transcarbamylase deficient mouse**

| Brain region | Control | *spf* mutant |
|---|---|---|
| Cerebral Cortex | 0.90 ± 0.03 | 0.74 ± 0.02* |
| Cerebellum | 1.52 ± 0.03 | 0.88 ± 0.07** |
| Striatum | 0.90 ± 0.06 | 0.70 ± 0.02* |
| Hippocampus | 0.79 ± 0.03 | 0.65 ± 0.06* |

Units: μmol of citrulline formed/10 min/mg protein. Each value is mean ± SE of 5 animals. *$P < 0.01$ and **$P < 0.001$ as compared to controls.

**Table 3. Distribution of [$^3$H]MK801 binding sites in brain regions of control and *spf* mutant mice**

| Brain region | Control | *spf* mutant | % Change over control |
|---|---|---|---|
| Frontal cortex | 1621 ± 4 | 1392 ± 17* | −14 |
| Parietal cortex | 1361 ± 13 | 1096 ± 23* | −19 |
| Striatal cortex | 1360 ± 9 | 1035 ± 15* | −24 |
| Caudate putamen | 1191 ± 4 | 832 ± 21* | −30 |
| CA1, hippocampus | 1823 ± 10 | 1451 ± 15* | −20 |
| CA2, hippocampus | 1801 ± 15 | 1503 ± 24* | −17 |
| CA3, hippocampus | 1310 ± 9 | 1161 ±10* | −11 |
| Dentate gyrus | 1094 ± 26 | 738 ± 35* | −33 |
| Inferior colliculus | 1000 ± 24 | 469 ± 12* | −53 |
| Superior colliculus | 1161 ±42 | 634 ± 15* | −45 |
| Anterior dorsal thalamus | 981 ± 17 | 767 ± 23* | −22 |
| Anterior ventral thalamus | 926 ± 20 | 702 ± 10* | −24 |
| Ventero posterior thalamus | 874 ± 7 | 593 ± 9* | −32 |
| Substantia nigra | 900 ± 10 | 616 ± 20* | −32 |
| Globus pallidus | 515 ± 11 | 333 ± 13* | −35 |
| Amygdala | 801 ± 18 | 826 ± 18 | +3 |
| Pons medulla | 701 ± 31 | 444 ± 11* | −37 |
| Medial geniculate nucleus | 1158 ± 12 | 757 ± 21* | −35 |
| Hypothalamus | 714± 17 | 584 ± 27* | −18 |
| Cerebellum | 514 ± 16 | 358 ±10* | −30 |
| Medial septum | 918 ± 29 | 794 ± 27 | −14 |

Values represent mean ± SE (pmol/g wet weight of tissue) specific binding of [$^3$H]MK801 in control and *spf* mutant mice. n = 5. *$P < 0.001$ *vs* control.

The brain ammonia levels were elevated significantly in *spf* mutant mice as compared to control mice (Table 1). The hepatic OTC activity levels of *spf* mutant mice were approximately one-tenth of the activity levels of control mice (Table 1). The cerebral levels of all three amino acids i.e. arginine, ornithine and citrulline were significantly lower in OTC deficient mutant mice as compared to normal controls (Table 1).

NOS activities were significantly decreased in the cerebral cortex, striatum, cerebellum and hippocampus of the OTC deficient mutant mice as compared to control mice (Table 2).

The highest density of NMDA sensitive binding sites was observed in CA1 and CA2 regions of the hippocampus (Table 3). Among different cortical areas, a higher density of MK801 binding sites was observed in the frontal cortex. Assessment of the density of [³H]MK801 binding sites revealed a significant difference between control and *spf* mutant mice (Table 3). Statistically significant decrease in densities of [³H]MK801 binding sites were observed in cortical areas, caudate-putamen, hippocampal regions, thalamic areas, inferior and superior colliculi, cerebellum and pons of congenitally hyperammonemic *spf* mutant mice as compared to normal control mice (Table 3).

## Discussion

Results of the present study show that *spf* mice are chronically hyperammonemic (Table 1) exhibiting only one-tenth of the control hepatic OTC activity (Table 1), which agrees well with our earlier reports [30,31]. The cerebral levels of arginine, ornithine and citrulline are significantly lower as compared to normal control mice (Table 1). Results of the present study demonstrate significant decreases in the activities of constitutive NOS in four different brain regions of *spf* mice when compared to normal controls (Table 2). The sustained decrease of arginine observed in the blood and brain of *spf* mutant mice might have an effect on the synthesis of cerebral NO. It has been suggested that one way of inhibiting NOS in biological systems is to limit the supply of one of its substrates or cofactors[19]. Exogenous administration of L-arginine has been shown to enhance NO production in brain slices[36]. Apart from producing NO from L-arginine, it has been shown that brain NOS can also generate oxygen-derived free radicals when L-arginine or $H_4$biopterin levels are suboptimal[11,23]. Since the involvement of oxygen radicals is attributed as an important causative factor in so many toxic processes, this phenomenon could have an important pathophysiological implication in the hyperammonemic condition.

The activity of the brain constitutive NOS depends on $Ca^{2+}$/calmodulin[19]. This dependence on $Ca^{2+}$ makes NOS a possible target for the post-synaptic alteration in $Ca^{2+}$ concentrations. In CNS, the activation of glutamate receptors, mainly of NMDA subtype leads to an influx of $Ca^{2+}$. Hence, we made an attempt to measure the changes in the densities of NMDA binding sites using [³H]MK801 as a ligand. The relative distribution of [³H]MK801 binding sites estimated in different brain regions of control mouse was consistent with the reported distribution of NMDA receptors[17,25,38]. The results of present study reveal a generalized decrease in the densities of [³H]MK801 binding sites in various brain regions of congenitally hyperammonemic *spf* mutant mice (Table 3). Similar reduced densities of NMDA binding sites were observed in the brains of chronically hyperammonemic portacaval shunted (PCS) rats[25]. Injection of acute and subacute doses of ammonium salts to the rats also resulted in specific decreases in the NMDA subclass of glutamate receptors[29]. The exact mechanism by which this reduction in NMDA subclass of glutamate receptor occurs in hyperammonemic conditions is not clear. It has been proposed that this could probably be due to the down-regulation of NMDA receptors[25]. Though the total glutamate content in various brain regions is either unchanged or decreased in hyperammonemic conditions, the extracellular concentration of glutamate has been shown to increase in hyperammonemic PCS rats[6,39]. The possibility of certain membrane proteins being masked by ammonium ions has also been suggested[29].

Alternatively, there might be an endogenous excitotoxin other than glutamate, such as quinolinic acid, a tryptophan metabolite that acts specifically at the NMDA receptor[37]. In support of this Batshaw *et al.* [2] have demonstrated a significant increase (2–10 fold) in CSF levels of quinolinic acid in children with congenital hyperammonemias. Increased brain levels of quinolinic acid have been demonstrated both in acute hyperammonemic as well as in PCS rats. Whatever may be the mechanism, the reduced glutamatergic neurotransmission might inhibit glutamate-induced calcium entry in the post-synaptic neuron and subsequent activation of NOS. This would lead to the decreased formation of cGMP. Similar type of synaptic inhibition has been suggested as the main causative factor underlying the common neurological abnormalities occurring in liver failure, Reye's syndrome and certain inborn errors of urea cycle enzymes[9].

In conclusion, the results of the present study may indicate that during chronic hyperammonemia due to congenital OTC deficiency, the decreased densities of NMDA binding sites and sustained low levels of cerebral arginine would result in a decreased production of NO. Since, NO serves as an important messenger molecule in the regulation of cell function and cellular communications, this observed decrease in NOS may effect the neuronal integrity in this metabolic disorder.

## Acknowledgments

The studies described were funded by The Medical Research Council of Canada (MT-9124 to IAQ and PG 11118 to RFB) UIA, NFWO (grants 3.0044.92 and 3.0064.93), Boon Bunge Foundation, the United Fund of Belgium and the Antwerp OCMW Medical Research Foundation). Part of this work was presented at the Annual Meeting of Society of Neuroscience, Miami, FL, 1994; Abstr. No. 343.13.

# References

1. Batshaw, M.L. (1994): Inborn errors of urea synthesis. *Ann. Neurol.* **35**, 133–141.

2. Batshaw, M.L., Robinson, M.B., Hyland, K., Djali, S. & Heyes, M.P. (1993): Quinolinic acid in children with congenital hyperammonemia. *Ann. Neurol.* **34,** 676–681.

3. Bredt, D.S. & Snyder, S.H. (1990): Isolation of nitric oxide synthetase, a calmodulin requiring enzyme. *Proc. Natl. Acad. Sci.* (USA) **87**, 682–685.

4. Bredt, D.S., Hwang, P. M. & Snyder, S.H. (1990): Localization of nitric oxide synthase indicating a neural role for nitric oxide. *Nature* **347**, 768–770.

5. Butterworth, R. F. (1992): Evidence that hepatic encephalopathy results from a defects of glutamatergic synaptic regulation. *Mol. Neuropharmacol.* **2**, 229–232.

6. Butterworth, R.F., Le, O., Lavoie, J. & Szerb, J.C. (1991): Effect of porta-caval anastomosis on electrically stimulated release of glutamate from rat hippocampal slices. *J. Neurochem.* **56**, 1481-84.

7. Calne, D.B. (1994): Glutamate and epilepsy: theories and therapeutics. *Neurology* **44** (Suppl 8), S5-S6.

8. Cathelineau, L. (1979): L'hyperammoniemie dans la pathologie pediatrique. *Arch. Franc. Pediatr.* **36**, 724-735.

9. Clark, I.A., Rockett, K.A. & Cowden, W.B. (1992): Possible central role of nitric oxide in conditions clinically similar to cerebral malaria. *Lancet* **340**, 894–896.

10. Collingridge, G.L. & Singer, W. (1990): Excitatory amino acid receptors and synaptic plasticity. *Trends. Pharmacol. Sci.* **11**, 290–296.

11. Culcasi, M., Lafon-Cazal, M., Pietri, S. & Bockaert, J. (1994): Glutamate receptors induce a burst of superoxide via activation of nitric oxide synthase in arginine-depleted neurons. *J. Biol. Chem.* **269**, 12589–12593.

12. Dawson, T. M. & Snyder, S. H. (1994): Gases as biological messengers: nitric oxide and carbon monoxide in the brain. *J. Neurosci.* **14**, 5147–5159.

13. DeMars, R., LeVan, S.L., Trend, B.L. & Russel, L.B. (1976): Abnormal ornithine carbamyltransferase in mice having the sparse-fur mutation. *Proc. Natl. Acad. Sci. (USA)* **73**, 1693–1697.

14. Dienst, S.G. (1961): An ion exchange method for plasma ammonia concentration. *J. Lab. Clin. Med.* **58**, 149–155.

15. Forstermann, U., Gorsky, L. D., Pollock, J. S., Schmidt, H.H.H.W., Heller, M. & Murad, F. (1990): Regional distribution of EDRF/NO-synthesizing enzyme(s) in rat brain. *Biochem. Biophys. Res. Commn.* **168**, 727–732.

16. Garthwaite, J. (1991): Glutamate, nitric oxide and cell-cell signalling in the nervous system. *Trends Neurosci.* **14,** 60–67.

17. Greenamyre, J.T., Olson, J.M.M., Penney. J.B. & Yound, A.B. (1985): Autoradiographic characterization of N-methyl-D-aspartate-, quisqualate-, and kainate- sensitive glutamate binding sites. *J. Pharmacol. Exp. Ther.* **233**, 254–263.

18.  Inoue, I., Gushiken, T., Kobayashi, K. & Saheki, T. (1987): Accumulation of large neutral amino acids in the brain of sparse-fur mice at hyperammonemic state. *Biochem. Med. Metab. Biol.* **38**, 378–386.

19.  Knowles, R. G. & Moncada, S. (1994): Nitric oxide synthases in mammals. *Biochem. J.* **298**, 249–258.

20.  Kornfeld, M., Woodfin, B.M., Papile, L., Davis, L E. & Bernard, L.R. (1985): Neuropathology of ornithine carbamyl transferase deficiency. *Acta Neuropathol (Berl).* **65**, 261–264.

21.  Levin, B., Oberholzer, V.G. & Sinclair, R.L. (1969): Biochemical investigation of hyperammonemia. *Lancet.* **2**, 170–174.

22.  Lowry, O. H., Rosebrough, N. J., Farr, A. L. & Randall, R. J. (1951): Protein measurement with the Folin-Phenol reagent. *J Biol. Chem.* **193**, 265–275.

23.  Mayer, B., John, M., Heinzel, B., Werner, E.R., Wachter, H., Schultz, G. & Bohme, E. (1991): Brain nitric oxide synthase is a biopterin- and flavin-containing multi-functional oxido-reductase. *FEBS Lett.* **288**, 187–191.

24.  Pei, H. (1994): The use of cation exchange resin in analytical research of amino acids. Dissertation to obtain the degree of Master in Biomedical Sciences,University of Antwerp (VIA), Belgium.

25.  Peterson, C., Giguere, J-F., Cotman, C.W. & Butterworth, R.F. (1990): Selective loss of N-methyl-D-aspartate-sensitive L-[$^3$H]glutamate binding sites in rat brain following portacaval anastomosis. *J. Neurochem .* **55**, 386–390.

26.  Qureshi, I.A., Letarte, J. & Ouellet, R. (1979): Ornithine transcarbamylase deficiency in mutant mice. 1. Studies on the characterization of enzyme defect and suitability as animal model of human disease. *Pediatr. Res.* **13**, 807–811.

27.  Qureshi, I.A., Marescau, B., Levy, M., De Deyn, P.P., Letarte, J. & Lowenthal, A. (1989): Serum and Urinary guandino compounds in "sparse-fur" mutant mice with ornithine transcarbamylase deficiency. In: *Guanidines 2.* Eds. A. Mori, B.D. Cohen and H. Koide, pp. 45–51. New York: Plenum press.

28.  Qureshi, I.A., Ratnakumari, L , Marescau, B. & De Deyn, P.P. (1994): Guanidino compounds in the brain and peripheral organs of hyperammonemic mice. In:*Guanidino Compounds: 2,* eds. P.P. De Deyn, B. Marescau, I.A. Qureshi. A. Mori, pp. 335-340. London: John Libbey.

29.  Rao, V.L.R., Agarwal, A.K. & Murthy, Ch. R.K. (1991): Ammonia-induced alterations in glutamate and muscimol binding to cerebellar synaptic membranes. *Neurosci. Lett.* **130**, 251–254.

30.  Ratnakumari, L., Qureshi, I.A. & Butterworth, R.F. (1992): Effects of congenital hyperammonemia on the cerebral and hepatic levels of the intermediates of energy metabolism in *spf* mice. *Biochem. Biophys. Res. Commn.* **184**, 746–751.

31.  Ratnakumari, L., Qureshi, I.A. & Butterworth, R.F. (1993): Effect of sodium benzoate on cerebral and hepatic energy metabolites in *spf* mice with congenital hypermmonemia. *Biochem. Pharmacol.* **45**, 137–146.

32.  Rothman, S.M & Olney, J.W. (1987): Excitotoxicity and the NMDA receptor. *Trends Neurosci .* **10**:299–301.

33.  Seiler, N., Grauffel, C., Daune-Anglard, G., Sarhan, S. & Knodgen, B. (1994): Decreased hyperammonaemia and orotic aciduria due to inactivation of ornithine aminotransferase in mice with a hereditary abnormal ornithine carbamoyltransferase. *J. Inher. Metab. Dis.* **17**, 691–703.

34.  Sidman, R.L., Angevine, J.B. & Taber Pierce, E. (1971): Atlas of the mouse brain and spinal cord. Harvard university Press, Cambridge, Massachusetts.

35.  Snyder, S.H. & Bredt, D.S. (1992): Biological roles of nitric oxide. *Sci. Am.* (May) , 68–77.

36.  Southam, E., East, S.J. & Garthwaite, J. (1991): Excitatory amino acid receptors coupled to nitric oxide/cyclic GMP pathway in rat cerebellum during development. *J. Neurochem .* **56**, 2072–2081.

37.  Stone, T.W. & Perkins, M.N. (1981): Quinolinic acid: a potent endogenous excitant at amino acid receptors in CNS. *Eur. J. Pharmacol.* **72**, 411–412.

38.  Subramaniam, S. & McGonigle, P. (1991): Quantitative autoradiographic characterization of the binding of (+)-5-methyl-10, 11-dihydro-5H-dibenzo[a,d]cyclohepten-5, 10-imine [$^3$H]MK-801 in rat brain: Regional effects of polyamines. *J. Pharmacol. Exp. Therp.* **256**, 811–819.

39.  Tossman, U., Delin, A., Eriksson, L.S. & Ungerstedt, U. (1987): Brain cortical amino acids measured by intracerebral dialysis in portacaval shunted rats. *Neurochem. Res.* **12**, 265–2.

*Guanidino Compounds : 2*, eds. by P.P. De Deyn, B. Marescau, I.A. Qureshi and A. Mori.
©1997 John Libbey & Company Ltd., pp. 17–20.

# Chapter 3

# Cerebral nitric oxide synthase activities in experimental hepatic encephalopathy

## V.L. RAGHAVENDRA RAO, R.M. AUDET and R.F. BUTTERWORTH

*Neuroscience Research Unit, Hôpital Saint-Luc (University of Montreal), Montreal, QC, Canada H2X 3J4*

## Summary

Hepatic encephalopathy is a neuropsychiatric syndrome resulting from chronic liver disease and characterized by inversion of sleep patterns progressing through stupor and coma. Increased release and concomitantly decreased binding of glutamate to NMDA receptors was reported previously in the brains of portacaval-shunted rats. As nitric oxide production was shown to be coupled with NMDA receptor activation, we studied the activities of nitric oxide synthase and the rate of transport of L-[$^3$H]arginine (precursor for nitric oxide) in the brains of portacaval-shunted rats in comparison with sham-operated controls. Activities of nitric oxide synthase as well as the transport of L-[$^3$H]arginine were elevated significantly in the brains of shunted rats. Increased production of nitric oxide may be of pathophysiological significance in the modulation of altered cerebral blood flow and neurological manifestations of hepatic encephalopathy.

## Introduction

Nitric oxide (NO) is a gaseous, highly reactive and diffusible molecule implicated in numerous biological functions such as learning, long-term depression, long-term potentiation, feeding and sexual behaviour, as well as neurotransmitter release and in the modulation of receptors[6,8,15]. In brain, NO is generated from the terminal guanidino group of L-arginine by NADPH-dependent nitric oxide synthase (NOS). Neurons are shown to contain the constitutive, Ca$^{2+}$/calmodulin-dependent isoform of NOS[5].

Glutamate, the major excitatory amino acid neurotransmitter in brain, stimulates the formation of cGMP in the cerebellum[2]. Activation of excitatory amino acid receptors, particularly those of the NMDA subtype, triggers Ca$^{2+}$-dependent synthesis of NO by NOS, which stimulates the enzyme guanylyl cyclase leading to the production of cGMP[29]. It has been suggested that glutamatergic synaptic dysfunction may be a causal factor responsible for the neuropsychiatric symptoms observed in hepatic encephalopathy[3,24]. Previous investigations have shown that the depolarization-induced release of glutamate was elevated and the binding of glutamate to NMDA receptors was decreased in brains of portacaval-shunted rats[3,19] as well as in acute hyperammonemia[21,23]. In order to evaluate the involvement of the glutamatergic signal transduction system in hepatic encephalopathy, we studied the activities of NOS in the brains of portacaval-shunted rats.

Exogenous L-arginine has been shown to enhance NO synthesis in brain slices[29] and the availability of the precursor L-arginine for the synthesis of NO appears to be dependent upon a specific uptake mechanism in neurons[31]. Consequently, we also studied the transport of L-[³H]arginine into synaptosomes prepared from the cerebral cortex and cerebellum of portacaval-shunted rats.

## Methods

L-[2,3-³H]arginine (specific activity, 43.5 Ci/mmol) was purchased from DuPont-New England Nuclear, Mississauga, Ont., Canada and its purity was confirmed by HPLC. (6R)-5,6,7,8-tetrahydro-L-biopterin dihydrochloride was obtained from Alexis Corporation, San Diego, CA, USA. L-Arginine, NADPH, HEPES, EDTA, dithiothreitol, calmodulin, sucrose, glucose, Dowex 50W X8 (200–400 mesh) cation exchange resin and Ficoll-400 were purchased from Sigma Chemical Co., MO, USA.

All surgical procedures were conducted and the animals were cared for according to the Principles of the Guide for Care and Use of Experimental Animals, vol. 2 (1984) (Canadian Council on Animal Care, Ottawa).

Portacaval-shunting was performed in male, Sprague-Dawley rats (175–200 g) by the method of Lee and Fisher[12], as described earlier[27]. Sham-operated rats served as controls. Animals were killed by decapitation 4 weeks after surgery.

NOS activity was determined in cerebellar or cerebral cortical cytosolic preparations by the method of Bredt and Snyder[2] as modified by Rao et al.[26] For measuring L-[³H]arginine transport, synaptosomes were prepared from fresh, unfrozen brain tissue by the Ficoll-400 discontinuous density gradient method of Cotman[4] as modified by Rao & Murthy[22]. L-[³H]arginine transport was measured in fresh synaptosomes by the method of Westergaard et al.[31] as modified by Rao et al.[26] Protein content was determined by the method of Lowry et al.[13].

## Results

NOS activity was significantly increased in the cerebral cortex (+65 per cent, $P < 0.01$) and cerebellum (+54 per cent, $P < 0.01$) of portacaval-shunted rats compared to sham operated control rats (Table 1). L-[³H]arginine transport was also elevated significantly in the synaptosomes isolated from cerebral cortex (by 51 per cent, $P < 0.01$) and cerebellum (by 52 per cent, $P < 0.01$) of portacaval-shunted rats compared to sham-operated control rats (Table 1).

**Table 1. Activities of nitric oxide synthase(NOS) and L-[³H]arginine transport in the brains of portacaval-shunted and sham-operated rats**

|  | Cerebral cortex | Cerebellum |
|---|---|---|
| NOS activity | | |
|   portacaval-shunted | 131.0 ± 11.7* | 243.8 ± 17.1* |
|   sham-operated | 79.3 ± 12.8 | 158.8 ± 13.2 |
| L-[³H]arginine transport | | |
|   portacaval-shunted | 209.4 ± 28.4* | 104.9 ± 8.3 |
|   sham-operated | 138.4 ± 37.3 | 69.3 ± 5.9 |

Units: pmol/mg protein/min. Each value is mean ± SE of 13 separate experiments for NOS activities and 6 separate experiments for L-[³H]arginine transport. Statistics: *$P < 0.01$ compared to sham-operated controls by Student's $t$-test.

## Discussion

Recent studies have shown that NO modulates both post-synaptic NMDA receptor function and pre-synaptic glutamate release by feed-back mechanisms[8,15]. NO in its nitrosonium ion (NO⁺) form can react with the thiol groups of NMDA receptors to block the ion currents through them[15]. Further, NO produced in the post-synaptic terminals may diffuse back into pre-synaptic terminals where it can stimulate the release of glutamate[8]. Consistent with these actions of NO, our present finding of elevated activities of NOS may lead to increased production of NO that may be responsible for the increased *in vivo* and *in vitro* release of glutamate[3,24] and a concomitant reduction in NMDA receptor densities observed previously in the brains of portacaval-shunted rats[19].

NO is known to activate, guanylyl cyclase in vascular smooth muscle cells and pericytes and thus may modulate cerebral blood flow (CBF)[9,30]. Possible sources of NO include cerebrovascular endothelium, parenchymal neurons adjacent to glia and perivascular nerves originating from parasymphathetic cranial ganglia[17,18]. Elevated CBF was reported previously in portacaval-shunted rats as well as in patients following portacaval anastomosis surgery[1,7]. It has been suggested that the hyperdynamic circulation observed in liver cirrhosis is mediated by NO, and administration of N[G]-nitro-L-arginine (L-NOARG; an NOS inhibitor) was shown to ameliorate portal-systemic shunting in portal hypertensive rats[11]. Though the precise neurochemical mechanisms for the neuropsychiatric disturbances observed in hepatic encephalopathy are not known, a number of studies suggest ammonia as the chief culprit[24]. It is interesting to note that a recent study showed that treatment of animals with L-NOARG protects the animals from hyperammonaemia and also prevents the biochemical changes due to ammonia toxicity[10].

Westergaard *et al.*[31] showed that L-arginine availability for the production of NO in brain was regulated by the uptake of this amino acid. Increased uptake of L-[³H]arginine into synaptosomes prepared from the brains of portacaval-shunted rats observed in the present study could help to sustain increased NO production in these rats. Mans *et al.*[14] recently reported that plasma and brain L-arginine levels are significantly decreased in acute hepatic encephalopathy. These authors speculated that decreased cerebral L-arginine levels are due to increased demand for arginine of NO synthesis.

Pathophysiological mechanisms proposed to explain hepatic encephalopathy have focused in recent years on alterations of neurotransmitter function in CNS. Such mechanisms include modifications of glutamatergic[3,24], and monoaminergic[16,20,25] function as well as the presence of endogenous benzodiazepines[28]. Results of the present study suggests that increased activities of NOS in brain following portacaval anastomosis could be of pathophysiological significance in modulating the CBF and neurological disturbances characteristic of hepatic encephalopathy.

**Acknowledgements**

Funded by the Medical Research Council (MRC) of Canada. VLRR is an MRC, Canada post-doctoral fellow.

## References

1.   Bianchi-Porro, G., Maiola, A.T. & Della-Porta, P. (1969): Cerebral blood flow and metabolism in hepatic cirrhosis before and after portacaval shunt operation. *Gut* **10**, 894–897.

2.   Bredt, D.S. & Snyder, S.H. (1989): Nitric oxide mediates glutamate-linked enhancement of cGMP levels in the cerebellum. *Proc. Natl. Acad. Sci. USA* **86**, 9030–9033.

3.   Butterworth, R.F., Le, O., Lavoie, J. & Szerb, J.C. (1991): Effect of portacaval anastomosis on electrically-stimulated release of glutamate from rat hippocampal slices. *J. Neurochem.* **56**, 1481–1484.

4.   Cotman, C.W. (1974): Isolation of synaptosomal and synaptic plasma membrane fractions. In: *Methods in enzymol*, Vol. 31, eds. S. Fleischer & L. Packer, pp. 445–452, New York: Academic Press.

5.   Dawson, T.M. & Snyder, S.H. (1994): Gases as biological messengers: nitric oxide and carbon monoxide in the brain. *J. Neurosci.* **14**, 5147–5159.

6.   Garthwaite, J. (1995): Neural nitric oxide signalling. *Trends Neurosci.* **18**, 51–52.

7.   Gjedde, A., Lockwood, A.H., Duffy, T.E. & Plum, F. (1978): Cerebral blood flow and metabolism in chronically hyperammonemic rats: effect of an acute ammonia challange. *Ann. Neurol.* **3**, 325–330.

8.   Guevara-Guzman, R., Emson, P.C. & Kendrick, K.M. (1994): Modulation of *in vivo* striatal transmitter release by nitric oxide and cyclic GMP. *J. Neurochem.* **62**, 807–810.

9.   Iadecola, C., Pelligrino, D.A., Moskowitz, M.A. & Lassen, N.A. (1994): Nitric oxide synthase inhibition and cerebrovascular regulation. *J. Cereb. Blood Flow Metab.* **14**, 175–192.

10.  Kosenko, E., Kaminsky, Y., Grau, E., Minana, M.-D., Grisolia, S. & Felipo, V. (1995): Nitroarginine, an inhibitor of nitric oxide synthetase, attenuates ammonia toxicity and ammonia-induced alterations in brain metabolism. *Neurochem. Res.* **20,** 381–386.

11.  Lee, F-Y., Colombato, L.A., Albillos, A. & Groszmann, R.J. (1993): Administration of $N^W$-nitro-L-arginine ameliorates portal-systemic shunting in portal-hypertensive rats. *Gastroenterology* **105,** 1464–1470.

12.  Lee, S.H. & Fisher, B. (1961): Portacaval shunt in the rat. *Surgery* **50,** 668–672.

13.  Lowry, O.H., Rosebrough, N.J., Farr, A.L. & Randall, R.J. (1951): Protein measurements with Folin-phenol reagent. *J. Biol. Chem.* **193,** 265–275.

14.  Mans, A.M., DeJoseph, M.R. & Hawkins, R.A. (1994): Metabolic abnormalities and grade of encephalopathy in acute hepatic failure. *J. Neurochem.* **63,** 1829–1838.

15.  Manzoni, O. & Bockaert, J. (1993): Nitric oxide synthase activity endogenously modulates NMDA receptors. *J. Neurochem.* **61,** 368–370.

16.  Mousseau, D.D., Perney, P., Layrargues, G.P. & Butterworth, R.F. (1993): Selective loss of palliadal dopamine $D_2$ receptor density in hepatic encephalopathy. *Neusroci. Lett.* **162,** 192–196.

17.  Murphy, S., Simmons, M.L. Agullo, L., Garcia, A., Feinstein, D.L., Galea, E., Reis, D.J., Minc-Golomb, D. & Schwartz, J.P. (1993): Synthesis of nitric oxide in CNS glial cells. *Trends Neurosci.* **16,** 323–328.

18.  Nozaki, K., Moskowitz, M.A., Maynard, K.I., Koketsu, N., Dawson, T.M., Bredt, D.S. & Snyder, S.H. (1993): Possible origin and distribution of immunoreactive nitric oxide synthase-containing nerve fibers in cerebral arteries. *J. Cereb. Blood Flow Metab.* **13,** 70–79.

19.  Peterson, C., Giguére, J.F., Cotman, C.W. & Butterworth, R.F. (1990): Selective loss of N-methyl-D-aspartate sensitive L-[$^3$H]glutamate binding sites in rat brain following portacaval anastomosis. *J. Neurochem.* **55,** 386–390.

20.  Rao, V.L.R. & Butterworth, R.F. (1994): Alterations of [$^3$H]8-OH-DPAT and [$^3$H]ketanserin binding sites in autopsied brain tissue from cirrhotic patients with hepatic encephalopathy. *Neurosci. Lett.* **182,** 69–72.

21.  Rao, V.L.R. & Murthy, Ch.R.K. (1991): Hyperammonemic alterations in the uptake and release of glutamate and aspartate by rat cerebellar preparations. *Neurosci. Lett.* **130,** 49–52.

22.  Rao, V.L.R. & Murthy, Ch.R.K. (1993): Uptake and metabolism of glutamate and aspartate by astroglial and neuronal preparations of rat cerebellum. *Neurochem. Res.* **18,** 647–654.

23.  Rao, V.L.R., Agrawal, A.K. & Murthy, Ch.R.K. (1991): Ammonia-induced alterations in glutamate and muscimol binding to cerebellar synaptic membranes. *Neurosci. Lett.* **130,** 251–254.

24.  Rao, V.L.R., Murthy, Ch.R.K. & Butterworth, R.F. (1992): Glutamatergic synaptic dysfunction in hyperammonemic syndromes. *Metab. Brain Dis.* **7,** 1–20.

25.  Rao, V.L.R., Giguere, J.F., Layrargues, G.P. & Butterworth, R.F. (1993): Increased activities of $MAO_A$ and $MAO_B$ in autopsied brain tissue from cirrhotic patients with hepatic encephalopathy. *Brain Res.* **621,** 349–352.

26.  Rao, V.L.R., Audet, R.M. & Butterworth, R.F. (1995a): Increased nitric oxide synthase activities and L-[$^3$H]arginine uptake in brain following portacaval anastomosis. *J. Neurochem.* **65,** 677–681..

27.  Rao, V.L.R., Audet, R.M. & Butterworth, R.F. (1995b): Selective alterations of extracellular brain amino acids in relation to function in experimental portal-systemic encephalopathy: Results of an *in vivo* microdialysis study. *J. Neurochem.* **65,** 1221–1228..

28.  Rothstein, J.D., McKhann, G., Guarneri, P., Barbaccia, M.L., Guidotti, A. & Costa, E. (1989): Cerebrospinal fluid content of diazepam binding inhibitor in chronic hepatic encephalopathy. *Ann. Neurol.* **26,** 57–62.

29.  Southam, E., East, S.J. & Garthwaite, J. (1992): Excitatory amino acid receptors coupled to nitric oxide/cyclic GMP pathway in rat cerebellum during development. *J. Neurochem.* **56,** 2072–2081.

30.  Wang, Q., Pauson, O.B. & Lassen, N.A. (1992): Effect of nitric oxide blockade by $N^G$-nitro-L-arginine on cerebral blood flow response to changes in carbon dioxide tension. *J. Cereb. Blood Flow Metab.* **12,** 947–953.

31.  Westergaard, N., Beart, P.M. & Schousboe, A. (1993): Transport of L-[$^3$H]arginine in cultured neurons: Characteristics and inhibition by nitric oxide synthase inhibitors. *J. Neurochem.* **61,** 364–367.

*Guanidino Compounds : 2*, eds. by P.P. De Deyn, B. Marescau, I.A. Qureshi and A. Mori.
©1997 John Libbey & Company Ltd., pp. 21–29.

# Chapter 4

# Nitric oxide is an important factor of autoregulation of retinal and choroidal blood flow in newborns

[1]Pierre HARDY, [1]Robert SEGAL,[2]Daya R. VARMA, [1]Sylvain CHEMTOB

[1]*Departments of Pediatrics, Ophthalmology and Pharmacology, Research Center of Hôpital Ste. Justine, 3175, Côte Sainte-Catherine, Montréal, Québec, Canada; and the* [2]*Department of Pharmacology and Therapeutics, McGill University, Montréal, Québec, Canada*

## Summary

We studied the role of nitric oxide (NO) in autoregulation of retinal blood flow (RBF), choroidal blood flow (ChBF), and $O_2$ delivery to the retina and the choroid during an acute change in ocular perfusion pressure (OPP) and during hyperoxia in newborn pigs (1–5 days old); blood flow was measured by the microsphere technique. Changes in perfusion pressure were produced by inflating balloon-tipped catheters placed at the aortic root and isthmus to cause hypotension and hypertension respectively. Hyperoxia was induced with 100 per cent $O_2$. Animals were either treated with the NO synthase inhibitors, $N^G$-nitro-L-arginine methyl ester (L-NAME, 1 mg/kg followed by 50 µg/kg/min; n = 15) or with saline (n = 14). In saline-treated animals, retinal oxygen delivery was constant only within a range of 30 to 80 mm Hg OPP ($R$ = 0.03, $P > 0.9$) and choroidal oxygen delivery increased as a function of OPP ($\tau = 0.51 - 0.62$, $P < 0.01$). L-NAME caused retinal and choroidal oxygen delivery to be maintained constant from 30 to 146 mm Hg (the highest OPP studied; [$R < 0.3$, $P > 0.15$]); correspondingly retinal and choroidal vascular resistance rose with increased OPP in L-NAME-treated pigs but not after saline ($P < 0.01$). In saline-treated animals, hyperoxia caused oxygen delivery to the retina to decrease but did not for the choroid. After L-NAME, oxygen delivery to the choroid significantly decreased in response to hyperoxia. Elevated OPP as well as hyperoxia caused increases in products of peroxidation in the choroid, which were prevented by L-NAME. The data suggest that released NO prevents adequate retinal and choroidal vasoconstriction in response to acute hypertension and hyperoxia in the newborn leading to increased retinal and choroidal $O_2$ delivery and consequently to increased free radical generation. These findings provide a new mechanism for the increased susceptibility of the newborn to increased $O_2$ tensions which predisposes to retinopathy of prematurity.

## Introduction

The newborn exhibits a limited ability to autoregulate blood flow to the retina and mainly the choroid (the major source of oxygen to the retina) in response to an acute rise in perfusion pressure[6], in contrast to the adult[1,16]. In the newborn, retinal blood flow (RBF) is autoregulated between 45 and 85 mm Hg[6], whereas in adults, the range of autoregulation is much wider, from 45 to greater than 145 mm Hg. When blood pressure is modestly increased, RBF varies as a function of perfusion pressure and this results in a rapid and marked increase in $O_2$ delivery and reactive oxygen species to the retina in the newborn. The mechanisms that govern the limits of perfusion pressure over which RBF and choroidal blood flow (ChBF) are autoregulated, especially in the newborn, remain largely unknown.

We have previously suggested that cyclooxygenase products, which include prostaglandins and free radicals, exert a significant role in the autoregulation of ocular blood flow[6,13]. It has been proposed that the reactive oxygen species, nitric oxide (NO), is released from the vasculature during an increase in perfusion pressure[27], and has been implicated in the coronary blood flow autoregulation in isolated hearts[25]. Moreover the role of NO in setting the perfusion pressure limits of RBF and ChBF autoregulation of the newborn has also been recently studied[13a].

It is however important to distinguish the relative role of perfusion pressures from that of increased tissue oxygenation. In the newborn, hyperoxia leads to retinal vasoconstriction[23,24]. On the other hand, the response of the choroid to hyperoxia is unknown in the newborn. Furthermore, the mechanisms responsible for the vasomotor responses to increased $O_2$ concentration are not known.

We therefore present studies based on the hypothesis that an increase in perfusion pressure causes a release of NO which contributes to the narrow range of RBF and ChBF during an acute change of perfusion pressure; and extend this role of NO in autoregulation to that in response to hyperoxia. For this purpose, we determined the effects of the NO synthase inhibitor, $N^G$-nitro-L-arginine methyl ester (L-NAME), on RBF and ChBF, during acute changes in perfusion pressure and during hyperoxia. In addition, we examined if the NOS inhibitor, L-NAME, would prevent an expected increase in peroxides during an acute change in perfusion pressure and hyperoxia by measuring two indices of peroxidation, malondialdehyde (MDA) and hydroperoxides in choroidal tissues.

## Materials and methods

### Animals and surgical preparation

Newborn pigs (1–5 day old) weighing 1.2–2.2 kg were used in this study according to a protocol approved by the Animal Care Committee of the Research Center of Ste. Justine Hospital. RBF and ChBF were studied in 29 animals as previously described[5,13]. Briefly, animals were anesthetized with 1.5 per cent halothane for tracheotomy and the catheterization of various blood vessels. The left subclavian artery was catheterized with a polyethylene catheter for the withdrawal of blood samples including reference samples. A similar catheter was placed into the left ventricle via the right subclavian artery for the injection of radiolabeled microspheres to measure blood flow and another one in the descending thoracic aorta via a femoral artery for continuous blood pressure (BP) recording by means of a Statham pressure transducer connected to a Gould multichannel recorder (TA240). To study the effects of changes in blood pressure on autoregulation a silicone coated balloon-tipped catheter (Berman Angiocath) was positioned in the distal thoracic descending aorta via a femoral artery; inflation of this balloon produces hypertension in the aortic arch. A second balloon-tipped catheter was placed at the root of the aorta via the right common carotid artery, and inflated to produce hypotension in the aortic arch. A polyethylene catheter was placed in the femoral vein for intravenous (i.v.) administration of drugs. In order to measure intraocular pressure (IOP), a 27 gauge butterfly needle attached to a catheter was introduced in the anterior chamber of the eye through the cornea and the site of entry was sealed with cyanoacrylate glue. Animals were ventilated by means of a Harvard small animal respirator with a gas mixture of 21 per cent $O_2$ and 79 per cent $N_2$. Halothane was discontinued after surgery; animals were maintained on α-chloralose (bolus i.v. injection of 50 mg/kg followed by infusion of 10 mg/kg/h) and paralyzed with pancuronium (0.1 mg/kg i.v.). Body temperature was maintained at 38 °C with an overhead radiant lamp and the animals were allowed to recover from the surgery for 2 h before starting the experiments.

### Experimental protocols

*Autoregulatory response to changes in perfusion pressure:*

Animals were assigned to receive i.v. either saline (1.5 ml; n = 7) or L-NAME (1 mg/kg followed by 50 µg/kg/min; n = 8). Protocols used to study the response of RBF and ChBF to changes in perfusion pressure have previously been reported[6,13a]. Briefly basal RBF and ChBF were recorded 45 min after the saline or drug injections and 10 min later one of the two balloon-tipped catheters was inflated to produce hypotension or hypertension. Once a steady state BP was achieved (within 30 to 40 s of balloon inflation), RBF and ChBF were determined again. The balloons were deflated after the blood flow measurements and animals were allowed to recover for 40 min. At the end of this recovery period a second baseline RBF and ChBF was recorded and, 10 min later, final measurements of RBF and ChBF were made after inflating the other balloon-tipped catheter. Hence each animal was subjected to one hypotensive and one hypertensive episode performed in a random order; for each treatment group of animals BP was scaled at intervals of 5 ± 1.8 mm Hg to cover a range of 5 to 146 mm Hg.

*Autoregulatory response to hyperoxia*

Animals were assigned to receive i.v. either saline (1.5 ml; n = 7) or L-NAME (1 mg/kg followed by 50 µg/kg/min; n = 7). Blood flow to the retina and choroid, arterial blood gases, were measured before and 45 min after piglets were exposed to 100 per cent $O_2$.

**Retinal and choroidal blood flow measurements**

RBF and ChBF were determined using the radionuclide-labelled microsphere technique as previously described in detail[6,13a]; approximately $10^6$ microspheres (15 µm diameter) labeled with [141]Ce, [95]Nb, [46]Sc and [113]Sn (New England Nuclear, Boston, MA) were injected in a random sequence into the left ventricle. Immediately after completion of injection of microspheres, blood samples were withdrawn from the left subclavian artery to determine blood gases, $O_2$ content (ABL 300; Radiometer, Copenhagen, Denmark), and hemoglobin concentrations. After the experiment, animals were killed with pentobarbital (120 mg/kg i.v.), the location of catheters was verified and the eyes were removed.

Radioactivity in the retina, choroid and reference blood samples was counted in a gamma scintillation counter (Cobra II, Canberra Packard, CA). Per cent interference between radionuclides was subtracted. Blood flow (ml/min/100 g) was calculated as [c.p.m/g of tissue × reference blood withdrawal rate] / [c.p.m in the reference blood]. Oxygen delivery to the retina and choroid was calculated as RBF and ChBF × arterial $O_2$ content (ml/min/100 g)[12]. Retinal and choroidal vascular resistances (mm Hg/ml/min/100 g) were calculated by dividing the ocular perfusion pressure (OPP; mean BP minus intraocular pressure [IOP]) by RBF and ChBF, respectively.

**Measurements of malondialdehyde and hydroperoxides in the choroid**

Forty additional piglets were prepared as described above to examine if the nitric oxide synthase inhibitor L-NAME would prevent an expected increase in peroxides during an acute elevation of the perfusion pressure and during hyperoxia in the newborn. For this purpose we measured MDA and hydroperoxide concentrations, both indices of peroxidation, in animals treated with L-NAME or saline. For the hypertension study, the animals were killed either at an average OPP of 90 mm Hg or immediately after increasing OPP to 125 ± 6 mm Hg (for 2 min, as per blood flow studies), a level greater than the upper OPP limit of autoregulation of RBF and ChBF of newborn pigs[13]; this enabled comparable increases in OPP in all treatment groups. For the hyperoxia study, animals were killed before or 45 min after exposure to 100 per cent $O_2$. Animals were killed by rapid i.v. injection of 120 mg/kg of pentobarbital sodium, and liquid $N_2$ was immediately thereafter poured on each eye. The tissues were then removed and stored at −80 °C until the assay was performed, within 2 weeks of storage, an interval during which no significant changes in the concentration of these products occur[7,13]. On the day of the assay, the choroid was thawed on ice and suspended in a cold buffer (pH 7.4) of the following composition: 5 mM Tris-HCl, 0.67 mM acetylsalicylic acid, 0.5 mM EGTA and 100 µM butylated hydroxytoluene. Tissues were homogenized and centrifuged at 1000 × *g* for 10 min to remove non-disrupted cells and nuclei. The supernatant was used to measure proteins[3], MDA and hydroperoxides. MDA and hydroperoxides were measured respectively by the thiobarbituric acid reaction and oxidation of $FeCl_2$, as we previously described in detail[7,13].

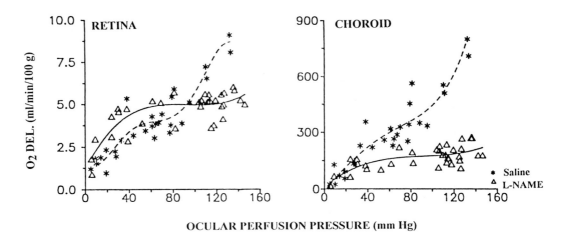

Fig. 1. Retinal and choroidal $O_2$ delivery autoregulation as a function of ocular perfusion pressure in newborn pigs treated with $N^G$-nitro-L-arginine methyl ester (L-NAME) (1 mg/kg followed by 50 µg/kg/min; n = 8) or saline (n = 7). Each animal was subjected to one hypotensive and one hypertensive episode to cover for each treatment group a range of ocular perfusion pressure from 5 to 146 mm Hg. In saline-treated animals the best fit regression line was a third-order polynomial for retinal $O_2$ delivery ($R^2 = 0.85$, $P < 0.001$) and a second-order polynomial for choroidal $O_2$ delivery ($R^2 = 0.87$, $P < 0.05$). In L-NAME-treated pigs a third-order polynomial regression fitted best the points for retinal and choroidal $O_2$ delivery ($R^2 = 0.64$ and 0.56, respectively, $P < 0.001$).

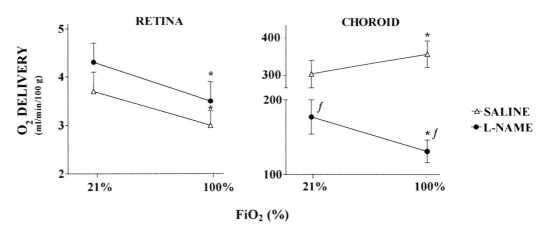

Fig. 2. Retinal and choroidal $O_2$ delivery as a function of $FiO_2$ (%) in newborn pigs treated with saline (n = 7) or $N^G$-nitro-L-arginine methyl ester (L-NAME, n= 7; 1 mg/kg followed by 50 µg/kg/min). *$P < 0.05$ compared to value at 21% $O_2$. ƒ $P < 0.05$ compared to corresponding value in saline-treated animals.

## Chemicals

$N^G$-nitro-L-arginine methyl ester, malonaldehyde bisdimethyl acetal, t-butyl-hydroperoxide, xylenol orange, butylated hydroxytoluene were purchased from Sigma Chemical Co., St. Louis, MO. Radionuclide microspheres were purchased from New England Nuclear, Boston, MA. All other chemicals were purchased from Fisher, Montréal, Québec.

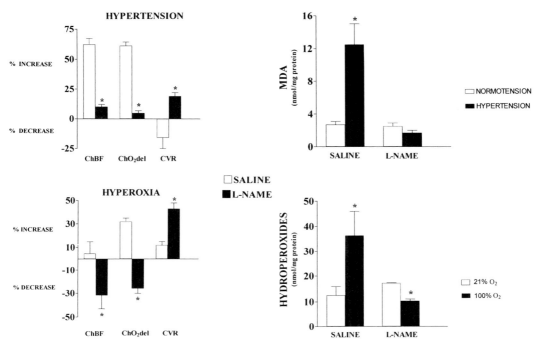

Fig. 3. Per cent change in choroidal blood flow (ChBF), choroidal $O_2$ delivery (ChO2del), choroidal vascular resistance (CVR) during hypertension (when ocular perfusion pressure is increased from an average of 90 (basal) to 125 mm Hg (high)) and hyperoxia (FiO2 = 100 per cent for 45 min) in newborn pigs treated with saline (n = 7 for each experimental protocol) or $N^G$-nitro-L-arginine methyl ester (L-NAME, n = 8 and 7 for each experimental protocol; 1 mg/kg followed by 50 μg/kg/min). * P < 0.05 compared to values after saline.

Fig. 4. Choroidal levels of MDA and hydroperoxides. Animals were pre-treated with saline (n = 7 for each experimental protocol) or $N^G$-nitro-L-arginine methyl ester (L-NAME, n = 7 for each experimental protocol; 1 mg/kg followed by 50 μg/kg/min) and exposed to an acute hypertension (when ocular perfusion pressure is increased from an average of 90 (basal) to 125 mm Hg (high)) or hyperoxia (FiO2 = 100 per cent for 45 min). * P < 0.05 compared to normotension or 21 per cent $O_2$.

## Statistical Analysis

Data were analyzed by paired Student's-*t*-test, analysis of variance for repeated measures and comparison among means tests, and by linear and non-linear correlation and regression analysis, as previously described in detail[5,6,13,13a]. Statistical significance was set at $P < 0.05$.

## Results

### Stability of Preparations

During the study on perfusion pressure changes, arterial pH, $pO_2$, $pCO_2$ and IOP remained stable throughout the course of an experiment (Table 1); RBF and ChBF also did not differ between the first and second baselines. L-NAME caused an increase in basal BP and consequently OPP, and a decrease in ChBF at baseline. During the hyperoxia study, pH, $pO_2$, $pCO_2$, heart rate, MBP, IOP, remained stable during respiration in room air (Table 2). Ventilation with 100 per cent $O_2$ increased arterial blood $pO_2$ and $O_2$ content. L-NAME caused an increase in basal blood pressure and consequently OPP.

**Table 1. Arterial blood pH and gases, mean blood pressure (MBP), intraocular pressure (IOP), ocular perfusion pressure (OPP), retinal blood flow (RBF) and choroidal blood flow (ChBF) at baseline, and during hypotension in newborn pigs treated with saline or $N^G$-nitro-L-arginine methyl ester (L-NAME)**

| Saline | First baseline | Second baseline | Hypotension (OPP < 30 mm Hg) | Hypertension (OPP > 80 mm Hg) |
|---|---|---|---|---|
| pH | $7.43 \pm 0.02$ | $7.41 \pm 0.03$ | $7.45 \pm 0.04$ | $7.42 \pm 0.02$ |
| pO2 | $128 \pm 4$ | $124 \pm 8$ | $121 \pm 14$ | $124 \pm 6$ |
| pCO2 | $36 \pm 1$ | $38 \pm 2$ | $34 \pm 3$ | $36 \pm 1$ |
| MBP | $70 \pm 6$ | $76 \pm 6$ | NP | NP |
| IOP | $13 \pm 2$ | $14 \pm 2$ | $12 \pm 3$ | $16 \pm 1$ |
| OPP | $58 \pm 6$ | $62 \pm 6$ | NP | NP |
| RBF | $35 \pm 2$ | $33 \pm 2$ | NP | NP |
| ChBF | $3186 \pm 251$ | $3291 \pm 159$ | NP | NP |
| **L-NAME** | | | | |
| pH | $7.38 \pm 0.01$ | $7.35 \pm 0.06$ | $7.35 \pm 0.04$ | $7.36 \pm 0.03$ |
| pO2 | $104 \pm 7$ | $109 \pm 14$ | $100 \pm 13$ | $109 \pm 16$ |
| pCO2 | $46 \pm 3$ | $45 \pm 4$ | $43 \pm 6$ | $42 \pm 1$ |
| MBP | $132 \pm 6^*$ | $126 \pm 3^*$ | NP | NP |
| IOP | $10 \pm 1$ | $13 \pm 1$ | $13 \pm 1$ | $13 \pm 3$ |
| OPP | $122 \pm 5^*$ | $113 \pm 4^*$ | NP | NP |
| RBF | $37 \pm 7$ | $42 \pm 10$ | NP | NP |
| ChBF | $942 \pm 86^*$ | $938 \pm 105^*$ | NP | NP |

Values are mean $\pm$ SEM; RBF and ChBF are expressed in ml/min/100 g, and all other values in mm Hg except for pH. Hypertension and hypotension are defined as the upper and lower OPP limits of RBF autoregulation in saline-treated animals. NP denotes not presented; see Fig. 1. *$P < 0.05$ compared to corresponding value in saline-treated animals.

## Effects of L-NAME on perfusion pressure-induced autoregulation

Retinal and choroidal oxygen delivery plotted as a function of OPP are shown in Fig. 1. In saline-treated animals retinal and choroidal $O_2$ delivery correlated non-linearly with OPP ( $\tau = 0.51$ and 0.62, respectively, $P < 0.01$). Retinal $O_2$ delivery was constant between 30 and 80 mm Hg of OPP (R = 0.03, $P > 0.9$, based on polynomial and LOWESS curves), and varied with OPP above and below this range (respectively, R = 0.97, $P < 0.01$; and R = 0.67, $P < 0.05$; Fig. 1). Choroidal $O_2$ delivery increased as a function of OPP over the entire range of OPP studied ($\tau = 0.62$, $P < 0.01$). Second- and third-order polynomial regressions fitted best all points for choroidal and retinal $O_2$ delivery respectively ($R^2 = 0.87$–0.85, $P < 0.001$).

After treatment with L-NAME, oxygen delivery to the eye remained constant between 30 and 146 mm Hg OPP (R < 0.1, $P > 0.9$). Third-order polynomial regressions fitted best all points for choroidal and retinal $O_2$ delivery respectively ($R^2 = 0.56$–0.64, $P < 0.001$).

L-NAME did not affect retinal and choroidal $O_2$ delivery when OPP was reduced below 30 mm Hg. Moreover, when OPP was raised to its highest levels (> 90 mm Hg), choroidal $O_2$ delivery increased on average by 61–62 per cent after saline treatment and only by 5–10 per cent after L-NAME (saline versus L-NAME-treated, $P < 0.01$; Fig. 3). Above an OPP of 90 mm Hg CVR decreased by 16 per cent in saline-treated animals but increased by 19 per cent with L-NAME. Above an OPP of 80 mm Hg retinal and choroidal vascular resistance was unrelated to OPP in saline-treated animals ($R < 0.02$, $P > 0.85$) whereas it increased linearly with OPP in L-NAME-treated pigs (respectively, R = 0.84 and 0.92, $P < 0.001$). In all experiments changes in blood flow paralleled those in oxygen delivery to the retina and choroid.

**Table 2. Arterial blood pH and gases, oxygen content (O$_2$ ct), heart rate (HR), mean blood pressure (MBP), intraocular pressure (IOP) and ocular perfusion pressure (OPP) during normoxia (21 per cent O$_2$) and hyperoxia (100 per cent 0$_2$) in newborn pigs treated with saline or N$^G$-nitro-L-arginine methyl ester (L-NAME)**

| FiO2 | Saline 21% | Saline 100% | L-NAME 21% | L-NAME 100% |
|---|---|---|---|---|
| pH | $7.43 \pm 0.03$ | $7.43 \pm 0.02$ | $7.39 \pm 0.04$ | $7.42 \pm 0.03$ |
| pCO2 (mm Hg) | $41 \pm 1$ | $41 \pm 1$ | $38 \pm 2$ | $37 \pm 4$ |
| pO2 (mm Hg) | $125 \pm 4$ | $468 \pm 20$* | $108 \pm 14$ | $474 \pm 40$* |
| O2ct (Vol%) | $9.9 \pm 1.0$ | $12.4 \pm 1.1$* | $10.0 \pm 1.0$ | $12.3 \pm 2.0$* |
| HR (beats/min) | $206 \pm 14$ | $228 \pm 9$ | $191 \pm 17$ | $196 \pm 17$ |
| MPB (mm Hg) | $74 \pm 3$ | $80 \pm 3$ | $94 \pm 5$† | $93 \pm 5$† |
| IOP (mm Hg) | $11 \pm 1$ | $9 \pm 1$ | $12 \pm 2$ | $10 \pm 2$ |
| OPP (mm Hg) | $63 \pm 2$ | $71 \pm 2$ | $82 \pm 3$† | $83 \pm 3$† |

Values are mean $\pm$ SEM.
*$P < 0.05$ compared to corresponding value in FiO2 21 %.
†P $< 0.05$ compared to corresponding values in saline-treated animals.

### Effects of L-NAME on hyperoxia-induced autoregulation

At 21 per cent oxygen the nitric oxide synthase inhibitor, L-NAME, did not affect oxygen delivery to the retina (Fig.2) and RBF. In response to 100 per cent oxygen, retinal O$_2$ delivery and blood flow significantly decreased by approximatively 40 per cent and was associated with an increase in vascular resistance. L-NAME did not affect this response of the retina. In contrast to the retina, ChBF of saline-treated animals did not change and choroidal O$_2$ delivery increased by 32 per cent in response to hyperoxia (Figs. 2 and 3). L-NAME decreased choroidal O$_2$ delivery and blood flow by 40 per cent and 57 per cent, respectively during normoxia, and more importantly choroidal blood flow and O$_2$ delivery decreased further by 31 per cent during hyperoxia (Fig. 2).

### Hypertension- and hyperoxia-induced changes in products of peroxidation

In saline-treated animals the concentrations of MDA and hydroperoxides increased in the choroid (a purely vascular tissue) respectively when BP was acutely raised above the upper limit of RBF autoregulation and during hyperoxia. L-NAME prevented these increases in MDA and hydroperoxide levels (Fig. 4); baseline levels of these products were not affected by L-NAME.

### Discussion

Mechanisms that control autoregulation of blood flow to the eye remain largely unknown. NO has been implied in coronary and cerebral blood flow autoregulation[15,25]. More recently evidence from our laboratory suggested as well a role for NO in the response of RBF and ChBF to an acute elevation in perfusion pressure[13a]. However, from these studies it was unclear whether some of the effects of increased perfusion pressure resulted from increased tissue oxygenation. We therefore proceeded to examine if NO was also involved in the autoregulatory hemodynamic response to hyperoxia.

Our results indicate that:

(i) in response to an acute rise perfusion pressure, inhibition of NOS with L-NAME enhanced RBF and ChBF autoregulation and this was associated with a stable oxygen delivery over a wider range of perfusion pressure, 30–146 mm Hg;

(ii) in a similar manner, ChBF and oxygen delivery to the choroid, which hardly responded to hyperoxia in control animals, decreased in piglets treated with L-NAME. This enhanced ability to contain O$_2$ delivery to the eye after NOS inhibition seemed to be secondary to an increase in vascular resistance. Finally, restriction of O$_2$ delivery to the tissues was associated with inhibition of increased peroxidation.

Thus, our data imply that released NO prevents adequate vasoconstriction necessary during acute hypertension and hyperoxia, both of which lead to increased tissue $O_2$ delivery, and result in excess formation of reactive oxygen species arising from increased NOS activity or from other sources of free radicals. On the other hand, we have recently reported that increased cyclooxygenase activity could account for nearly all of the rise in hypertension-induced formation of peroxides[13]. With regards to hyperoxia the source for increased generation of free radicals remains to be determined but may be through leakage of electrons along the mitochondrial electron transport chain[11].

Although NO is involved in the upper limit of autoregulation of ocular blood flow, it does not seem to contribute to the RBF and ChBF response at the lower limit of perfusion pressure (Fig. 1), as reported by others[4,25,28]. A reduction in oxygen delivery to tissues during hypotension may decrease the $O_2$-dependent NO synthesis so that further inhibition may not be significantly induced by NO synthase inhibitors. NO also does not appear to contribute to resting RBF as opposed to ChBF (Table 1), as previously reported[8]. Several authors have suggested that the role of NO in the control of basal blood flow differs among tissues and species, such that for instance NO seems to be important in resting circulation to certain areas of the brain[14] and in the kidneys[22], but not in the heart[20]; this may explain in part the heterogeneity in endothelium-dependent responses of different blood vessels[26].

Our findings reveal that inhibition of NOS in the newborn animal uncovers a vasoconstrictor response to increased perfusion pressure and hyperoxia, similar to that observed in the untreated adult[1,9,16]. These data are consistent with increased NOS activity in the newborn[18]. However, the mechanisms responsible for the vasoconstriction due to acute hypertension and hyperoxia remain to be elucidated.

The data presented provide a new mechanism for the increased generation of free radicals in the eye of the newborn during conditions of acute hypertension and hyperoxia. We propose that NO exerts a significant contribution to the limited ability of the newborn choroid (major $O_2$ supply to the eye) to restrain $O_2$ delivery to the eye during an acute increase in perfusion pressure as well as during hyperoxia, by masking a vasoconstrictor response; the latter combined with the inherent decreased antioxidant activity in the newborn[2,17,19] facilitate peroxidation. Given the importance of increased retinal oxygenation and reactive oxygen species in the genesis of retinopathy of prematurity[10,21] our findings are of physiological and clinical relevance.

### Acknowledgments

The authors wish to thank Mrs. Hensy Fernandez for technical assistance. We also thank F. Menard Inc. (Ange-Gardien, Québec) for generous supply of newborn pigs. This work was supported by grants from the Medical Research Council of Canada, the Heart and Stroke Foundation of Québec, the Hospital for Sick Children Foundation, and the March of Dimes Birth Defects Foundation. P. Hardy is a recipient of a fellowship from the Medical Research Council of Canada.

## References

1. Alm, A. & Bill, A. (1972): The oxygen supply to the retina. II. Effects of high intraocular pressure and of increased arterial carbon dioxide tension on uveal and retinal blood flow in cats. *Acta Physiol. Scand.* **84**, 306–319.

2. Bougle, D., Vert, P., Reichart, E., Hartemeann, D., & Heng, E.L. (1982): Retinal superoxide dismutase activity in newborn kittens exposed to normobaric hyperoxia: effect of vitamin E. *Pediatr. Res.* **16**, 400–402.

3. Bradford, M.M. (1976): A rapid and sensitive method for the quantitation of microgram quantities of protein utilizing the principle of protein dye binding. *Anal. Biochem.* **72**, 248-254.

4. Buchanan, J.E. & Phillis, J.W. (1993): The role of nitric oxide in the regulation of cerebral blood flow. *Brain. Res.* **610**, 248–255.

5. Chemtob, S., Beharry, K., Rex, J., Varma, D.R. & Aranda, J.V. (1990): Changes in cerebrovascular prostaglandins and thromboxane as a function of systemic blood pressure: cerebral blood flow autoregulation of newborn piglets. *Circ. Res.* **67**, 674–682.

6. Chemtob, S., Beharry, K., Rex, J., Chatterjee, T., Varma, D.R. & Aranda, J.V. (1991): Ibuprofen enhances retinal and choroidal blood flow autoregulation in newborn piglets. *Invest. Ophthalmol. Vis. Sci.* **32**, 1799–1807.

7.  Chemtob, S., Roy, M-S., Abran, D., Fernandez, H. & Varma D.R. (1993): Prevention of post-asphyxial increase in lipid peroxides and retinal function deterioration in the newborn pig, by inhibition of cyclooxygenase activity and free radical generation. *Pediatr. Res.* **33**, 336–340.

8.  Deussen, A., Sonntag, M. & Vogel, R. (1993): L-arginine-derived nitric oxide: a major determinant of uveal blood flow. *Exp. Eye. Res.* **57**, 129–134.

9.  Flower, R.W., Fryczkowski, A.W. & McLeod, D.S. (1995): Variability in choriocapillaris blood flow distribution. *Invest. Ophthalmol. Vis. Sci.* **36**, 1247–1258.

10. Flynn, J.T., Bancalari, E. & Snyder, E.S. (1992): A cohort study of transcutaneous oxygen tension and the incidence and severity of retinopathy of prematurity. *New. Engl. J. Med.* **326**, 1050–1054.

11. Freeman, B.A. & Crapo, J.D. (1981): Hyperoxia increases oxygen radical production in rat lungs and lung mitochondria. *J. Biol. Chem.* **256**, 10986–10992.

12. Guyton, A.C. (1991): Overview of the circulation, and physics of pressure, flow, and resistance: Hemodynamics. In: *Textbook of Medical Physiology*, 8th edn., ed. A.C. Guyton, pp. 150–158. Philadelphia: W.B. Saunders.

13. Hardy, P., Abran, D., Li, D-Y., Fernandez, H., Varma, D.R. & Chemtob, S. (1994): Free radicals in retinal and choroidal blood flow autoregulation in the piglet: interaction with prostaglandins. *Invest. Ophthalmol. Vis. Sci.* **35**, 580–591.

13a. Hardy, P., Nuyt, A.M., Abran, D., St. Louis, J., Varma, D.R., & Chemtob, S. (1996): Nitric oxide in retinal and choroidal blood flow autoregulation in newborn pigs: interaction with prostaglandins. *Pediatr. Res.* **39**, 487–493.

14. Iadecola, C., Pelligrino, D.A., Moskowitz, M.A. & Lassen, N.A. (1994): Nitric oxide synthase inhibition and cerebrovascular regulation. *J. Cerebr. Blood Flow Metab.* **14**, 175–192.

15. Kelly, P.A.T., Thomas, C.L., Ritchie, I.M. & Arbuthnott, G.W. (1994): Cerebrovascular autoregulation in response to hypertension induced by N$^G$-nitro-L-arginine methyl ester. *Neuroscience.* **59**, 13–20.

16. Kiel, J.W. & Shepherd, A.P. (1992): Autoregulation of choroidal blood flow in the rabbit. *Investig. Ophthalmol. Vis. Sci.* **33**, 2399–2410.

17. Nielsen, J.C., Naashi, M.I., Anderson, R.E. (1988): The regional distribution of vitamin E and C in human adult and preterm infant retinas. *Invest. Ophthalmol. Vis. Sci.* **39**, (3) 22–26.

18. North A.J., Yuhanna, I.S., Zohre, G. & Shaul, P.W. (1995): Estrogen upregulates nitric oxide synthase gene expression in fetal pulmonary artery endothelial cells. *Pediatr. Res.* **37**, 31A.

19. Oliver, P.D. & Newsome, D.A. (1992): Mitochondrial superoxide dismutase in mature and developing human retinal pigment epithelium. *Invest. Ophthalmol. Vis. Sci.* **33**, 1909–1918.

20. Parent, R., Al-Obaidi, M. & Lavallee, M. (1993): Nitric oxide formation contributes to β-adrenergic dilation of resistance coronary vessels in conscious dogs. *Circ. Res.* **73**, 241–251.

21. Penn, J.S. (1990): Oxygen-induced retinopathy in the rat: Possible contribution of peroxidation reactions. *Doc. Ophthalmol.* **74**, 179–186.

22. Raij, L. (1993): Nitric Oxide and the Kidney. *Circulation* **87**, V26–V29.

23. Rici, B., (1987): Effects of hyperbaric, normobaric and hypobaric oxygen supplementation on retinal vessels in newborn rats: a preliminary study. *Exp. Eye. Res.* **44**, 459–464.

24. Stiris, T., Hansen, T.W.R., Odden, J-P., Mørkrid, L. & Bratlid, D. (1989): Effect of light and hyperoxia on ocular blood flow in the newborn piglet. *Biol. Neonate.* **55**, 191–196.

25. Ueeda, M., Silvia, S.K. & Olsson, R.A. (1992): Nitric oxide modulates coronary autoregulation in the guinea pig. *Circ. Res.* **70**, 1296–1303.

26. Vanhoutte, P.M. & Miller, V.M. (1985): Heterogeneity of endothelium-dependent responses in mammalian blood vessels. *J. Cardiovasc. Pharm.* **7**, S12–23.

27. Vargas, H.M., Ignarro, L.J. & Chaudhuri, G. (1990): Physiological release of nitric oxide is dependent on the level of vascular tone. *Eur. J. Pharmacol.* **190**, 393–397.

28. Wang, Q., Paulson, O.B. & Lassen, N.A. (1992): Is autoregulation of cerebral blood flow in rats influenced by nitro-L-arginine, a blocker of the synthesis of nitric oxide? *Acta Physiol. Scand.* **145**, 297–298.

*Guanidino Compounds : 2*, eds. by P.P. De Deyn, B. Marescau, I.A. Qureshi and A. Mori.
©1997 John Libbey & Company Ltd., pp. 31–36.

# Chapter 5

---

# Detection of nitric oxide from α-guanidinoglutaric acid by hydroxyl radicals using an electron spin resonance method with spin trap

---

## Midori HIRAMATSU and Makiko KOMATSU

*Division of Medical Science, Institute for Life Support Technology, Yamagata Technopolis Foundation,*
*2–2–1 Matsuei, Yamagata 990, Japan*

## Summary

Generation of nitric oxide from guanidino compounds was examined using electron spin resonance (ESR) spectrometry with spin trap, 2-phenyl-4,4,5,5-tetramethylimidazoline-1-oxyl 3-oxide (PTIO). 2-Phenyl-4,4,5,5-tetramethylimidazoline-1-oxyl (PTI) was formed after the reaction of nitric oxide with PTIO, and nine line signals of PTI were detected by an ESR spectrometer. PTI were observed in the solution of sodium nitrite with hydrochloric acid, and nitroprusside solution, respectively, suggesting that nitric oxide was generated in both solutions. Nine signals of PTI were observed in the solution of α-guanidinoglutaric acid with ferric chloride and hydrogen peroxide and the intensity increased in a dose-dependent manner. These results suggest that nitric oxide was generated from α-guanidinoglutaric acid through hydroxyl radicals.

## Introduction

There are many guanidino compounds in the brain and some of them, like guanidinoethanesulphonic acid[9], guanidinoacetic acid[4], α-guanidinoglutaric acid[13], γ-guanidinobutyric acid[3], N-acetylarginine[11], methylguanidinie[8] and homoarginine[14], induce epileptic discharges and/or convulsions in rats and cats. Recently nitric oxide has been reported to be generated from arginine by nitric oxide synthase, and nitric oxide generation is thought to derive from the amidino group[5]. Nitric oxide is considered as a second messenger in the brain and some papers have reported that nitric oxide may be related to seizures[7,12]. In the present study we examined generation of nitric oxide from other guanidino compounds using ×-band ESR spectrometry with the spin trap method[1].

## Experimental method

### Chemicals

2-Phenyl-4,4,5,5-tetramethylimidazoline-1-oxyl 3-oxide (PTIO) and 2-phenyl-4,4,5,5-tetramethylimidazoline-1-oxyl (PTI) was purchased from SIGMA Chemical Co. (St. Louis, MO, USA). All other chemicals and reagents were of the highest grade available.

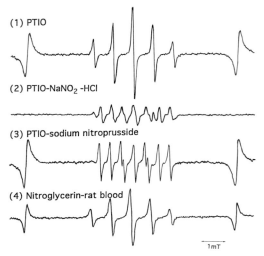

(1) PTIO

(2) PTIO-NaNO$_2$ -HCl

(3) PTIO-sodium nitroprusside

(4) Nitroglycerin-rat blood

1 mT

*Fig. 1. ESR spectra of PTIO and those obtained from the reaction mixture of PTIO plus system for NO generation. (1) Control, (2) sodium nitrite and hydrochloride, (3) sodium nitroprusside, and (4) nitroglycerin in the rat blood.*

## Animals

Male Wistar rats of 170–200 g were obtained from Funabashi Farm (Sendai, Japan). Animals were housed at 23 °C and 55 per cent humidity under 12 h cycle of light and dark (7:00 a.m. – 19:00 p.m. light). Animals were decapitated and the cerebral cortex was rapidly removed on an ice plate, and was used for analysis of nitric oxide.

## Analysis of nitric oxide

PTIO, sodium nitroprusside and sodium nitrite were dissolved in distilled water. 20 µl of 10 mM sodium nitroprusside and 180 µl of 3 µM PTI were mixed and were left for 30 min at room temperature. The mixture was then placed in a flat cell and the nitric oxide spin adduct of PTI in the mixture was measured by an ESR spectrometer (JES-RE1X ESR spectrometer, Tokyo, Japan). Conditions for analysis of nitric oxide were as follows: magnetic field, 335 ± 5 mT; response, 0.1 s; sweep time, 2 min; field modulation, 0.1 mT and room temperature.

## Results

PTIO shows five signal lines by ESR spectrometry, and when it reacts with nitric oxide, PTIO changes to PTI, which shows nine signal lines[1].

## (1) Sodium nitroprusside/saline

## (2) Sodium nitroprusside/rat brain homogenate

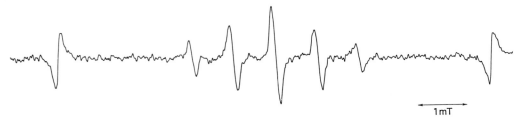

1 mT

*Fig. 2. ESR spectra of the mixture of PTIO, nitroprusside and rat brain homogenate. (1) Saline and (2) rat brain homogenate.*

**(1) NaNO$_2$-HCl / saline**

**(2) NaNO$_2$-HCl / rat brain homogenate**

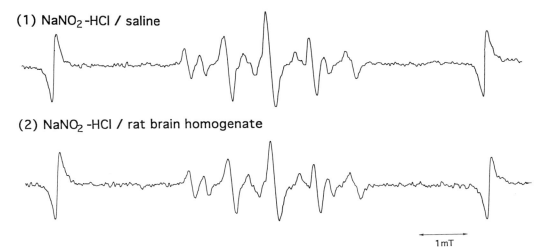

1mT

*Fig. 3. ESR spectra of the mixture of PTIO, sodium nitrite, hydrochloride and rat brain homogenate.*

The reaction mixture of PTIO, sodium nitrite and hydrochloric acid showed nine line signals of PTI (Figs. 1–2) and nitroprusside, which is a releaser of nitric oxide, also showed nine line signals (Fig. 1–3). That means PTIO reacted with nitric oxide generated from the system of sodium nitrite and hydrochloric acid and nitroprusside alone. However, no PTI formation was found in nitroglycerin in rat blood (Figs. 1–4).

All nine line signals of PTI were observed in the solution of nitroprusside; no PTI was found in the mixture of rat brain homogenate (Fig. 2). On the other hand, PTI was found in the mixture of sodium nitrite and hydrochloric acid and rat brain homogenate (Fig. 3).

The addition of ferric chloride and hydrogen peroxide to α-guanidinoglutaric acid solution induced generation of nine line signals of PTI but α-guanidinoglutaric acid solution alone did not. The addition of ferric chloride to α-guanidinoglutaric acid solution did not induce any nine line signals of PTI. The addition of hydrogen peroxide to the α-guanidinoglutaric acid solution generated nine line signals of PTI and the signal intensities were much higher in the use of addition of ferric chloride and hydrogen peroxide than those of hydrogen peroxide only (Fig. 4).

The signal intensity of PTI was examined with different concentrations of hydrogen peroxide and ferric chloride. The signal intensity of PTI increased in a dose-dependent manner (Fig. 5). Signal intensity of PTI from α-guanidinoglutaric acid solution with ferric chloride and hydrogen peroxide was increased by the concentration of α-guanidinoglutaric acid in a dose-dependent manner (Fig. 6).

On the other hand, no generation of nine line signals of PTI was observed from other guanidino compound solutions, such as methylguanidine, arginine and guanidinoacetic acid with ferric chloride and hydrogen peroxide (Fig. 7).

## Discussion

α-Guanidinoglutaric acid is a derivative of glutamic acid, in which a hydrogen of the amino group of glutamate is substituted by an amidino group (Fig. 8). The administration of cobalt powder on the sensori motor cortex of cat induced epileptic discharges[10] and intraventricular injection of it induced epileptic-like discharges in rat[3]. Though a very low level of α-guanidinoglutaric acid is present in the cat cortex, α-guanidinoglutaric acid level was much elevated in the cortex 24 h after the administration of cobalt powder in the sensori motor cortex of cat[10].

α-Guanidinoglutaric acid in aqueous solutions generated a lot of free radicals, such as hydroxyl radicals and carbon–centred radicals compared with other guanidino compounds (unpublished data).

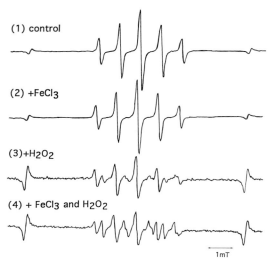

Fig. 4. ESR spectra of the mixture of PTIO, α–guanidinoglutaric acid and rat brain homogenate with (1) control, (2) ferric chloride, (3) hydrogen peroxide, (4) ferric chloride and hydrogen peroxide.

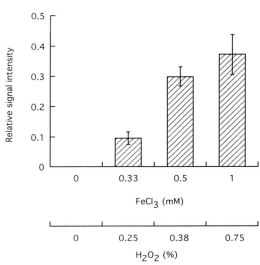

Fig. 5. Effect of ferric chloride and hydrogen peroxide on PTI formation in the mixture of PTIO, α-guanidinoglutaric acid and rat brain homogenate. Each value represents the mean ± SEM of 4 determinations.

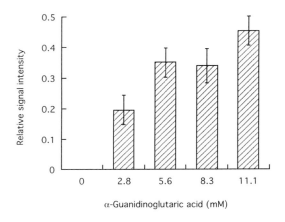

Fig. 6. Effect of α-guanidinoglutaric acid on PTI formation in the mixture of PTIO, ferric chloride, hydrogen peroxide and rat brain homogenate. Each value represents the mean ± SEM of 4 determinations.

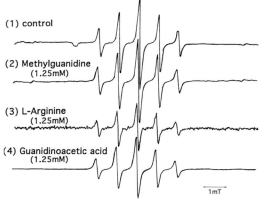

Fig. 7. ESR spectra of the mixture of PTIO, guanidino compounds, ferric chloride and hydrogen peroxide, and rat brain homogenate. (1) control, (2) methylguanidine, (3) L-arginine, (4) guanidinoacetic acid.

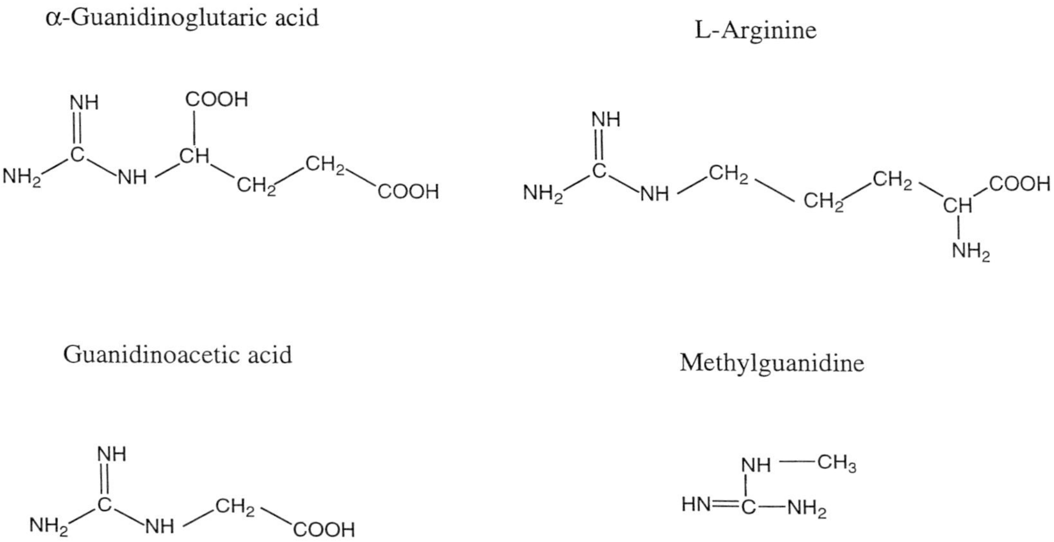

α-Guanidinoglutaric acid

L-Arginine

Guanidinoacetic acid

Methylguanidine

*Fig. 8. Chemical structure of guanidino compounds of α-guanidinoglutaric acid, L-arginine, guanidinoacetic acid and methylguanidine.*

Therefore the seizure mechanism for cobalt-induced epileptic discharges was thought to be due to the generation of free radicals by α-guanidinoglutaric acid application.

We could detect generated nitric oxide from the system of sodium nitrite and hydrochloric acid, and nitroprusside using ESR spectrometry with the spin trap PTIO. However, the generation of nitric oxide was not found in the solution of nitroglycerin and rat blood. Recently nitric oxide production from nitroglycerin and sodium nitrite was found in rat blood using ESR spectrometry with haemo-globin binding at 77 °K[6]. As the method for detection of nitric oxide formation was different between our spin trap method at room temperature and the haemoglobin binding method at 77 °K, discrepancy for detection of nitric oxide formation from nitroglycerin in rat blood may be due to differences in methodology.

We found that the scavenging activity for nitric oxide, generated from nitroprusside and not that from sodium nitrite, was present in the brain homogenate. The amount of nitric oxide generated from sodium nitrite and nitroprusside may be different.

In this experiment α-guanidinoglutaric acid was found to generate nitric oxide by hydroxyl radicals generated from ferric chloride and hydrogen peroxide using the spin trap PTIO (Fig. 9). In relation to convulsions, some papers have reported that, nitric oxide may decrease the frequency of appear-ance of amygdala kindling[12], delay the formation of early stage of kainate-limbic status epilepticus and render the seizure pattern worse[7]. On the other hand, kainate accelerates nitric oxide release[2]. There is a discrepancy for a role of nitric oxide in seizure mechanism. However, our results suggest that nitric oxide may be involved in hydroxyl radical generation in brain during cobalt-induced seizures in rat and cat.

## References

1.   Akaike, T., Yoshida, M., Miyamoto, Y., Sato, K., Kohno, M., Sasamoto, K., Miyazaki, K., Ueda, S. & Maeda, H. (1993): Antagonistic action of imidazolineoxyl N-oxides against endothelium-derived relaxing factor/ •NO through a radical reaction. *Biochemistry* **32,** 827–832.

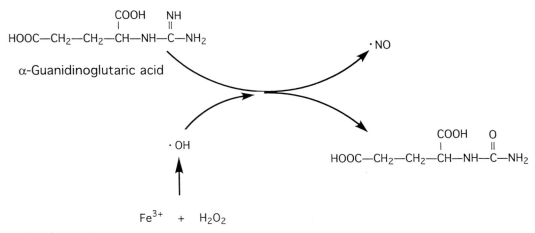

Fig. 9. *Possible mechanism for the formation of nitric oxide from α-guanidinoglutaric acid with ferric chloride and hydrogen peroxide.*

2.  Balcioglu, A. & Maher, T.J. (1993): Determination of kinic acid – induced release of nitric oxide using a novel hemoglobin trapping technique with microdialysis. *J. Neurochem.* **61,** 2311–2313.

3.  Jinnai, D., Sawai, A. & Mori, A. (1966): γ-Guanidinobutyric acid as a convulsive substance. *Nature* **212,** 617.

4.  Jinnai, D., Mori, A., Mukawa, J., Ohkusu, H., Hosotani, M., Mizuno, A. & Tye, L.C. (1969): Biological and physical studies on guanidino compounds induced convulsion. *Jpn. J. Brain Physiol.* **106,** 3668–3673.

5.  Klatt, P., Schmidt, K., Uray, G. & Mayer, B. (1993): Multiple catalytic functions of brain nitric oxide synthase. *J. Biol. Chem.* **268,** 14781–14787.

6.  Kohno, M., Masumizu, T. & Mori, A. (1995): ESR demonstration of nitric oxide production from nitroglycerin and sodium nitrite in the blood of rats. *Free Radical Biol. Med.* **18,** 451–457.

7.  Lerver-Natoli, M., Bockaert, J. & Rondouin, G. (1993): Role of nitric oxide in limbic epilepsy. *Epilepsia 34* (Suppl. 2), 99.

8.  Matsumoto, M., Kobayashi, K., Kishikawa, H. & Mori, A. (1976): Convulsive activity of methylguanidine in cats and rabbits. *IRCS Med. Sci.* **4,** 65.

9.  Mizuno, A., Mukawa, J., Kobayashi, K. & Mori, A. (1975): Convulsive activity of taurocyamine in cats and rabbits. *IRCS Med. Sci.* **3,** 385.

10. Mori, A., Watanabe, Y., Shindo, S., Akagi, M. & Hiramatsu, M. (1983) : α-guanidinoglutaric acid and epilepsy. In: *Urea Cycle Diseases*, eds. A. Lowenthal, A. Mori & B. Marescau, pp. 465–470. New York: Plenum.

11. Ohkusu, H. (1970): Isolation of N-acetyl-L-arginine from calf brain and convulsive seizure induced by this substance. *Osaka-Igakkai-Zasshi* **21,** 49–50.

12. Rondouin, G., Lerner-Natoli, M., Manzoni, O., Lafon-Cazal, M. & Bockaert, J. (1992): A nitric oxide (NO) synthase inhibitor accelerates amygdala kindling. *Neuro Report* **3,** 805–808.

13. Shiraga, H., Hiramatsu, M. & Mori, A. (1986): Convulsive activity of α-guanidinoglutaric acid and the possible involvement of 5-hydroxytryptamine in the α-guanidinoglutaric acid induced seizure mechanism. *J. Neurochem.* **47,** 1832–1836.

14. Yokoi, I., Toma, J. & Mori, A. (1984): The effect of homoarginine on the EEG of rats. *Neurochem. Pathol.* **2,** 295–300.

*Guanidino Compounds : 2*, eds. by P.P. De Deyn, B. Marescau, I.A. Qureshi and A. Mori.
©1997 John Libbey & Company Ltd., pp. 37–41.

# Chapter 6

# A new chemiluminescence detection method of nitric oxide and nitric oxide synthase activity

[1]Moto-o NAKAJIMA, [2]Yukiko KURASHIMA and [2]Hiroyasu ESWUMI

[1]*Research & Development Division, Kikkoman Corporation, 399 Noda, Noda-shi, Chiba-ken 278 and* [2]*National Cancer Center, Research Institute, East, 6–5–1 Kashiwanoha, Kashiwa-shi, Ciba-ken 277, Japan*

## Summary

Luminol and MCLA as chemiluminescence probes were recognized to emit strongly for hydrogen peroxide and superoxide, but were non–specific for peroxynitrite. We found however that the firefly luciferin chemiluminescence was specific for peroxynitrite. NO generated from SNAP was detected by firefly luciferin–hydrogen peroxide chemiluminescence. Peroxynitrite from SIN-1 was detected by only firefly luciferin chemiluminescence. These chemiluminescences disappeared by the addition of carboxy-PTIO as NO scavenger. We could detect the NO synthase activity by the firefly luciferin–hydrogen peroxide chemiluminescence method. However, this method is not yet sensitive enough as a routine assay method for measuring NO synthase activity.

## Introduction

Nitric oxide (NO) is generated from L-arginine by nitric oxide synthase. This NO appears to be responsible for cytotoxic effects of macrophages and neutrophilis[5], for vasodilatation mediated by endothelial cells[9] and for cell-to-cell communication in the nervous system[3]. It is extremely labile and easily oxidized by oxygen to $NO_2^-$ and $NO_3^-$, with a half-life of 6 s[6].

NO can be detected by a chemiluminescence assay method based on the reaction of NO with ozone[9], a spectrophotometric method[2], an electron spin resonance method[10] and a gas chromatography–mass spectrometry method[11]. However, due to the limitation of low sensitivity, non-specificity, and/or requirement for expensive instruments, none of these methods can be used for routine analysis in physiological solutions.

Generally, the chemiluminescence assay is considered as a useful method with a high sensitivity, and as a real-time assay for unstable radicals at low concentrations in physiological solutions. NO is reported to react with $O_2^-$ or $H_2O_2$ to form peroxynitrite[1,11], which is a stronger oxidizing species than $O_2^-$ or $H_2O_2$. Luminol and *Cypridina* luciferin analog (MCLA) chemiluminescence have been widely used to detect the production of reactive oxygen species. Recently, peroxynitrite-induced luminol or MCLA chemiluminescence methods were reported[4,7,12]. On the other hand, the firefly luciferin and luciferase bioluminescence has been widely used for ATP determination[8]. Here, we report that firefly luciferin reacts with peroxynitrite and emits a chemiluminescence without luciferase and ATP. We have extended this observation to detection of NO synthase.

## Materials and methods

### Apparatus

Chemiluminescence was measured with Lumat LB 9501 luminometer (Berthold, Germany). Measuring mode used was the ratemeter mode.

### Chemicals

Firefly luciferin was purchased from Sigma. S-nitroso-N-acetyl-D,L-penicillamine (SNAP), carboxy-2-phenyl-4,4,5,5-tetramethyl-imidazoline-3-oxide-1-oxyl (carboxy-PTIO), 3-morpholinosydnomine (SIN-1) were obtained from Dojindo Laboratories (Kumamoto, Japan), Luminol, $H_2O_2$ and xanthine were purchased from Wako Pure Chemical (Osaka, Japan), MCLA from Tokyo Kasei Kogyo (Tokyo, Japan). Xanthine oxidase was obtained from Boehringer Mannheim.

### NO synthase preparation

NO synthase activity was induced in the liver of male Sprague-Dawley rats by injection of heat-killed *P. acnes* and lipopolysaccharide. Rat livers were homogenized and centrifuged. The resulting supernatant was partially purified by 2′, 5′-ADP-agarose chromatography.

## Results and discussion

### Chemiluminescence for hydrogen peroxide or superoxide

For the generation of $O_2^-$ we used the xanthine–xanthine oxidase system. Table 1 shows that the luminol and MCLA chemiluminescence by $H_2O_2$ or $O_2^-$ were very strong. However, the firefly luciferin emitted a weak luminescence with $H_2O_2$ or $O_2^-$. On the other hand, luminol or MCLA

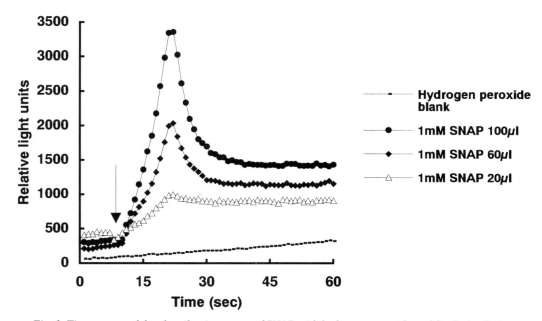

*Fig. 1. Time course of the chemiluminescence of SNAP with hydrogen peroxide and firefly luciferin.*

chemiluminescences by peroxynitrite were confirmed in our experiments. For detection of NO with luminol or MCLA chemiluminescence, we have to evaluate the effect of $H_2O_2$ or $O_2{}^-$. On the other hand, we found the firefly luciferin chemiluminescence with peroxynitrite. This chemiluminescence was specific for peroxynitrite. Therefore, firefly luciferin is expected to be excellent for the chemiluminescence probe against peroxynitrite.

**Table 1. Chemiluminescence of luminol, MCLA and firefly luciferin for hydrogen peroxide and superoxide**

| Probe | Hydrogen peroxide | | Superoxide |
|---|---|---|---|
| Luminol (1 μM) | 12879 | (relative light units) | 149322 |
| MCLA (1 μM) | 73913 | | 1403285 |
| Firefly luciferin (1 μM) | 41 | | 163 |

## Detection of NO from SNAP by firefly luciferin chemiluminescence

To examine the optimum conditions of chemiluminescence reaction for firefly luciferin and NO, we used SNAP solution as the NO generation reagent. Kolthoff's buffers were used for assay pH range 7.5 to 9.0 (0.1 M $KH_2PO_4$–0.05 M $Na_2B_4O_7$), 9.5 to 11 (0.05 M $Na_2CO_3$–0.05 M $Na_2B_4O_7$). SNAP, $H_2O_2$ and firefly luciferin concentrations were 1 mM, 50 mM and 1 μM, respectively. Optimum pH was 9.5. The emission intensity at pH 8.0 was only 8.9 per cent at optimum pH. Fig. 1 shows the time course of the chemiluminescence of SNAP with 100 mM $H_2O_2$ and 1 μM firefly luciferin. Chemiluminescence peaks were about 10 seconds after the addition of $H_2O_2$. Each peak height was proportional to the volume of SNAP solution. Figure 2 shows the calibration curve for SNAP obtained from chemiluminescence with 1 μM firefly luciferin and 10 mM $H_2O_2$. Linearity was obtained for SNAP concentrations less than 0.4 mM. To estimate the true amount of the NO generation from SNAP, we added the carboxy-PTIO solution as NO scavenger to the reaction mixture. We found that the firefly luciferin chemiluminescence for the NO generation from SNAP was abolished by addition of carboxy-PTIO at more than 10 μM. When 5 μM carboxy-PTIO was added, the NO generation started after a delay of about 100 s. Therefore, the firefly luciferin chemiluminescence with $H_2O_2$ from SNAP was confirmed to be caused by NO generation. We thought that 10 mM SNAP generated truly about 5 μM NO.

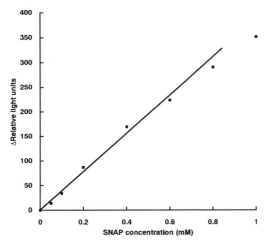

*Fig. 2. Calibration curve of nitric acid from SNAP by chemiluminescence with firefly luciferin and hydrogen peroxide.*

## Detection of peroxynitrite from SIN-1 by firefly luciferin chemiluminescence

SIN-1 is known to spontaneously generate peroxynitrite via $O_2{}^-$ and NO. The firefly luciferin with SIN-1 emitted luminescence without $H_2O_2$ (see Fig. 3). The chemiluminescence peak of firefly luciferin and SIN-1 reaction appeared 60s after the addition of SIN-1 and it was about 50s later than the peak of firefly luciferin and SNAP reaction. This delay of peak was thought to be due to different generation type of peroxynitrite from SIN-1. Linearity of the calibration curve for SIN-1 was obtained in the concentration range less than 2 mM. Figure 3 shows the disappearance of firefly chemiluminescence for NO by carboxy-PTIO addition. Addition of 2 μM carboxy-PTIO decreased to one-tenth the chemiluminescence intensity for SIN-1. The firefly luciferin chemiluminescence without $H_2O_2$ from SIN-1 was confirmed by NO generation.

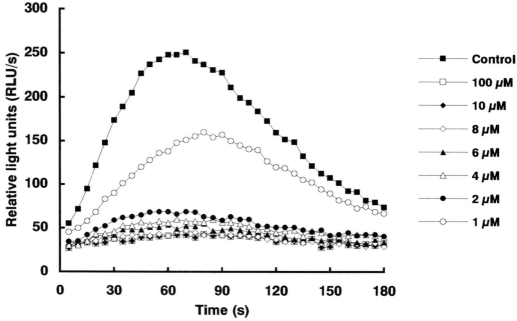

*Fig. 3. Effect of carboxy-PTIO on firefly luciferin chemiluminescence with 10 mM SIN-1 at pH 9.5. Each symbol shows the concentration of carboxy-PTIO.*

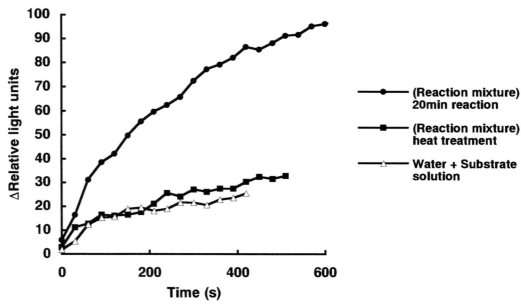

*Fig. 4. Chemiluminescence change of firefly luciferin and hydrogen peroxide on NO synthase activity.*

**Detection of NO synthase activity by firefly luciferin chemiluminescence**

We examined the application of firefly chemiluminescence to detection of NO synthase activity. Figure 4 shows the chemiluminescence change of firefly luciferin and $H_2O_2$ at pH 8.0 on inducible rat liver NO synthase activity. When the reaction mixture of substrate solution and NO synthase preparation was added to the firefly luciferin and $H_2O_2$ mixture, an increase of chemiluminescence intensity by NO generation was recognized. However, in the case of heat treated reaction mixture, chemiluminescence intensity did not increase similarly the control. This inducible NO synthase activity was highest at pH 7.73, but its activity at pH 8.0 was 58 per cent at optimum pH. NO synthase activity was inhibited by 98 per cent by the addition of $H_2O_2$, final concentration 0.04 per cent. On the other hand, the emission intensity at pH 8.0 was only 8.9 per cent at optimum pH as described previously. Consequently, the sensitivity of our NO detection was too low for the NO synthase assay.

## References

1.  Beckman, J.S., Beckman, T.W., Chen, J., Marshall, P.A. & Freeman, B.A. (1990): Apparent hydroxyl radical production by peroxynitrite: implication for endothelial injury from nitric oxide and superoxide. *Proc. Natl. Acad. Sci.* USA **87**, 1620–1624.

2.  Bredt, D.S. & Snyder, S.H. (1989): Nitric oxide mediates glutamate-linked enhancement of cGMP levels in cerebellum. *Proc. Natl. Acad. Sci.* USA **86**, 9030–9033.

3.  Bredt, D.S. & Snyder, S.H. (1990): Isolation of nitric oxide synthetase, a calmodulin- requiring enzyme. *Proc. Natl. Acad. Sci.* USA **87**, 682–685.

4.  Fukahori, M., Ichimori, K., Nakazawa, H. & Okino, H. (1992): Real-time monitoring of peroxynitrite generation in the reaction of superoxide and NO by MCLA chemiluminescence. In: *Oxygen radicals*, ed. K. Yagi *et al.*, pp. 207–210. Amsterdam: Elsevier.

5.  Hibbs, J.B., Taintor, R.R., Vavrin, Z. & Rachlin, E.M. (1988): Nitric oxide: a cytotoxic activated macrophage effector molecule. *Biochem. Biophys. Res. Commun.* **157**, 87–94

6.  Kelm, M., Feelish, M., Spahr, R., Piper, H.M., Noak, E. & Schuraderr, J. (1988): Quantitative and kinetic characterization of nitric oxide and EDRF released from cultured endothelial cells. *Biochem. Biophys. Res. Commun.* **154**, 236–241.

7.  Kikuchi, K., Hayakawa, H., Nagano, T., Hirata, Y., Sugimoto, T. & Hirobe, M. (1992): New method of detecting nitric oxide production. *Chem. Pharm. Bull. Tokyo,* **40**, 2233–2235.

8.  Leach, F. R. (1981): ATP determination with firefly luciferase. *J. Appl. Biochem.* **3**, 473–517.

9.  Palmer, R.M.J., Ferrige, A. G. & Moncada, S. (1987): Nitric oxide release accounts for the biological activity of endothelium-derived relaxing factor. *Nature* **327**, 524–526.

10.  Palmer, R.M.J., Ashton, D. S. & Moncada, S. (1988): Vascular endothelial cells synthesize nitric oxide from L-arginine. *Nature* **333**, 664–666.

11.  Petriconi, G. L. & Papee, H.M. (1966): Aqueous solutions of 'sodium pernitrite' from alkaline hydrogen peroxide and nitric oxide. *Can. J. Chem.* **44**, 977–980.

12.  Radi, R., Cosgrove, T.P., Beckman, J.S. & Freeman, B.A. (1993): Peroxynitrite-induced luminol chemiluminescence. *Biochem. J.* **290**, 51–57.

*Guanidino Compounds : 2*, eds. by P.P. De Deyn, B. Marescau, I.A. Qureshi and A. Mori.
©1997 John Libbey & Company Ltd., pp. 43–49.

# Chapter 7

# Arginine, a possible adaptogen?

V.S. SHUGALEY, L.V. MOGILNITSKAYA, A.A. ANANYAN, N.P. MILUTINA and
S.I. SADEKOVA

*Rostov State University, Department of Biochemistry, B. Sadovaya, 105, 344006, Rostov-on Don, Russia*

## Summary

The levels of arginine, urea and guanidino compounds were determined 1.5 and 3.5 h after the intraperitoneal injection of arginine (120 mg/100 g) in the blood, liver and brain of the rat. The highest concentrations of the blood arginine, liver urea and brain guanidino compounds were found 1.5 h after arginine administration. The reduction of free serum hemoglobin and the activation of erythrocyte superoxide dismutase (SOD) and catalase by the arginine were shown. The activity of cytochrome P-450 was increased *in vivo* and *in vitro* in liver microsomes after addition of exogenous arginine. There were no apparent effects of arginine on the liver and testicle microsomes. Erythrocytes diene conjugates and Shiff bases were significantly decreased after arginine injection as well as the microviscosity of erythrocytes and liver microsome membranes. The addition of arginine reduced the plasma lipid peroxidation initiated by ferrum ascorbate, whereas the formation of superoxide anion *in vitro* significantly decreased. These data suggested an important role of arginine in the regulation of metabolic pathways connected with the adaptation to unfavourable ambient conditions.

## Introduction

At the present time compounds of varying chemical nature and of different origin are described as adaptogens. A common feature is their biological activity, aimed at increasing the non-specific stability of an organism under unfavourable ecologic conditions.

Adaptogens improve the protective potential and bring about an adaptive restructuring of metabolism. The subject of adaptogens is being intensively studied nowadays, yet there is no classification of these substances so far. Moreover, the molecular mechanisms of their activity are unknown, and pharmacologists do not single out these substances as belonging to a certain group. The physiological criteria which enable us to classify a given compound as an adaptogen may not be called strict ones. The empirism of the two concepts, adaptogens and adaptation, is quite obvious. One cannot say these concepts have an absolutely definite meaning, and they have plenty of interpretations from the point of view of both physiology and biochemistry[10]. The chemical compounds which could function as such signals are primarily products of catabolism, naturally excreted into the environment. Such substances are the nitrogen-containing products of catabolism like ammonia, urea, amino acids and the derivatives, nitrous bases of purine and pyrimidine, and the product of their oxidation (uric acid). The presence of nitrogen in molecules was an additional factor of specificity which later made it possible for some of these to stand out as special regulators and receptor-active substances. In our laboratory we focus our attention on the biological activity of these catabolic products[6,7]. The phenomenon of the biological activity of urea, which was studied by Z.S. Gershenovich *et al.*[5,7], was the starting point of the research into the biological role of arginine. Among the nitrogen-containing catabolic products, arginine has a special place. On the one hand, it is an amino acid, that is, an active participant of primary metabolism, the structural component of all enzymes, dominating component of positively charged cation proteins (histones, macrophages cation proteins, the precursor of creatine phosphate and other phosphagens, the precursor of low-mol. wt regulating substances like polyamines, nitric oxide etc). On the other hand, arginine is related to nitrogen catabolism by means of urea synthesis. The characteristic feature of arginine in metabolism is its polyfunctionality, which seems to be an indispensable quality of all natural adaptogens[9].

## Methods

The contents of arginine, guanidino compounds, and urea were estimated, according to Gershenovich et al.[7] and Krichevskaya et al.[12]. The lysosomal and microsomal fractions were pretreated according to Milutina et al.[15], Mogilnitskaya et al.[17] and Shugaley et al.[18]. The activity of microsomal oxidation was measured using amidopyrine and aniline as substrates[2]. Free serum hemoglobin was estimated according to Milutina et al.[15]. Diene conjugates and Shiff bases were measured as products of lipid peroxidation (LP) from absorption at 233 nm and fluorescence at 440 nm[14,17]. The microviscosity of lipid–lipid and lipid–protein zones of microsomal membranes was estimated with pyrene as a fluorescent probe, using the relation of its eximere and monomere (F/F) at the excitation wavelength 334 nm and 286 nm respectively. As the microviscosity was reduced this coefficient increased. The activity of catalase and SOD was estimated according to Milutina et al.[14,15], Mogilnitskaya et al.[17] and Shugaley et al.[18].

## Results and discussion

We have studied the distribution in the tissues of arginine and the products of its transformation after it was injected.

**Table 1. The contents of arginine, urea and guanidino compounds in rat blood, liver and brain after intraperitoneal administration of arginine (120 mg/100 g)**

|  | Arginine mg/100g or 100 ml | Urea µM/g or ml | Guanidino compounds mg/100g or 100 ml |
|---|---|---|---|
| *Blood* |  |  |  |
| control | $2.03 \pm 0.3$ (n = 8) | $4.91 \pm 0.2$ (n =15) | $2.73 \pm 0.06$ (n = 8) |
| 1.5 h after administration | $13.35 \pm 1.24$ (n = 8) | $9.71 \pm 0.58$ (n = 12) | $4.07 \pm 0.35$ (n = 8) |
|  | + 557% $P < 0.001$ | + 40% $P < 0.001$ | + 91% $P < 0.001$ |
| 3.5 h after administration | $3.12 \pm 0.22$ (n = 8) | $6.87 \pm 0.42$ (n = 12) | $2.17 \pm 0.28$ (n = 8) |
|  | + 54% $P < 0.001$ | + 40% $P < 0.001$ | + 27% $P < 0.05$ |
| *Liver* |  |  |  |
| control | $1.27 \pm 0.1$ (n = 8) | $4.58 \pm 0.43$ (n = 10) | $2.67 \pm 0.18$ (n = 8) |
| 1.5 h after administration | $1.45 \pm 0.21$ (n = 8) | $9.39 \pm 0.22$ (n = 8) | $2.80 \pm 0.23$ (n = 8) |
|  | +14% $P > 0.2$ | + 104% $P < 0.001$ | + 5% $P > 0.5$ |
| 3.5 h after administration | $1.38 \pm 0.21$ (n = 8) | $9.10 \pm 0.47$ (n = 12) | $2.79 \pm 0.25$ (n = 8) |
|  | + 8% $P > 0.2$ | + 98% $P < 0.001$ | +4% $P > 0.2$ |
| *Brain* |  |  |  |
| control | $0.56 \pm 0.08$ (n = 8) | $2.47 \pm 0.1$ (n =18) | $0.68 \pm 0.07$ (n = 8) |
| 1.5 h after administration | $0.71 \pm 0.05$ (n = 8) | $4.24 \pm 0.18$ (n = 12) | $2.65 \pm 0.26$ (n = 8) |
|  | + 26% $P < 0.02$ | + 72% $P < 0.001$ | + 290% $P < 0.001$ |
| 3.5 h after administration | $0.78 \pm 0.09$ (n = 8) | $4.52 \pm 0.30$ (n = 9) | $1.61 \pm 0.12$ (n = 8) |
|  | + 40% $P < 0.05$ | + 83% $P < 0.001$ | + 137% $P < 0.001$ |

Values are given as mean ± SE, Student's *t*-test.

The highest concentration of arginine is observed in the blood one hour after intraperitonial injection (Table 1). The level of arginine is observed to become normal 24 h after administration. The principal place of transformation of the introduced arginine is the liver, where arginine is split into urea and ornithine by arginase. The concentrations of urea and ornithine are also observed to become higher. Arginine, while metabolized in the organism, is a source of a large number of other biologically active compounds such as urea, the polyamines spermine and spermidine[12], guanidino compounds[3,4] and, as recent works on the subject have shown, it is also a source of nitric oxide, which makes the adaptogenic properties more balanced, complex and broad. Indeed, we have demonstrated that introduction of arginine produces a stabilizing effect on cellular and intracellular membranes[1,14,17] (Table 2). The effect is observed in the experiments both *in vivo* and *in vitro* in the process of incubating cells and subcellular organelles in the presence of arginine in various concentrations.

**Table 2. The effect of arginine administration on the activity of rat brain and liver lysosomal acid proteinase (in µM N/100 mg protein/1 h)**

| | Distribution of acid poteinase | |
| --- | --- | --- |
| | In lysomes | Free activity |
| *Brain* | | |
| control | $54.40 \pm 4.00$ (n = 9) | $18.61 \pm 2.07$ (n = 9) |
| With arginine | $65.20 \pm 5.18$ ( n = 9) | $15.08 \pm 1.08$ (n = 8) |
| | + 20%  $P < 0.1$ | –23%  $P < 0.1$ |
| *Liver* | | |
| Control | $155.00 \pm 18.70$ (n = 9) | $13.81 \pm 1.29$ (n = 9) |
| With arginine | $156.10 \pm 10.00$ ( n = 9) | $16.24 \pm 1.61$ (n = 9) |
| | + 1% $P > 0.1$ | + 18%  $P > 0.1$ |

**Table 3. The effect of arginine on the content of free serum haemoglobin, superoxide dismutase (SOD) and catalase activity in erythrocytes**

| | Control | 1.5 h after administration | 3.5 h after administration |
| --- | --- | --- | --- |
| Free serum | 15.9± 1.76 | $13.58 \pm 0.56$ (n = 10) | $10.09 \pm 0.15$ (n = 10) |
| haemoglobin (mg %) | n=13 | –18%  $P < 0.1$ | –36%  $P <0.02$ |
| SOD | $15.5 \pm 2.3$ | | $28.8 \pm 3.54$( n = 11) |
| (U/mg Hb/min) | (n = 10) | | + 81%  $P < 0.001$ |
| Catalase | $50.4 \pm 7.36$ | | $71.5 \pm 10.2$ (n = 10) |
| (U/mg Hb/min) | (n = 10) | | + 42%  $P < 0.001$ |

The most detailed research has been done on the effect of arginine on the properties of erythrocytes (Table 3) microsomes and lysosomes. The effect of arginine on the subcellular organelles is not confined to structural modifications in membranes, but is also manifested on the level of enzymes. Arginine regulates the activity of lysosome and non-lysosome proteinases[16]. Arginine considerably activates (2–3 fold) the microsomal oxidation of cytochrome P-450 (Table 4), which may be of particular importance if one keeps in mind the detoxication and protective role of that system[17,19]. It has already been determined in the brain that arginine activates enzymes such as glutaminase and glutamate decarboxylase, the latter being directly related to the formation of GABA[12]. We have shown in various experiments that in the presence of arginine the concentration of LP products (diene conjugates, malonyl dialdehyde, Shiff bases) goes down (Table 5 and 6). This phenomenon has no organo-specific character and is observed in erythrocytes, liver, brain, lung, testicles etc.[17,18]. One cannot rule out the possibility that arginine regulates the principal enzymes of antioxidant protection such as SOD, catalase or peroxidase.

Thus, on the biochemical level the adaptogenic properties of arginine and the products of its transformation could be realized through several routes, namely through the activation of the antioxidation system, stabilization and regulation of lysosome apparatus activity, activation of the cytochrome P-450 system, and stabilization of cellular and subcellular membranes.

**Table 4. The effect of arginine on the activity of cytochrome P-450 in liver microsomes**

| | In vivo | | In vitro | | |
|---|---|---|---|---|---|
| | Control | Arginine | Control | Arginine (0.5 µM) | Arginine (1 µM) |
| Aniline hydroxylase (nM amino-phenol/min) | $0.32 \pm 0.09$ (n = 9) | $0.56 \pm 0.04$ + 74% (n = 9) $P < 0.001$ | $0.34 \pm 0.04$ (n = 10) | $0.42 \pm 0.04$ + 24% (n = 10) $P < 0.001$ | $0.35 \pm 0.04$ +2% (n = 10) $P < 0.1$ |
| Amidopyrine demethylase (nM formal-dehyde/mg protein/min) | $2.16 \pm 0.31$ (n = 10) | $7.13 \pm 0.76$ + 230% (n = 10) $P < 0.001$ | $4.28 \pm 0.66$ (n = 10) | $6.35 \pm 0.66$ + 48% (n = 10) $P < 0.001$ | $5.56 \pm 0.96$ + 30% (n = 10) $P < 0.02$ |

The consequences of arginine interference into the functioning of any of these routes may be quite substantial, because these metabolic links perform protective functions and are involved in the reciprocal reactions under all sorts of effects on an organism or cell. Therefore the protective properties of arginine are non-specific. Arginine takes part in protein biosynthesis and at the same time it is a source of regulators like nitric oxide, which is formed from arginine with the participation of the microsomal enzymes. Nitric oxide can interact with free radicals, thus displaying its antioxidant effect.

**Table 5. The effect of arginine administration on the contents of diene conjugates, Shiff bases and microviscosity of rat liver, testicle microsomes and erythrocyte membranes**

| | Liver | | Testicles | | Erythrocytes | |
|---|---|---|---|---|---|---|
| | Control | Arginine | Control | Arginine | Control | Arginine |
| Diene conjugates (µmol/mg lipid) | $9.06 \pm 0.93$ (n = 8) | $10.57 \pm 1.61$ +15% (n=8) $P > 0.2$ | $10.10 \pm 1.85$ (n = 8) | $8.27 \pm 2.62$ −26% (n=8) $P > 0.2$ | $5.59 \pm 0.34$ (n = 10) | $4.51 \pm 0.33$ −19% (n=10) $P < 0.02$ |
| Shif bases (U fluorescence/ mg lipid) | $2.21 \pm 0.22$ (n = 13) | $2.00 \pm 0.12$ −10% (n=13) $P > 0.5$ | $2.61 \pm 0.36$ (n = 8) | $2.18 \pm 0.45$ − 16% (n=8) $P > 0.2$ | $6.62 \pm 0.35$ (n = 6) | $5.72 \pm 0.25$ −15% (n=6) $P < 0.05$ |
| Microviscosity (with pyrene as fluorescent probe at 334 nm) | 2.47 (n = 10) | 1.61 $P < 0.001$ (n=40) | | | 1.82 (n = 9) | 1.59 $P < 0.001$ (n=9) |

Earlier we found that arginine (120 mg per 100 g weight) introduction into rats prior to the exposition to hyperoxia made the latent period before the beginning of hyperoxial convulsions 2 to 3 times longer[5,11,12]. The analogous dose of arginine produced an anti-hypothermal effect and facilitated quicker normalization of temperature after exposition to artificial hypothermia[13]. Introduction of

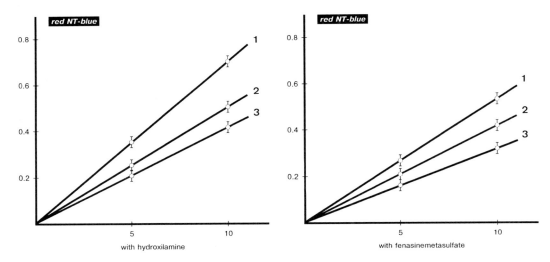

*Fig. 1. The inhibition of superoxide anion formation by arginine (systems with NADH-fenasinemetasulfate and hydroxilamine autooxidation with nitrotetrazole). 1 = Without arginine (n = 10); 2 = 0.5 μM arginine (inhibition 14–17%, P < 0.05); 3 = 1.0 μM arginine (inhibition 15-16%, P < 0.05).*

arginine had an anti-stress effect under the stress caused by cold and immobilization[1,13]. Arginine also displayed an anti-hypoxial effect which was manifested by the fact that animals lived 3 to 6 times longer under the acute hypoxia[17]; however the relationship between the dose and effect was not studied completely. The dose of 60 to 120 mg produced an approximately similar effect, the activity of 30 mg was low. Arginine is popularly used as an antidote, especially in the case of poisoning by ammonium salts[7,8]. Arginine produces a positive effect in the case of some forms of tumours in animals and human beings[7]. Arginine infusions also influence insulin and glucagon levels[20]. The above-mentioned facts make it possible to define arginine as belonging to biologically active compounds with a wide range of effects.

**Table 6. The effect of arginine on the plasma lipid peroxidation initiated by ferrum-ascorbate (in per cent to control)**

|  | Control | Ferrum-ascorbate induction LP | | |
|---|---|---|---|---|
|  |  | Without arginine | Arginine 0.5 μM | Arginine 1 μM |
| Diene conjugates (μmol/mg lipid) | 5.25 ± 0.32 (n = 11) | 7.50 ± 0.45 + 40%( n = 9) P < 0.001 | 3.91 ± 0.39 −48% (n = 9) P<0.001 | 4.87 ± 0.61 −35% (n = 10) P < 0.001 |
| Shift bases (U fluorescence/ mg lipid) | 1.33 ± 0.3 (n = 10) | 3.55 ± 0.70 +167% (n = 9) P < 0.001 | 1.73 ± 0.26 −50% (n = 11) P < 0.05 | 1.78 ± 0.28 −51% (n = 11) P < 0.05 |

The next stage of our work is to establish the mechanisms of the adaptogenic effect of arginine. In the experiments on animals we estimate their stability and resistance to environmental change. It was demonstrated in the model systems *in vitro* that arginine cut down the formation of superoxide radicals (Fig. 1) and some products of LP, which means that arginine displays an antioxidant effect when its concentrations are close to physiological. The antioxidant properties of arginine and the products of

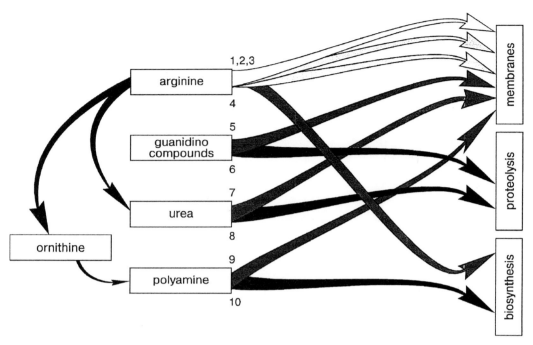

*Fig. 2. Metabolic pathways of protective effect of arginine under unfavourable
ambient condition. 1 = Activation of microsomes; 2 = Activation of
mitochondria; 3,5,7,9 = Antioxidant effect, inhibition of lipid peroxidation;
4 = Activation of hormones (insulin and others); 6,8 = Inhibition of proteolyses;
10 = Regulation of protein and nucleic acid biosynthesis.*

its transformation, urea and polyamines, which are antioxidants as well, may account for the fact that arginine **affects the molecular level**, and its effect should be stronger in the *in vivo* conditions compared with those *in vitro* (Table 4). The molecular and submolecular mechanisms of anti-radical and anti-peroxide effect of arginine are still unclear. One cannot rule out the possibility that the ability of arginine to produce nitric oxide makes a special contribution to these properties. One does not know yet the range of such activity, but it may be quite broad. Studying the ways through which arginine affects the activity of cytochrome P-450 seems to be one of the promising directions of research. One may assume that arginine may be a natural stimulator of the immune system at the level of cell immunity. Arginine may take part in optimizing the functions of leucocytes and macrophages by regulating antioxidation of lysosome and microsome systems of cells. Figure 2 suggests some protective effects of arginine and related compounds.

## References

1.  Ananyan, A.A., Milutina, N.P., Sadekova, S.I., Mogilnitskaya, L.V. & Shugaley, V.S. (1993): Protective effects of arginine on cell membranes in high altitude rats. In: *The III World congress of the international society for adaptive medicine*, Abstracts. Adaptation. pp. 77. Tokyo, Japan.

2.  Arinc, E. & Jescan, M. (1983): Comparative studies of sheep liver and lung microsomal aniline-4-hydroxylase. *Comp. Biochem. Physiol.* **74,** 2, 151–158.

3.  Colombo, J.P. (1992): Argininaemia: clinical and biochemical aspects. In: *Guanidino compounds in biology and medicine*, eds. P.P. De Deyn, B. Marescau, V. Stalon & I.A. Qureshi, pp. 343–348. London: John Libbey and Company Ltd.

4.   De Deyn, P.P., Marescau, B., Cuyckens, J.J., Van Gorp, L., Lowenthal, A. & De Potter, W.P. (1987): Guanidino compounds in serum and cerebrospinal fluid of non-dialyzed patients with renal insufficiency. *Clin. Chim. Acta* **169,** 81–88.

5.   Gershenovich, Z.S. & Krichevskaya, A.A. (1960): A protective effect of arginine under hyperoxia. *Biochimia* **5,** 790–792.

6.   Gershenovich, Z.S., Krichevskaya, A.A. & Shugaley, VS. (1969): The arginase and guanidinobutyrate-ureahydrolase of rat brain. *Enzymol. Biol. Chlin.* **10,** 181–186.

7.   Gershenovich, Z.S., Krichevskaya, A.A., Lukash, A.I. & Shugaley, V.S. (1970): *Urea in living organisms,* p. 90. Rostov-on Don.

8.   Goodman, M.W., Zieve, L., Konstantinides, F.N. & Cerra, F.V. (1984): Mechanism of arginine protection against ammonia intoxication in the rat. *Am. J. Physiol.* **247,** 3, 290–295.

9.   Guarniery, C., Fussy, F. & Fawelly, O. (1979): RNA and protein synthesis in rat brain during exercise, Effect of arginine and some phosphorylated aminoacids. *Pharmacology J.* **19,** 1, 51–56.

10.   Kaplan, E.J. (1990): *The optimization of the adaptation process.* pp. 150. Moscow: Nauka.

11.   Krichevskaya, A.A., Lukash, A.I. & Bronovitskaya, Z.G. (1980): *Biochemical mechanisms of oxygen intoxication,* pp. 120. Rostov-on Don.

12.   Krichevskaya, A.A., Shugaley, V.S. & Tsvetnenko, E.Z. (1981): Brain and liver arginase and polyamines in the mechanism of arginine- protective effect under hyperoxia. *Byulleten Exper. Biol. Medicine* **4,** 445–447.

13.   Krichevskaya, A.A., Shugaley, V.S., Ananyan, A.A. & Zygova, I.G. (1985): Protective effect of arginine on cold stress. *Voprosy meditsinskoi khimii* **6,** 50–53.

14.   Milutina, N.P., Ananyan, A.A. & Shugaley, V.S. (1990): Arginine antiradical and antioxidant effect and its influence on lipid peroxidation during hypoxia. *Byulleten Exper. Biol. Medicine* **9,** 263–265.

15.   Milutina, N.P., Ananyan, A.A., Sapozhnikov, V.M. & Novikova, E.I. (1992): The activity of lipid peroxidation and the functioning of erythrocyte membranes under long hyperbaric conditions. *Byulleten. Exper. Biol. Medicine* **5,** 674–676.

16.   Mogilnitskaya, L.V., Shugaley, V.S. & Sukhinina, I.V. (1987): Influence of arginine on the activity of proteolytic enzymes of the brain and liver of rats. *Ukrainskii Biochim Zhurnal* **1,** 15–19.

17.   Mogilnitskaya, L.V., An Fan, Baranova, N.Ju. & Shugaley, V.S. (1992): Influence of arginine on the erythrocyte membranes under hypoxia. *Byulleten Exper. Biol. Medicine* **5,** 497–498.

18.   Shugaley, V.S., Ananyan, A.A., Milutina, N.P. & Chin Kim Thi Thoa (1991): Regulation of cytochrome P-450 activity and peroxidation of lipids in rats liver and testicles under hypoxia by arginine. *Voprosy Meditsinskoi Khimii* **4,** 51–54.

19.   Shugaley, V.S., Sadekova, S.I., Ananyan, A.A. & Milutina, N.P. (1992): Oxygen and Cytochrome P-450 activity on rat liver (1992): Cytochrome P-450 Biochemistry and Biophysics. In: *Proceedings of the 7th international conference,* Inco-TNC Joint stock company, pp. 158–161.

20.   Walter, R.M., Gold, E.M., Michas, S.A. & Ensinck, J.W. (1980): Portal and peripheric vein concentration of insulin and glucagon after arginine infusion. *Metabolism* **11,** 1037–1040.

# Section II
## Hyperargininemia and arginase

*Guanidino Compounds : 2*, eds. by P.P. De Deyn, B. Marescau, I.A. Qureshi and A. Mori.
©1997 John Libbey & Company Ltd., pp. 53–69.

# Chapter 8

# Hyperargininemia: a treatable inborn error of metabolism?

P.P. DE DEYN, B. MARESCAU, I.A. QURESHI, S.D. CEDERBAUM, M. LAMBERT, R. CERONE, N. CHAMOLES, N. SPÉCOLA, J.V. LEONARD, R. GATTI, GREEN, S.S. KANG, N. MIZUTANI, I. REZVANI, S.E. SNYDERMAN, H.G. TERHEGGEN, M. YOSHINO, B. APPEL, J.J. MARTIN, ROTH, A.L. BEAUDET, L. VILARINHO, E. HIRSCH, K. JAKOBS, M.S. VAN DER KNAAP, H. NAITO, B.A. PICKUT, S.K. SHAPIRA, A. FUCHSHUBER, B. ROTH and K. HYLAN

*Hyperargininemia Research Group, Laboratory of Neurochemistry and Behavior, University of Antwerp, Born-Bunge Foundation, Universiteitsplein 1, 2610 Wilrijk*

## Summary

Hyperargininemia is an inborn error of the urea cycle characterized by arginase type I deficiency. The authors present data regarding all 55 patients presently diagnosed world-wide. Data on 20 hitherto unpublished cases are included in this review and emphasis is put on the proposed pathogenesis of the neurological complications and the presumable treatable nature of this autosomal recessive disorder. The disease is heterogenous at the genotypic level encompassing a variety of mutations.

The clinical presentation of hyperargininemia is unlike this observed in other urea cycle enzymopathies. Usually, early infancy is without symptoms other than irritability. Typically, episodes of irritability, poor appetite and periodical vomiting and lethargy, in some cases progressing to seizures and coma, develop on weaning from breast feeding or on changing from formula to cow's milk. After an initial normal or only slightly delayed psychomotor development, motor and mental deterioration is observed at ages varying from 3 months to 4 years. The neurodevelopmental outcome ranges from normal to severe mental retardation and pronounced pyramidal tract signs. Upper motor neuron involvement was present in 80 per cent of patients. In the majority (49 per cent), pyramidal tract signs are restricted to the lower extremities with scissoring and tiptoe gait as most characteristic feature. Seizures were observed in 66 per cent of cases and occurred in the absence of hyperammonemia. Failure to thrive was observed in 66 per cent of cases and occasional findings were ataxia, athetosis, spinal deformities, microcephaly and hepatomegaly. MRI showed the presence in white matter of high-signal intensit(y)(ies) on T2 weighted images.

Biochemical findings show an accumulation of arginine in serum as the hallmark of the disease. Arginase deficiency in red and white blood cells, liver and stratum corneum confirms the diagnosis. Argininuria was only present in 68 per cent and lysine-cystinuria was present in 51 per cent of patients. Hyperammonemia is only periodically observed here, less pronounced and even absent in certain cases. Furthermore, the important accumulation of secondary catabolites of arginine, guanidino compounds, is demonstrated in biological fluids and tissues including brain.

The authors suggest that the extreme degree of spasticity and the frequent epileptic symptomatology, could be caused by guanidino compounds which are proven experimental toxins. Moreover, clinical observations suggest that hyperargininemia might be another treatable error of metabolism. The treatment of choice is proposed to be the administration of an essential amino acid mixture without arginine.

All patients with mental retardation/deterioration, spasticity and epilepsy should be screened by plasma guanidino compound or plasma amino acid analysis.

## Introduction

The catabolism of amino acids and protein releases high amounts of nitrogen in the form of ammonia, a toxic metabolite that must be converted in a nontoxic, readily excretable form. Mammals excrete nitrogen primarily as urea, the end-product of the Krebs-Henseliet urea cycle. This cycle consists of five steps in which two molecules of ammonia are combined with $CO_2$ to form urea. Inborn errors involving deficiency of the second and fourth enzymes of this pathway, ornithine transcarbamylase and argininosuccinic acid lyase, have been most frequently described while deficiencies of argininosuccinic acid synthetase and carbamyl phosphate synthetase, the third and the first enzyme respectively, have been reported less frequently.

Deficiency of the fifth enzyme of the urea cycle, arginase, has been diagnosed in at least 54 children of 45 families scattered all over the world. In this paper, we present a review of the clinical and biochemical findings on 35 published and 20 new hyperargininemic patients. The overview is based on the original papers, the medical records of the patients and in most cases on clinical neurological observations made by one board-certified neurologist. In addition, we present determinations of guanidino compounds together with arginine in the biological fluids of most of the patients.

The first probable case of hyperargininemia was reported by Peralto Serrano in 1965[44]. The propositus was a Spanish male who developed convulsions in early infancy and presented with hypertonicity, hyperreflexia, severe psychomotor retardation, discoloration of the scalp and hepatomegaly. Argininuria was noted and arginine was markedly elevated in blood and cerebrospinal fluid. However, arginase activity was not assayed. In 1969, Terheggen et al.[58], reported the first two fully documented cases. These two sisters from German origin, exhibited periodic vomiting, anorexia, lethargy, epilepsy, mental retardation and spasticity. Amino acid analysis revealed the existence of a cystine-lysine-ornithine-argininuria and in blood, a hyperargininemia and hyperammonemia was found. A deficiency of arginase in the erythrocytes confirmed the diagnosis[59,60].

Since then, 32 more cases have been described and at least 20 more hyperargininemic patients have been diagnosed (See Table 1). The disease is not tied to a geographical distribution or a particular ethnic group. Hyperargininemia was diagnosed in Western and Southern Europe, Canada, United States, Central and South America, Japan and Australia. Hyperargininemia was diagnosed in patients with the following geographical origin: Argentina (n = 2), Australia (n = 1), Cambodia (n = 1), Canada (French )(n = 5), France (n = 1), Germany (n = 5), Guatemala (n = 1), Hong-Kong (n = 1), India (Hindu) (n = 1), Italy (n = 4), Japan (n = 5), Marocco (n = 2), Mexico (n = 3), Oman (n = 1), Pakistan (n = 2), Portugal (n = 4), Puerto Rico (n = 5), Saoudi Arabia (n = 1), Senegal (n = 2), Spain (n = 2), United States of America (n = 5; Jewish: 2, Hispanic: 2, African:1), Uruguay (n = 2).

The sex distribution, girls and boys being as frequently affected (F = 22, M = 33), the occurrence of affected siblings[38,53,59,60] and reduced enzyme activity in parents indicated hyperargininemia to be an autosomal recessive trait[10,11,12,38,46,53,58,59,60]. Consanguinity was reported in 15 of the 46 families.

## Clinical features

The clinical picture of 'full-blown' argininemia is quite uniform and unlike that observed in other urea cycle enzymopathies. A summary of the clinical and biochemical findings in all patients studied are shown in Tables 2 and 3. In all but three patients, early infancy was without symptoms other than irritability. While the clinical presentation is a delayed one, there are only a couple of cases neonatal presentation, with or without hepatic disturbances. More typically, in almost all patients, episodes of irritability, poor appetite and periodical vomiting and lethargy, in some cases progressing to seizures and coma, developed on weaning from breast feeding or on changing from formula to cow's milk. After an initial normal or only slightly delayed psychomotor development, motor and mental deterioration is observed at ages varying from 3 months to 4 years. The neurodevelopmental outcome ranges from normal or close-to-normal intellectual and motor function to severe mental deterioration

## Table 1. Demographic data on all hyperargininemic patients

| Case number and initials | Sex | Year of birth | Deceased or alive | Consanguinity | Ethnicity | Reference |
|---|---|---|---|---|---|---|
| 1 (?.?.) | M | 1963 (?) | D | + | Spanish | 44 |
| 2 (A.W.) | F | 1964 | A | + | German | 58,59,60 |
| 3 (M.W.) | F | 1968 | A | + | German | 58,59,60 |
| 4 (I.W.) | F | 1971 | A | + | German | 57 |
| 5 (D.R.) | M | 1969 | A | – | Puerto Rican | 53 |
| 6 (N.R.) | F | 1969 | A | – | Puerto Rican | 53 |
| 7 (J.R.) | M | 1976 | A | – | Puerto Rican | 54 |
| 8 (M.O.) | M | 1968 | D 1990 | – | Ashkenazi Jewish | 10,11 |
| 9 (R.U.) | F | 1962 | A ??? | – | Puerto Rican | 12,13 |
| 10 (M.U.) | M | 1969 | D 1989 | – | Puerto Rican | 12,13 |
| 11 (S.T.) | M | 1972 | A | – | Guatemalan | 3 |
| 12 (T.G.) | F | | A | | Mexican | cederb unpub |
| 13 (G.M.) | F | 1971 | A | – | Mexican | 38 |
| 14 (L.C.) | F | 1965 | A | + | French Canadian | 46 |
| 15 (F.F.L.) | M | 1977 | A | – | French Canadian | 45 |
| 16 (A.R.) | F | 1980 | A | + | French Canadian | 32 |
| 17 (P.R.) | M | 1987 | A | – | French Canadian | Lamb-Qur unpub |
| 18 (D.B.) | M | 1990 | A | – | French Canadian | Lamb-Our unpub |
| 19 (A.Y.) | F | 1972 | A ? | – | Japanese | 68 |
| 20 (K.Y.) | M | 1978 | D 1992 | + | Japanese | 40 |
| 21 (N.S.) | M | 1960 | A | – | Japanese | Mizutani Unpub |
| 22 (D.C.) | F | 1970 | A | – | Hispanic | 31 |
| 23 (?.?.) | M | 1983? | D | ? | German? | 19 |
| 24 (P.C.) | F | 1978 | A | – | Uruguay | Chamoles unpub |
| 25 (C.C.) | F | 1989 | A | – | Uruguay | Chamoles unpub |
| 26 (S.O.) | M | ≤ 1989 | A | – | Argentinian | Chamoles unpub |
| 27 (?.?.) | M | 1976 | D | – | Japanese | 51 |
| 28 (?.?.) | M | 1986? | D ? | ? | Spanish | 30 |
| 29 (T.D.) | F | 1972 | A | – | Australian | 29 |
| 30 (F.B.) | M | 1978? | D 1988 | – | Italian | 1 |
| 31 (R.M.) | M | 1983 | A | – | Italian | 22 |
| 32 (E.D.D.) | F | 1974 | D | – | Italian | 50 |
| 33 (B.R.) | M | 1982 | A | – | Jewish | Rezvani unpub |
| 34 (S.W.) | M | 1986 | D 1989 | +? | German | 20 |
| 35 (F.H.) | F | 1969 | A | – | French | 14 |
| 36 (L.B.) | ? | | A? | + | Saoudi | Levy & Bruslow unpub |
| 37 (Y.S.) | M | 1984 | A | – | Japanese | Naito unpub |
| 38 (?.?.) | F | ?? | A? | + | Hindu | 21 |
| 39 (W.M.) | M | 1984 | A | + | Pakistani | 4 |
| 40 (FG.OF) | M | 1984 | A | – | Portugese | 67 |
| 41 (S.S.) | F | 1992 | A | – | Portugese | Vilarinho unpub |
| 42 (R.L.) | M | 1985 | A | + | Hong Kong (Chinese) | Green, Rylance & Brice unpub |
| 43 (D.L.) | M | | D | + | Cambodian | 25 |
| 44 (A.M.) | F | 1993 | A | – | Italian | 8 |
| 45 (J.L.) | M | 1984 | A | – | Argentinian | Chamoles unpub |
| 46 (I.H.) | F | 1986 | A | + | Pakistani | Leonard, unpub |
| 47(M.AL-B) | M | 1992 | A | + | Omani | Leonard, unpub |
| 48 E.G. | M | 1982 | A | – | African-American | 52 |
| 49 A.G. | M | 1986 | A | – | Mexican | 18 |
| 50 A.P. | M | 1986 | A | + | Portugese | Abs. by I. Tavares de Almeida |
| 51 I.K. | M | 1984 | A | + | Senegalese | 7 |
| 52 A.K. | M | 1991 | A | + | Senegalese | 7 |
| 53 J.L. | F | 1992 | A | – | Portugese | Vilarhino unpub |
| 54 F.A. | M | 1988 | A | + | Maroccan | van der Knaap & Jakobs unpub |
| 55 K.A. | M | 1993 | A | + | Maroccan | van der Knaap & Jakobs unpub |

and pronounced pyramidal tract signs. At least 3 patients (No. 7, 17 and 25), aged over 4 years by now and treated from birth with an arginine free essential amino acid mixture, remained free of any degree of deterioration of the mental and motor functions. In addition, patients 40, 44, 52 and 53 had a normal neurological examination.

Varying degrees of upper motor neuron involvement were present in 80 per cent of patients and resulted in spastic paraparesis, paraplegia, quadriparesis or quadriplegia. In 49 per cent of patients, pyramidal tract signs remained restricted to the lower extremities. The neurologic examination revealed hyperactive deep tendon reflexes, spread of reflexes, clonus, extensor plantar reflexes and

# Table 2. Clinical findings in all diagnosed hyperargininemic patients

| Case number and initials | Pyramidal tract sign (spasticity) | Cognitive impairment | Age of onset of psychomot or deterioration | Seizures | Ataxia | EEG abnormalities | Cerebral atrophy | Periodic vomiting | Failure to thrive | Microcephaly | Spinal deformity | Hepatomegaly | Therapy (+ response when well documented) |
|---|---|---|---|---|---|---|---|---|---|---|---|---|---|
| 1 (?.?.) | | | neon? | + (GTC) | | + | | | | | | + | |
| 2 (A.W.) | ++:U.L+ | ++ | 0y3mth | + (+) (GTC) | (+) | + | + | + | – | – | – | + | PR |
| 3 (M.W.) | ++:L | ++ | 3y | +(+) F,GTC | – | + | + | + | – | – | + | – | PR |
| 4 (I.W.) | ++:L | ++ | 0y5mth | +(+)F,GTC | – | – | NA | – | – | – | – | – | PR |
| 5 (D.R.) | +:L | – | 0y4mth | (+) | (+) | + | – | + | + | – | – | + | PR,EAA(good, well impr.) |
| 6 (N.R.) | +:L | + | 1y6mth | +(+) | (+) | – | + | + | – | – | + | – | PR,EAA(good, well impr.) |
| 7 (J.R.) | – | – | – | – | – | – | – | – | + (±) | – | – | – | PR,EAA(ex., norm.) |
| 8 (M.O.) | ++:L;+:U | ++ | 1y6mth | +AS(GT,SE) | – | + | PA | + | + | – | + | – | PR,SB,PA |
| 9 (R.U.) | ++:L | ++ | 3y | +(GTC) | – | + | + | – | + | – | + | – | PR,SB,EAA |
| 10 (M.U.) | ++:L;+:U | ++ | 2y6mth | (+)(GTC) | – | + | + | – | + | – | + | (+) | PR,SB,EAA |
| 11 (S.T.) | +:L;±:U | ± | 4y | – | – | + | NA | + | + | – | – | (+) | PR,SB |
| 12 (T.G.) | +L? | + | 1y | – | – | – | – | – | – | + | – | – | PR,SB,EAA? |
| 13 (G.M.) | ++:L | +++ | 1y6mth | – | – | + | – | – | + | + | – | – | PR,EAA? |
| 14 (L.C.) | ++:L;+:U | + | 4y | (+) | – | + | NA | + | + | – | – | – | |
| 15 (F.F.L.) | +:L | + | 3y | – | (+) | + | NA | – | – | – | + | – | PR,SB,L-Carn,EAA (no determination outcome) |
| 16 (A.R.) | +:L | + | 15mth | – | – | + | – | + | – | – | – | + | PR,SB,L-Carn,EAA (well impr.52–74) |
| 17 (P.R.) | – | – | screening | – | – | – | – | – | + | – | – | – | PR,SB,EAA |
| 18 (D.B.) | ++:L | + | 0y10mth | + | – | + | – | – | – | – | – | – | PR,SB,EAA (impr.) |
| 19 (A.Y.) | ++:L.U | +++ | 0y4mth | +(+) | – | + | ++ | + | + | – | – | – | PR,EAA (unimpr.) |
| 20 (K.Y.) | ++:L;+:U | +++ | 1y6mth | +(+) | – | (+) | ++ | + | – | – | – | – | PR,SB,PA,EAA(det.) |
| 21 (N.S.) | ++:L,+:U | ++ | 3mth | +(+)GTC | – | + | + | + | + | – | + | – | PR,SB |
| 22 (D.C.) | ++:L | + | 2y | +(+)F,GTC | (+) | + | NA | + | + | – | – | – | LO |
| 23 (?.?.) | NA | ++ | NA | –? | – | NA | NA | NA | ? | – | NA | – | PR,EAA |
| 24 (P.C.) | ++:L | + | 2y6mth | +F,GTC | – | + | + | – | + | – | + | – | PR,SB |
| 25 (C.C.) | – | – | – | – | – | – | NA | – | – | – | – | – | PR,SB,EAA-carn. |
| 26 (S.O.) | ++:L;U | + | 2y8mth | +,F,GT | – | + | – | + | + | + | – | – | PR,SB-carn. |
| 27 (?.?.) | ++:L | NA? | 3y | +GTC | – | NA | + | + | + | – | NA | + | PR,EAA,PRBC |
| 28 (?.?.) | ? | NA? | neon? | (+) | – | NA | NA | + | ? | – | NA | + | |
| 29 (T.D.) | ++:L;+:U | +++ | motor: 2ycogn:8y | +AS | – | +(ME) | NA | – | + | – | + | – | PR,SB,EAA(good response) |
| 30 (F.B.) | ++:L | +++ | 10mth | +(GTC) | – | + | – | + | – | – | – | + | PR |
| 31 (R.M.) | +:L | + | 2y | +(GTC) | – | + | NA | + | – | – | – | – | PR,SB,EAA,carn. |
| 32 (E.D.D.) | ++:L | + | 1y6mth | – | – | + | NA | + | + | – | – | – | PR,SB,EAA,PD,PRCB |
| 33 (B.R.) | +:L | ++ | 6mth | + | – | – | – | + | – | – | – | – | PR,EAA |
| 34 (S.W.) | +:L | + | 1y | – | – | + | – | + | + | – | – | –? | PR,EAA,HD,HF |
| 35 (F.H.) | ++:L;++:U | ++ | 1y3mth | (+)F,GT,GTC,SE | – | + | + | + | + | – | – | – | ±PR |
| 36 (L.B.) | NA | | | – | | | | | | | | | |
| 37 (Y.S.) | ++:L;+U | +++ | 9mth | + (GTC) | – | + | + | + | + | + | + | – | PR,PRBC(deteriorated) |
| 38 (?.?.) | –(hypot) | + | 2y? | + | – | NA | NA | ? | ? | ? | ? | + | NA |
| 39 (W.M.) | ++:L;++:U | ++ | 2y | +(GTC,AS) | (+) | + | + | + | + | + | – | – | PR,EAA |
| 40 (FG.OF) | – | – | 3y6mth | +(F) | – | + | – | + | – | – | – | – | PR,SB,EAA |
| 41 (S.S.) | – | ± | 6–9mth(?) | – | – | NA | NA | – | + | – | – | + | PR,SB,EAA |
| 42 (R.L.) | ++:L;+U | +++ | 6–9mth | – | – | + | NA | – | + | – | – | – | SB,PB |
| 43 (D.L.) | – | | neon | – | – | – | – | – | – | – | – | + | PR,SB |
| 44 (A.M.) | – | – | prenat | – | – | – | – | – | – | – | – | – | PR,EAA |
| 45 (J.L.) | ++L,+U | ++ | 2y | +AS | + | + | – | + | + | + | + | – | PR,SB |
| 46 (I.H.) | +++:L;+:U | – | motor: 4y normIQ | – | – | NA | + | + | ± | – | + | – | PR,SB,EAA(good response) |
| 47 (M.ALB) | +++:L;+:U | + | 5mth | +GTC | + | + | + | + | – | – | – | + | PR,SB(good) |
| 48 E.G. | +++:L | + | 5y | +GTC | – | + | – | – | – | – | – | + | PR,phenylbutyrate (partly impr.) |
| 49 A.G. | +:L | + | 0y6mth | +AS | – | + | – | – | – | – | – | – | PR,EAA |
| 50 A.P. | +++:L;+:U | ++ | 1y5mth | +FC | + | +(E) | + | – | – | – | – | – | PR,EAA |
| 51 I.K. | +++:L +U | +++ | 3y | – | – | + | + | + | – | – | – | – | PR |
| 52 A.K. | – | – | – | – | – | ± | – | + | – | – | – | – | EAA,SB |
| 53 J.L. | ++:U+++:L | – | 18mth | – | – | – | – | – | – | – | – | + | PR,EAA |
| 54 F.A. | +:L | +++ | 0y6mth | +(GTC) | – | + | + | + | + | + | – | – | PR,(EAA(poor compliance) |
| 55 K.A. | – | – | – | – | – | ? | ± | + | – | – | – | – | PR, (EAA) |

Spasticity: absent = –; moderate = +; severe = ++. U=upper extremities; L= lower extremities. Cognitive impairment: normal intelligence = –; borderline =±; mild impairment =+; moderate impairment; severe impairment = +++.
Findings indicated between brackets refer to hyperammonemic episodes or protein load.
Therapy: PR= protein restriction; SB = sodium benzoate; EAA = essential amino acid mixture; PA = phenylacetate; PB = phenylbutyrate; LO = lysine+ ornithine supplementation; PD= peritoneal dialysis; HD = hemodialysis; PRBC = packed red blood cells. Seizure types: F=focal, GTC = generalized tonic-clonic; GM = generalized myoclonic; AS = absence seizures; SE = Grand Mal status epilepticus. NA: Not available.

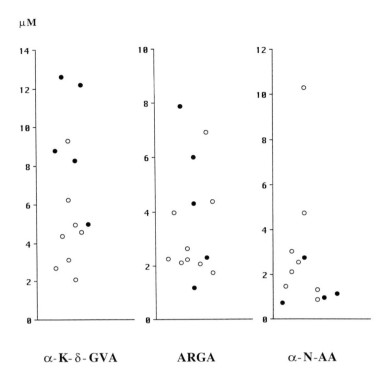

μM

α-K-δ-GVA          ARGA          α-N-AA

*Fig. 1.Serum levels (mM) of α-keto-δ-guanidinovaleric acid
(α-K-δ-GVA), argininic acid (ARGA) and α-N-acetylarginine
(α-N-AA) in untreated (closed circles) and protein-restricted
(open circles) hyperargininemic patients.The upper normal
limit for α-K-δ-GVA was 0.250 μM and for ARGA and α-N-
AA, respectively 0.115 and 0.310 μM.*

Hoffmann-Trömmer signs. A disuse atrophy was observed in several patients. The most characteristic feature reflecting spasticity of the lower extremities, was a scissoring or tiptoe gait. In several patients with heel cord, hamstring or adductor contractures, surgical release was required. Ten patients (No. 7,17,25,38,40,41,43,44,52,53) were devoid of any pyramidal tract signs. Non-persistent pseudobulbar palsy with excessive drooling, reduced gag reflex and poorly coordinated swallowing was observed in 2 patients (No. 9,10) and pseudobulbar palsy was also reported in propositus No. 29.

Ataxia was observed in 9 patients (Nos. 2,5,6,15,22,40,46,48,51) out of whom 6 presented ataxia only intermittently and linked to hyperammonemic episodes. Intermittent athetosis was reported in one patient (No. 4). In addition, fine tremors were observed in one patient during infusion of amino acids (No. 17). Seizures were observed in 33 out of 55 patients; in 28 of these patients, epileptic symptoms occurred in the absence of hyperammonemia. Clinical types of epilepsy were simple focal epilepsy (n = 7), complex focal epilepsy (n = 1), generalized tonic-clonic seizures (n = 18), generalized tonic seizures (n = 3) and generalized absence seizures (n = 5). At least two patients presented with primary generalized tonic-clonic status epilepticus (No. 8,35). Remarkable hyperactivity was reported in two patients before dietary treatment (No. 5,6). Occasional findings were spinal deformities, such as scoliosis and lordosis, which were found in 11 patients. Microcephalia was present in 4 patients. A

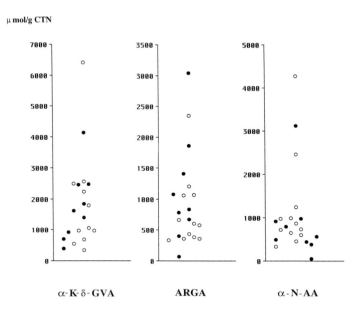

Fig. 2. Urinary excretion levels ( μmol/g creatinine) of α-keto-
δ-guanidinovaleric acid (α-K-δ-GVA), argininic acid (ARGA)
and α-N-acetylarginine (α-N-AA) in untreated (closed circles)
and protein-restricted (open circles) hyperargininemic pa-
tients.The upper normal limit for α-K-S-GVA was 35  μmol/g
creatinine, for ARGA and_-N-AA it was respectively 25 and
65 μmol/g creatinine.

non-persistent hepatomegaly was observed in 7 of the studied patients and a failure to thrive was observed in 25 out of 55 patients.

Electroencephalographic recordings were abnormal in most patients, showing diffuse slowing compatible with a metabolic encephalopathy and epileptic graphoelements in more than 50 per cent of patients of whom EEGs were available. Electromyographic examination of upper and lower extremities showed no evidence of a myopathic or neurogenic process (No. 8,11). Sensory and motor nerve conduction velocities were within the normal limits in these patients and a light microscopic exam of a sural nerve biopsy of patient No. 8 did not demonstrate significant changes. Pneumoencephalography and/or CAT-scan of the brain were performed in 10 patients showing normal findings in 4 subjects and cerebral atrophy in 6 children; the atrophy varying from mild subcortical atrophy to pronounced cortical and subcortical atrophy. Nuclear magnetic imaging of the brain was performed in at least 5 patients. This showed variable degrees of cerebral and cerebellar atrophy and the presence of high-signal intensities on $T_2$ weighted images in the periventricular and more peripheral white matter.

Liver biopsy showed swollen hepatocytes in a normal lobular configuration with minimal signs of interportal fibrosis (No. 9,10,13,19,28). Electronmicroscopic and histochemical studies revealed multi-focal hydropic changes, increased cellular glycogen, normal mitochondria and dilated endoplasmic reticulum (No. 9,13,19). Autopsy in patient No. 28 revealed portal fibrosis with ductular

## Table 3. Laboratory data in all hyperargininemic patients

| | Increased liver transaminases | Blood ammonia (µg/dl) | Serum arginine (µM) | Serum α-keto-δ-GVA (µM) | Serum ArgA (µM) | CSF arginine (µM) | CSF α-keto-δ-GVA (µM) | CSF Arg A (µM) | Aminoaciduria | α-keto-d-GVA in urine (µmoles/grCTN) | Arg A in urine (µmoles/gr CTN) | Arginase activity |
|---|---|---|---|---|---|---|---|---|---|---|---|---|
| ) | + | 392–671 | 640 | | | 53.3 | | | A,C,Ct,L,O | (2230) | (355) | E(O) |
| .) | + | 165–317 | 996 | | | 94.8 | | | A,C,Ct,L,O | (6410) | (1069) | E(O) |
| | – | 237–590 | 252–1579 | | | 113 | | | A,C,Ct,L,O | (2554) | (360) | E(5.9) |
| .) | – | nl–±500 | 445–540 | (1.25)–(2.47) | (2.15)–(2.80) | 94.5 | (<0.018) | (0.04)–(0.10) | A,C,L,O | (30.3)–(183) | (102)–(139) | E(O) |
| | + | nl–±200 | 471–572 | (3.17) | (3.05) | 98.3 | | | A,C,L,O | (257) | (171) | E(O) |
| | – | nl–(118) | (80)–402 | (1.38)–(4.69) | (2.38)–(5.28) | NA | (<0.018) | (0.08) | – | (475) | (344) | E(5) |
| ) | + | nl–340 | 192–625 | (5.68) | (1.85) | 70.1 | | | (A,C,L,Ho) | (326) | (290) | E(1) |
| | – | nl–258 | (60)–786–(913) | (4.72) | (2.74) | 85–98 | | | A,C,L,O | (285) | (400) | E(0.5)L(1.6) |
| J.) | – | nl–132 | (40)–591–(677) | (3.39) | (3.21) | 69 | | | A,C,L,O | (30)–(190) | (276)–(353) | E(0.1) |
| .) | + | 180 | 200–672 | | | 73 | | | – | | | E(0) |
| 1.) | – | nl–171 | 544–1391 | (2.94) | (3.12) | 88 | | | A,Ct,L,As | (2487) | (2351) | E(5),L,SC |
| .) | – | 100–183 | 600–1200 | (0.50)–(4.83) | (0.35)–(3.32) | 16,4–? | ND | (0.08) | A,C,Ct,L,Or | (<DL)–(702) | (40)–(670) | E(1) |
| ,L.) | – | nl (=?) | 663 | (1.40)–8.28 | (1.02)–788 | 57,7 | (0.074)–(0.08) | (0.09)–(0.15) | C,Ct,L,Or | (207)–4823 | (155)–4079 | E(O) |
| .) | – | nl–602 | (120)–895 | (0.51)–(6.05) | (0.30)–(3.64) | 61 | (<0.025)–(0.140) | (0.08)–(0.100) | – | (<DL)–(708 | (28.1)–(445) | E(O) |
| .) | – | nl(=?) | (33)–625 | (0.26)–(4.96) | (0.27)–(3.96) | NA | NA | NA | A,C,L,Or | (54.9)–(969) | (362)–(1206) | E(O) |
| .) | + | (28)–118 | (43)–505 | NA | NA | NA | NA | NA | A,C,Ct,L,Or | | | E(O) |
| ') | – | nl–960 | 178–1115 | (3.50) | (2.14) | 110 | | | Or | (867) | (590) | E(<0.5)L(0) |
| ') | + | 239–400 | (301)–604 | (2.94) | (2.20) | 30.7–85 | (0.208) | (0.110) | A,Or | (829) | (1146) | E(O) |
| .) | – | nl–215 | 300–778 | | | ? | | | –? | | | E(O) |
| .) | – | nl–132 | (424)–478–638 | (4.66) | (2.76) | 65 | | | A,Or | (913) | (502) | E(O) |
| ) | – | 700 | 695 | | | NA | | | NA | | | E(O) |
| .) | – | nl(+?) incr | 498–666 | (4.57)–(7.89) | (2.08)–(3.35) | – | | | A,O,Glut | (1064)–(1387) | (581)–(813) | E(0.1) |
| .) | – | (nl) | (169)–649 | (2.83)–12.6 | (0.61)–2.30 | – | | | A,O,Glut,L,C | (511)–1838 | (128)–404 | E(0.1) |
| .) | + | 107 | 556–769 | (2.85) | (1.25)–2.93 | – | | | – | 658 | 658 | E(32) |
| ) | + | 50–330 | 115–534 | | | 69 | | | A,L,Or | | | E(0) |
| ) | + | 338–440 | 170 | | | NA | | | C,O,Or | | | L(1–3) |
| .) | NA | NA | 161–663 | | | 36–63 | (0.241)–(0.362) | (0.100)–(0.230) | NA | | | E(0) |
| .) | – | nl–(310) | 886 | (6.25) | (4.39) | NA | | | (A),L | (676) | (437) | E(5) |
| 1.) | + | nl–(=80) | 239–837 | (9.29) | (6.93) | 94 | (0.280) | (0.240) | A,Or,Ct,C | (1793) | (1055) | E(2) |
| .D.) | + | 118–435 | 754 | (3.69) | (2.02) | NA | | | Or | (619) | (229) | E(O) |
| .) | + | nl–900 | (80)–500 | | | NA | | | Or | (144) | (94.4) | E(O) |
| '.) | + | nl–350 | 150–600 | (1.12)–(24) | (0.97)–(7.81) | 186 | (2.51) | (1.98) | – | (483) | (356) | E(O) |
| .) | + | (67)–nl–652? | ?–501 | 6.85 | 4.04 | 191 | 0.408 | 0.310 | A,C,O,Or,L? | 1456 | 833 | E(<1) |
| .) | ? | ? | ? | | | ? | | | ? | | | ? |
| .) | | 369–682 | ?–682 | | | NA?? | | | A,C,L | | | E(0) |
| ) | | | 483–(826) | | | NA | | | A,C,O,L | | | NA |
| M.) | – | (15)–104 | (54)–907 | (0.650)–(4.38) | (0.57)–7.58 | 78 | | | A,C,L | (151)–1400 | (193)–1243 | E(3) |
| OF) | – | nl–355 | (198)–923 | (1.85)–(3.15) | (1.38)–(2.23) | NA | | | A,C,Or,L | (100)–2474 | (243)–675 | E(3) |
| ) | + | ?–255 | 350–1756 | 8.77? | 6.02 | NA | | | A,C,Ct,L,O | 1626 | 1407 | E(0.5) |
| .) | + | nl(+?) | (480)–817 | (2.09) | (2.12) | NA | NA | NA | A | (343) | (329) | E(O) |
| ) | + | 126–447 | 500–1305 | | | NA | | | – | | | E(<1),L(3) |
| 1.) | + | nl(+?) | 140 | (1.36)–5.08 | (0.96)–1.60 | - | | | – | (384)–1365 | (71.1)–(177) | NA |
| .) | | 208 | 691 | | | 96 | | | –(ND) | | | E(30) |
| ) | ++ | 167 | 161–774 | | | 63–287 | | | NA | | | E(O) |
| LB) | ++ | 136,(6mth) | 109–786 | | | NA | | | C,O,A;L | | | E(O) |
| | | 22–95?? | 53–334 | | | NA | | | normal | | | E(0.04) |
| | | 25–36?? | 312–592 | | | NA | | | ? | | | E(0.03) |
| | + | 47???? | 502 | | | 92 | | | C,dib aa,Or | | | E(O) |
| | – | (90)–136 | (264)–554 | NA | NA | NA | NA | NA | A,Or | NA | NA | E(<1) |
| | – | nl–187 | 88–200 | NA | NA | NA | NA | NA | NA | NA | NA | E(4) |
| | + | 76 | 464 | ? | ? | NA | NA | NA | A,C,L,O,Or | NA | | E(4) |
| | – | (22)–180 | (164)–475 | NA | NA | NA | NA | NA | A,C,Ct,L,O | NA | | E(0) |
| | – | nl(45) | 82–96 | NA | NA | NA | NA | NA | – | | | E(0) |

eviations: NA = not available; A = arginine; C = cystine; Ct = citrulline; Ho = homocitrulline; L = lysine; O = ornithine; Or = orotic acid; ythrocyte, L = liver;  SC = stratum corneum. Normal values: Erythrocyte arginase activity = 2100–12.400 µmoles/h/gHb. Ammonia blood < 80 µg/dl; serum arginine = 40–130 µM; CSF arginine = 6–30 µM. Data and/or findings indicated between brackets refer to ammonemic episodes or protein load.

μM

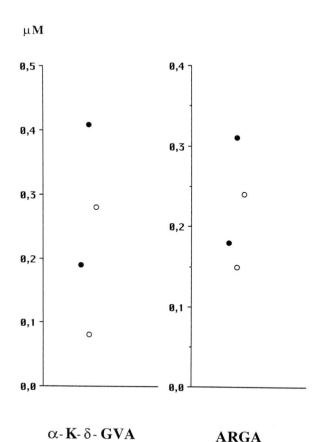

## α- **K**- δ- **GVA**     **ARGA**

*Fig. 3. Cerebrospinal fluid levels (μM) of α-keto-δ-guanidino-valeric acid (α-K-δ-GVA) and argininic acid (ARGA) in un-treated (closed circles) and protein-restricted (open circles) hyperargininemic patients.In controls the α-K-δ-GVA levels were lower than the detection limit (< 0.025 μM).The upper normal limit for ARGA was 0.070 μM.*

proliferation, macrovascular steatosis and megamitochondria in the hepatocytes. Autoptic examination in patient No. 34 revealed veno-occlusive hepatic disease with sinusoidal fibrosis, steatosis and intrahepatic cholestasis. Hyperargininemic patient No. 41 presented with a neonatal presentation of persistant jaundice and hepatosplenomegaly. Liver biopsy showed a picture of hepatic cirrhosis with formation of regeneratative nodules and marked cholestasis. Postmortem examation of the liver in patient No. 43 demonstrated micronodular liver cirrhosis. Microscopic examination of the liver and premortem biopsies demonstrated the evolving consequences of chronic cholestasis. In the latter case, there might have been an interfering hepatitis B infection.

Extensive post-mortem neuropathological examinations have been performed in patient No. 34[20] and will be the subject of a future publication. No obvious alterations were found; the maturation of the brain in this 3 year old boy was normal. Brain stem, spinal cord and cauda equina were normal but for a pallor of the myelin in the crossed pyramidal tracts. The latter phenomenon might hypothetically correlate with the pyramidal tract signs observed in the lower extremities of this individual.

# Biochemical findings

## General

Accumulation of arginine, the substrate proximal to the metabolic block, is the hallmark of the disease as it is found in all patients. In the blood, arginine increases 5 to 20 fold as compared to controls. In 25 patients arginine levels were determined in the CSF and found to be increased 2 to 10 fold.

Argininuria and increased excretion of the other dibasic amino acids together with cystine was present in approximately two thirds of the patients. In one of the patients (No. 8), a modest dibasic aminoaciduria occurred only after a very high protein meal[10]. The lysine-cystinuria is due to competitive inhibition of tubular reabsorption of these amino acids by excess arginine as demonstrated by van Sande *et al.*[66]. In 17 out of 47 tested patients, urinary orotic acid excretion was greatly increased. The orotic aciduria is rather puzzling, especially since it sometimes occurs in the absence of hyperammonemia[13,42,47]. Excessive orotic acid excretion is a characteristic of ornithine transcarbamylase deficiency due to an excess of mitochondrial carbamyl phosphate going to the extramitochondrial pyrimidine pathway[33] and is also reported in citrullinemia[5,6]. In hyperargininemia however, the metabolic block would be too far removed to cause an accumulation of carbamyl phosphate due to a simple feedback inhibition. Therefore, Bachman and Colombo[2] suggested that the cause of increased pyrimidine synthesis in argininemia is the stimulating effect of arginine on N-acetylglutamate synthetase which results in an increase of N-acetylglutamate that activates carbamyl phosphate synthetase. Carbamyl phosphate would thus accumulate and be channeled into pyrimidine synthesis.

The hyperammonemia, usually found in all urea cycle diseases, is only periodically observed here, less pronounced and even absent in certain cases. In at least 8 patients, hyperammonemia was never observed. One patient with otherwise normal ammonia levels developed hyperammonemic episodes since the age of 16, associated with the menses. In the other patients febrile illnesses were the most frequent conditions leading to hyperammonemia.

The disease is characterized by a deficiency of arginase (type I) that is predominantly found in the cytosol of liver and red blood cells. Arginase isozyme(s) that differs from the liver-type enzyme in catalytic, molecular, and immunological properties, is present in the kidney, small intestine, brain and lactating mammary gland[23,28,48,49]. In all 50 tested patients, the liver-type arginase deficiency was demonstrated in red blood cells. The residual arginase activity was either below detection limit or very low (≤ 5 per cent of normal) in all but two patients (No. 26,45). The virtual absence of liver arginase activity was demonstrated in all seven tested patients. Arginase in leukocytes in patients No. 2–4[37], 9 and 10[12,13], 30[1] and Endres' propositus[19] and in stratum corneum in No. 13[38] were also shown to be deficient. The residual erythrocyte arginase activities in parents approximated 50 per cent of normal values.

A human liver-derived arginase cDNA clone has been produced and the gene assigned to band q23 on the long arm of chromosome 6[18,27,55,56]. Human liver-type arginase consists of 322 amino acid residues, and the estimated molecular mass is 34,732. The arginase gene has 11. 5 kb, including eight exons[56].

Molecular genetic data have been generated with regard to patients 2,5,6,7,16,17,19,20,21,29,36,37,39 and 48. Several interesting papers demonstrated a series of mutations illustrating the molecular heterogeneity in mutant arginase alleles[24,26,64,65]. Based on an in vitro expression test, the mutations could be considered either severe or moderate[65]. Uchino et al. moreover suggested a molecular basis for the clinical response to dietary treatment in hyperargininemia[65].

## Guanidino compounds

Part of the results presented here were previously reported[35]. We analysed the following 13 guanidino compounds in serum, urine and or cerebrospinal fluid of treated and untreated patients: arginine, homoarginine, creatine, creatinine, α-keto-δ-guanidinovaleric acid, guanidinosuccinic acid, guanidi-

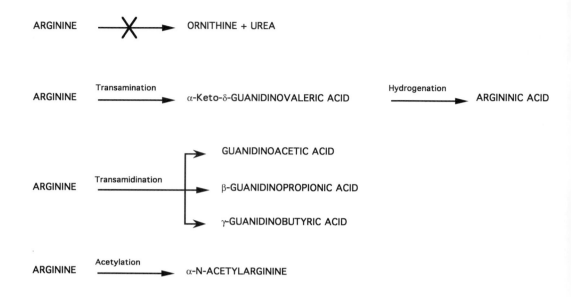

Fig. 4. Proposed catablic pathways of arginine in hyperargininemic patients.

noacetic acid, α-N-acetylarginine, β-guanidinopropionic acid, argininic acid, γ-guanidinobutyric acid, guanidine and methylguanidine. Data are represented in Table 3 and figures 1–3.

Investigating the possible usefulness of guanidino compound analysis as a complementary diagnostic parameter for hyperargininemia, we focussed on the group of untreated and protein-restricted hyperargininemic patients. We considered them as one group since hyperargininemic patients, as do patients with other urea cycle disorders, mostly impose a restricted protein diet upon themselves. The serum levels of β-guanidinopropionic acid and methylguanidine were below the detection limit in all patients of this group (n = 14) and comparable to control values. The serum levels of creatinine and guanidine of most of the patients of this group were within the normal range. The serum creatine levels fell within the normal levels for 50 per cent of these patients while the others had increased (7 to 50 per cent) concentrations. Only one patient had serum creatine levels which were 2. 5 times higher than the upper normal limit. The serum arginine and homoarginine levels in this group were roughly equally increased ranging from 2.5 up to 10 times for arginine and from 3 up to 13 times for homoarginine. The guanidino compounds which are most increased are α-keto-N-guanidinovaleric acid (8 up to 60 times), argininic acid (10 up to 90 times) and α-N-acetylarginine (2 up to 30 times the upper normal limit)(Fig. 1). The serum guanidinoacetic acid levels are also increased in untreated and protein restricted hyperargininemic patients, however to a smaller extent. Some patients have levels just above the upper normal limit, others have increased levels up to 4 times. The serum γ-guanidinobutyric acid levels ranged from lower than the detection limit (like in most of the controls) up to 32 times the upper normal limit. It must be stressed that γ-guanidinobutyric acid levels can be overestimated since this guanidino compound can be formed by chemical oxidation from α-keto-δ-guanidinovaleric acid[63]. However the serum guanidinosuccinic acid levels of most (80 per cent) of the studied patients in this group are lower than control values.

The pattern of the urinary excretion levels of the guanidino compounds in the studied untreated and protein restricted patients (n = 20) roughly reflects the picture observed in serum. However, it should be stressed that 20 per cent of the studied untreated and protein restricted patients have urinary arginine excretion levels which fall within the normal range. The others have excretion levels which were increased from 30 per cent up to 540 times the upper normal limit. Also 50 per cent of these patients have normal homoarginine excretion levels. The other have excretion levels ranging from the upper normal limit up to 15 times this limit. The urinary excretion levels of β-guanidinopropionic acid, guanidine and methylguanidine from most of the patients were comparable to controls. Like in serum, the urinary excretion levels which are most increased are those of α-keto-δ-guanidinovaleric acid (up to 180 times), argininic acid (up to 100 times) and α-N-acetylarginine (up to 70 times the upper normal limit)(Fig. 2). The excretion levels of creatine in untreated and protein restricted patients ranged from 2 up to 14 times the upper normal limit. Respectively twenty and 10 per cent of these patients had normal guanidinoacetic acid and γ-guanidinobutyric acid excretion levels. The other patients have increased excretion levels of guanidinoacetic acid (ranging from 20 per cent up to 9 times the upper normal limit) and γ-guanidinobutyric acid (ranging from 30 per cent up to 19 times the upper normal limit). Twenty five per cent of patients have low normal excretion levels of guanidinosuccinic acid while the excretion levels were decreased for the other patients.

Unfortunately, we could only analyse the guanidino compound levels in cerebrospinal fluid of 4 untreated or protein restricted hyperargininemic patients. The cerebrospinal fluid levels of creatine, guanidinoacetic acid, γ-guanidinopropionic acid, creatinine, γ-guanidinobutyric acid, guanidine and methylguanidine were within the normal range, except for one untreated hyperargininemic patient who had increased levels of γ-guanidinobutyric acid (10 times), guanidine (4 times) and methylguanidine (6 times the upper normal limit). The increase of arginine ranged from 2 up to 5 times the upper normal limit. Three out of the four patients had homoarginine levels just above the upper normal limit; only one patient had increased levels up to 3 times. In one untreated hyperarginininemic patient the levels of α-keto-δ-guanidinovaleric acid, argininic acid and α-N-acetylarginine were most increased respectively up to 16, 4.5 and 8 times the upper normal limit. The increase of α-keto-δ-guanidinovaleric acid in the four patients ranged from 3 up to 16 times the upper normal limit, the argininic acid levels from 2 up to 4. 5 times (Fig. 3).

Follow-up of serum guanidino compound levels during therapy demonstrated that the serum arginine levels could be lowered to the upper limit of the control range through restriction of daily arginine intake (protein restriction and supplementation of essential amino acids without arginine). However, the decrease and the normalization of the serum arginine levels could certainly not normalize the serum levels of α-keto-δ-guanidinovaleric acid, argininic acid, guanidinoacetic acid and α-N-acetylarginine which remained increased.

Quantitative determination of guanidino compounds in the biological fluids of hyperargininemic patients clearly shows that some guanidino compounds increased to an important extent. As a consequence of the arginase deficiency the serum arginine and homoarginine levels are increased. The secondary catabolic pathway which is most activated in hyperargininemic patients is the one with the formation of α-keto-δ-guanidinovaleric acid (Fig. 4). This formation is probably catalyzed by a transaminase. The increased activity of the hydrogenation of α-keto-δ-guanidinovaleric acid to argininic acid is also remarkable. The increase of the transamidination activity involved in the biosynthesis of guanidinoacetic acid, γ-guanidinobutyric acid and β-guanidinopropionic acid in hyperargininemic patients is less pronounced. It should also be mentioned that the transamidination of arginine with γ-guanidinobutyric acid formation can always be overestimated in hyperargininemic patients. Indeed, α-keto-δ-guanidinovaleric acid is an unstable product and degrades partially to γ-guanidinobutyric acid by chemical oxidation[63]. Therefore the time between sampling and analysis must be as short as possible in order to prevent overestimation of the δ−guanidinobutyric acid and an underestimation of the α-keto-δ-guanidinovaleric acid levels.

The biosynthesis of guanidinosuccinic acid is related to urea[41]. Hyperargininemic patients have a decreased urea cycle activity, they have also decreased levels of guanidinosuccinic acid in their biological fluids.

We also determined the guanidino compound levels in different brain regions of one hyperargininemic boy (patient No. 34). The values were compared with those obtained from autopsy material of one age-matched control individual. In addition, we determined the brain guanidino compound levels in different regions of an adult hyperargininemic patient and compared them with those of adult controls (n = 5). Although the arginine levels found in both patients were higher than those found in controls (in some regions the increase is 100 per cent), the increases of α-keto-δ-guanidinovaleric acid, argininic acid and homoarginine are clearer. The increase of guanidinoacetic acid in most regions ranged from 70 to 100 per cent. The levels of the other guanidino compounds in brain regions of hyperargininemic patients were comparable with controls. The guanidinosuccinic acid levels were lower than the control levels or low normal.

## Pathogenesis of neurological complications

Undoubtedly, hyperammonemia could underlie the episodic vomiting, anorexia, irritability, ataxia, convulsions or coma presenting in hyperargininemia[9]. However, considering the pathogenesis of the extreme degree of spasticity (which has either not been or has extremely rarely been observed in other forms of hyperammonemia) and the frequent epileptic symptomatology (often not temporally related to hyperammonemic episodes), one might question whether these neurological complications result simply from ammonia intoxication or are due to the accumulation of arginine and/or its metabolites. Results of therapeutical trials support this view; both Terheggen et al. [57] and Cederbaum et al.[10] were successful in controlling the hyperammonemia with protein restriction, but the plasma arginine levels remained elevated and no clinical improvement was observed. Moreover, several guanidino compounds, found to be increased in the biological fluids of hyperargininemic patients have been shown to be in vitro and/or in vivo neurotoxins. α-keto-δ-guanidinovaleric acid was demonstrated to be a convulsant in rabbit[36]. After topical application of α-keto-δ-guanidinovaleric acid on the sensory motor cortex, tonic-clonic seizures were evoked and epileptiform discharges were recorded on the electroencephalogram[36]. The convulsant effect of this metabolite and the other accumulated guanidino compounds, could contribute to the epileptic symptomatology observed in hyperargininemia. Indeed, α-keto-δ-guanidinovaleric acid and the other 'hyperargininemic guanidino compounds' were shown to inhibit inhibitory amino acid responses on mouse neurons in cell culture; a characteristic of many convulsants[15,16,17]. Extensive investigations indicated that argininic acid was the most potent in reducing inhibitory amino acid responses, followed in decreasing potency by α-keto-δ-guanidino-valeric acid, homoarginine and arginine[17]. Hypothetically, an inhibition of transketolase activity, as has been observed for some guanidino compounds that are increased in uremia[34] could produce demyelinization resulting in upper motor neuron signs.

## Therapy

Enzyme replacement constitutes the only causal treatment of inborn errors of metabolism but has not been successful in hyperargininemia. Michels & Beaudet[38] and Sakiyama et al. [51] made an attempt to replace the deficient enzyme by administration of packed red blood cells. While a small immediate decrease in serum arginine was obtained, no significant clinical changes occurred. In 1987, Mizutani et al.[39] performed an erythrocyte exchange transfusion in one patient who was in addition treated with a low-protein diet, an essential amino acid mixture and sodium benzoate. This treatment resulted in a three months normalization of ammonia levels and serum arginine concentrations. On the other hand, CSF concentrations of arginine remained unchanged in spite of the exchange transfusion. An attempt at gene replacement by intravenous injection of the Shope papilloma virus in three patients (No. 3–5) was unsuccessful[62].

Other therapeutic measures have been applied with varying success. Protein restriction alone is unsatisfactory, the plasma arginine levels remaining high without improvement of the clinical picture[10,12,53,56,61]. Lysine supplementation has been tried not only to augment argininuria but also in the hope that lysine might compete with arginine for uptake in brain, thus lowering brain arginine levels[31,38,43].

The first successful dietary therapy was reported by Snyderman *et al.* [54] who started therapy shortly after birth in a sibling (No. 7) of two previous patients. An essential amino acid mixture, excluding arginine but including tyrosine and cystine, was fed at 2g/kg/day as the only source of amino acids for the first four months of life. Plasma arginine has remained normal and physical and neurologic development, with exception of a somewhat delayed growth, are normal still to date. In two other patients from the same kindred, the same treatment was started from the age of four years on and almost totally prevented episodic hyperammonemia and resulted in a significant improvement in mental capacity and a normal growth. Plasma arginine levels in these patients fell to a value two to three times normal. Three older patients (patients No. 9,10,13) have been treated in the same manner for some years; plasma arginine has fallen to 2 to 3 fold normal and some clinical improvement has been achieved[13]. This treatment is currently prescribed in all newly diagnosed cases, and sometimes in combination with sodium benzoate. Lambert *et al.* [32] described intellectual and motor improvement related to changes in biochemical data under the influence of dietary treatment.

In addition, alternative pathways for the excretion of waste nitrogen as substitutes for the defective ureagenic pathway have been used. The administration of sodium benzoate diverts ammonium nitrogen from the defective urea pathway to hippurate synthesis by way of the glycine cleavage complex. Qureshi *et al.*[45,46] were the first to use this treatment and achieved successful biochemical control with disappearance of the orotic aciduria.

Although the above mentioned therapeutic effects obtained by Snyderman *et al.* and Qureshi *et al.* are promising, one should remain cautious. Indeed, the molecular deficits in different cases of arginase-deficiency are possibly quite heterogenous and therefore it is hard to generalize findings to all the hyperargininemic patients.

## Concluding remarks

Hyperargininemic patients, unlike other urea cycle patients, present typical neurological complications such as pronounced signs of upper motor neuron disease and frequent epileptic symtomatology non-temporally related to hyperammonemia. The hyperammonemia is not persistent and less pronounced than in other urea cycle disorders and even absent in certain hyperargininemic patients. Substantial suggestive evidence supports the view that ammonia is not the sole toxic agent in this disease but that arginine and/or its metabolites, the guanidino compounds might contribute to at least some of the neurological complications of this disease.

The clinical diagnosis may be difficult. A normal urinary amino acid chromatogram in several patients indicates that this most frequently used criterion for the identification of inborn errors of metabolism may be inadequate. Neurologic disorders with progressive spasticity, seizures and mental deterioration may require a study of plasma amino acids. A review of the basic aminoacidurias (cystinuria) in retarded populations may uncover more cases of this disease.

In view of the possibility of the false negative diagnosis of hyperargininemia, the true incidence of this disease is unknown. It may not be as rare as once thought as is illustrated by the diagnosis of multiple families by three expert research groups.

An early diagnosis is imperative because hyperargininemia may be another disorder in which early nutritional management is successful in controlling the detrimental effects.

## Acknowledgments

A. Mori and M. Hiramatsu for their practical help in localizing and examining the Japanese hyperargininemic patients. This work was supported by the Flemish Ministry of Education, the Baron Bogaert-Scheid Fund, Born-Bunge Foundation, Medical Research Foundation OCMW Antwerp, University of Antwerp, Neurosearch Antwerp, the United Fund of Belgium and the NFWO grants N¡ 3. 0044. 92 and 3. 0064. 93.

# APPENDIX

## Institutional affiliations of each author

P. P. De Deyn, B. Marescau and J. J. Martin, University of Antwerp (Antwerp); P. P. De Deyn, B. Pickut and B. Appel, Algemeen Ziekenhuis Middelheim (Antwerp); I. A. Qureshi and M. Lambert, Hôpital Sainte Justine (Montreal); S.D. Cederbaum, University of California (Los Angeles); R. Cerone and R. Gatti, Istituto G. Gaslini (Genova); N. Chamoles and N. Spécola; Foundation for the Study of Metabolic Diseases (Buenos Aires); J.V. Leonard, Institute of Child Health (London); S.S. Kang, Rush Presbytarian-St. Luke's Medical Center (Chicago); N. Mizutani, University of Nagoya (Nagoya); I. Rezvani, St. Christopher's Hospital for Children (Philadelphia); S.E. Snyderman, New York University Medical Center (New York); H.G. Terheggen, Mainzweg 16 (Cologne); A. Fuch-shuber and B. Roth, University of Cologne (Cologne); L. Vilarinho, Instituto de Genetica Medica (Porto); E. Hirsch, Hôpital Civil (Strasbourg); H. Hyland, Institute for Child Health and the Hospital for Sick Children (London); K. Jakobs and M.S. Van der Knaap, Free University Hospital (Amsterdam). A.L. Beaudet and S.K. Shapira, Baylor College of Medicine (Houston); H. Naito, National Childrens' Hospital (Tokyo).

# References

1. Antonozzi, I., Leuzzi, V., Radice, M.L. & Porro, G. (1986): Iperargininemia: descrizione di un caso. *Riv. Ital. Ped.* **12**, 270–274.

2. Bachmann, C. & Colombo, J.P. (1980): Diagnostic value of orotic acid excretion in heritable disorders of the urea cycle and in hyperammonemia due to organic acidurias. *Eur. J. Pediatr.* **134**, 109–113.

3. Bernar, J., Hanson, R.A., Kern, R., Phoenix, B., Shaw, K.N.F. & Cederbaum, S.D. (1986): Arginase deficiency in a 12-year-old boy with mild impairment of intellectual function. *J. Pediatr.* **108**, 432–435.

4. Brockstedt, M., Smit, L.M.E., de Grauw, A.J.C., van der Klei-van Moorsel, J.M. & Jakobs, C. (1990): A new case of hyperargininemia: neurological and biochemical findings prior to and during dietary treatment. *Eur. J. Pediatr.* **149**, 341–343.

5. Brusilow, S., Batshaw, M., Robinson, B., Sherwood, G. & Walser, M. (1977): The mechanism of hyperammonemia in citrullinemia. *Pediatr. Res.* **11**, 453.

6. Buist, N.R.M., Kennaway, N.G., Hepburn, C.A., Strandholm J.J. & Ramburg, D.A. (1974): Citrullinemia: investigation and treatment over a 4 year period. *J. Pediatr.* **85**, 208–214.

7. Candito, M., Bebin, B., VianeyãSaban, C., Rabier, D., Bekri, S., Sebag, F., Chambon, P. & Kamoun, P. (1993): Arginase deficiency in two brothers. *J. Inher. Metab. Dis.* **16**, 1054–1056.

8. Caruso, U., Cerone, R., Schaffino, M.C., Minniti, G., Romano, C., Gatti, R., Filocamo, M. & Colombo, J.P. (1994): Prenatal diagnosis of argininemia: experience on two pregnancies in the same family. *International Pediatrics* **9**, 77.

9. Cathelineau, L. (1979): L'hyperammonièmie dans la pathologie pédiatrique. *Arch. Franç Pédiat.* **36**, 724–735.

10. Cederbaum, S.D. & Shaw, K.N.F. (1977): Hyperargininemia. *J. Pediatr.* **90**, 569–573.

11. Cederbaum, S.D., Shaw, K.N.F., Spector, E.B. & Snodgrass, P.J. (1976): Hyperargininemia with arginase deficiency in two siblings. Fifth International Congress of Human Genetics, Mexico City. *Excerpta Med. Int. Congr. Ser.* **397**, 27.

12. Cederbaum, S.D., Shaw, K.N.F., Spector, E.B., Verity, M.A., Snodgrass, P.J. & Sugarman, G.I. (1979): Hyperargininemia with arginase deficiency. *Ped. Res.* **13**, 827–833.

13. Cederbaum, S.D., Moedjono, S.J., Shaw, K.N.F., Carter, M., Naylor, E. & Walser, M. (1982): Treatment of hyperargininemia due to arginase deficiency with a chemically defined diet. *J. Inher. Metab. Dis.* **5**, 95–99.

14.   Christmann, D., Hirsh, E., Mutschler, V., Collard, M., Marescaux, C. & Colombo J.P. (1990): Argininimie congénitale diagnostique tardivement de l'occasion de la prescription de valproate de sodium. *Rev. Neurol.* **146,** 764–766.

15.   De Deyn, P.P., D'Hooge, R., Marescau, B. & Pei, Y.Q. (1992): Chemical models of epilepsy and their applicability in the development of anticonvulsants. *Epilepsy Res.* **12,** 87–110.

16.   De Deyn, P.P., Marescau, B. & Macdonald, R.L. (1988): Effects of α-keto-δ-guanidinovaleric acid on inhibitory amino acid responses on mouse neurons in cell culture. *Brain Res.* **449,** 54–60.

17.   De Deyn, P.P., Marescau, B. & Macdonald, R.L. (1991): Guanidino compounds that are increased in hyperargininemia inhibit GABA and glycine responses on mouse neurons in cell culture. *Epilepsy Res.* **8,** 134–141.

18.   Dizikes, G.J., Grody, W.W., Kern, R.M. & Cederbaum, S.D. (1985): Isolation and study of DNA clones for human liver arginase. *Am. J. Hum. Genet.* **37,** A152.

19.   Endres, W., Schaller, R. & Shin, Y.S. (1984): Diagnosis and treatment of argininaemia. Characteristics of arginase in human erythrocytes and tissues. *J. Inher. Metab. Dis.* **7,** 8.

20.   Fuchshuber, A., Marescau, B., Roth, B., De Deyn, P.P., Sprenger H-J & Michalk, D.V. (1993): Haemodialysis and continuous veno-venous haemofiltration in a patient with hyperargininaemia and acute renal failure. *J. Inher. Metab. Dis.* **16,** 909–910.

21.   Gambhir, P.S., Phadke, M.A., Khedkar, V.A., Padalkar, J.A., Joshi, A.S., Limaye, A.S., Bhate, S.M. (1989): Argininemia. *Indian Pediatrics* **26,** 1260–1262.

22.   Gatti, R., Cerone, R., Caruso, U., Schiaffino, M.C. & Ciccone. O. (1993): Biochemical diagnosis and follow-up in a new Italian patient with hyperargininemia. *J. Inher. Metab. Dis.* **16,**1050.

23.   Glass, R.D. & Knox, W.E. (1973): Arginase isozymes of rat mammary gland, liver and other tissues. *J. Biol. Chem.* **248,** 5785–5789.

24.   Grody, W.W., Klein, D., Dodson, A.E., Kern, R.M., Wissmann, P.B., Goodman, B.K., Bassand, P., Marescau, B., Soo-Sang Kang, Leonard, J.V. & Cederbaum, S.D. (1992): Molecular genetic study of human arginase deficiency. *Am. J Hum. Genet.* **50,** 1281–290.

25.   Grody, W.W., Argyle, C., Kern, R.M., Dizikes, G.J., Spector, E.B., Strickland, A.D., Klein, D, & Cederbaum, S.D. (1989): Differential expression of the two human arginase genes in hyperargininemia. Enzymatic, pathologic, and moleular analysis. *J. Clin. Invest.* **83,** 602–609.

26.   Haraguchi, Y., Aparicio, J.M., Takiguchi, M., Akaboshi, I., Yoshino, M., Mori, M. & Matsuda, I. (1990): Molecular basis of argininemia.Identification of two discrete frame-shift deletions in the liver-type arginase gene. *J. Clin. Invest.* **86,** 347–350.

27.   Haraguchi, Y, Takiguchi, M, Amaya, Y., Kawamoto, S, Matsuda, I. & Mori, M. (1987): Molecular cloning and nucleotide sequence of cDNA for human liver arginase. *Proc. Natl. Acad. Sci.* **84,** 412–415.

28.   Herzfeld, A. & Raper S.M. (1976): The heterogeneity of arginase in rat tissues. *Biochem. J.* **153,** 469–478.

29.   Hyland, K., Smith, I., Clayton, P.T. & Leonard, J.V. (1986): Impaired neurotransmitter amine metabolism in arginase deficiency. *J. Neurol. Neurosurg. Psychiatr.* **49,** 1188–1189.

30.   Jorda, A., Rubio, V., Portolés, M., Vilas, J. & Garcia-Pino, J. (1986): A new case of arginase in a Spanish male. *J. Inher. Metab.* **9,** 393–397.

31.   Kang, S.S., Wong, P.W.K. & Melyn, M.A. (1983): Hyperargininemia: effect of ornithine and lysine supplemetation. *J. Pediatr.* **103,** 763–765.

32.   Lambert, M.A., Marescau, B., Desjardins, M., Laberge, M., Dhondt, J.āL., Dallaire, L., De Deyn, P.P. & Qureshi, I.A. (1991): Hyperargininemia: intellectual and motor improvement related to changes in biochemical data. *J. Pediatr.* **118,** 420–424.

33.   Levin, B., Oberholzer, V.G. & Sinclair, L. (1969): Biochemical investigation of hyperammonemia. *Lancet* 2,**170,** 174.

34.   Lonergan, E.T., Semar, M., Sterzel, R.B., Treser, G., Needle, M.A., Voyles, L. & Lange, K. (1971): Erythrocyte transketolase activity in dialyzed patients: a reversible metabolic lesion in uremia. *N. Engl. J. Med.* **284,** 1399–1403.

35. Marescau, B., De Deyn, P.P., Lowenthal, A., Qureshi, I.A., Antonozzi, I., Bachmann, C., Cederbaum, S.D., Cerone, R., Chamoles, N., Colombo, J.P., Hyland, K., Gatti, R., Kang, S.S., Letarte, J., Lambert, M., Mizutani, N., Possemiers, I., Rezvani, I., Snyderman, S.E., Terheggen, H.G. & Yoshino, M. (1990): Guanidino compound analysis as a complementary diagnostic parameter for hyperargininemia: Follow-up of guanidino compound levels during therapy. *Pediatr. Res.* **27,** 297–303.

36. Marescau, B., Hiramatsu, M. & Mori, A. (1983):-Keto-guanidinovaleric acid-induced epileptiform discharges in rabbits. *Neurochem. Pathol.* **1,** 203–209.

37. Marescau, B., Pintens, J., Lowenthal, A., Terheggen, H.G. & Adriaenssens, K. (1979) Arginase and free amino acids in hyperargininemia. *J. Clin. Chem. Clin. Biochem.* **17,** 211–217.

38. Michels, V.V. & Beaudet, A.L. (1978): Arginase deficiency in multiple tissues in argininemia. *Clin. Genet.* **13,** 61–67.

39. Mizutani, N., Hayakawa, C., Maehara, M. & Watanabe, K. (1987): Enzyme replacement therapy in a patient with hyperargininemia. *Tohoku J. Exp. Med.* **151,** 301–307.

40. Mizutani, N., Maehara, M., Hayakawa, C., Kato, T., Watanabe, K. & Suzuki, S. (1983): Hyperargininemia: clinical course and treatment with sodium benzoate and phenylacetic acid. *Brain & Development* **5,** 555–563.

41. Natelson, S. & Sherwin, J.E. (1979): Proposed mechanism for urea nitrogen reutilization: relationship between urea and proposed guanidine cycles. *Clin. Chem.* **25,** 1343–1344.

42. Naylor, E.W. & Cederbaum, S.D. (1981): Urinary pyrimidine excretion in arginase deficiency. *J. Inher. Metab. Dis.* **4,** 207–210.

43. Pardridge, W.M. (1977): Lysine supplementation in hyperargininemia. *J. Pediatr.* **91,** 1032–1033.

44. Peralta Serrano, A. (1965): Argininuria, convulsiones y oligofrenia. Un nuevo error innata del metabolismo? *Rev. Clin. Esp.* **97,** 176–184.

45. Qureshi, I.A., Letarte, J., Ouellet, R., Larochelle, J. & Lemieux, B. (1983): A new French–Canadian family affected by hyperargininemia. *J. Inher. Metab. Dis.* **6,** 179–182.

46. Qureshi, I.A., Letarte, J., Ouellet, R., Lelièvre, M. & Laberge, C. (1981): Ammonia metabolism in a family affected by hyperargininemia. *Diabète et Métabolisme* (Paris) **7,** 5–11.

47. Qureshi, I.A., Letarte, J., Ouellet, R. & Lelièvre, M. (1981): Sodium benzoate therapy and dietary control in hyperargininemia. *Pediatr. Res.* **15,** 6–38.

48. Ratner, S., Morell, H. & Carvalho, E. (1960): Enzymes of arginase metabolism in brain. *Arch. Biochem. Biophys.* **91,** 280–289.

49. Reddi, P.K., Knox, W.E. & Herzfeld, A. (1975): Types of arginase in rat tissues. *Enzyme* (Basel). **20,** 305–310.

50. Romano C., Pescetto T., Caruso U., Cerone R. & Caffarena G. (1987): Clinical and biochemical studies on a new patient affected by argininemia. *Persp. Inher. Metab. Dis.* **7,** 19–22.

51. Sakiyama, T., Nakabayashi, H., Kondo, Y., Shimizu, H., Kodama, S. & Kitagawa T. (1982): Argininemia: clinical course and trial of enzyme replacement therapy. *Biomedicine and therapeutics* **8,** 907–910.

52. Scheuerle, A.E., McVie, R., Beaudet, A.L. & Shapira, S.K. (1993): Arginase deficiency presenting as cerebral palsy. *Pediatrics* **91,** 995–996.

53. Snyderman, S.E., Sansaricq, C., Chen, W.J., Norton, P.M. & Phansalkar, S.V. (1977): Argininemia. *J. Pediatr.* **90,** 563–568.

54. Snyderman, S.E., Sansaricq, C., Norton, P.M. & Goldstein, F. (1979): Argininemia treated from birth. *J. Pediatr.* **95,** 61–63.

55. Sparkes, R.S., Dizikis, G.J., Klisak, I., Grody, W.W., Mohandas, T., Heinzmann, C., Zollman, S., Lusis, A.J. & Cederbaum, S.D. (1986): The gene for human liver arginase (Arg1) is assigned to chromosome band 6q23. *Am. J. Hum. Genet.* **39,** 186–193.

56. Takiguchi, M., Haraguchi, Y. & Mori, M. (1988): Human liver–type arginase gene: structure of the gene and analysis of the promotor region. *Nucleic Acids Res.* **16,** 8789–8802.

57. Terheggen, H.G., Lowenthal, A., Lavinha, F. & Colombo, J.P. (1975): Familial hyperargininaemia. *Arch. Dis. Child.* **50,** 57–62.

58. Terheggen, H.G., Schwenk, A., Lowenthal, A., van Sande, M. & Colombo, J.P. (1969): Argininaemia with arginase deficiency. *Lancet* **2,** 748–749.

59.  Terheggen, H.G., Schwenk, A., Lowenthal, A., van Sande, M. & Colombo J.P. (1970): Hyperargininämie mit Arginasedefekt eine neue familiäre Stoffwechselstörung. II. Biochemische Untersuchungen. *Z. Kinderheilk.* **107,** 313–323.

60.  Terheggen, H.G., Schwenk, A., Lowenthal, A., van Sande, M. & Colombo, J.P. (1970): Hyperargininämie mit Arginasedefekt eine neue familiäre Stoffwechselstörung. I. Klinische Befunde. *Z. Kinderheilk.* **107,** 298–312.

61.  Terheggen, H.G., Lavinha, F., Colombo, J.P., van Sande, M. & Lowenthal, A. (1972): Familial hyperargininemia. *J. Genet. Hum.* **20,** 69–84.

62.  Terheggen, H.G., Lowenthal, A., Lavinha, F., Colombo, J.P. & Rogers, S. (1975): Unsuccessful trial of gene replacement in arginase deficiency. *Z. Kinderheilk.* **119,** 1–3.

63.  Thoai, N.V.& Roche, J. (1960): Dérivés guanidiques biologiques. In: *Progress in the chemistry of organic natural products*, vol 18, ed. L. Zechmeister, pp. 83–121. Vienna: Springer-Verlag.

64.  Uchino, T., Haraguchi, Y., Aparicio, J.M., Mizutani, N., Higashikawa, M., Naitoh, H., Mori, M. & Matsuda, I. (1992): Three novel mutations in the liver-type arginase gene in three unrelated Japanese patients with argininemia. *Am. J. Hum. Genet.* **51,** 1406–1412.

65.  Uchino, T., Snyderman, S.E., Lambert, M., Qureshi, I.A., Shapira, S.K., Sansaricq, C., Smit L.M., Jakobs, C.& Matsuda, I. (1995): Molecular basis of phenotypic variation in patients with argininemia. *Human Genetics.***96,** 255–260.

66.  van Sande, M., Terheggen, H.G., Clara R., Leroy J.G. & Lowenthal A. (1971): Lysine cystine pattern associated with neurological disorders. In: *Inherited disorders of sulfur metabolism*, eds. N.A.J. Carson & D.N. Raine, pp. 85–112. Edinburgh: Churchill Livingstone.

67.  Vilarinho, L., Senra, V., Vilarinho, A., Barbosa, C., Parvy, P., Rabier, D. & Kamoun, P. (1990): A new case of argininaemia without spastic diplegia in a Portuguese male. *J. Inher. Metab. Dis.* **13,** 751–753.

68.  Yoshino, M., Kubota, K., Yoshida, I., Murakami, T. & Yamashita, F. (1982): Argininemia: report of a new case and mechanisms of orotic aciduria and hyperammonemia. In: *Urea cycle disease*, eds. A. Lowenthal , A. Mori & B. Marescau, pp. 121–125. New York: Plenum Press.

*Guanidino Compounds : 2*, eds. by P.P. De Deyn, B. Marescau, I.A. Qureshi and A. Mori.
©1997 John Libbey & Company Ltd., pp. 71–76.

# Chapter 9

## Hyperargininemia: pre-natal diagnosis and treatment from birth

R. CERONE, U. CARUSO, A. BARABINO, R. GATTI, [1]C. JAKOBS,
[2]I. JACQUEMYN, [3]B. MARESCAU and [3]P.P. DE DEYN

*University Department of Pediatrics, Division of Pediatrics, G. Gaslini Institute, Largo G. Gaslini 5, 16147 Genova-Quarto, Italy, [1]Department of Pediatrics, Free University Amsterdam, Acad. Ziekenhuis, De Boele-laan 1117, Amsterdam, The Netherlands, [2]Department of obstetrics and gynaecology-AZM, Lindendreef 1, 2018 Antwerp, Belgium, and [3]Laboratory of Neurochemistry and Behavior, Born Bunge Foundation, University of Antwerp, Universiteitsplein 1, 2610 Wilrijk, Antwerp, Belgium*

### Summary

We report the first experience on pre-natal diagnosis of hyperargininemia and the second experience on treatment from birth. Amino acids and arginase activity measured in red blood cells from foetal blood indicated a probable affection of the foetus by arginase deficiency. The diagnosis was confirmed in the newborn. 12 hours after birth human milk was started; at the age of 4 months arginine-free essential amino acid mixture was introduced. The follow-up of the child until 18 months of age is presented here; mental and physical development are normal until now. We conclude that arginase deficiency can be diagnosed pre-natally and that neonatal treatment can prevent hyperammonaemic episodes. In addition, we demonstrated that specific guanidino compound patterns in blood from umbilical cord and amniotic fluid can contribute to the pre-natal diagnosis of this treatable inborn error of metabolism.

### Introduction

Arginase, the last enzyme of the urea cycle, catalyses the formation of urea and ornithine from arginine. The ornithine is recycled by ornithine transcarbamylase, which catalyzes the formation of citrulline. The activity of the next two enzymes, argininosuccinic synthetase and argininosuccinic lyase, lead to the production of a new molecule of arginine. The cycle also produces a molecule of aspartic acid which can react with carbamyl phosphate (when it accumulates as a result of a metabolic impairment) producing orotic acid. Arginase deficiency is an autosomal recessively transmitted disorder presenting with accumulation of arginine in biological fluids, hyperammonemia and orotic aciduria. Levels of some secondary catabolites of arginine, guanidino compounds, are increased in plasma and massively excreted in the urine of these patients.

The clinical picture is characterized by varying degrees of mental retardation, epilepsy and progressive spasticity. Some patients have recurrent episodes of vomiting and irritability. Since the first clinical and biochemical description[5] about 26 cases have been reported[3] and only one experience on treatment from birth has been published[4]. At present, data on approximately 54 patients have been compiled by De Deyn and co-workers.

To our knowledge, we report the first experience on pre-natal diagnosis of hyperargininaemia. Arginase activity determination allowed the pre-natal diagnosis and was supported by the increased level of arginine. Additional determinations of guanidino compounds on umbilical plasma and

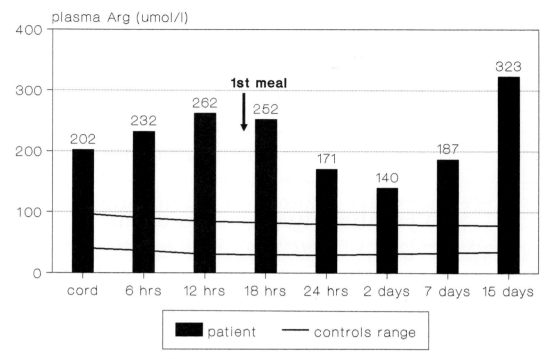

*Fig. 1. Arginine during the first 15 days of life.*

amniotic fluid pre-elevated at birth, indicate that determination of these compounds could further substantiate the pre-natal diagnosis of this treatable inborn error of metabolism.

## Case report and material

The patient (A.M.), a 18 month-old girl, is the third child of healthy, non-consanguineous Italian parents. Her elder brother with known hyperargininaemia has been previously reported[2]. At the 21st gestational week of the propositus, 1 ml of foetal blood was collected by ultrasound-guided puncture from umbilical cord. Plasma was immediately separated from erythrocytes (RBC) which were washed with saline solution. Amino acids and guanidino compounds were quantified in plasma, and arginase activity was measured in RBC.

The pregnancy was uneventful. Apgar scores were 7 and 8 at 1 and 5 minutes, respectively. The birth weight was 3670 g and head circumference 35 cm. Clinical examination showed no abnormalities. Ammonia, amino acids and guanidino compounds were also quantified in plasma and arginase activity was measured in RBC obtained at birth.

The infant was given glucose-water for the first 12 h of life, then human milk was started. At the age of 4 months arginine-free essential amino acid mixture was introduced. After 6 months of life the diet was supplemented with fruits, vegetables and protein-free products. The natural protein intake never exceeded 7 g/day. Sodium benzoate or other drugs were never administrated.

## Results and discussion

Amino acid analysis in foetal plasma revealed a marked elevation of arginine (192 µM) against the reference values (31; 51 µM), indicating that the foetus (Pregn III, Table 1) was probably affected.

These data were supported by the result of arginase activity measurement in foetal RBC: 4.3 per cent of the lower control value was found (Table 1).

**Table 1. Plasma arginine (μM) and arginase activity in foetuses, newborns and mother**

|  | Foetuses umbilical cord | Umbilical cord birth | Newborns | | Mother | |
|---|---|---|---|---|---|---|
|  | Arginine | Arginase | Arginine | Arginase | Arginine | Arginase |
| Pregn II | 98 | 1504 | NA | 63 | 2553 | 59 |
| Pregn III | 192 | 135 | 202 | 262 | NA | 79 |
| Controls | 31; 51 | 3127; 3305 | 41–97 | 31–85 | 2100–12400 | 60–128 |

Arginase activity is expressed as μmol of product/60 min /g Hb; NA = not assessed.

The same procedure was performed[1] in another pregnancy at risk for argininemia from the same family (= Pregn II, Table 1).The foetal plasma arginine levels of Pregn II were slightly higher than the upper control value. The arginase activity in RBC of Pregn II was 48 per cent of the lower control reference value, suggesting that Pregn II was a heterozygote. The plasma arginine level of the newborn Pregn II and the arginase activity in RBC fell into the normal range.

The plasma arginine levels at birth of Pregn III were elevated (202 μM for umbilical cord and 232 μM 6 h after birth) (Table 1 and Fig. 1). Arginase activity in RBC was absent (Table 1), while ammonia levels remained within normal range. The following 15 days plasma ammonia levels were normal. Plasma arginine levels were maintained between 220 and 323 μM (Fig. 1).

Next to arginine, other guanidino compounds were determined in amniotic fluid at birth, plasma and urine at birth and the first days after birth. The determinations in amniotic fluid at birth clearly demonstrate increased arginine levels (Table 2). The increase of homoarginine are from the same order. Both metabolites can normally be hydrolysed by arginase. The keto analogue (α-keto-δ-guanidinovaleric acid) and the hydroxylated analogue (argininic acid) of arginine are the catabolites of arginine which are most increased in the amniotic fluid. Similar results were obtained in plasma, furthermore the clear increase in plasma of homoarginine is remarkable and many times higher than the increase of arginine (Table 3). The urinary excretion of arginine and the guanidino compounds in the propositus depends on protein intake (Table 4), as has been seen earlier in other patients with arginase deficiency[3].

**Table 2. Guanidino compound levels (M) in amniotic fluid of controls and propositus**

| Guanidino compound | Controls 1618 wks gest | Controls 31–35 wks gest | Propositus (A.M.) at birth |
|---|---|---|---|
| α-keto-δ-guanidinovaleric acid | 0.123 ± 0.050 | 0.205 ± 0.060 | 2.12 |
| guanidinosuccinic acid | 0.199 ± 0.092 | 0.292 ± 0.094 | 0.15 |
| guanidinoacetic acid | 2.56 ± 0.402 | 2.83 ± 0.716 | 6.06 |
| Argininic acid | < 0.015–0.130 | 0.100 ± 0.044 | 0.63 |
| γ-guanidino acid | < 0.013–0.07 | < 0.013–0.14 | 0.24 |
| Arginine | 55.2 ± 15.7 | 9.76 ± 1.98 | 37.0 |
| Homoarginine | 4.97 ± 2.19 | 0.737 ± 0.474 | 2.65 |

The propositus never presented hyperammonaemia or seizures on follow-up until 18 months of age. During this period plasma arginine levels were maintained under 350 μM. Her body weight was 10 kg (10th centile) and length 81 cm (50th centile). She was physically, neurologically and mentally normal. The diet provides 7 g/day of natural protein and 100 Kcal/Kg/day: arginine free essential amino acids mixture (Glycinex®) is supplied to provide the total protein need.

Table 3. Guanidino compound levels (M) in plasma of controls and propositus

| Guanidino compound | Controls (5 days) | Propositus (time after birth) | | |
|---|---|---|---|---|
| | | 6 h | 24 h | 5 days |
| α-keto-δ-guanidinovaleric acid | < 0.035 | 4.97 | 5.08 | 1.36 |
| guanidinosuccinic acid | < 0.025–0.093 | 0.18 | 0.14 | 0.04 |
| guanidinoacetic acid | 0.770–0.137 | 2.52 | 1.71 | 1.14 |
| Argininic acid | < 0.025–0.040 | 1.16 | 1.60 | 0.96 |
| γ-guanidino acid | < 0.025 | 0.09 | 0.14 | 0.03 |
| Arginine | 51.0 ± 11.3 | 232 | 171 | 140 |
| Homoarginine | 0.992 ± 0.598 | 29.1 | 29.5 | 33.6 |

The propositus never presented hyperammonaemia or seizures on follow-up. With regard to early treatment our results are similar to the data reported previously[4]. Protein restriction with administration of an arginine-free essential amino acid mixture has been effective in controlling plasma ammonia levels; however the plasma arginine levels showed only a few decreases. An early diagnosis of hyperargininemia is important since the disease seems to be treatable by dietary measures as illustrated also in the case described by Snyderman *et al.* However, one needs to be cautious since there might be a genetic heterogeneity co-determining the clinical outcome in these patients.

Our presented experience demonstrates clearly the possibility of pre-natal diagnosis of argininemia by measurement of arginase activity in foetal RBC and by quantitation of arginine in plasma from foetal blood. We underline the relevance that the foetal arginine levels could be informative for diagnostic purposes, however confirmation by measurement of arginase activity in RBC is necessary.

The guanidino compound determinations on neonatal amniotic fluid and plasma showed typical increases of the secondary catabolytes of arginine which were more pronounced than the particular increases of arginine. Therefore we suggest to determine in the future, next to the arginine levels, also the guanidino compound levels in foetal blood combined with the arginase activity in RBC in high risk pregnancies (parents known heterozygote).

Table 4. Guanidino compound levels (μmol/g creatinine) in urine of controls and propositus

| Guanidino compound | Controls | Propositus (time after birth) | |
|---|---|---|---|
| | | 1 day | 10 days (after fed milk) |
| α-keto-δ-guanidinovaleric acid | < DL–30 | 384 | 1365 |
| guanidosuccinic acid | 25–100 | 14.3 | 4.90 |
| guanidoacetic acid | 250–850 | 552 | 1187 |
| Argininic acid | 5–20 | 71.0 | 177 |
| γ-guanidino acid | 5–30 | 89.5 | 81.1 |
| Arginine | 10–60 | 80.3 | 130 |
| Homoarginine | < DL–10 | 18.7 | 21.9 |

## Acknowledgement

The authors thank Professor J.P. Colombo for arginase activity assay and Professor F. Carnevale for follow-up of the patient.

# References

1.   Caruso, U., Cerone, R., Schaffino, M.C., Minniti, G., Romano, C., Gatti, R. & Filocamo, M., Colombo, J.P. (1994): Prenatal diagnosis of argininemia: experience on two pregnancies in the same family. *Int. Pediatr.* **9,** 77.

2.   Gatti, R., Cerone, R., Caruso, U., Schiaffino, M.C. & Ciccone, O. (1993): Biochemical diagnosis and follow-up in a new Italian patient with hyperargininemia. *J. Inher. Metab. Dis.* **16,** 1050–1051.

3.   Marescau, B., De Deyn, P.P., Lowenthal, A., Qureshi, I.A., Antonozzi, I., Bachmann, C., Cederbaum, S.D., Cerone, R., Chamoles, N., Colombo, J.P., Hyland, K., Gatti, R., Kang, S.S., Letarte, J., Lambert, M., Mizutani, N., Possemiers, I., Rezvani, I., Snyderman, S.E., Terheggen, H.G. & Yoshino, M. (1990): Guanidino compounds analysis as a complementary diagnostic parameter for hyperargininemia: follow-up of guanidino compound levels during therapy. *Pediatr. Res.* **27,** 297–303.

4.   Snyderman, S.E., Sansaricq O., Norton, P.M. & Goldstein, F. (1979): Argininemia treated from birth. *J. Pediatr.* **95,** 61–63.

5.   Terheggen, H.G., Schwenk, A., Lowenthal, A., Van Sande, M. & Colombo, J.P. (1969): Argininemia with arginase deficiency. *Lancet,* **2,** 748–749.

*Guanidino Compounds : 2*, eds. by P.P. De Deyn, B. Marescau, I.A. Qureshi and A. Mori.
©1997 John Libbey & Company Ltd., pp. 77–85.

# Chapter 10

# Distribution of arginase along the mammalian nephron

Olivier LEVILLAIN and Annette Hus-CITHAREL

*Laboratoire de Physiologie Cellulaire, C.N.R.S. URA 219; Collège de France, 75231 Paris Cedex 05, France*

## Summary

Several years ago, an arginase activity was discovered in mammalian kidney but the precise localization and physiological roles of this enzyme remain unknown. Because of the high cellular heterogeneity of the kidney, the different nephron segments were microdissected in order to localize and quantify urea production. The nephron segments were incubated with 216 µM L-[guanido-$^{14}$C] arginine and the [$^{14}$C] urea resulting from intracellular arginase activity was released in the incubating medium which contained purified urease. [$^{14}$C] urea was immediately split into ammonia and $^{14}$CO$_2$ which was trapped in KOH and counted by liquid scintillation. The results indicated that the distribution pattern of arginase along the nephron was specific to the species studied. A constant finding was that the highest rate of urea was produced in the straight proximal tubule with an increasing cortico-medullary gradient. Arginase was found in the proximal convoluted tubule of most species. This enzyme was also present in the thin descending limbs of rats and *Meriones shawi* and in the thick ascending limb of rabbits. A high production of urea increasing from the cortex towards the medulla was observed in the collecting duct of the rat. The physiological role of arginase might be: (1) production of urea for the urinary concentrating mechanism or for the 'guanidine cycle'; (2) production of ornithine for either polyamine synthesis via ornithine decarboxylase or glutamate via ornithine amino-transferase; (3) modulation of the availability of arginine which is a substrate for nitric oxide synthase; this way arginase could modulate vascular tone, filtration rate and renal blood flow.

## Introduction

The ornithine cycle which was discovered in the liver by Krebs[13] consists of five enzymatic steps, the last one being catalysed by arginase (L-arginine urea hydrolase, EC 3.5.3.1) which hydrolyses L-arginine into urea and ornithine. Arginine is either synthesized in hepatocytes by the condensation of ornithine with bicarbonate, ammonia and the amino group of aspartate or comes from the food or a breakdown of proteins. It has been established that significant amounts of urea are synthesized only in the liver because (1) hepatocytes contain the five enzymes of the ornithine cycle; (2) arginase activity is higher in the liver[8] than in non-hepatic tissues such as the kidney, brain, spleen, intestine and submaxillary gland[8]. Urea produced by hepatocytes is released into the bloodstream and excreted in the urine.

At high plasma and cell levels, urea is a toxic end-product of amino acid catabolism. However, urea plays an important role in the urinary concentrating mechanism[2]. In the interstitium and the tubular lumen of the renal medulla, high concentrations of urea have been measured. Urea has been demonstrated to be concentrated and recycled leading to an important osmotic gradient at the tip of the papilla. Actually, all the mechanisms involved in these processes are not yet completely understood, leading us to hypothesize that an intrarenal production of urea might be one of the unknown factors which contribute to build or to maintain the gradient of urea when the kidney concentrates the urine. An intrarenal production of urea involves the presence of an arginase activity and the availability of the precursor, L-arginine.

Two possible sources of arginine have been reported: (1) the blood which contains arginine continuously supplies the needs of renal cells; (2) in the kidney, the three subsegments of the proximal tubule[19,21] produce arginine from citrulline and aspartate[5,7,26,27,30].

Concerning the enzymatic machinery needed, it has been reported that the kidney contains an arginase activity[1,6,8,12,26–29,31,33,34,38] which is bound to the mitochondrial fraction[12,33]. In the renal tissue, two isoenzymes of arginase have been identified[26,27,33] and characterized by measuring their immunological, biochemical and catalytical properties which differ from the two hepatic isoenzymes[26]. One of the most important differences is that, in the liver, arginase activity is about 1000-fold higher than in the kidney[8]. However, these experiments were performed on renal tissue homogenate. Because of the high cellular heterogeneity of the kidney[14,25], it is not excluded that arginase is localized only in one of the different nephron segments containing a high arginase activity. Thus, the characteristics of the renal production of urea and ornithine should be reconsidered.

This work was designed to localize arginase activity along the nephron of several mammals, to know whether or not the kidney produces urea and to quantify this production on pieces of freshly microdissected nephron segment. The experiments were performed in different mammalian species whose capacity to concentrate urine is either high (i.e. rat, mouse and *Meriones shawi*) or low (i.e. rabbit and guinea pig).

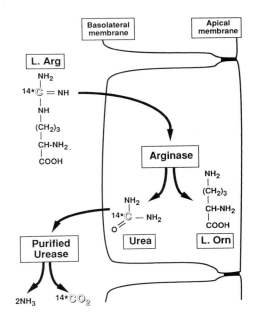

Fig. 1. The principle of urea measurement is based on the specificity of arginase activity. L-[guanido-$^{14}$C]arginine (L-Arg) enters the cells by facilitated diffusion. In the presence of an intracellular arginase, the labelled arginine is cleaved into ornithine (L-Orn) and $^{14}$C urea. Labelled urea diffuses through the membrane and is hydrolysed into $^{14}CO_2$ in the presence of purified urease. $^{14}CO_2$ is trapped in KOH. The labelled carbon of arginine is indicated on the panel.

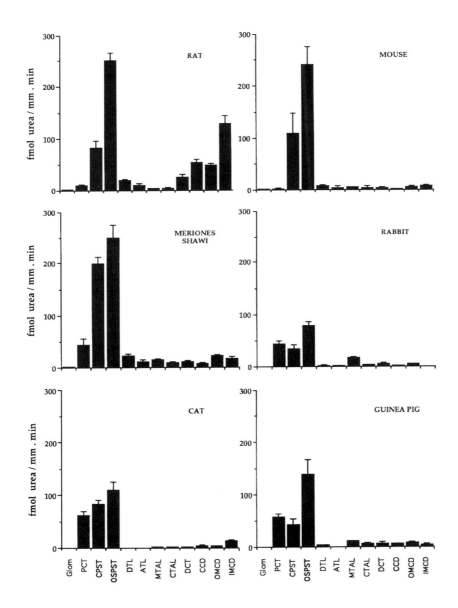

*Fig. 2. Distribution of urea and ornithine production and pattern of arginase activity along the nephron of several mammals. Rats (n = 5–9), mice (n = 5), rabbits (n = 5), Meriones shawi (n = 3–6), guinea pigs (n = 5) and cats (n = 5). Abbreviations of the different nephron segments (see 'Material and Methods' section). Fresh microdissected tubules were incubated with 216 μM L-[guanido-*[14]*C]arginine. Results are mean ± SEM. Reproduced with permission from* Am. J. Physiol.

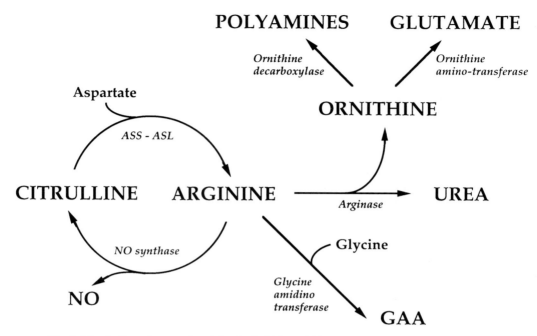

Fig. 3. Metabolic pathways of arginine in the kidney. ASS – ASL: argininosuccinate synthase, argininosuccinate lyase, NO = nitric oxide, GAA = guanidinoacetic acid.

## Material and methods

### Animals

Adult male Sprague-Dawley rats (150 g body weight (BW)), OF-A Swiss mice (35 g BW), New Zealand rabbits (1.5 kg BW), *Meriones shawi*, a desert rodent which lives in Tunisia (150–200 g BW), guinea pigs (500 g BW) and female cats (2–2.5 kg BW) were used for the different experiments. Animals had free access to tap water and their respective chow.

### Chemicals and solutions

The medium used for the metabolic incubation and for the collagenase perfusion (MIP) consisted of a basal salt solution in which metabolic substrates were added. Salts were as follows (in mM): NaCl: 137, KCl: 5, $KH_2PO_4$: 0.44, $MgCl_2$: 1, $MgSO_4$: 0.8, $Na_2HPO_4$: 0.33, $CaCl_2$: 1 and N-2-hydroxy-ethylpiperazine-N'-2-ethanesulfonic acid (HEPES): 20 as well as 0.1 per cent bovine serum albumin (BSA), vitamins mixture 1 per cent and dextran 6 per cent. Energetic substrates for the metabolic needs of the different nephron segments were as follows (in mM): glucose 5, lactate 5, acetate 10, pyruvate 1, glutamine 2. The pH was adjusted to 7.40 with NaOH and the osmotic pressure was 330–350 mosmol/kg $H_2O$.

Salts and most chemicals were purchased from Sigma. Collagenase from *Clostridium histolyticum* (CLS II, 151 U/mg) was bought from Worthington (Freehold, NJ, USA). Urease from jack bean (124 U/mg) was from Boehringer (Mannheim, FRG). Vitamins (100X MBE VT 00) were from Eurobio (Paris, France). Liquid scintillation counting mixture (Aqueous Counting Scintillant ACS II) was from Amersham International (Buckinghamshire, UK). L-[guanido-$^{14}$C]arginine, 1.85 GBq/mmol (50 mCi/mmol) was from the Commissariat à l'Energie Atomique (Saclay, France).

## Preparation and microdissection of the kidney

Animals were anaesthetized by intraperitoneal injection of Nembutal (pentobarbital 6 per cent, Clin Midy, Abbott, Ingelheim, Germany). A catheter was introduced in the abdominal aorta and the left kidney was perfused with 0.9 per cent cold NaCl to wash out the blood and reduce the cellular functions, and then with a collagenase Hanks solution (MIP) containing salts and energetic substrates to meet the metabolic needs of the different nephron segments[10,15,18,20]. The kidney was rapidly removed and sliced along the cortico-papillary axis. The small pyramids obtained were incubated at 30 °C in the same Hank's solution containing 1 mg/ml collagenase, bubbled with oxygen. The incubating period lasted 15–20 min (microdissection of tubules from the cortex and the outer medulla) or 60–90 min (microdissection of tubules from the inner medulla). Pyramids were rinsed with Hank's solution and microdissection was performed by hand at 4 °C under stereomicroscopic control.

Pieces of tubules were dissected and identified with the anatomical and morphological criteria used for rodents[14]: glomerulus (Glom); proximal convoluted tubule (PCT); straight parts of the proximal tubule were taken in the medullary rays of the cortex (CPST) and from the outer stripe of the outer medulla (OSPST); descending thin limb (DTL); ascending thin limb (ATL); cortical and medullary thick ascending limb, (CTAL) and (MTAL) respectively; distal convoluted tubule (DCT); cortical collecting duct (CCD); outer medullary collecting duct (OMCD); inner medullary collecting duct (IMCD).

Glomeruli and tubules were transferred onto a siliconized hollow glass slide coated with BSA in 0.5 µl Hank's solution and tightly sealed with another cover glass slide. Pieces of tubules were photographed in order to determine the tubular length. Samples were kept at 4 °C for 1–2 h until the metabolic incubation.

## Principle of urea measurement

The micromethod is based on the use of a radiolabelled precursor: L-[guanido $^{14}$C]arginine. In the presence of an intracellular arginase activity, labelled arginine is hydrolysed into ornithine and [$^{14}$C] urea. Labelled urea diffuses out of the cells into the incubating medium in which we have previously added purified urease (EC 3. 5. 1. 5), see Fig. 1. The [$^{14}$C] urea is cleaved into 2 $NH_3$ and $^{14}CO_2$ which is continuously trapped by KOH during the metabolic incubation.

In detail, metabolic incubation was started by adding to the 0.5 µl droplet which contains the tubule(s), another 0.5 µl droplet of MIP containing L-[guanidino $^{14}$C]arginine (400 Bq = 10.8 nCi per sample) and urease (124 mU/sample). The final concentration of arginine (without carrier) was 216 µM, a concentration close to physiological plasma concentration[32]. The samples were sealed by a glass cover slide containing a 2 µl droplet iso-osmotic KOH solution and immersed in a water bath at 37 °C for 70 min. At the end of the incubation, KOH was removed and transferred into a glass vial and the $^{14}CO_2$ trapped was counted by liquid scintillation.

The amount of urea produced during incubation of individual nephron segments was deduced from the amount of $^{14}CO_2$ counted in KOH. Blank was measured in samples containing labelled arginine but no tubule. Results were expressed in femtomoles urea per minute and per millimeter tubular length.

## Determination of amino acid concentration

Arterial blood was sampled in the abdominal aorta and transferred into lithium heparinized vacutainer tubes and maintained at 4 °C before centrifugation. One part of the blood was immediately deproteinized with sulfosalysilic acid (SSA) in the proportion of 200 µl/ml blood and the other part of the blood was centrifuged at 4000 × $g$ for 20 min at 4 °C. The plasma obtained was deproteinized with SSA in the proportion of 100 µl/ml plasma. Amino acids were measured in SSA deproteinized samples by ion exchange chromatography on a Beckman 6300 AA analyser using the manufacturer's methodology.

## Results

### Urea production and arginase localization

The results presented in Fig. 2 depict the pattern of urea and ornithine production along the nephron of several mammalian species. This pattern also reflects the distribution of arginase activity.

Considering the sites of urea and ornithine production and the amounts of urea produced in each nephron segment, we observe and conclude that the pattern of urea/ornithine production and arginase activity is species specific.

In rats, the different nephron segments are able to hydrolyse arginine but the main sites of urea and ornithine production are the pars recta of the proximal tubule and the whole collecting duct. In both structures, an increasing cortico-medullary gradient of urea/ornithine production has been observed[18]. Lower amounts of urea/ornithine are produced in the thin descending and ascending limb of Henle's loop[18].

In mice, the only site of urea and ornithine production is the pars recta of the proximal tubule; this production increases from the cortical towards the medullary portion of this nephron segment[20].

In *Meriones shawi*, the different nephron segments produce urea and ornithine from arginine, but the main site of arginine hydrolysis is the proximal straight tubule. An increasing gradient of urea and ornithine production is observed from the end of the proximal convoluted tubule towards the OSPST. In more detailed studies, we demonstrated that urea production was higher in OSPST and TDL of deep nephrons than in superficial nephrons[10].

In rabbits, urea and ornithine are produced in the proximal convoluted tubule, the cortical and medullary pars recta of the proximal tubule and in another nephron segment: the medullary thick ascending limb[20].

In cats, arginine hydrolysis occurs in the whole proximal tubule with an increasing gradient of urea and ornithine production from the PCT towards the OSPST. In this species, a small but significant amount of urea is produced in the inner medullary collecting duct[15].

In guinea pigs, the main site of urea and ornithine production is the whole proximal tubule[15].

In the dog kidney, urea production has been measured in tubular suspensions prepared from the cortex, the outer and the inner medulla. The cortical zone which was enriched in PCT segments showed the highest level of urea production[37].

### Blood and plasma arginine concentration

In normal adult mice, rabbits, cats and *Meriones shawi*, the blood concentration of arginine was (mean $\pm$ SEM in $\mu$M, (n)): $186 \pm 6$ (5); $321 \pm 62$ (3); $97 \pm 13$ (5); $98 \pm 9$ (4), respectively.

In normal adult rats, mice, rabbits, cats, plasma concentration of arginine was (mean $\pm$ SEM in $\mu$M, (n)): $179 \pm 9$ (9); $103 \pm 8$ (8); $214 \pm 30$ (6); $70 \pm 13$ (5), respectively.

The arginine concentration in blood of *Meriones shawi* and cats is about 2–3 fold lower than that used for the metabolic incubation (216 $\mu$M). This implicates that urea production *in vivo*, in the tubules of these species might be lower than that reported in Fig. 2.

## Discussion

The presented experiments were performed in order to know whether or not the kidney is able to produce urea and to identify the nephron structures containing an arginase activity.

Our results clearly indicate that the kidney of several mammals contains an arginase activity and produces urea when arginine is the only substrate. In contrast to the liver, the kidney does not require the ornithine cycle to produce urea. Arginase activity and its distribution along the nephron are heterogenous and vary from one species to another. Consequently, the previous works reporting that renal arginase activity from homogenate is 1000-fold lower than that of the liver should be reevaluated[12]. So, a higher specific activity of the renal arginase is obtained when measured on each nephron segment, in contrast to kidney homogenate[12,26,31,34]. It has been reported that the kidney contains two arginase isoenzymes: $A_1$ and $A_4$, according to the new nomenclature and IUPAC-IUB recommendations[27]. Arginase $A_1$, the major form, is bound to the mitochondrial fraction[6,33] and $A_4$, the minor form, is a cytosolic enzyme[33]. These two isoenzymes also differ by their immunological properties and their specific activity in both human ($A_1$: 1000 and $A_4$: 30 µmol/min per mg protein)[38] and rat kidney ($A_1$: 1054 and $A_4$: 30–40 µmol/min per mg protein)[27,33]. It should be underlined that arginase $A_1$ activity is in the same range of magnitude as that of the liver arginase isoenzyme $A_5$ (human: 2500 and rat: 3500 µmol/min per mg protein), the main arginase isoform of the liver[27,38].

Based on these results and the anatomical organization of the nephron, we propose two hypotheses concerning the possible localization of arginase isoforms. On the one hand, the mitochondrial arginase $A_1$, which represents about 80–90 per cent of the renal arginase activity[26,33], might be localized in the proximal tubule which is quantitatively the most abundant structure in the kidney and which also possesses large amounts of mitochondria[25]. In contrast, the cytosolic arginase $A_4$ could be localized in the collecting duct, mainly the inner medullary collecting duct which contains very few mitochondria. On the other hand, arginases $A_1$ and $A_4$ could be localized in the same nephron structure, $A_1$ being always more important than $A_4$ or the ratio $A_1/A_4$ could vary depending on the nephron segment (i.e. $A_1 > A_4$ in the proximal tubule). Because of the different properties of renal arginase isoenzymes, it is conceivable that their physiological roles also differ.

Actually, the physiological roles of arginase and the products resulting from arginine breakdown in kidney remain unknown, but we propose three hypotheses.

**First hypothesis: role of arginase in the urinary concentrating mechanism  (Fig. 3.)**

The amount of urea produced by OSPST could contribute to build the gradient of urea in the renal medulla (i.e. after diuresis). This hypothesis is supported by the following arguments: (1) it has been reported that, in the straight proximal tubule, urea is preferentially transported into the tubular lumen[11]; (2) urea is produced at the entry of the nephron structures involved in the urinary concentrating mechanism (just before the medulla); (3) mammalian species which bring their urine osmolality to high osmotic pressures (rat, mouse, *Meriones shawi*) produce the highest amounts of urea (250 fmol/min per mm) whereas species such as rabbits and guinea pigs produce 3–4 fold lower amounts of urea and their urine osmolality is low; in *Meriones shawi*, we demonstrated that OSPST and TDL of deep nephrons produce more urea than those of superficial nephrons[10]. This hypothesis should be tested experimentally.

**Second hypothesis: ornithine might be used for polyamines and ATP synthesis  (Fig. 3.)**

Ornithine could be the most important product resulting from arginine hydrolysis. Ornithine enters the polyamine pathway in which ornithine decarboxylase (ODC, EC 4. 1. 1. 17) is the key enzyme[24,35]. (see Fig. 3). It is known that the kidney contains a very high ornithine decarboxylase activity[24,35]. In addition to the physiological roles of polyamines in cell growth[35], new biochemical functions have been discovered for these compounds[3,4,22]. It has been recently proposed that nephrotoxicity of aminoglycosides could result from their interaction with the polyamine pathway, through an inhibition of the renal ODC activity[9]. So, polyamines would contribute to maintain cell integrity and functions, act as hormones, modulate enzyme activities, and polyamines produced in the kidney could also be available for other tissues. We have recently demonstrated in rat kidney, that ODC activity is mainly localized in the three subsegments of the proximal tubule, but mainly in the proximal convoluted tubule[15,16]. In another metabolic pathway whose key enzyme is ornithine amino-transferase (OAT, EC 2. 6. 1. 13), ornithine is transaminated to produce γ-glutamyl semialdehyde which in turn is oxidized and deaminated before entering the Krebs – cycle to be oxidating decarboxylated and to provide ATP (see Fig. 3). This pathway has also been demonstrated in isolated fresh nephron segments, the highest OAT activity being found in the proximal tubule (mainly in the OSPST)[15,16].

## Third hypothesis: local control of nitric oxide synthase activity by arginase

Arginine is the common substrate of several metabolic pathways (i.e. synthesis of urea, nitric oxide (NO), guanidino compounds, see Fig. 3). Thus, arginase which consumes arginine, could modulate arginine availability for the other enzymes using arginine as substrate. If a cell possesses both arginase and NO synthase activity (i.e. rat collecting duct)[36], a modulation of the former[17,34] or an induction of the latter[23] would influence preferentially the fate of arginine towards either production of urea plus ornithine or nitric oxide plus citrulline. An alternative possibility is that the two enzymes are localized in different but neighbouring nephron segments (i.e. OSPST and MTAL)[23]. In this case, arginase would decrease the extracellular concentration of arginine and indirectly modulate the production of NO and the vascular tone.

In conclusion, this work demonstrates the presence of an arginase activity in specific nephron segments and a net production of urea. In all species studied, the highest rate of urea production was found in the OSPST, suggesting a high arginase activity in this segment. Actually, the physiological role of the renal urea production remains unclear. In contrast, ornithine may be involved in the polyamine pathway or in the production of energy (ATP).

## Acknowledgement

O. Levillain received a scholarship from the Foundation de la Recherche Médicale.

# References

1.  Aperia, A., Broberger, O., Larsson, A. & Snellman, K. (1979): Studies of renal urea cycle enzymes. I. Renal concentrating ability and urea enzymes in the rat during protein deprivation. *Scand. J. Clin. Lab. Invest.* **39,** 329–336.

2.  Berliner, R.W. (1953): Mechanism of urine concentration. *Kidney Int.* **22,** 202–211.

3.  Charlton, J.A. & Baylis, P.H. (1989): Stimulation of ornithine decarboxylase activity by arginine vasopressin in the rat medullary thick ascending limb of Henle's loop. *J. Endocrinol.* **120,** 195–199.

4.  Charlton, J.A. & Baylis, P.H. (1990): Stimulation of Na+/K+–ATPase activity by polyamines in the rat medullary cells of the thick ascending limb of Henle's loop. *J. Endocrinol.* **127,** 377–382.

5.  Dhanakoti, S.N., Brosnan, J.T., Herzberg, G.R. & Brosnan, M.E. (1990): Renal arginine synthesis: studies *in vitro* and *in vivo. Am. J. Physiol.* **259,** E437–E442.

6.  Dhanakoti, S.N., Brosnan, M.E., Herzberg, G.R. & Brosnan, J.T. (1992): Cellular and subcellular localization of enzymes of arginine metabolism in rat kidney. *Biochem. J.* **282,** 369–375.

7.  Featherston, W.R., Rogers, Q.R. & Freedland, R.A. (1973): Relative importance of kidney and liver in synthesis of arginine by the rat. *Am. J. Physiol.* **224,** 127–129.

8.  Gasiorowska, I., Porembska, Z., Jachimowicz, J. & Mochnacka, I. (1970): Isoenzymes of renal arginase in rat tissues. *Acta Biochim. Pol.* **17,** 19–30.

9.  Henley, III.C.M., Mahran, L.G. & Chacht, J.S. (1988): Inhibition of renal ornithine decarboxylase by aminoglycoside antibiotics *in vitro. Biochem. Pharmacol.* **37,** 1679–1682.

10. Hus-Citharel, A., Levillain, O. & Morel, F. (1995): Site of arginine synthesis and urea production along the nephron of a desert rodent species, *Meriones shawi. Pflügers Arch.* **429,** 485–493.

11. Kawamura, S. & Kokko, J.P. (1976): Urea secretion by the straight segment of the proximal tubule. *J. Clin. Invest.* **58,** 604–612.

12. Kaysen, G.A. & Strecker, H.J. (1973). Purification and properties of arginase of rat kidney. *Biochem. J.* **133,** 779–788.

13. Krebs, H.A. & Henseleit, K. (1932): Untersuchungen über die Harnstoffbildung im Tierkörper. *Z. Physiol. Chem.* **210,** 33–66.

14. Kriz, W. (1981): Structural organisation of the renal medulla: comparative and functional aspects. *Am. J. Physiol.* **241,** R3–R16.

15. Levillain, O. (1992): Métabolisme de la citrulline, de l'arginine et de l'ornithine le long du néphron de plusieurs mammifères. Thesis, Université Paris 7, France.

16. Levillain, O., Hus-Citharel, A. & Morel, F. (1993): Localization of ornithine decarboxylation and oxidation along the rat nephron. *J. Am. Soc. Nephrol.* **4**, 891.

17. Levillain, O., Hus-Citharel, A. & Morel, F. (1994): Amino acids affect urea production in the rat collecting duct. *Pflügers Arch.* **426**, 481–490.

18. Levillain, O., Hus-Citharel, A., Morel, F. & Bankir, L. (1989): Production of urea from arginine in pars recta and collecting duct of the rat kidney. *Renal Physiol. Biochem.* **12**, 302–312.

19. Levillain, O., Hus-Citharel, A., Morel, F. & Bankir, L. (1990): Localization of arginine synthesis along rat nephron. *Am. J. Physiol.* **259**, F916–F923.

20. Levillain, O., Hus-Citharel, A., Morel, F. & Bankir, L. (1992): Localization of urea and ornithine production along mouse and rabbit nephrons: functional significance. *Am. J. Physiol.* **263**, F878–F885.

21. Levillain, O., Hus-Citharel, A., Morel, F. & Bankir, L. (1993). Arginine synthesis in mouse and rabbit nephron: localization and functional significance. *Am. J. Physiol.* **264**, F1038–F1045.

22. Manteuffel-Cymborowska, M., Chmurzynska, W. & Grzelakowska-Sztabert, B. (1991): Ornithine decarboxylase induction in mouse kidney as indicator of renal damage. Differential nephrotoxic effect of anticancer antifolate drugs. *Cancer Lett.* **59**, 237–241.

23. Morrissey, J.J., McCracken, R., Kaneto, H., Vehaskari, M., Montani, D. & Klahr, S. (1994): Location of an inductible nitric oxide synthase mRNA in the normal kidney. *Kidney Int.* **45**, 998–1005.

24. Pegg, A.E. & McCann, P.P. (1982): Polyamines metabolism and function. *Am. J. Physiol.* **243**, C212–C221.

25. Pfaller, W. (1982): Structure function correlation on rat kidney. Quantitative correlation of structure and function in the normal and injured kidney. *Adv. Anat. Embryol. Cell Biol.* **70**, 1–106.

26. Porembska, Z., Baranczyk, A. & Jachimowicz, J. (1971): Arginase isoenzymes in liver and kidney of some mammals. *Acta Biochim. Pol.* **18**, 77–85.

27. Porembska, Z. & Zamecka, E. (1984): Immunological properties of rat arginases. *Acta Biochim. Pol.* **31**, 223–227.

28. Rabinowitz, L., Gunther, R.A., Shoji, E.S., Freedland, R.A. & Avery, E.H. (1973): Effects of high and low protein diets on sheep renal function and metabolism. *Kidney Int.* **4**, 188–207.

29. Ratner, S. (1973): Enzymes of arginine and urea synthesis. *Adv. Enzymol.* **39**, 1–90.

30. Ratner, S. & Petrack, B. (1953): The mechanism of arginine synthesis from citrulline in kidney. *J. Biol. Chem.* **200**, 175–185.

31. Robinson, R.R. & Schmidt-Nielsen, B. (1963): Distribution of arginase within the kidneys of several vertebrate species. *J. Cell. Comp. Physiol.* **62**, 147–157.

32. Scharff, R. & Wool, I.G. (1964): Concentration of amino acids in rat muscle and plasma. *Nature* **202**, 603–604.

33. Skrzypek-Osiecka, I., Robin, Y. & Porembska, Z. (1983): Purification of rat kidney arginases A1 and A4 and their subcellular distribution. *Acta Biochem. Pol.* **30**, 83–92.

34. Snellman, K., Aperia, A. & Broberger, O. (1979): Studies of renal urea cycle enzymes. II. Human renal arginase activity and location of the adaptive changes of renal arginase in the protein deprived rat. *Scand. J. Clin. Lab. Invest.* **39**, 337–342.

35. Tabor, C.W. & Tabor, H. (1984): Polyamines. *Ann. Rev. Biochem.* **53**, 749–790.

36. Terada, Y., Tomita, K., Nonoguchi, H. & Marumo, F. (1992): Polymerase chain reaction localization of constitutive nitric oxide synthase and soluble guanylate cyclase messenger RNAs in microdissected rat nephron segments. *J. Clin. Invest.* **90**, 659–665.

37. Vinay, P., Levillain, O. & Bankir, L. (1991): Urea synthesis along the dog nephron. *J. Am. Soc. Nephrol.* **2**, (Abrst.) 728.

38. Zamecka, E. & Porembska, Z. (1988): Five forms of arginase in human tissues. *Biochem. Med. Metab. Biol.* **39**, 258–266.

*Guanidino Compounds : 2*, eds. by P.P. De Deyn, B. Marescau, I.A. Qureshi and A. Mori.
©1997 John Libbey & Company Ltd., pp. 87–91.

# Chapter 11

# The effect of hyperbaric oxygen exposure on brain arginase and arginine: glycine amidinotransferase activity

[1]Takehiko ITO, [2]Katsumi YUFU, [1]Rei EDAMATSU, [2]Hidehiko YATSUZUKA, [2]Hidenori HASHIMOTO and [1]Akitane MORI

[1]*Department of Neuroscience, Institute of Molecular and Cellular Medicine, [2]Department of Anesthesiology and Resuscitology, Okayama University Medical School, 2–5–1 Shikata-cho, Okayama, 700 Japan*

## Summary

Our previous study on the effect of hyperbaric oxygen (HBO) at 3 atmosphere absolute for 2 h revealed a marked increase of arginine in rat brains. Also, guanidinoacetic acid levels were increased. This study aimed to explore the possible mechanism(s) of these effects. Activities of two enzymes of arginine metabolism, arginase (E.C. 3.5.3.1) and arginine:glycine amidinotransferase (GAT E.C. 2.1.4.1.) were measured in several brain regions. A discrete distribution of arginase was observed in the brain. Following HBO, arginase activity in the cerebral cortex and midbrain was decreased, while GAT activity was decreased in the pons-medulla oblongata. These data suggest that decrease in arginase activity was observed activity in the cerebral cortex without alteration in GAT activity was, at least in part, responsible for the increase of arginine and guanidinoacetic acid contents.

## Introduction

Hyperbaric oxygen (HBO) exposure to animals has been utilized in studies of seizures and in the field of hyperbaric medicine[10], because it causes many changes in the activities of enzymes and in the contents of bioreactive substances *in vivo*[2]. However, many former experiments seem to be performed at convulsive doses of HBO rather than at clinically relevant doses. In order to determine effects of HBO under clinically relevant conditions, we previously tried to investigate the effect of HBO on rat brain under relatively mild HBO exposure[4]. At the same time, we wanted to know changes that may represent a preconvulsive state induced by high oxygen pressure. We were particularly interested in the metabolism of guanidino compounds, since many of them, such as methylguanidine[6] and arginine[7], are convulsants or proconvulsants. Our previous data show increases, of arginine, glutamine, taurine and guanidinoacetic acid (GAA) contents, while glycine, aspartate and glutamate content were not changed. The increase in arginine content was marked. In addition, an increase in the level of thiobarbituric acid reactive substances, carbon centred radicals and SOD-like activity was found[4]. These results suggested that reactive oxygen species are involved in the increased contents of several guanidino compounds contents. However, the affected pathway of the metabolism of guanidino compounds remained unclear.

In this study, we attempted to study the effect of HBO exposure on two key enzymes of guanidino compound metabolism, namely arginase (E.C. 3.5.3.1) and arginine:glycine-amidinotransferase (E.C. 2.1.4.1) (GAT). The former enzyme forms equimolecular urea and ornithine from arginine, while the latter forms equimolecular ornithine and GAA from glycine and arginine. The results obtained are discussed in relation to our former data on the effect of HBO exposure on arginine metabolism as well as to the nitric oxide synthase (NOS) system.

## Materials and methods

### Animals

Seven-week-old male Sprague-Dawley rats (Japan Charles River, Tokyo) were used. Rats were housed in a standard air conditioned and light–dark controlled room before the experiment. They were fed rat chow (MF, Oriental Yeast Company, Tokyo) and water *ad libitum*.

### HBO procedure

HBO procedure was carried out according to previously described method[4].

### Sample preparation

Immediately after HBO or sham procedure, all rats were anaesthetized with ether and their abdominal aorta was cannulated with a 22 gauge needle. 2 ml of aortic blood was aspirated. After perfusing rats with cold saline solution (approximately 4 °C), they were decapitated and the brain was removed. The obtained brain was dissected into seven regions according to the method of Glowinski and Iversen[3]. The liver and kidney were also sampled. All samples were kept at −80 °C until analysis.

### Arginase activity assay

Arginase activity was measured according to the method of Colombo and Konarska[1]. For brain, liver and kidney, samples were homogenized using a glass-Teflon homogenizer with 5 mM Tris buffer adjusted to pH 9.5 containing 1 mM of manganese chloride. 50 μl of homogenate was mixed with 50 μl of 10 mM manganese chloride solution. For the erythrocytes, 50 μl of rinsed and packed cells were haemolysed by adding 30 volumes of the Tris buffer. Then 50 μl of this hemolysate was mixed with 50 μl of 10 μM manganese chloride solution. These mixtures were preincubated at 55 °C for 20 min to activate arginase. Then 300 μl of 0.1 M carbonate buffer adjusted to pH 9.5 was added. Enzymatic reaction was initiated by adding 100 μl of 0.1 M arginine solution. The reaction mixture was incubated exactly at 37 °C for 10 min, 30 min or 2 h for the liver, brain and kidney or hemolysate respectively. The reaction was terminated by adding 1.5 ml of glacial acetic acid. 500 ml of 140 mM ninhydrin reagent was added and the mixture boiled for an hour in a water bath. After cooling the mixture to room temperature, absorbance was read at 515 nm. A zero and reagent blanks were always included among the samples to cancel preexisting substrate and product in the samples. Arginase activities were expressed as μmol ornithine formed per g tissue per hour except for erythrocyte samples; these were expressed as μmol ornithine formed per g haemoglobin per hour.

### GAT activity assay

GAT activities in the cerebral cortex, cerebellum and pons-medulla oblongata were measured according to the method of Van Pilsum *et al.*[9] with some modifications. Each sample was homogenized in 0.1 M phosphate buffer adjusted to pH 7.4. Then, 250 μl of arginine–glycine substrate solution (A-G solution) or arginine solution (A solution) was added to 250 μl of homogenate. The mixture was incubated at 37 °C for 3 h. The reaction was terminated by adding 1.5 ml of 0.6 M perchloric acid solution. The mixture was centrifuged at 2650 × g for 10 min. 1 ml of the supernatant was carefully removed and combined with 3 ml of ninhydrin reagent. The mixture was boiled for 25 min in a water bath. After cooling the samples to room temperature, absorbance was read at 505 nm.

GAT activity was defined by subtracting the absorbances of two tubes containing A-G solution and A solution, and expressed as nmol ornithine formed per g tissue per hour.

## Measurement of haemoglobin content

Haemoglobin content in the haemolysate was measured using a CO-Oxymeter (Instrumental Laboratory, USA).

## Statistical analysis

Significance of differences was evaluated by Student's $t$-test when two groups were compared; one way ANOVA followed by Bonferroni's multiple comparison test was performed when more than two groups were compared.

## Results

Arginase activity was measured in seven different regions in the brain as well as in the liver, kidney and erythrocytes (Table 1). The highest arginase activity in the brain was detected in the hypothalamus while the lowest was found in the striatum. The activities were significantly different from region to region except for the combination of midbrain and cerebellum. HBO exposure significantly decreased arginase activity in the cerebral cortex and midbrain, while no significant alteration was detected in other brain regions or in the liver and kidney. In contrast, there was a significant increase in the erythrocyte arginase activity of the HBO group. GAT activity was measured in three brain regions, namely the cerebral cortex, cerebellum and pons-medulla oblongata (Table 2). While there was no significant difference in GAT activity in the cerebral cortex and cerebellum, GAT activity in pons-medulla oblongata increased significantly.

### Table 1. Arginase activity[*] in brain, liver, kidney and erythrocytes

|  | Control | HBO |
|---|---|---|
| Region |  |  |
| *Brain* |  |  |
| cerebral cortex | $50.4 \pm 1.3$ | $48.4 \pm 1.0$a |
| midbrain | $73.4 \pm 2.4$ | $70.0 \pm 2.7$b |
| hippocampus | $82.7 \pm 6.8$ | $80.6 \pm 4.9$ |
| striatum | $26.9 \pm 3.3$ | $27.3 \pm 3.2$ |
| hypothalamus | $106.1 \pm 5.6$ | $107.0 \pm 2.3$ |
| cerebellum | $73.3 \pm 7.3$ | $70.3 \pm 7.4$ |
| pons-medulla oblongata | $41.0 \pm 6.2$ | $47.0 \pm 4.7$ |
| *Liver* | $3.16 \pm 0.60 \times 10^3$ | $2.99 \pm 0.50 \times 10^3$ |
| *Kidney* | $1.24 \pm 0.99 \times 10^2$ | $1.37 \pm 0.16 \times 10^2$ |
| *Erthrocytes*[**] | $7.34 \pm 9.53$ | $18.8 \pm 7.0$b |

Each value represents mean $\pm$ SD of 7 or 8 rats.
Significance: (a)$P < 0.01$ *vs* control, (b) $P < 0.05$ *vs* control.
[*]µmol ornithine formed/g tissue or [**]haemoglobin/hour

## Discussion

Our former study revealed that HBO exposure to rats increased arginine and guanidinoacetic acid contents in brain[4]. The mechanism underlying this phenomenon seemed to be complex, but we postulated the changes of arginase activity could be involved. As arginase cleaves arginine to ornithine, the decrease in arginase activity may result in the increase of arginine content. Present data showed significant decrease in the arginase activity in the cerebral cortex and midbrain, while those in other regions examined did not show significant alterations. Though our experimental conditions were at subconvulsive doses, parameters of oxidative stress (lipid peroxidation, free radical content and SOD-like activity) increased significantly[4]. Since reactive oxygen species are generated in the brain, the decrease of arginase activity seemed to be the result of direct oxidation of the enzyme *in vivo*. The decrease in arginase activity, however, showed regional differences in the susceptibility to oxidative stress and suggested regional differences in the response to oxidative stress.

**Table 2. GAT activity in brain (nmol ornithine formed/g tissue/hour)**

|  | Control | HBO |
|---|---|---|
| Region of the brain |  |  |
| cerebral cortex | 45.3 ± 47.8 | 21.0 ± 15.4 |
| cerebellum | 91.5 ± 61.3 | 73.0 ± 39.6 |
| pons-medulla oblongata | 234 ± 91 | 107 ± 26a |

Each value represents mean ± SD of 7 or 8 rats.
Significance: (a) $P < 0.05$ *vs* control.

Other enzymes that contribute to arginine metabolism are GAT and NOS. Unaltered GAT activity except in pons-medulla oblongata seemed to give a reason to the increase in guanidinoacetic acid together with the increase in arginine level.

The involvement of NOS system in HBO-induced seizures is studied by Zhang *et al.*[11]. HBO exposure at 6 ATA for 20 min is performed with or without N-ω-nitro-L-arginine (LNNA), an NOS inhibitor. Glutamate, aspartate and GABA concentrations are decreased by HBO without LNNA, while arginine and glutamine concentration are increased. Interestingly, the rise in arginine levels is suppressed by LNNA administration with HBO, giving the rats protection against HBO induced seizures. Co-administration of arginine with LNNA diminishes this protective effect. Zhang *et al.*, suggest that glutamate and norepinephrine release is decreased by lowered NO release, which leads to less arginine production via $NH_3$ generation by an unknown pathway. They propose this mechanism as a protective system.

On the other hand, our former experiments also revealed increases in arginine and glutamine contents, while there was no change in glutamate and GABA contents[4]. The discrepancy between their result and ours may be explained by the difference in HBO conditions. It is not likely that under milder HBO conditions, GABA or glutamate levels alter to the extent they demonstrated. Our hypothesis was that susceptibility of glutamate dehydrogenase and glutaminase to high oxygen tension is higher than that of glutamine synthase, which shifts the reaction to produce more glutamine[2,5].

The major discrepancy between the two series of results is that, even though arginase and NOS are both arginine cleaving enzymes, inhibition of arginase resulted in increased arginine content while inhibition of NOS causes a decrease. This contradiction may be explained by the arginine level being controlled by the two enzymes. NOS seems to control arginine content through indirect mechanisms via amino acid metabolism[9] whereas arginase may directly regulate arginine content.

As arginine is a precursor for NO synthesis, arginine level modifications, may result in altered NO production[8]. In other words, arginase activity may be closely related to NO synthesis. In fact it has been reported in animals that they become resistant to HBO when arginase activity is increased[5].

There was no change in liver and kidney arginase levels, probably because these organs are less susceptible to HBO under this condition. The increase of arginase in erythrocytes seemed hard to explain because blood cells are apparently exposed to highest oxygen concentrations in the body. Response of arginase to oxygen might be different among various organs or might be biphasic, but this fact still needs further investigation.

In conclusion, HBO at 3 ATA for 2 h decreased arginase activity in the cerebral cortex and midbrain. GAT activity decreased in pons-medulla oblongata. These changes seemed to correlate, at least in part, with increased arginine and guanidinoacetic acid levels. The modulation of arginase activity may alter NO synthesis via the arginine level, but more experiments are needed to confirm this hypothesis.

## References

1.  Colombo, J.P. & Konarska, L. (1984): Arginase. In: *Methods in enzymatic analysis*, Vol. 4, ed. H.U., Bergmeiyer, pp. 285–294. Weinheim: Verlag Chemie.

2.  Davies, H.C. & Davies, R.E. (1965): Biochemical aspects of oxygen poisoning. In: *Handbook of physiology*, Vol. 2, section 3, eds. W.O., Fenn & H., Rahn, pp. 1047–1058. Washington: Am. Physiol. Soc.

3.  Glowinski, J. & Iversen, L. (1966): Regional studies of catecholamines in the rat brain. *J. Neurochem.* **13,** 655–669.

4.  Itoh, T., Yufu, K., Edamatsu, R., Yatsuzuka, H. & Mori, A. (1992): The effect of hyperbaric oxygen on the guanidino compounds in rat brain. In: *Guanidino compounds in biology and medicine*, eds. P.P. De Deyn, B. Marescau, V. Stalon & I.A. Qureshi, pp. 403–408. London: John Libbey.

5.  Krichevskaia, A.A., Shugalei, V.S., Tsvetnenko, E.Z. & Ananian, A.A. (1978): Protective effect of arginine in hyperoxia. Activity of cerebral glutaminase and glutamate decarboxylase (translation). *Vopr. Med.* **24,** 42–46.

6.  Matsumoto, M., Kobayashi, K., Kishikawa, H. & Mori, A. (1976): Convulsive activity of methyguanidine in cats and rabbits. *ICRS Med. Sci.* **4,** 65.

7.  Mollace, V., Baggetta, G. & Nistico, G (1991): Evidence that L-arginine possesses proconvulsant effects mediated through nitric oxide. *Neuro. Report* **2,** 269–272.

8.  Nakamura, H., Saheki, T. & Nakagawa, S. (1990): Differential cellular localization of enzymes of L-arginine metabolism in the rat brain. *Brain Res.* **530,** 108–112.

9.  Van Pilsum, J.F., Taylor, D., Zakis, B. & McCormic, P. (1970): Simplified assay for transamidinase activities of rat kidney homogenates. *Anal. Biochem.* **35,** 277–286.

10. Wood, J.D. (1972): Systemic oxygen derangements. In: *Experimental models of epilepsy*, eds. D.P. Purpura, J.K. Penry, D.B. Tower, D.M., Woodbury & R.D. Walter, pp. 461–476. New York: Raven Press.

11. Zhang, J., Su, Y., Oury, T.D. & Piantadoshi, C.A. (1993): Cerebral amino acid, norepinephrine and nitric oxide metabolism in CNS oxygen toxicity. *Brain Res.* **606,** 56–62.

# Section III

# Biochemistry-metabolism of the creatine-creatinine, phosphocreatine biosynthesis pathway

*Guanidino Compounds : 2*, eds. by P.P. De Deyn, B. Marescau, I.A. Qureshi and A. Mori.
©1997 John Libbey & Company Ltd., pp.95–102.

# Chapter 12

---

# The role of growth hormone and insulin in rat kidney creatine transport

## Thomas BLOOM and John F. VAN PILSUM

*Department of Biochemistry, Medical School, University of Minnesota, Minneapolis, Minnesota 55455, USA*

## Summary

Both creatine and growth hormone regulate the expression of the rat kidney L-arginine:glycine amidinotransferase (transamidinase) gene, and further the action of each effector is diminished by the presence of the other[12]. A possible explanation for the interrelationship between the two effectors could be that creatine acts as the direct intracellular mediator of the gene expression with its transport in the kidney regulated by growth hormone. The role of growth hormone and of insulin in the regulation of creatine transport in kidney was investigated since both hormones have been reported to regulate amino acid and creatine transport in a variety of tissues.

An assay for the rate of creatine uptake by rat kidney cortex slices was developed. The $K_m$ and $V_{max}$ for creatine uptake was 30 mM and 14 µmol/g kidney/10 min, respectively. No significant differences were found in the uptake of creatine by kidney slices from hypophysectomized rats maintained with and without growth hormone or from diabetic or non-diabetic rats. The interrelationship between the two effectors on transamidinase gene expression is therefore not explained by a regulation of creatine transport by growth hormone. Creatine is suggested to regulate the secretion of an extra-renal factor, possibly a hormone, which inhibits the induction of the transamidinase gene by growth hormone.

## Introduction

Creatine plays a major role in vertebrate energy metabolism and both endogenous and exogenous creatine can be utilized for this purpose. Muscle, liver and kidney from rats have been reported to accumulate creatine from the diet[7]. The creatine requirements can also be met by the stepwise synthesis of creatine in the kidneys (and pancreas) and liver[4,5]. Block & Schoenheimer suggested in 1939 that when exogenous creatine is given the endogenous production of creatine is proportionally retarded or inhibited[2]. The feed-back repression of the transamidinase gene expression by creatine in the diet[12] is evidence that creatine synthesis is repressed by exogenous creatine. The body, therefore, must possess some system that detects variations in blood levels of creatine and thereby adjusts its synthetic capacity for creatine accordingly.

L-arginine:glycine amidinotransferase (transamidinase) catalyses the first of two steps in creatine synthesis in mammals[4,5] and is thought to be the rate controlling step[6]. Transamidinase activities have been found in significant amounts only in the kidney and pancreas of mammals[24,30,31]. Guanidinoacetic acid, formed by the action of transamidinase in kidney and pancreas, is transported to the liver where it is methylated by S-adenosylmethionine guanidinoacetate N-methyltransferase to form creatine. Kidney transamidinase activities have been reported to be altered greatly in a variety of physiological states. For example, kidneys from intact rats fed creatine-supplemented diets had only ~ 20 per cent

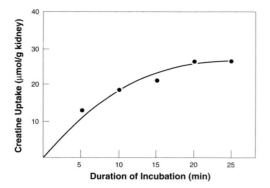

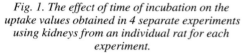

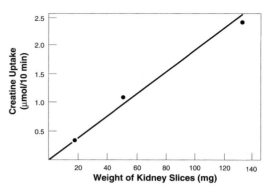

Fig. 1. The effect of time of incubation on the uptake values obtained in 4 separate experiments using kidneys from an individual rat for each experiment.

Fig. 2. The effect of mass of kidney slices in the incubation flask on the uptake of creatine: ten minute incubation times and a creatine concentration of 60 mM was used.

of the transamidinase activities as kidneys from similar rats fed diets not supplemented with creatine[8,25,26,32,33]. Kidneys from hypophysectomized rats also have been found to have ~20 per cent of the transamidinase activities of kidneys from intact rats. Hypophysectomized rats given injections of growth hormone had similar transamidinase activities as intact rats[22]. Both the repression of transamidinase activities by creatine in the diet and the induction of transamidinase activities by growth hormone in hypophysectomized rats have been reported to be at the pre-translational level using a cDNA hybridization technique[12]. An excellent correlation between transamidinase activities and relative amounts of transamidinase mRNA was observed in the hypophysectomized rats fed either a creatine-free diet or a creatine supplemented diet and maintained both with and without growth hormone administration.

Creatine is a catabolically inert molecule[3,27] and therefore may act as a direct effector or mediator of transamidinase gene expression in the proximal kidney tubules. Transamidinase is localized in the mitochondria of the proximal tubules[16] and if creatine acts as the direct mediator of transamidinase gene expression its concentration in the tubule cells should affect the gene expression process. Evidence to this effect has been reported previously. An inverse correlation between kidney and blood concentrations of creatine and kidney transamidinase activities has been found in intact rats fed protein-free diets or complete diets supplemented with varying amounts of creatine[28]. A good correlation also was found between the low transamidinase activities and the high levels of creatine in the kidneys and blood of rats that had received excessive doses of thyroid or adrenal steroid hormones[23]. The decline in transamidinase activities in kidneys in the first 3 days after hypophysec- tomy was correlated with an increase in the levels of creatine in the blood and kidneys[22].

Evidence, however, has been reported that low levels of transamidinase activities are not always correlated with high amounts of creatine in the kidneys. Rats had normal levels of creatine in their blood and kidneys 7 days after hypophysectomy while their transamidinase activities remained at the low levels found 3 days after hypophysectomy. The low transamidinase activities of hypophysec- tomized rats were returned to normal levels only after the administration of growth hormone[22]. Conversely, the low transamidinase activities of intact rats fed creatine supplemented diets returned to normal after the removal of creatine from the diet[33]. Thus, the repression of transamidinase activities by creatine and the induction of activities by growth hormone in hypophysectomized rats seemed to occur by distinct mechanisms.

Evidence now has been obtained for an interrelationship between creatine and growth hormone in the regulation of transamidinase activities and gene expression since the action of each effector was diminished by the presence of the other[12]. A possible explanation for the interrelationship between

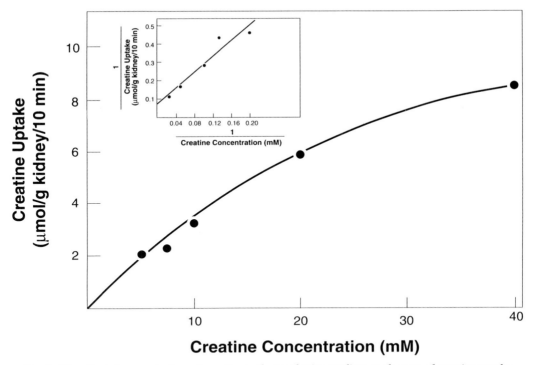

*Fig. 3 . The effect of concentration of creatine in the incubation medium on the rate of creatine uptake. Ten minute incubation periods were used.*

the two effectors on transamidinase gene expression could be a regulation of creatine transport in the kidney by growth hormone. If creatine acts as the intracellular mediator of transamidinase gene expression in the proximal tubule, a stimulation of creatine transport into the tubule cells by growth hormone might explain the diminution by creatine of the induction of gene expression by growth hormone. A growth hormone regulation of creatine transport in kidney seemed possible since growth hormone has been reported to regulate amino acid transport in a variety of tissues and has been implicated in creatine uptake by rat skeletal muscle[20]. A sodium-dependent creatine transporter in rabbit tissues, including kidney, has been discovered recently[11]. A transport system for creatine in the kidney which functions to transfer the creatine and other solutes such as amino acids, etc. from the ultra-filtrate of plasma in the proximal tubules back into the blood was reported in 1943 by Pitts[18]. Solutes such as creatine are considered to enter the proximal tubule cells by a carrier-mediated, sodium-coupled transport mechanism. The solutes leave the tubules and enter the peritubular space by means of a carrier-mediated, passive facilitated diffusion process[1]. The rate limiting step in the transfer of the solutes to the blood from the ultra-filtrate of plasma is considered to be the carrier-mediated, sodium-coupled transport into the kidney tubule cells.

The rates of the uptake of creatine by kidney cortex slices from hypophysectomized rats maintained with and without growth hormone administration and from diabetic and non-diabetic rats were determined since insulin like growth hormone has long been known to regulate amino acid transport in a variety of tissues. Any differences in the rate of creatine uptake by the kidney cortex slices between the hormone-deficient and the hormone-intact rats would be assumed to be differences in the availability or the activity of the creatine transporter as the result of the hormone deficiency.

## Materials and methods

[1–[14]C]creatine hydrate, specific activity 41 μCi/mg was purchased from Amersham Corporation, Arlington Heights, Illinois, and purified by thin layer chromatography in n-butanol:water:acetic acid (12:5:3). Non-labelled creatine was added to the solutions of the 1–[14]C labeled creatine to obtain the desired molarity for the incubation medium. The incubation medium was Krebs-ringer phosphate Buffer (KRP):$KH_2PO_4$, $Na_2HPO_4$, NaCl, KCl, $MgSO_4$, $CaCl_2$ at mM concentrations of 1.3, 6.3, 127.3, 5.1, 1.3, and 1.2, respectively, pH 7.4. The incubation medium, also contained 10 mg lactic acid per 100 ml, and the labelled plus non-labeled creatine dissolved in the KRP buffer.

Male hypophysectomized Sprague-Dawley rats, weighing ~130 g, were purchased from Harlan Laboratories, Madison, Wisconsin. The rats were fed a complete purified creatine-free diet[28] for the duration of the experiment. One half of the rats were given 200 μg/100 g body weight of recombinant human growth hormone in physiological saline by subcutaneous injection in the dorsal neck region daily for 14 days. The recombinant growth hormone was a generous gift from Genentech, South San Francisco, CA. The average weight of the rats maintained with and without growth hormone injections after 14 days was 120 and 160 g, respectively.

For studies on the effect of insulin in males, Sprague-Dawley rats (Harlan Sprague-Dawley, Madison, WI, USA). The animals were housed in individual cages and given free access to the tap water and standard rat chow (Purina laboratory chow #5001, RFG PET@ Supply company, Plymouth, MN) for the duration of the experiment. At 6 weeks of age (170–190 g), rats were made diabetic by the intravenous injection of 52 mg/kg of streptozotocin (Zanazar brand of Streptozotocin, sterile powder, Upjohn Corp., Kalamazoo, MI) reconstituted with Hanks balanced salt solution (HBBS, Catalog #H8264) from pH 7.2–7.6 to pH 4.5–5.0. Age-matched littermates served as controls for the diabetic animals. Induction of the diabetic state was confirmed on fifth day post-injection by blood glucose determination in non-fasting animals using the glucose oxidase method (Beckman Glucose Analyzer, Beckman Instruments, Inc., Fullerton, CA.). Diabetes was defined as a blood glucose level in excess of 500 mg/dl. Body weights were determined weekly, blood glucose levels were determined at 4 weeks, after diabetes induction and on the day before termination of these experiments.

Creatine uptake was assessed by a modification of the procedure of Rosenberg et al.[19] used for determining the uptake of amino acids by rat kidney cortex slices. Rats were killed by decapitation, their kidneys excized, placed on ice and the capsule removed. The kidneys were bisected transversely and immediately transferred to ice cold KRP buffer. Kidney slices (0.5 mm thick) were made in a Campden vibroslice apparatus with the kidney immersed in the cold KRP buffer. The first slice was discarded and the second and third slices, containing the cortex region, were preincubated in 2 ml of KRP at room temperature for 12 min, gently blotted between filter paper and placed in 10 ml flasks containing 1.0 ml of the incubation medium. Incubation was performed with gentle shaking for 10 min at 37 °C. The slices were removed from the incubation mixture with a small metal spatula and quickly immersed and removed from two 10 ml volumes of KRP at room temperature. After the slices were removed from the second KRP wash, they were blotted gently with filter paper, weighed, and transferred to 12 ml centrifuge tubes containing 1.0 ml distilled water. The tubes containing the kidney slices in water were immersed in a boiling water bath for 10 min, cooled to room temperature and centrifuged at 2700 r.p.m in a clinical centrifuge for 5 min. Two 0.25 ml aliquots of the water extract of the kidney slices were removed and added to vials containing 5 ml of Beckman Ready-Solv H.P. scintillation fluid for the determination of the radioactivity in a Beckman model LS3801 Scintillation Counter. The creatine uptake by the slices was expressed as μmol/g kidney/10 min. The creatine uptake by kidney slices without incubation at 37 °C was used as an 'uptake blank' and subtracted from all values of incubated samples.

## Results and discussion

### Development of the method for measuring creatine uptake by rat kidney cortex slices

The uptake of creatine as a function of time of incubation is shown in Fig. 1. The uptake was approximately linear with respect to time for incubation periods of 5 to 10 min. A 10 min incubation was chosen over the 5 min period in order to facilitate processing of multiple incubation samples in a single experiment. The uptake of creatine as a function of kidney mass is shown in Fig. 2. An excellent correlation between kidney mass and creatine uptake was observed. The uptake as a function of the concentration of creatine in the incubation mixture is shown in Fig. 3. The $K_m$ was determined to be 30 mM and the $V_{max}$ to be 14 µmol/g kidney/10 min.

The cortex is the site of the proximal tubules which are the only cells in the kidney known to possess transamidinase[16]. Rosenberg *et al.*, reported that the amount of non-tubular tissue in the kidney cortex slices is small and stated that the uptakes of amino acids by the cortex slices should therefore correlate well with the processes taking place in the renal tubular system of the intact animal[19]. The rate of creatine uptake by the kidney cortex slices in the present investigation was assumed to be a measure of the creatine transported by the rate-limiting, carrier-mediated, sodium-coupled transport system of the kidney tubules for the following reasons: (1) no circulating blood system was present to remove the creatine from the tubule cells; (2) the large concentration of creatine in the incubation media should inhibit the passive, facilitated diffusion of creatine from the tubule cells.

The kinetic data that were obtained in our experiments are compared with the kinetic data reported by Fitch and Shields[9] on the uptake of creatine by strips of rabbit skeletal muscle. The $K_m$ for creatine uptake by kidney slices of 30 mM is 60-fold greater than the reported $K_m$ for muscle of 0.5 mM. For comparative purposes the $V_{max}$ for creatine uptake by kidney slices of 14 µmol/g kidney/10 min has been converted to 168 mmol/liter intracellular $H_2O$/hr, the units used to express $V_{max}$ for skeletal muscle. The intracellular water was assumed to be 50 per cent of the kidney weight for this calculation, thus, the $V_{max}$ for creatine uptake by kidney is 280 fold greater than the reported $V_{max}$ for muscle of 0.6 mmol/l intracellular $H_2O$ per h.

Little information on the effect of hormones on creatine transport has been published. The rate of transport of creatine into muscle was inhibited in hyperthyroidism[10]. A role of growth hormone in the regulation of creatine uptake in rat skeletal muscle has been suggested by Tan and Ungar[20]. Hypophysectomized rats maintained with and without growth hormone and sham-hypophysec-tomized rats were given intraperitoneal injections of [14]C-creatine. The amounts of [14]C-creatine isolated from the muscle of hypophysectomized rats maintained with and without growth hormone were 73 and 48 per cent, respectively, of the values found in the sham-hypophysectomized rats. A role of insulin in the regulation of creatine uptake in rat skeletal muscle also has been reported[13,14].

### Creatine uptake by kidney slices from hypophysectomized rats and from diabetic non-diabetic rate

The uptake of creatine by kidney cortex slices from hypophysectomized  rats, maintained with and without growth hormone injections and from the diabetic and non-diabetic rats are shown in Table 1. Incubation was at 37 °C for 10 min with 20 mM creatine concentration. There were no significant differences in the uptake of creatine by kidney slices from hypophysectomized rats maintained with and without growth hormone or from slices from diabetic or non-diabetic rats. Neither growth hormone nor insulin are concluded to be implicated in the regulation of creatine transport in rat kidney cortex.

A growth hormone induction of transamidinase gene expression seems likely to be a direct action of the hormone at the site of the kidney tubules. That creatine acts in the kidney tubules as the intracellular mediator of the gene expression also seemed possible since creatine inhibited the induction of transamidinase gene expression by growth hormone. The lack of an effect of growth hormone on the uptake of creatine by the kidney cortex slices does not support the concept that the interrelationship between the two effectors of the gene expression is at the creatine transport level. We suggest that the diminution of each effector's action on transamidinase gene expression by the presence of the other is best explained by an indirect action of creatine on the mediation of kidney transamidinase gene expression.

**Table I. The uptake of creatine by kidney cortex slices from hypophysectomized rats maintained with and without growth hormone and from diabetic and non-diabetic rats**

| Group | Rats | Creatine uptake (µmol/ g kidney/10 min) Mean ± SD |
|---|---|---|
| Hypophysectomized | 4 | 6.45 ± 0.8 |
| Hypophysectomized + growth hormone | 4 | 7.05 ± 0.8 |
| Non-diabetic | 2 | 4.30 ± 0.2 |
| Diabetic | 4 | 3.90 ± 0.3 |

The levels of creatine in the blood could regulate the secretion or metabolism of an extrarenal factor (such as a hormone) that counteracts or inhibits the action of growth hormone on the induction of transamidinase gene expression in the kidney tubules. A creatine regulation of growth hormone secretion seems unlikely since creatine supplementation of diets has no effect on the weight gains of intact rats or hypophysectomized rats maintained with or without growth hormone administration. Also, the creatine repression of transamidinase gene expression was observed in hypophysectomized rats both with and without growth hormone administration. A creatine regulation of the secretion of any pituitary or pituitary axis hormone is not likely since the effect of creatine on transamidinase gene expression was observed in hypophysectomized rats. Since both creatine and growth hormone were found to regulate transamidinase gene expression at the pre-translational level, it seems likely that a growth hormone induction of transamidinase gene expression would be opposed by the action of some other factor rather than by the direct action of creatine. For example, a possible mediator of creatine on the gene expression could be estradiol.

Female rats have 50 per cent of the kidney transamidinase activities found in male rats[15,29]. Intact male rats given injections of as little as 1 µg estradiol/day for 8 days had significantly lower transamidinase activities than the non-injected rats[29]. Hypophysectomized rats maintained both with and without growth hormone administration and given injections of estradiol also had lower transamidinase activities than the rats not given estradiol. Further, estradiol and growth hormone inhibited each other's action on the transamidinase activities of hypophysectomized rats. Thus, the interrelationships between growth hormone and estradiol on transamidinase activities are not unlike that of growth hormone and creatine. The similarity between creatine and estradiol in their interrelationship with growth hormone in the regulation of transamidinase activities is support for an extrarenal factor responding to alterations in blood creatine levels. That this putative extrarenal factor or system that detects changes in blood creatine concentrations could be estradiol seems possible. The concentration of creatine in the plasma of human females is two-fold greater than in males[17]. The amount of creatinine excreted in the urine should be proportional to the creatine requirement of the body. Female humans excrete ~75 per cent as much creatinine per kilogram body wt/day as males, thus indicating that females synthesize creatine at a lower rate than males. The possibility that blood creatine levels and estrogen metabolism or secretion are interrelated in their effects on transamidinase gene expression has not been investigated. The action of estrogen on the expression of a large number of genes has been reported[21] and the action of oestrogen on the expression of the transamidinase gene now needs to be investigated.

We have concluded that the possibility of creatine acting directly as the intracellular regulator of transamidinase gene expression is less likely than its acting indirectly via an extrarenal factor. Furthermore, an antithetical regulation of transamidinase gene expression seems much more likely to involve, for example, two hormones rather than one hormone and a nutrient or metabolite. The lack of a regulation of creatine transport in the kidney by growth hormone and the similar trends of action of creatine and estradiol in their effect on the action of growth hormone on transamidinase activity is offered as support for the suggestion that creatine acts indirectly on transamidinase gene expression.

## Acknowledgements

This work was supported by a grant from the Minneosta Medical Foundation. We thank: (1) Dr. Michael S. Mauer and Silvia K. Rosen of the Department of Pediatrics, the Medical School, University of Minnesota for preparing the diabetic rats; (2) Genentech (South Francisco, CA) for providing us with recombinant human growth hormone; (3) Drs. James Koerner and Robert Roon for allowing us to do the experimental work in their research laboratory in the Department of Biochemistry at the University of Minnesota; (4) Drs. Frank Ungar, Howard Towle, and Nelson Goldberg for their reviews of the manuscript during its preparation.

# References

1.   Berry, C.A. & Rector, F.C. (1991): Renal transport of glucose, amino acids, sodium, chloride, and water. In: *The kidney*, eds. B.M. Brenner & F.C. Rector, pp. 245–282. Philadelphia, London, Toronto, Montreal, Sydney, Tokyo: W.B. Saunders Company.

2.   Block, K. & Schoenheimer, R. (1939): Studies in protein metabolism XI. The metabolic relation of creatine and creatinine studied with isotopic nitrogen. *J. Biol. Chem.* **131,** 111–119.

3.   Block, K., Schoenheimer, R. & Rittenberg, D. (1941): Rate of formation and disappearance of body creatine in normal animals. *J. Biol. Chem.* **138,** 155–166.

4.   Borsook, H. & Dubnoff, J.W. (1940): The formation of creatine from glycocyamine in the liver. *J. Biol. Chem.* **132,** 559–574.

5.   Borsook H. & Dubnoff, J.W. (1941): The formation of glycocyamine in animal tissues. *J. Biol. Chem.* **138,** 389–403.

6.   Carlson, M. & Van Pilsum, J.F. (1973): S-adenosylmethionine: guanidinoacetate N-methyltransferase activities in livers from rats with hormonal deficiencies or excesses. *Proc. Soc. Exp. Biol. Med.* **143,** 1256–1259.

7.   Chanutin, A. (1930): Studies on the creatine and nitrogen content of the whole rat after the feeding of a variety of diets and after nephrectomy. *J. Biol. Chem.* **89,** 765–774.

8.   Fitch, C.D., Hsu, C. & Dinning, J.S. (1960): Some factors affecting kidney transamidinase activity in rats. *J. Biol. Chem.* **235,** 2362–2364.

9.   Fitch, C.D. & Shields, R.P. (1966): Creatine metabolism in skeletal muscle I creatine movement across muscle membranes. *J. Biol. Chem.* **241,** 3611–3614.

10.   Fitch, C.D., Coker, R. & Dinning, J.S. (1960): Metabolism of creatine-1-$^{14}$C by vitamin E-deficient and hyperthyroid rats. *Am. J. Physiol.* **198,** 1232–1234.

11.   Guimbal, C. & Kilimann, M.W. (1993): A Na$^+$-dependent creatine transporter in rabbit brain, muscle, heart, and kidney. *J. Biol. Chem.* **268,** 8418–8421.

12.   Guthmiller, P., Van Pilsum, J.F., Boen, J.R. & McGuire, D.M. (1994): Cloning and sequencing of rat kidney L-arginine:glycine amidinotransferase: studies on the mechanism of regulation by growth hormone and creatine. *J. Biol. Chem.* **269,** 17556–17560.

13.   Haugland, R.B. & Chang, D.T. (1975): Insulin effect on creatine transport in skeletal muscle. *Proc. Soc. Exp. Biol. Med.* **148,** 1–4.

14.   Koszalka, T.R. & Andrew, C.L. (1972): Effect of insulin on the uptake of creatine-1-$^{14}$C by skeletal muscle in normal and X-irradiated rats. *Proc. Soc. Exp. Biol. Med.* **139,** 1265–1271.

15.   Krisko, I. & Walker, J.B. (1966): Influence of sex hormones on amidinotransferase levels. Metabolic control of creatine bioynthesis. *Acta Endocrinologica* **58,** 655–662.

16.   McGuire, D.M., Gross, M.D., Elde, R.P. & Van Pilsum, J.F. (1986): Localization of L-arginine:glycine amidinotransferase protein in rat tissues by immunofluorescence microscopy. *J. Histochem. and Cytochem.* **34,** 429–435.

17. Painter, P.C., Cope, J.Y. & Smith, J.L. (1994): Reference intervals. In: *Clinical chemistry*, C.A. Burtis & E.R. Ashwood, eds. pp. 2175–2217, Philadephia, London, Toronto, Montreal, Sydney, Tokyo: W.B. Saunders Company.

18. Pitts, R.F. (1943): A renal reabsorptive mechanism in the dog common to glycin and creatine. *Am. J. Physiol.* **140,** 156–167.

19. Rosenberg, L.E., Blair, A. & Segal, S. (1961): Transport of amino acids by slices of rat-kidney cortex. *Biochim. Biophys. Acta* **54,** 479–488.

20. Tan, A.W.H. & Ungar, F. (1979): Growth hormone effects on creatine uptake by muscle in the hypophysectomized rat. *Mol. Cell. Biochem.* **25,** 67–77.

21. Tsai, M.J. & O'Malley, B.W. (1994): *Mechanisms of Steroid Hormone Regulation of Gene Transcription.* Austin: R.G. Landes Co.

22. Ungar, F. & Van Pilsum, J.F. (1966): Hormonal regulation of rat kidney transamidinase; effect of growth hormone in the hypophysectomized rat. *Endocrinology* **78,** 1238–1247.

23. Ungar, F. & Van Pilsum, J.F. (1966): Effects of adrenal steroids and thyroid hormone on creatine and kidney transamidinase in the rat. *Endocrinology* **79,** 1143–1148.

24. Van Pilsum, J.F., Stephens, G.C. & Taylor, D. (1972): Distribution of creatine guanidinoacetate and the enzymes for their biosynthesis in the animal kingdom. *Biochem. J.* **126,** 325–345.

25. Van Pilsum, J.F. & Canfield, T.M. (1962): Transamidinase activities, *in vitro*, of kidneys from rats fed diets supplemented with nitrogen- containing compounds. *J. Biol. Chem.* **237,** 2574–2577.

26. Van Pilsum, J.F., Olsen, B., Taylor, D., Rozycki, T. & Pierce, J.C. (1963): Transamidiniase activities, *in vitro*, of tissues from various mammals and from rats fed protein-free, creatine-supplemented and normal diets. *Arch. Biochem. Biophys.* **100,** 520–524.

27. Van Pilsum, J.F. & Warhol, R.M. (1963): The fate of large doses of creatine injected intraperitoneally into normal rats. *Clin. Chem.* **9,** 347–350.

28. Van Pilsum, J.F., Taylor, D. & Boen, J.R. (1967): Evidence that creatine may be one factor in the low transamidinase activities of kidneys from protein-depleted rats. *J. Nutr.* **91,** 383–390.

29. Van Pilsum, J.F. & Ungar, F. (1968): Effect of castration and steroid sex hormones on rat kidney transamidinase. *Arch. Biochem. Biophys.* **124,** 372–379.

30. Walker, J.B. (1958): Role for pancreas in biosynthesis of creatine. *Proc. Soc. Exp. Biol. Med.* **98,** 7–9.

31. Walker, J.B. & Walker, M.S. (1959): Formation of creatine from guanidinoacetate in pancreas. *Proc. Soc. Exp. Biol. Med.* **101,** 807–809.

32. Walker, J.B. (1960): Metabolic control of creatine biosynthesis. I. Effect of dietary creatine. *J. Biol. Chem.* **235,** 2357–2361.

33. Walker, J.B. (1961): Metabolic control of creatine biosynthesis II. Restoration of transamidinase activity following creatine repression. *J. Biol. Chem.* **236,** 493–498.

*Guanidino Compounds : 2*, eds. by P.P. De Deyn, B. Marescau, I.A. Qureshi and A. Mori.
©1997 John Libbey & Company Ltd., pp. 103–109.

# Chapter 13

# The localization of L-arginine: glycine amidinotransferase in rat pancreas

John F. VAN PILSUM, Robert L. SORENSON, Larry STOUT and T. Clark BRELJE

*Departments of Biochemistry and of Cell Biology and Neuroanatomy, University of Minnesota, Minneapolis, MN 55455–0347, USA*

## Summary

A role of L-arginine:glycine amidinotransferase, commonly called transamidinase, in glucagon secretion seemed possible in view of the previous report that transamidinase immunoreactivity was localized in the α-cells of rat pancreas. Glucagon and ornithine secretion by rat pancreata perfused with arginine and glycine or with canavanine and glycine were determined. Both substrate solutions stimulated glucagon secretion; only arginine plus glycine stimulated ornithine secretion. The chronology of transamidinase immunoreactivity appearance was investigated in pancreata from foetal, neonatal, and adult rats. Transamidinase was detected only in the acinar cells of all rats. Transamidinase enzyme activities of homogenates of entire pancreas and of isolated acinar tissue were similar to each other and to those of the pancreas perfused with arginine plus glycine. No transamidinase activity was detected in homogenates of islet cells. We have concluded that transamidinase is localized exclusively in acinar tissue and further, transamidinase immunoreactivity appears in acinar tissue of foetal rats as early as 16 days of age. Pancreatic transamidinase is suggested to have a major role in furnishing ornithine and guanidinoacetic acid directly to the liver via the portal vein. A role of transamidinase in the regulation of glucagon secretion seems unlikely since the islet cells do not have access to acinar cell secretory products.

## Introduction

The first step in the synthesis of creatine in mammals is the conversion of glycine to guanidinoacetic acid, catalysed by L-arginine: glycine amidinotransferase (transamidinase, E C 2.1.4.1) in the kidney and pancreas[15]. The cellular localization of transamidinase, as determined by immunofluorescence microscopy, has been reported previously[10]. Purified homogeneous rat kidney transamidinase was used to generate rabbit polyclonal[11] and mouse monoclonal[3] antibodies for those studies. Transamidinase immunoreactivity was detected in the proximal tubules of the rat kidney and the α-cells of the rat pancreas using rabbit polyclonal antibodies[10]. The localization of the enzyme in the proximal tubules of the kidney was confirmed by the same technique using mouse monoclonal antibody; however, no transamidinase immunoreactivity was detected in the α-cells of the pancreas by this technique. Further, no transamidinase immunoreactivity was detected in the acinar cells of the rat pancreas with either the rabbit polyclonal or mouse monoclonal antibodies. It was therefore suggested that pancreas and kidney transamidinase proteins were not exactly identical with respect to the immunological determinant recognized by the monoclonal antibody. Also, multiple forms of rat kidney transamidinase have been reported[4]; therefore, transamidinase was apparently confined to the α-cells of pancreas. The fact that the pancreas transamidinase activity, similar to that of kidney, was confined to a cell type represented only 0.3 per cent of the pancreas tissue[2] was of little concern to us at this time.

The purpose of this investigation was to determine the chronology of the appearance of α-cells in pancreata from foetal, neonatal, and adult rats using as a marker transamidinase immunoreactivity by immunofluorescence microscopy. Polyclonal guinea pig anti-transamidinase was used and transamidinase immunoreactivity was detected only in the acinar cells of pancreata of all ages of rats. Therefore the localization of transamidinase in pancreatic tissue from rats was re-examined.

## Material and methods

### L-arginine: glycine amidinotransferase immunohistochemistry

Polyclonal guinea pig anti-transamidinase sera was prepared using the rat kidney enzyme purified by a modified procedure[11]. The modification of the isolation procedure included chromatography on Sephadex G-150 and HPLC on hydroxyapatite. 200 μg of purified transamidinase in Freund's complete adjuvant was injected into multiple sites in guinea pigs. Three biweekly injections were given and the animals were bled two weeks after the last injection. Booster injections were given with emulsion of Freund's incomplete adjuvant and 200 μg of purified protein. Specificity of the antisera to transamidinase was determined by Western blot analysis of a rat kidney homogenate. Only a single 42 kd band was detected by the transamidinase antisera.

Pancreata were obtained from 16-day foetal, 5-day neonatal and adult Sprague-Dawley rats. The pancreata from the foetal and neonatal rats were fixed by immersion in 4 per cent paraformaldehyde for 24 h. The adult pancreata were perfusion-fixed with 4 per cent paraformaldehyde and post-fixed by immersion for an additional 24 h. The tissue was then extensively washed and stored in 30 per cent sucrose for 3–5 days at 4 °C before sectioning. Sections, 22 μm thick, were cut on a cryostat and thaw-mounted onto gelatin-coated slides. The sections were incubated with guinea-pig anti-transamidinase serum (1:100 dilution in phosphate buffered saline (PBS) containing 0.3 per cent Triton X-100) for 24 h. Tissue was washed eight times over a period of four hours in PBS and then incubated with 10 per cent normal donkey serum for 1 h. After additional washes, the sections were incubated with fluorescein isothiocyanate (FITC) labeled donkey anti-guinea-pig serum (Jackson ImmunoResearch, West Grove, PA) (1:100 dilution in PBS-Triton) for 24 h. The tissue was washed six times in PBS over a period of 3 h. The stained sections were cover-slipped using PBS-glycerine containing p-phenylenediamine to reduce photo-bleaching of fluorescein[6]. Immunostained pancreatic sections were examined with an MRC-600 Confocal Imaging System (Bio-Rad Life Science, Hercules, CA) mounted on an Olympus BH-2 epifluoresence microscope.

### Isolated pancreas perfusion

The isolated pancreas perfusion procedure used in this study was a modification of that of Loubatieres, et al.[8] as previously reported[14,1]. Pancreata were obtained from pentobarbital-anaesthetized, 150–250 g Sprague-Dawley male rats. The perfusate was Krebs-Ringer bicarbonate solution containing 15 mM HEPES, 3.8 per cent dextran, 0.25 per cent BSA, 10 mg/dl soybean trypsin inhibitor and 2.8 mM glucose. The medium was gassed continuously with 95 per cent $O_2$/5 per cent $CO_2$ and maintained at 37 °C. The isolated pancreata were perfused with either 2.5 mM canavanine and 2.5 mM glycine or 2.5 mM arginine and 2.5 mM glycine in the standard perfusion buffer, solutions that should produce the reaction products canaline and guanidinoacetic acid or ornithine and guanidinoacetic acid, respectively. The inlet for the vascular perfusion was the coeliac artery, and the outlet was through the severed portal vein. The effluent fractions containing the pancreatic secretions were collected in small test tubes. The perfusion rate was 1.3 ml/min and fractions were collected every 2 min. The pancreata were perfused for a 20 min equilibration period prior to the onset of media collection for analysis.

**Determination of transamidinase activity**

The method used was based on the production of ornithine using arginine and glycine as the substrate[16]. One unit of transamidinase activity is equal to 1 µmol of ornithine produced per hour.

*Acinar and islet tissue*

The pancreas was first subjected to the procedure used for isolation of islets[7]. Thus, the acinar tissue was essentially devoid of islet tissue and the islet was free of contaminating acinar tissue. One per cent aqueous homogenates of the acinar and of the islet tissue were used for determination of transamidinase activity.

*Perfused pancreata*

Ornithine production and release was determined by measuring the amount of ornithine released into the perfusate by the pancreata perfused with the substrates (arginine and glycine or canavanine and glycine). The ornithine production is reported as nmol ornithine/ml of perfusate and is an indicator of the transamidinase activity in the intact pancreas. Glucagon was also monitored during the pancreatic perfusion and its presence determined by radioimmunoassay as previously described[14].

## Results and discussion

**Immunohistochemistry of L-Arginine: glycine amidinotransferase in the rat pancreas**

The immunofluorescence staining of pancreata from foetal, neonatal and adult rats are shown in Fig. 1. Transamidinase immunoreactivity was identified in only the acinar tissue of pancreata of all groups of rats and, further, was confined to the mitochondria. Magri *et al.*[9] had reported previously that transamidinase was confined to the mitochondria of kidney. A few cells of unknown origin in the foetal tissue, scattered throughout the acinar and endocrine pancreata, had a diffuse staining pattern. These cells were not observed in the pancreata from the neonatal or adult rats. Transamidinase immunoreactivity was not observed in islets of pancreata from neonatal or adult rats.

We are unable to explain the staining of the α-cells and not the acinar tissue with the rabbit anti-transamidinase in the previous investigation[10] in contrast to the staining of the acinar tissue and not the α-cells with the guinea-pig anti-transamidinase in the present investigation. The transamidinase used in the guinea pig antibody preparation is believed to have been more pure that the enzyme used in the rabbit antibody preparation, in view of the HPLC chromatography purification step.

**L-*arginine: glycine amidinotransferase activity determinations of pancreata homogenates and of homogenates of isolated islet and acinar tissue***

The enzyme activities of the various homogenates are shown in Table 1. The activity of the isolated acinar tissue was believed to be greater than the whole pancreata homogenates because connective tissue is absent in the isolated acinar tissue. No transamidinase activity was found in the homogenates of the islet tissues (i.e. less than 0.5 units/g).

**Table 1.** L-arginine:glycine amidinotransferase activities of homogenates of whole pancreas and of isolated acinar and islet tissue (mean ± SEM)

| Tissue | Transamiddinase Activity (units/g) |
|---|---|
| Whole pancreas (n = 11) | 13.7 ± 0.7 |
| Isolated acinar tissue (n = 2) | 16.8 ± 2.0 |
| Isolated islet tissue (n = 2) | Not detectable |

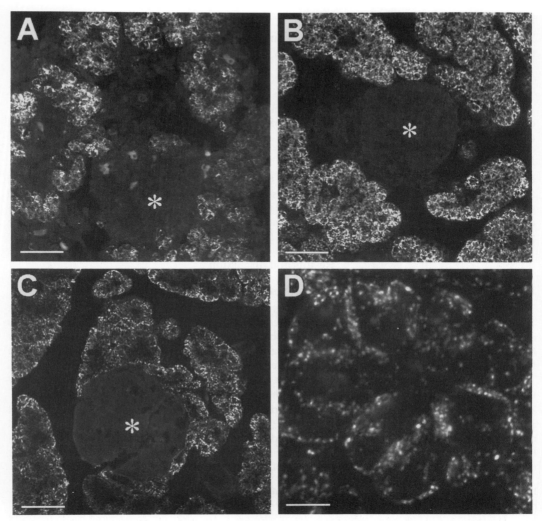

*Fig. 1. Immunohistochemical localization of L-arginine:glycine amidinotransferase in the pancreas using a guinea-pig antibody against HPLC purified transamidinase. A. 16 day foetal rat pancreas (scale bar = 50 μm); B. Five day neonatal rat pancreas (scale bar = 50 μm); C. Adult rat pancreas (scale bar = 50 μm); D. Adult rat pancreas (scale bar = 10 μm). Immunostaining of the transamidinase was associated with acinar cells and not with any of the islet cells. The immunostaining pattern is characteristic of that associated with mitochondria. The location of islets in the photomicrographs is indicated with an asterisk.*

## Production of ornithine by intact isolated pancreata perfused with substrates of L-arginine: glycine amidinotransferase (i.e. arginine and glycine or canavanine and glycine)

The secretions of ornithine and of glucagon by the perfused isolated pancreata are shown in Fig. 2. Ornithine was produced only when arginine and glycine were included in the perfusion medium. There was a rapid increase in ornithine production when arginine and glycine were included in the media. The rate of production increased throughout the perfusion period and declined towards basal levels after removal of the amino acids from the medium. Ornithine production reached a peak level

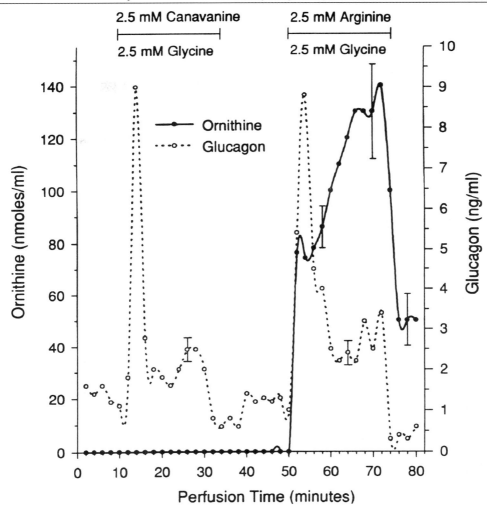

*Fig. 2. Production of ornithine by the isolated perfused pancreas. Ornithine was only produced when the substrates, arginine and glycine, are available for metabolism by* L-*arginine:glycine amidinotransferase. In contrast, glucagon secretion responds to both sets of amino acids and is similarly secreted into the portal vein. The data represent the mean ± SEM of five pancreatic perfusion experiments.*

of $182 \pm 26$ nmol/min. The average wet weight of the pancreata was $535 \pm 30$ mg. The specific activity based on ornithine production during the 20 min treatment period ($270 \pm 29$ nmol ornithine/min/g pancreas) is equivalent to $16.4 \pm 1.8$ units/g of pancreas. The specific activity based on ornithine production at its peak level during the experiment ($336 \pm 35$ nmol ornithine/min/g pancreas) is equivalent to $20.0 \pm 2.1$ units/g pancreas.

Glucagon secretion was also determined in the pancreatic perfusate, since islet hormone secretion and their secretion profiles in this preparation are well characterized. In addition, glucagon secretion responds to amino acids used in this study and served as an indicator of the efficacy of the perfused pancreas preparation. Both treatment with canavanine and glycine and treatment with arginine and glycine resulted in a prominent first-phase glucagon release and a smaller second-phase glucagon release.

In summary, transamidinase is present in high concentrations in the pancreas, where it is primarily if not exclusively found in acinar cells. The production of ornithine and guanidinoacetic acid is now concluded to be restricted to the acinar cells of the pancreas, the location of L-arginine: glycine amidinotransferase. The proximal convoluted tubules comprise approximately 30 per cent of the organ mass of the kidney[13]. If the enzyme were confined to the α-cells as originally concluded, its concentration in the α-cells would have been over 100-fold greater than in the proximal tubules of the kidney, an unlikely possibility. The localization of the enzyme in the acinar cells of the pancreas and in the proximal tubules of the kidney means that approximately equal concentrations of the enzyme (i.e. units/g tissue) are present in both tissues.

The physiological significance of the localization of transamidinase in the acinar cells of the pancreas and the proximal tubules of the kidney is not known other than their being the site of synthesis of guanidinoacetic acid. The localization of transamidinase in these cells constitutes a highly concentrated package of the enzyme which could have as its function the ability to deliver a high concentration of guanidinoacetic acid in the blood to the liver. The blood concentration of guanidinoacetic acid should influence its rate of conversion to creatine by S-adenosylmethionine guanidinoacetate N-methyltransferase. Therefore, the pancreas may play a major role in furnishing guanidinoacetic acid to the liver in that blood from the pancreas enters the portal vein of the liver directly. In contrast, blood from the kidney is circulated throughout the body prior to its entering the liver as arterial blood. Approximately 80 per cent of the blood that enters the liver is venous blood, the route of entry of the pancreatic secretory products into the liver. In view of the fact that the islet cells of the pancreas do not have access to the acinar cell secretory products, we have concluded that a role of L-arginine: glycine amidinotransferase in the secretion of glucagon seems unlikely.

In addition to the probable major role of pancreatic transamidinase in providing ornithine and guanidinoacetic acid to the liver, guanidinoacetic acid has been suggested to have an anti-hyperglycaemic effect[12]. The concentration of guanidinoacetic acid required to elicit an hyperglycaemic effect is substantially higher than its concentration in serum, however, it is of interest that pancreatic transamidinase activity has been reported to be regulated by insulin[5].

## Acknowledgements

This work was supported by NIH Grant DK33655. We thank Dr. D.M. McGuire, Department of Biological Sciences, St. Cloud State University, St. Cloud, MN 55301 for the guinea-pig antibody used in this investigation.

# References

1. Brelje, T.C. & Sorenson, R.L. (1988): Nutrient and hormonal regulation of the threshold of glucose-stimulated insulin secretion in isolated rat pancreases. *Endocrinology* **123**, 1582–1590.

2. Erlandsen, S.L., Hegre, O.D., Parsons, J.A., McEvoy, R.C. & Elde, R.P. (1976): Pancreatic islet cell hormones: Distribution of cell types in the islet and evidence for the presence of somatostatin and gastrin within the D cell. *J. Histochem. Cytochem.* **24**, 883–897.

3. Gross, M.D., McGuire, D.M. & Van Pilsum, J.F. (1985): The production and characterization of two monoclonal antibodies to rat kidney L-arginine:glycine amidinotransferase. *Hybridoma* **4**, 257–269.

4. Gross, M.D., Simon, A.M., Jenny, R.J., Gray, E.D., McGuire, D.M. & Van Pilsum, J.F. (1988): Multiple forms of rat kidney L-arginine: glycine amidinotransferase. *J. Nutr.* **118**, 1403–1409.

5. Hirata, M. (1989): Study of impaired metabolism of guanidinoacetic acid in uremia. The compensatory role of the pancreas in guanidinoacetic acid synthesis. *Nippon Jinzo Gakkai Shi.* **9**, 951–961.

6. Johnson, G.D. & Araujo, G.M. de C.N. (1981): A simple method of reducing the fading of immunofluorescence during microscopy. *J. Immunol. Methods* **43**, 349–350.

7. Lindall, A., Steffes, M. & Sorenson, R.L. (1969): Immunoassayable insulin content of subcellular fractions of rat islets. *Endocrinology* **85**, 218–223.

8. Loubatieres A., Mariani, M.M., Ribes, G., de Malbosc, H. & Chapal, J. (1969): Etude expérimentale d'un nouveau sulfamide hypoglycémiant particulièrement actif, le HB 419 on glibenclamaide: 1' action betacytotrope et insulino-sécrétrice. *Diabetologia* **5**, 1–10.

9. Magri, E., Baldoni G. & Grazi, E. (1975): On the biosynthesis of creatine. Intramitochondrial localization of transamidinase from rat kidney. *FEBS Lett.* **55(1)**, 91–93.

10. McGuire, D.M., Gross, M.D., Elde, R.P. & Van Pilsum, J.F. (1986): Localization of L-arginine: glycine amidinotransferase protein in rat tissues by immunofluorscence microscopy. *J. Histochem. Cytochem.* **34**, 429–435.

11. McGuire, D.M., Tormanen, C.D., Segal, I.S. & Van Pilsum, J.F. (1980): The effect of growth hormone and thyroxine on the amount of L- arginine:glycine amidinotransferase in kidneys of hypophysectomized rats. *J. Biol. Chem.* **255**, 1152–1159.

12. Meglasson, M.D., Wilson, J.M., Yu, J.H., Robinson, D.D., Wyse, B.M. & de Souza, C.J. (1993): Antihyperglycemic action of guanidinoalkanoic acids: 3 guanidinopropionic acid ameliorates hyperglycemia in diabetic KKA$^y$ and C57BL6 ob/ob mice and increases glucose disappearance in Rhesus monkeys. *J. Pharm. Exp. Therap.* **266**, 1454–1462.

13. Pfaller, W., Gstraunthaler, G. & Kotanko, P. (1985): Compartments and surfaces in renal cells. In: *Renal biochemistry; cells, membranes, molecules*, ed. R.K.H. Kinne, pp. 2–57, Amsterdam, New York, Oxford: Elsevier Science Publishers B.V.

14. Sorenson, R.L., Lindell, D.V. & Elde, R.P. (1980): Glucose stimulation of somatostatin and insulin release from the isolated, perfused rat pancreas. *Diabetes* **29**, 747–751.

15. Van Pilsum, J.F., Stephens, G.C. & Taylor, D. (1972): Distribution of creatine, guanidinoacetic acid and the enzymes for their biosynthesis in the animal kingdom. *Biochem. J.* **126**, 325–345.

16. Van Pilsum, J.F., Taylor, D., Zakis, B. & McCormick, P. (1970): Simplified assay for transamidinase activities of rat kidney homogenates. *Anal. Biochem.* **35**, 277–286.

*Guanidino Compounds : 2*, eds. by P.P. De Deyn, B. Marescau, I.A. Qureshi and A. Mori.
©1997 John Libbey & Company Ltd., pp. 111–119.

# Chapter 14

# Sequence comparison and functional analysis of amidinotranferases from eukaryotes and prokaryotes

John F. VAN PILSUM, David H. SHERMAN, Theresa V. LINE, Ajay BEDEKAR, and Lilliam AYALA

*Department of Biochemistry and Department of Microbiology and Biological Process Technology Institute, University of Minnesota, Twin Cities, MN 55455–0347, USA*

## Summary

Amidinotransferases are involved in creatine and neuroactive guanidino compound synthesis in vertebrates and in antibiotic synthesis in the gram positive soil bacteria *Streptomyces*. Amidinotransferases from four species of *Streptomyces* have been found to have overall amino acid sequence identities and similarities with the rat kidney amidinotransferase ranging from 30–39 and 47–58 per cent, respectively. The *Streptomyces* amidinotransferase genes have been found to have overall nucleotide sequence identities with the rat kidney gene of 46–48 per cent. We have now done a computer assisted analysis of the comparative structures of the amidinotranferase genes and enzymes of various species of *Streptomyces* and rat kidney. Identities of 70, 57, and 80 per cent between the *Streptomyces griseus strB* and rat kidney genes were found near the 5′ end, midway, and near the 3′ ends, isd as the α, β, and γ regions of high identity, respectively. Identities of 75, 52, and 80 per cent between the *Streptomyces griseus* and rat enzymes were found near the N-terminal, midway, and near the COOH-terminal ends, isd as α′, β′, and γ′ regions of high identity, respectively. Evidence was obtained that rat kidney and pancreas L- arginine:glycine amidinotransferases were able to catalyze the amidination of a derivative of inosamine, a step in the synthesis of streptomycin by *Streptomyces griseus*. We therefore, have concluded that the amidinotransferase genes and enzymes are a remarkable example of conservation of structure and function through evolution, from prokaryotes to higher eukaryotes.

## Introduction

Over 100 naturally occurring organic compounds containing the guanidine moiety, called guanidino compounds, have been identified in the plant and animal kingdoms[13]. The guanidino compounds are synthesized by transamidination reactions, catalyzed by amidinotransferases, commonly called transamidinases. Transamidinases catalyze the reversible transfer of an amidine group between arginine or canavanine, and a variety of compounds containing an amino group. The first step in creatine synthesis in vertebrates is the transfer of the amidine group of arginine to the amino group of glycine, catalyzed by L-arginine: glycine amidinotransferase[15]. One step in the synthesis of streptomycin by *Streptomyces* is the transfer of the amidine group of arginine to inosamine phosphate, catalyzed by L-arginine:inosamine phosphate amidinotransferase[20].

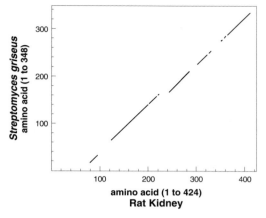

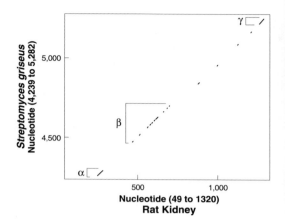

Fig. 1. A DOTPLOT computer analysis comparing the amino acid sequences of the amidinotransferases from rat kidney and streptomyces griseus strB.The N-terminal amino acids of the enzymes are designated as amino acid residue 1. The amino acid residues from 1 to 424 in the rat kidney and from 1 to 348 in the Streptomyces amidinotransferases, respectively, are compared. Density: 397.73; COMPARE, window: 30; stringency: 14.

Fig. 2. A DOTPLOT computer analysis comparing the nucleotide sequences of the amidinotransferase genes from rat kidney and Streptomyces griseus strB. The 5' end of the genes were designated as nucleotide residue 1. The nucleotide residues 4,239 to 5,282 of the Streptomyces griseus strB gene was the location of its amidinotransferase coding. The nucleotide residues 49 to 1,320 of the rat kidney gene were the location of the amidinotransferase coding which exclude the coding of its leader nucleotide sequence. The density was 1188.64; COMPARE, window: 21; stringency:14.

A variety of compounds have been tested previously for their abilities to be amindinated by the *Streptomyces* and rat kidney amidinotransferases. Watanabe *et al.,* has reported that purified homogeneous rat kidney L-arginine:glycine amidinotransferase catalyzed the amidination of a variety of compounds containing an amino group to form neuroactive guanidino compounds[21]. Walker has reported that both inosamine-P amidinotransferase from *Streptomyces* and glycine amidinotransferase from rat kidney catalyzed the amidination of the ornithine analogs glycylglycine and 1,4 diamino-butylphosphonic acid[17]. However, the *Streptomyces* amidinotransferase was unable to utilize glycine, the amidine acceptor for rat kidney amidinotransferase[18]. Here we report the ability of rat kidney amidinotransferase to utilize the amidine acceptor from *Streptomyces*.

## Materials and methods

### Protein and nucleotide sequence analysis

Sequences were found using the programs TFASTA and BLASTP on the GenBank DNA sequence database. The alignment of closely related sequences was done using PILEUP with its default settings of 3.0 for GapWeight and 0.1 for GapLengthWeight. Nucleic acid sequence comparison of the rat kidney gene to each of the 7 *Streptomyces* genes was done using COMPARE with a window of 21 and stringency of 14.0. Protein sequence comparison was done using COMPARE with a window of 30 and stringency of 14.0. DOTPLOT graphs were made from the output files of COMPARE. The codonpreference graphs were produced using a software package by IntelliGenetics, Inc.: GENEWORKS. For both of the graphs, a window size of 50 was used.

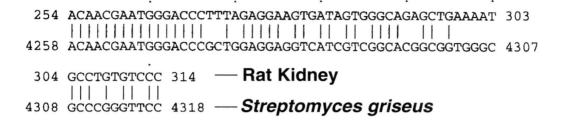

254 ACAACGAATGGGACCCTTTAGAGGAAGTGATAGTGGGCAGAGCTGAAAAT 303
    ||||||||||||||||| | ||||| || || || |||| || |
4258 ACAACGAATGGGACCCGCTGGAGGAGGTCATCGTCGGCACGGCGGTGGGC 4307

304 GCCTGTGTCCC 314    — **Rat Kidney**
    ||| | || ||
4308 GCCCGGGTTCC 4318   — *Streptomyces griseus*

per cent identity = 70.49

*Fig. 3. The nucleotide sequences of the α-identity regions of rat kidney and Streptomyces griseus strB amidinotransferase genes.The nucleotide sequences of the rat kidney amidinotransferase gene from residue 254 to 314 and the Streptomyces griseus strB amidinotransferase gene from residue 4258 to 4318 are shown.*

### Rat kidney and pancreas L-arginine: glycine amidinotransferase preparation

Rat kidney L-arginine: glycine amidinotransferase was purified by a modification of the procedure of McGuire *et al.*[9]. The modified procedure involved chromatography on Sephadex G-150, phenyl Sepharose and hydroxyapatite agarose gel. The purified kidney enzyme appeared homogeneous by SDS gel electrophoresis. The preparation of pancreas transamidinase did not appear homogeneous by SDS gel electrophoresis. The concentration of enzyme protein in the solution used in the assay was ~0.5 mg/ml.

### *Streptomyces griseus* amidine acceptor preparation

The amidine acceptor preparation was a heat-treated sonicate of the mycelia of *Streptomyces griseus* grown in the presence of 1 per cent myo-inositol according to the procedure of Walker[19].

### Assay for the ability of rat L-arginine:glycine amidinotransferase to amidinate the amidine acceptor from *Streptomyces griseus*

The heat treated sonicate of the mycelia of *Streptomyces griseus* was incubated in the presence of arginine and the production of ornithine was determined. The complete incubation mixture contained: 40 μl arginine substrate; 10 μl extract of *Streptomyces griseus* mycelia; 20 μl rat kidney enzyme; and 30 μl Tris-EDTA[19]. Incubation was 1 h at 37 °C and 150 μl perchloric acid and 750 μl ninhydrin solution were added and the mixture heated in a boiling water bath for 25 min and the OD at 516nm recorded[16].

### Results and discussion

### Sequence comparison of amidinotransferase genes and enzymes from *Streptomyces* and rat kidney

The overall per centage amino acid sequence identities and similarities of the *Streptomyces* amidinotransferases and the per cent overall identities of the nucleotide sequences of their genes to those of rat kidney amidinotransferase are listed in Table 1. The deduced amino acid sequences of the *Streptomyces* enzymes had similarities and identities with the rat kidney enzyme ranging from 47–58 and 34–40 per cent, respectively. The nucleotide identities of all the *Streptomyces* genes with the rat kidney genes were 47–48 per cent. All of the *Streptomyces* species listed in Table 1 produce streptomycin, except *S. bluensis*, which produces bluensomycin. Both streptomycin and bluensomycin are guanidino compounds.

**Table 1. A comparison of the structures of the amidinotransferases genes and enzymes with rat kidney amidinotransferase**

| Streptomyces | % Deduced amino acid similarity (identitiy) | | % Nucleotide identity |
|---|---|---|---|
| S griseus StrB[2] | 58 | 40 | 47 |
| S glaucescens strB1[5] | 47 | 38 | 48 |
| S griseus SPH[14] | 56 | 39 | 47 |
| S bluenisis[6] | 47 | 37 | 47 |
| S galbus[7] | 56 | 37 | 47 |
| S griseus strB1[11] | 57 | 36 | 47 |
| S glaucescens strB2[8] | 51 | 34 | 48 |

DOTPLOT computer analysis comparing the amino acid sequences of each *Streptomyces* amidino-transferase with that of rat kidney were performed. Large areas of identities plus similarities for each *Streptomyces* enzyme with the rat kidney enzyme were found. A DOTPLOT comparison of the amino acid sequence similarities and identities of *Streptomyces griseus strB* amidinotransferase with the rat

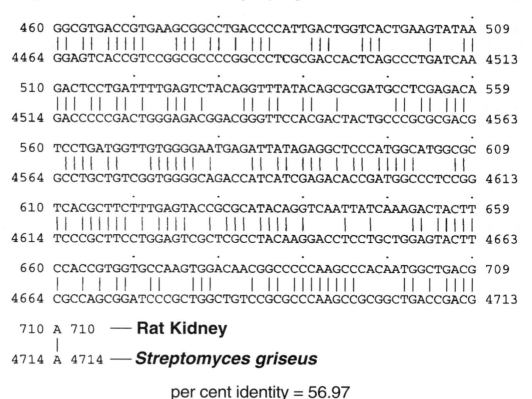

per cent identity = 56.97

*Fig. 4. The nucleotide sequence of the β-identity regions of rat kidney and Streptomyces griseus strB amidinotransferase genes.The nucleotide sequence of the rat kidney amidinotransferase gene from residue 460 to 710 and the nucleotide sequence of the Streptomyces griseus strB amidinotransferase gene from residue 4464 to 4714 are shown.*

```
1247  CCCTAGGAGGAGGCTTCCACTGCTGGACCTGCGACGTCCGCCGC 1290  —— Rat Kidney
      |||| || || |||||||||||||  ||||  |||||| || |||
5212  CCCTCGGGGGCGGCTTCCACTGCGCGACCCTCGACGTGCGGCGC 5255  —— Streptomyces griseus
```

per cent identity = 79.54

*Fig. 5. The nucleotide sequences of theg γ-identity regions of rat kidney and Streptomyces griseus strB amidinotransferase genes.The nucleotide sequence of the rat kidney amidinotransferase gene from residue 1247 to 1290 and the nucleotide sequence of the Streptomyces griseus strB gene from residue 5212 to 5255 are shown.*

```
YNEWDPLEEVIVGRAENACV — Rat Kidney
.||||||||||||| | .| |
HNEWDPLEEVIVGTAVGARV — Streptomyces griseus
```

per cent identity = 75.0
per cent similarity = 75.0

*Fig. 6. The amino acid sequences of the α′-identity regions of rat kidney and Streptomyces griseus strB amidinotransferases. Two dots between the amino acid residues of the two enzymes indicates similar amino acids.*

```
GVTVKRPDPIDWSLKYKTPDFESTGLYSAMPRDILMVVGNEIIEAPMAWR
||||:||:| | |  .||||:|..|:..  ||| |: ||..|||.|||:|
GVTVRRPGPRDHSALIKTPDWETDGFHDYCPRDGLLSVGQTIIETPMALR

SRFFEYRAYRSIIKDYFHRGAKWTTAPKPTMAD  —— Rat Kidney
|||:|  ||:.:: :|| .|.:| .|||| :.|
SRFLESLAYKDLLLEYFASGSRWLSAPKPRLTD  —— Streptomyces griseus
```

per cent identity = 51.8
per cent similarity = 67.5

*Fig. 7. The amino acid sequences of the β′-identity regions of rat kidney and Streptomyces griseus strB amidinotransferases.*

kidney enzyme is shown in Fig. 1. A DOTPLOT comparison of the nucleotide sequence identities of *Streptomyces griseus strB* amidinotransferase gene with that of kidney enzyme is shown in Fig. 2. Three regions of identity between the 2 genes were found; near the 5′ end, midway, and near the 3′ end of the genes, isd as the α, β, and γ high identity regions. Similar DOTPLOT comparisons between the 7 other *Streptomyces* amidinotransferases listed in Table 1 and the rat kidney amidinotransferase were obtained. *Streptomyces griseus strB* was the most similar to rat kidney; therefore, only further comparative studies with these two species are described in this report. The nucleotide sequences of the *Streptomyces griseus strB* and rat kidney genes in the α, β, and γ regions of high identity are shown in Figs. 3, 4, and 5. The per cent identities of the nucleotide sequences in the α, ß, and γ regions of the two genes were 70, 57, and 80 per cent, respectively.

The amino acid sequences of the *Streptomyces griseus strB* and rat kidney amidinotransferase corresponding to the α, β, and γ regions of their genes are shown in Figs. 6, 7, and 8. The two enzymes had amino acid sequence identities of 75, 52, and 80 per cent, respectively in the α′, β′, and γ′. Thus, excellent correlation between the identities of the nucleotide and amino acid sequences were found in all three high identity regions of *Streptomyces griseus strB* and rat kidney amidinotransferase genes and their enzymes.

The entire amino acid sequences of all the enzymes from the *Streptomyces* and rat kidney were aligned side by side. In general, good correlation of amino acid sequences of all eight enzymes were found. The amino acid sequences of segments of enzymes in the α′ region of high identity are shown in Table 2.

**Table 2. A comparison of the amino acid sequences of the amidinotransferases from the various species of *Streptomyces* with that of rat kidney in the α′ region of identity**

| Species | Amino acid residues | | | | | | | | | | | | | | | | | | | |
| --- | --- | --- | --- | --- | --- | --- | --- | --- | --- | --- | --- | --- | --- | --- | --- | --- | --- | --- | --- | --- |
| *S griseus StrB* | N | E | W | D | L | E | E | V | I | I | V | G | T | A | V | G | A | R | V | P |
| *S glaucescens strB1* | N | E | W | D | P | L | E | E | I | I | V | G | T | A | V | G | A | R | V | P |
| *S griseus SPH* | N | E | W | D | P | L | E | E | I | I | V | G | T | A | Q | G | A | R | V | P |
| *S bluenisis* | . | . | . | . | . | . | E | E | I | I | V | G | T | A | Q | G | A | R | V | P |
| *S galbus* | . | . | . | . | . | . | E | E | I | I | D | G | T | A | V | G | A | R | V | P |
| *S griseus strB2* | N | E | W | D | P | L | E | E | V | V | V | G | T | A | R | R | P | . | C | A |
| *S glaucescens strB2* | T | E | W | D | P | L | E | E | I | V | V | G | T | A | V | G | S | . | . | . |
| Rat Kidney | N | E | W | D | P | L | E | E | V | I | V | G | R | A | E | N | A | C | V | P |

Streptomyees are reported to have a average G+C base content of 70–80 per cent while eukaryotes have an average G+C content of 50 per cent. The per cent G+C bases in the α, β, and γ regions of identity of the two genes are shown in Table 3. The γ region of the rat kidney gene that has a G+C content approaching the G+C content of the *Streptomyces* gene is the γ region.

The codon frequencies of the *Streptomyces griseus strB* and the rat kidney are shown in Figs. 9 and 10 respectively. Protein-coding bacterial genes seem to process a codon usage indicative of their overall base composition[1]. *Streptomyces*, with a high overall G+C base content, has been reported to have a widely varying G+C base composition dependent on codon position. The codonpreference plot (Fig. 9) shows that ~90 per cent of the third position codon bases in *Streptomyces griseus strB* are G+C. This concurs with previously published codonpreference data of other *Streptomyces* genes[3]. As in the case of the *Streptomyces griseus strB* gene, the third codon position of the rat kidney gene (Fig. 10) has the highest overall G+C base composition relative to the other two codon positions. Quantitatively, however, the rat gene third codon position G+C distribution is significantly lower than in its *Streptomyces* counterpart; only ~60 per cent of the third position codon bases are G+C. We think it noteworthy that 3 separate third position codon peaks of high G+C base content appear in the α, β, and γ regions of high identity between the *Streptomyces griseus strB* and rat kidney genes.

SLGGGFHCWTCDVRR — **Rat Kidney**

·||||||| | ||||

TLGGGFHCATLDVRR — *Streptomyces griseus*

per cent identity = 80.0

per cent similarity =80.0

*Fig. 8. The amino acid sequence of the γ-identity regions of rat kidney and Streptomyces griseus strB amidinotransferases.*

**Table 3. The per cent of G+C base preference in the third position of the codons in the α, β, and γ regions of the amidinotransferases from rat kidney and from *Streptomyces***

| | % of G + C bases | |
| --- | --- | --- |
| Region | Rat | *Streptomyces* |
| α | 45 | 90 |
| β | 56 | 96 |
| γ | 77 | 86 |

**Functional analysis of amidinotransferase amidino acceptor specificity**

The amidinotransferase from *Streptomyces* has been reported to be unable to utilize glycine, the amidine acceptor for the rat kidney amidinotransferase[18]. The rat kidney amidinotransferase has been reported to utilize a variety of amidine acceptors[10,21,–23], therefore its ability to utilize inosamine phosphate, the acceptor for the *Streptomyces* enzyme, was investigated. Purified homogeneous rat kidney and a partially purified rat pancreas L-arginine: glycine amidinotransferase were incubated with arginine and a heat-treated sonicate of mycelia of *Streptomyces griseus*. Ornithine production was used as a measure of the ability of the rat enzyme to transfer the amidino group from arginine to the inosamine phosphate. No ornithine was found in the absence of the extract of *Streptomyces griseus* mycelia. The ornithine formed in the presence of the extract of mycelia was directly proportional to the duration of the incubation, the amount of kidney enzyme, and the amount of the extract of the mycelia, respectively. No ornithine was formed with a dialyzed preparation of the mycelia extract. The formation of ornithine was confirmed by incubation with ornithine carbamyltransferase in the presence of carbamyl phosphate. The absence of glycine in the mycelia extract, acting as a possible amidine acceptor for the kidney enzyme, was established by the lack of guanidinoacetic acid formed in the incubation. A determination of the kinetic constants for the rat kidney amidinotransferase utilization of inosamine phosphate was not possible in view of the unknown concentration of the amidine acceptor in the mycelia extract.

In addition to the above described evidence that rat kidney L-arginine: glycine amidinotransferase can utilize the amidine acceptor from *Streptomyces griseus*, additional reasons can be listed to support this concept. Watanabe *et al.* have reported that purified homogenous rat kidney amidinotransferase catalyzed the amidination of ethanolamine, 4-aminobutyric acid, lysine, 5-aminovaleric acid, 3-aminopropionic acid, and taurine[21]. The regions of high identity between *Streptomyces* and rat kidney

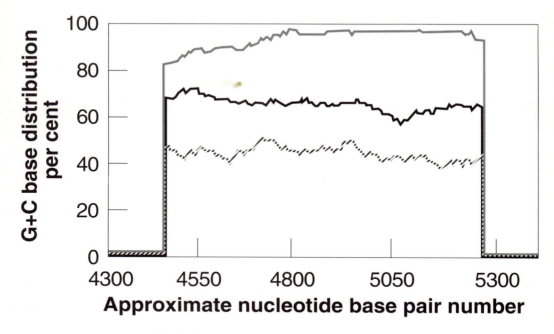

*Fig. 9. Streptomyces griseus strB gene codon usage.*
*The G+C base composition at the first, second, and third codon positions are locally averaged throughout the nucleotide sequence using a codonpreference algorithm. The topmost line represents the G+C Base distribution in the third codon position; the dark line represents the first codon position; the faint line represents the second codon position. The overall G+C base content is 68 per cent.*

amidinotransferases also support the concept that the rat enzyme should be able to use the amidine acceptor from *Streptomyces*.

The high degree of conservation of structure and function of the amidinotransferases between prokaryotes and eukaryotes that we report here also is believed to have implications for the active site of the amidinotransferases. The active sites of these enzymes could well be located in the regions of high identity of amino acid sequences between the enzymes. The degree of conservation of structure and function of the amidinotransferases is also believed to support the evolutionary aspects of the distribution of L-arginine: glycine amidinotransferase in the animal kingdom. This enzyme has been reported previously to have evolved in the vertebrates with the lamprey[15]. The presence of creatine in most species of vertebrates investigated was accounted for by the presence of the L-arginine:glycine amidinotransferase and N-methyl guanidinoacetate methyltransferase. Vertebrates lacking these enzymes that contained creatine in their bodies were suggested to obtain creatine via their diet, i.e. the sharks and rays. A repression by creatine of the expression of the gene for L-arginine:glycine amidinotransferase has been reported in rats[4].

Whether the gene for amidinotransferase has been transferred to the lamprey and vertebrates from bacteria or evolved with the lamprey cannot be answered at this time. We suggest, however, that the ability to transamidinate amino compounds evolved at least as early as the *Streptomyces* bacteria, and this ability has been conserved through evolution, perhaps by horizontal gene transfer, to vertebrates. The ability of the vertebrate, but not the *Streptomyces* amidinotransferases, to utilize glycine as an amidine acceptor may represent a random or direct mutation. In any case, transamidination is a pivotal reaction for both prokaryotes and eukaryotes. We consider the fact that the high G+C base nucleotide composition of rat kidney gene in the regions of high identities with the *Streptomyces* genes signifies

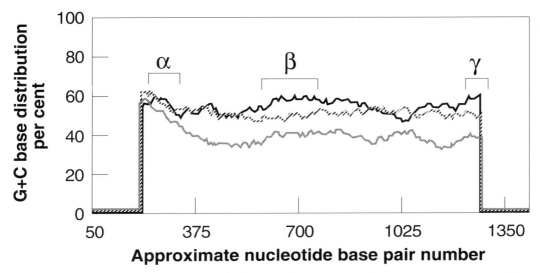

*Fig. 10. Rat kidney gene codon usage.The dark, topmost line represents the G+C base distribution in the third codon position. The third codon position peaks are found in the regions that are homologous to the Streptomyces griseus strB gene: α, β, and γ. The overall G+C base content is 51 per cent.*

that the rat gene has retained much of the primitive form of the gene found in *Streptomyces*. The gene clusters and evolutionary aspects of *Streptomyces* species have been discussed by Retzloff *et. al.*[12].

## References

1.  Bibb, M.J., Findlay, P.R. & Johnson, M.W. (1984): The relationships between base composition and codon usage in bacterial genes and reliable identification of protein-coding sequences. *Gene* **30,**157–166.

2.  Distler, J., Ebert, A., Mansouri, K., Pissowotzki, K., Stockman, M. & Piepersberg, W. (1987): Gene cluster for streptomycin biosynthesis in *Streptomyces griseus*: nucleotide sequence of three genes and analysis of transcriptional activity. *Nucleic Acids Res.* **15,** 8041–8056. EMBL/GenBank/DDBJ database accession number Y00459.

3.  Gray, G. & Thomson, C.J. (1983): Nucleotide sequence of a streptomycete aminoglycoside phosphotransferase gene and its relationships to phosphotransferases encoded by resistant plasmids. *Proc. Natl. Acad. Sci. USA* **80,** 5190–5194.

4.  Guthmiller, P., Van Pilsum, J.F., Boen, J.R. & McGuire, D. (1994): Cloning and sequencing of rat kidney L-arginine:glycine amidinotransferase: studies on the mechanism of regulation by growth hormone and creatine. *J. Biol. Chem.* **269,** 17556–17560.

5.  Mayer, G. (1994): Submitted to the EMBL/GenBank/DDBJ database accession number X78974.

6.  Mayer, G. (1994): Submitted to the EMBL/GenBank/DDBJ database accession number X78972.

7.  Mayer, G. (1994): Submitted to the EMBL/GenBank/DDBJ database accession number X78973.

8.  Mayer, G. (1994): Submitted to the EMBL/GenBank/DDBJ database accession number X78975.

9.  McGuire, D.M., Tormanen, C.D., Segal, I.S. & Van Pilsum, J.F. (1980): The effect of growth hormone and thyroxine on the amount of L-arginine:glycine amidinotransferase in kidneys of hypophysectomized rats. *J. Biol. Chem.* **255,** 1152–1159.

10. Pisano, J.J., Abraham, D. & Udenfriend, S. (1963): Biosynthesis and disposition of γ-guanidinobutyric acid in mammalian tissues. *Arch. Biochem. Biophys.* **100,** 323–329.

11. Pissowotzki, K., Manouri, K. & Piepersberg, W. (1991): Genetics of streptomycin production in *Streptomyces griseus*: molecular structure and putative function of genes strELMB2N. *Mol. Gen. Genet.* **231,** 113–123. EMBL/GenBank/DDBJ database accession number X62567.

12.  Retzloff, L., Mayer, G., Beyer, S., Ahlert, J., Verseck, S., Distler, J. & Piepersberg, W. (1993): Streptomycin production in *Streptomyces*: a progress report. In: *Industrial Microorganisms*: basic and appled molecular genetics, eds. R.H. Baltz, G.D., Hegeman & P.L. Skatrud, pp. 183–193. Washington D.C: American Society for Microbiology.

13.  Robin, Y. & Marescau, B. (1985): Natural guanidino compounds; a review. In: *Guanidines*, eds. A. Mori, B.D. Cohen & A. Lowenthal, pp. 383–438. New York and London: Plenum Press.

14.  Tohyama, H., Okami, Y. & Umezawa, H. (1987): Nucleotide sequence of the streptomycinphospho-transferase and amidinotransferase genes from *Streptomyces griseus*. *Nucleic Acids Res*. **15**, 1819–1833. EMBL/GenBank/DDBJ database accession number X05045.

15.  Van Pilsum, J.F., Stephens, G.C. & Taylor, D. (1972): Distribution of creatine, guanidinoacetate and the enzymes for their biosynthesis in the animal kingdom. *Biochem. J.* **126**, 325–345.

16.  Van Pilsum, J.F., Taylor, D., Zakis, B. & McCormick, P. (1970): Simplified assay for transamidinase activities of rat kidney homogenates. *Anal. Biochem.* **35**, 277–286.

17.  Walker, J.B. (1971): Enzymatic reactions involved in streptomycin biosynthesis and metabolism. *Lloydia* **34**, 363–371.

18.  Walker, J.B. (1958): Further studies on the mechanism of transamidinase action: Transamidination in *Streptomyces griseus*. *J. Biol. Chem.* **231**, 1–9.

19.  Walker, J.B. (1975): Pathways of biosynthesis of guanidinated inositol moieties of streptomycin and bluensomycin. In: *Methods in enzymology*, ed. J.H. Hash. pp. 429–470. New York, San Francisco, London: Academic Press.

20.  Walker, J.B. (1974): Biosynthesis of the monoguanidinated inositol moiety of *Bluensomycin*, a possible evolutionary precursor of streptomycin. *J. Biol. Chem.* **249**, 2397–2404.

21.  Watanabe, Y., Van Pilsum, J.F., Yokoi, I. & Mori, A. (1994): Synthesis of neuroactive guanidino compounds by rat kidney L-arginine:glycine amidinotransferase. *Life Sci.* **55**, 351–358.

22.  Watanabe, Y., Yokoi, I. & Mori, A. (1987): Biosynthesis of 2-guanidinoethanol. In: *Guanidines*, eds. A. Mori, B.D. Cohen & A. Lowenthal, pp. 53–60. New York and London: Plenum Press.

23.  Watanabe, Y., Yokoi, I., Watanabe, S., Sugi, H. & Mori, A. (1988): Formation of 2-guanidinoethanol by a transamidination reaction from arginine and ethanolamine by the rat kidney and pancreas. *Life Sci.* **43**, 295–302.

*Guanidino Compounds : 2*, eds. by P.P. De Deyn, B. Marescau, I.A. Qureshi and A. Mori.
©1997 John Libbey & Company Ltd., pp. 121–130.

# Chapter 15

# Guanidinoacetate methyltransferase: identification of the *S*-adenosylmethionine-binding site by affinity labelling and site-directed mutagenesis

Kiyoshi KONISHI, Yoshimi TAKATA, Tomoharu GOMI, Hirofumi OGAWA, and Motoji FUJIOKA

*Department of Biochemistry, Toyama Medical and Pharmaceutical University Faculty of Medicine, Sugitani, Toyama 930–01, Japan*

## Summary

The *S*-adenosylmethionine (AdoMet)-binding region of rat guanidinoacetate methyltransferase was probed by photo-affinity labelling and site-directed mutagenesis. Ultraviolet irradiation of guanidinoacetate methyltransferase in the presence of radiolabelled AdoMet resulted in a time-dependent covalent attachment of radioactivity to the protein. The rate of photolabelling was saturable with AdoMet, and the concentration of AdoMet yielding half-maximal rate was in excellent agreement with the dissociation constant determined by the equilibrium dialysis method. Low concentrations of competitive inhibitors, *S*-adenosylhomocysteine and sinefungin, were effective in preventing photoincorporation of AdoMet. These results indicate that AdoMet bound at the active site is fixed to the protein by photochemical reaction. By peptide analysis of the AdoMet-labelled guanidinoacetate methyltransferase, Tyr136 was found to carry AdoMet. Most low molecular weight methyltransferases have, in the middle portion of the sequence, an aspartate residue that is preceded by a hydrophobic amino acid and followed successively by a small neutral and a hydrophobic residue. In guanidinoacetate methyltransferase Asp129 and Asp134 satisfy the requirement. To test the role of the conserved aspartate residues, we performed amino acid replacement studies. Whereas amino acid substitutions at position 129 had little or no effect on catalytic activity, mutation of Asp134 produced profound effects on $k_{cat}$ and $K_m$. Reductions in $k_{cat}/K_m^{AdoMet}$ of $10^3$- and $10^5$-fold were observed for the Glu134 and Asn134 mutants, respectively. Guanidinoacetate methyltransferase binds AdoMet first, followed by guanidinoacetate. Therefore, a large decrease of $k_{cat}/K_m^{AdoMet}$ observed upon mutation indicates a crucial role of Asp134 in binding AdoMet. The data described here are consistent with the contention that the region of guanidinoacetate methyltransferase comprising residues 134–136 forms part of the AdoMet-binding site.

## Introduction

Guanidinoacetate methyltransferase (GAMT) catalyzes the last step of creatine biosynthesis by transferring the methyl group of *S*-adenosylmethionine (AdoMet) to guanidinoacetate. Creatine is reversibly phosphorylated by ATP to creatine phosphate which serves as an energy reservoir. Thus creatine phosphate is abundant in such tissues as skeletal muscle, cardiac muscle, brain and testis, where sudden changes of cellular energy are great. Creatine phosphate is sponta-neously converted to creatinine which is excreted in the urine. In humans as much as 1–2 g of creatine must be renewed daily, and it is estimated from nutritional studies that the GAMT-catalyzed reaction consumes over 70 per cent of the total methionine metabolized through the transsulfuration/homo-cysteine conservation pathway[12]. S-Adenosylhomocysteine (AdoHcy), the product of all transmethy-lation reactions, is cleaved to adenosine and homocysteine by AdoHcy hydrolase. AdoHcy is the sole source of homocysteine in mammals. Normally, about one-third of the homocysteine formed is used for the synthesis of cysteine and the remaining two-thirds are recycled to methionine via the 5-methyltetrahydrofolate-dependent reaction. Since the cellular folate coenzymes exist largely in the form of 5-methyl derivatives, the latter reaction is essential for regeneration of tetrahydrofolate which in turn is utilized for the synthesis of methylene- and methenyltetrahydrofolate. In view of the preponderance of GAMT over other methyltransferases in converting AdoMet to AdoHcy, this enzyme is considered to play a crucial role in the metabolism of one-carbon compounds and eventually in the biosynthesis of nucleotides in addition to the role of providing creatine.

Rat liver GAMT is a simple, monomeric protein with a MW of 26,000[14]. The enzyme consists of 235 amino acid residues, and has an acetylated N-terminus[13]. The GAMT-catalyzed reaction is essentially irreversible and not inhibited by creatine. Initial velocity, product (AdoHcy) and dead-end (sinefun-gin) inhibition studies are consistent with a sequential ordered mechanism in which AdoMet is the first substrate to bind to the enzyme[19]. Although the crystal structure of GAMT is not known, previous chemical modification[4,9,17], amino acid replacement[16], and limited proteolysis studies[5] have revealed some structural features for the enzyme. The N-terminal segment of about 25 residues is exposed to solvent and the portion comprising residues 15–20 is highly flexible. Cys15, Cys90, and Cys219 are juxtaposed in the three-dimensional structure, and Cys15 can form a disulfide with Cys90 or Cys219 when incubated with a stoichiometric amount of 5,5'-dithiobis(2-nitrobenzoate). Removal of the N-terminal segment of ~20 residues by proteolysis or disulfide cross-linking of Cys15 with either Cys90 or Cys219 results in a large or complete loss of activity. Whereas these results suggest that the integrity or flexibility of the N-terminal region is crucial for catalysis, the region is apparently distant from the active site because the modified enzymes retain the capacity to bind substrates. The present article summarizes the results of work aimed at elucidating the AdoMet-binding site of rat GAMT.

## Materials and methods

### Enzyme

GAMT used in this study was a recombinant rat enzyme produced in *Escherichia coli* JM109 transformed with plasmid pUCGAT9-1 that contained the coding sequence of rat GAMT cDNA[13]. The enzyme was structurally and functionally indistinguishable from the liver enzyme except that it has no N-terminal blocking group. cDNAs encoding mutant GAMTs were prepared by oligonucleo-tide-directed mutagenesis of pUCGAT9-1, and mutant enzymes were prepared using Mutan-K site-directed mutagenesis kit (Takara Shuzo, Kyoto, Japan)[19] which is based on the method of Kunkel *et al*[10]. All mutant enzymes as well as the recombinant wild-type enzyme were purified to homogene-ity[19].

### Enzyme assay

The GAMT activity was determined by monitoring the formation of inosine in a coupled assay with AdoHcy hydrolase and adenosine deaminase. The inosine was determined by HPLC on a TSK ODS 120T column (0.46 × 25 cm) as described[19].

### Photolabelling of GAMT with [*methyl*-³H]AdoMet

GAMT was photolabelled with [*methyl*-³H]AdoMet (1.2 × 10⁴ cpm/nmol) in 0.1 M Tris-HCl (pH 8.0)/1 mM EDTA/1 mM dithiothreitol. 50 µl aliquots of the reaction mixture were placed in the wells

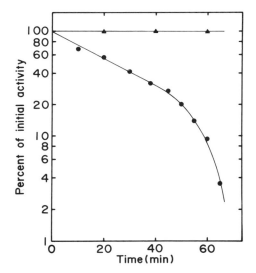

*Fig. 1. Inactivation of GAMT by UV irradiation. GAMT (17 μM) was UV irradiated in 0.1 M Tris-HCl (pH 8.0), containing 1 mM EDTA and 1 mM dithiothreitol. At the times indicated, aliquots (10–20 μl) were withdrawn from the incubation mixture and assayed for residual activity. The enzyme not irradiated (▲) lost no activity. Adapted with permission from Takata and Fujioka[18]. Copyright [1992] American Chemical Society.*

of a multititre plate on ice and irradiated at a distance of 7 cm from a 15W germicidal lamp. Incorporation of radioactivity into the enzyme protein was determined after isolation of the protein by gel column centrifugation[15].

## Results and discussion

### Photoaffinity labelling of guanidinoacetate methyltransferase

Ultraviolet (UV) irradiation of GAMT caused a progressive loss of enzyme activity (Fig. 1). The inactivation did not follow a simple decay kinetics and was not influenced by the presence of AdoMet and its analogues. No detectable activity remained after a 70 min irradiation. The inactivation is not due to UV-induced cross-linking or fragmentation of the enzyme; SDS-polyacrylamide gel electrophoresis showed the presence of only one protein band with MW 26,000 at least for this time period (data not shown). GAMT that had been irradiated for 60 min, despite being virtually inactive, normally bound AdoMet; an equilibrium dialysis study indicated that it binds a stoichiometric amount of AdoMet with a dissociation constant of 0.085 mM that was essentially identical to that of the untreated enzyme. These results indicate that the UV-induced changes that cause inactivation exert no influence on the active site geometry.

### Table 1. Effect of AdoMet analogues on photolabelling

| Compound | Concentration (μM) | Inhibition of photolabelling (%) |
|---|---|---|
| AdoHcy | 5 | 25 |
| | 50 | 79 |
| | 200 | 96 |
| Sinefungin | 160 | 62 |
| | 480 | 84 |
| | 960 | 99 |
| ATP | 500 | 33 |
| | 1000 | 44 |

GAMT (17 μM) was UV-irradiated with 0.1 mM [*methyl*-³H]AdoMet in the presence and absence of the compounds indicated. The per cent inhibition was obtained by comparing the radioactivity found in the presence and absence of the test compound. Adapted with permission from Takata and Fujioka[18]. Copyright [1992] American Chemical Society.

Exposure of GAMT to UV irradiation in the presence of [*methyl*-³H]AdoMet resulted in a time- and concentration-dependent covalent attachment of radioactivity to the protein. Figure 2 shows a plot of radioactivity incorporation as a function of [*methyl*-³H]AdoMet concentration. The data conform to Eqn 1 and shows that a reversible E·AdoMet complex is formed before covalent modification. A $K_d$

value of $0.10 \pm 0.01$ mM was obtained from this experiment, and this supports the idea that the AdoMet bound at the active site of GAMT undergoes photochemical reaction to form a covalent adduct. Furthermore, photolabelling was effectively prevented by low concentrations of competitive inhibitors AdoHcy and sinefungin, but not by ATP which was not bound to GAMT (Table 1).

**Table 2. Amino acid sequence of radioactive peptide**

| Cycle | Residue | pmol |
|---|---|---|
| 1 | Asp | 351 |
| 2 | Thr | 167 |
| 3 | X | |
| 4 | Pro | 23 |
| 5 | Leu | 9 |
| 6 | Ser | 11 |
| 7 | Glu | 6 |
| 8 | Glu | 6 |
| 9 | Thr | 3 |
| 10 | Trp | 2 |

Adapted with permission from Takata & Fujioka[18]. Copyright [1992] American Chemical Society.

**Table 3. Alignment of amino acid sequences of mammalian AdoMet-dependent methyltransferases**

| Enzyme[a] | Sequence | Position | Reference |
|---|---|---|---|
| Guanidinoacetate MT (rat liver) | HFDGILY | 127–133 | b |
| | LYDTYPL | 132–138 | |
| Hydroxyindole MT (bovine pineal gland) | FKDALPE | 238–244 | c |
| Phenylethanolamine MT (bovine adrenal medulla) | PADALVS | 174–180 | d |
| Glycine MT (rat liver) | GFDAVIC | 131–137 | e |
| D-Asp/L-isoAsp MT (bovine brain) | PYDAIHV | 151–157 | f |
| Catechol MT (rat liver) | DVDTLDM | 181–187 | g |
| Histamine MT (rat kidney) | KWDFIHM | 136–142 | h |

[a]MT, methyltransferase; [b]Ogawa, H., Date, T., Gomi, T., Konishi, K., Pitot, H.C., Cantoni, G. L. & Fujioka, M. (1988): *Proc. Natl. Acad. Sci.* USA 85, 694–698; [c]Ishida, I., Obinata, M. & Deguchi, T. (1987): *J. Biol. Chem.* 262, 2895–2899; [d]Baetge, E. E., Suh, Y. H. & Joh, T.H. (1986): *Proc. Natl. Acad. Sci.* USA. 83, 5454–5458; [e]Ogawa, H., Konishi, K., Takata, Y., Nakashima, H. & Fujioka, M. (1987): *Eur. J. Biochem.* 168, 141–151; [f]Henzel, W.J., Stults, J. T., Hsu, C.-A. & Aswad, D.W. (1989) *J. Biol. Chem.* 264, 15905–15911; [g]Salminen, M., Lundstrom, K., Tilgman, C., Savolainen, R. Kalkkinen, N. & Ulmanen, I. (1990): *Gene (Amst.)* 93, 241–247; [h]Takemura, M., Tanaka, T., Taguchi, Y., Imamura, I., Mizuguchi, H., Kurida, M., Fukui, H., Yamatodani, A. & Wada, H. (1992): *J. Biol. Chem.* 267, 15687–15691. Adapted with permission from Takata *et al.*[19]. Copyright [1994]. American Society for Biochemistry and Molecular Biology, Inc.

To locate the sites photolabelled with AdoMet on the polypeptide chain, GAMT that had been UV-irradiated with [*methyl*-³H]AdoMet for 60 min was digested with chymotrypsin, and radioactive peptides were isolated by reverse-phase HPLC. The effluent was monitored by absorbance at 220 nm and collected in 1-min fractions. About 41 per cent of the total radioactivity eluted was recovered in a single fraction collected between 40 and 41 min (Fig. 3). This fraction contained three small peaks of absorbance (a, b, and c), of which only peak b was radioactive. Rechromatography of peak b in a different solvent system gave a single radioactive peptide (Fig. 4). Automated Edman degradation indicated that this was the peptide comprising residues 134–143 with a modified Tyr136 (Table 2). In order to gain insight into the structure of the modified residue, the decapeptide was subjected to amino acid analysis by the phenylisothiocyanate precolumn derivatization method[1]. In addition to the phenylthiocarbamoyl derivatives of the component amino acids and methionine, a peak suggestive of the presence of adenine was found (Fig. 5). Thus, although the chemical nature is not known, it appears that the entire molecule of AdoMet is attached to Tyr136. The presence of the methionine

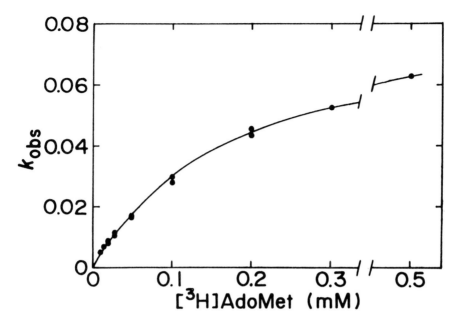

*Fig. 2. Concentration dependence of photolabelling of GAMT with [methyl-$^3$H]AdoMet. GAMT (35 μM) was exposed to UV light as described in Fig. 1 in the presence of various concentrations of [methyl-$^3$H]AdoMet for 30 min. The data were fitted to Eq. 1,*

$$k_{obs} = \frac{k}{1 + K_d / [\text{AdoMet}]} \tag{1}$$

*which was derived from Eq. 2.*

$$\text{E + AdoMet} \overset{K_d}{\longleftrightarrow} \text{E·AdoMet} \overset{k}{\longrightarrow} \text{X} \tag{2}$$

*k is expressed as mol of [$^3$H]AdoMet bound/mol enzyme in 30 min.*
*Reprinted with permission from Takata and Fujioka[18].*
*Copyright [1992] American Chemical Society.*

moiety of AdoMet in the photoadduct is also evident from the fact that the tyrosine is radiolabelled with [*carboxyl*-$^{14}$C]AdoMet (data not shown).

### Site-directed mutagenesis at the AdoMet-binding region

Most mammalian AdoMet-dependent methyltransferases possess in the middle portion of the sequence an aspartate residue that is preceded by a hydrophobic amino acid and followed successively by a small neutral and a hydrophobic amino acid. Rat GAMT has two such aspartates at positions 129 and 134 (Table 3). Since these residues are near the tyrosine that is photoaffinity-labelled with AdoMet, it is anticipated that either Asp129 or Asp134 is within the AdoMet-binding pocket of the enzyme. To probe the role of the conserved aspartate, we carried out amino acid replacement studies. Replacement of Asp129 with either an asparagine or alanine resulted in a functional enzyme, indicating that this residue is not involved in the catalytic activity of the enzyme. However, mutation of Asp134 produced profound effects on kinetic parameters (Table 4). Conversion of the residue to a glutamate decreased $k_{cat}$ about 3-fold and increased $K_m$ for AdoMet and guanidinoacetate 160- and 80-fold, respectively. Alteration to an asparagine resulted in further reduction in $k_{cat}$, but the $K_m$ values were not greatly different from those of D134E. (Mutant enzymes are designated by the one-letter symbol of the residue being changed, followed by the sequence number and the symbol of the substituted amino acid.) The D134A mutant had no catalytic activity and was not able to bind AdoMet as determined by the equilibrium dialysis method.

**Table 4. Apparent kinetic constants of wild-type and mutant GAMT**

| Enzyme | | AdoMet | | Guanidinoacetate | |
|---|---|---|---|---|---|
| | $k_{cat}$ (min$^{-1}$) | $k_m$ ($\mu$M) | $k_{cat}/K_m{}^a$ | $K_m$ ($\mu$M) | $k_{cat}/K_m{}^a$ |
| Wild Type | 4.86 ± 0.21 | 6.76 ± 0.59 | | 25.14 ± 2.32 | |
| D129N | 4.33 ± 0.16 | 4.93 ± 0.73 | 1.22 | 29.06 ± 1.30 | 0.77 |
| D129A | 2.71 ± 0.04 | 5.94 ± 0.26 | 0.63 | 14.72 ± 0.50 | 0.95 |
| Y133F | 3.92 ± 0.08 | 4.59 ± 0.31 | 1.19 | 12.58 ± 1.96 | 1.62 |
| D134E | 1.68 ± 0.05 | $1.10 ± 0.11 \times 10^3$ | $2.12 \times 10^{-3}$ | $1.94 ± 0.16 \times 10^3$ | $4.49 \times 10^{-3}$ |
| D135N | 0.04 ± 0.002 | $5.71 ± 0.69 \times 10^3$ | $9.74 \times 10^{-3}$ | $1.22 ± 0.11 \times 10^3$ | $1.70 \times 10^{-4}$ |
| T135A | 3.54 ± 0.04 | 5.41 ± 0.21 | 0.91 | 10.73 ± 1.17 | 1.71 |
| Y136F | 5.73 ± 0.06 | 6.47 ± 0.17 | 0.57 | 61.23 ± 0.39 | 0.49 |

[a]Relative to the wild type enzyme. Adapted with permission from Takata *et al.*[19]. Copyright [1994]. American Society for Biochemistry and Molecular Biology, Inc.

All mutant enzymes including the Asp134 mutants had UV absorption spectra and far- and near-UV CD spectra indistinguishable from those of wild-type. Fluorescence emission spectra ($\lambda$ max = 337 nm; excitation at 280 nm) were also identical except for D134A that had the same $\lambda$max but a 7.5 per cent decreased fluorescence intensity. Thus, the large impairments of activity of the Asp134 mutants are not the results of gross conformational changes, and it is suggested that Asp134 has a crucial role in catalysis and/or substrate binding. The fact that little changes in activity are observed when the residues surrounding the aspartate are mutated also indicate a specific role of Asp134.

The GAMT-catalyzed reaction follows an ordered Bi Bi mechanism and may be represented as,

$$E \xrightleftharpoons[k_2]{k_1[\text{AdoMet}]} \text{E–AdoMet} \xrightleftharpoons[k_4]{k_3\,[\text{GAA}]} \text{E} \cdot \text{AdoMet} \cdot \text{GAA} \xrightarrow{k_5} \text{E} \cdot \text{AdoHcy} \cdot \text{Cr} \xrightarrow{k_6} \text{E}$$

(3)

where GAA is guanidinoacetate and Cr is creatine. The steady-state kinetic parameters for the mechanism are expressed in terms of individual rate constants as,

$$K_{cat} = \frac{k_5\,k_6}{k_5 + k_6}$$

(4)

$$K_m{}^{\text{AdoMet}} = \frac{k_5 + k_6}{k_1\,(k_5 + k_6)}$$

(5)

$$K_m{}^{\text{GAA}} = \frac{k_6\,(k_4 + k_5)}{k_3\,(k_5 + k_6)}$$

(6)

Therefore, $K_{cat}/K_m^{\text{AdoMet}}$ represents the association rate constant of AdoMet and GAMT. Thus, only minor changes such as addition of one extra methylene and deprivation of negative charge at residue

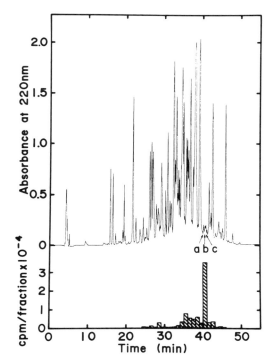

*Fig. 3. Isolation of chymotryptic peptide photolabelled with [methyl-³H]AdoMet. GAMT (42 μM) was exposed to UV light in the presence of 0.2 mM [methyl-³H]AdoMet (1.2 × 10⁵ cpm./nmol) for 60 min. The resulting enzyme was S-carboxymethylated with iodoacetate and digested with chymotrypsin. The digest was fractionated on a TSK ODS 120T column with a gradient of acetonitrile on 0.05 per cent trifluoroacetic acid. The concentration of acetonitrile was varied from 0 to 48 per cent over a period of 50 min starting at 10 min. The flow rate was 0.8 ml/min. The effluent was monitored by absorbance at 220 nm and collected in 1-min fractions. Aliquots from each fraction were assayed for radioactivity. The shaded bars indicate the total radioactivity found in each fraction. Reprinted with permission from Takata and Fujioka[18]. Copyright [1992] American Chemical Society.*

134 greatly reduce the enzyme's capacity to bind AdoMet. As calculated from the magnitude of the change in $k_{cat}/K_m$ values, a change of Asp134 to glutamate is accompanied by a loss of binding energy of about 4 kcal/mol, and a change to asparagine causes a loss of about 7 kcal/mol. GAMT would certainly provide multiple sites of interaction, and these values suggest that something more than merely a salt bridge or hydrogen bonding interaction is lost when Asp134 is mutated to glutamate or asparagine.

Amino acid replacements at residue 134 also reduced the $k_{cat}/K_m^{GAA}$ value greatly. Since guanidino acetate binds only to the E·AdoMet complex, the simplest explanation would be that improper binding of AdoMet in turn perturbs the geometry of guanidinoacetate binding site so as to cause weaker binding. However, since guanidinoacetate is the last substrate in an ordered mechanism, the situation is more complex. The expression for $k_{cat}/K_m^{GAA}$ includes the unimolecular rate constant for isomerization of the central complex ($k_5$) and, in the absence of the knowledge about the rate-limiting steps in the wild-type and mutant enzymes, it is not possible to determine whether the amino acid replacement interferes with binding of guanidinoacetate or whether it causes impairment of catalysis.

It has been recognized that most small molecule methyltransferases and protein carboxyl methyltransferases have common sequence motifs in addition to the one described above[3,7]. Kagan & Clarke[8] have recently extended this observation to a broader group of methyltransferases and other enzymes that bind AdoMet and/or AdoHcy (AdoMet synthetase, AdoMet decarboxylase, and AdoHcy hydrolase). They termed three regions of sequence similarity motifs I, II, and III, from the N-terminal side. Motifs I and III are found in 82 and 73 per cent, respectively, of the known non-DNA methyltransferase sequences, and motif II occurs in 55 per cent of the sequences. More than half of the enzymes have all three regions. These motifs are usually separated by comparable intervals on the polypeptide chain. GAMT has all three motifs, and the residues mutagenized in the present study are those present in motif II.

The crystal structures of a DNA methyltransferase, *Hha*I DNA methyltransferase[2], and a small molecule methyltransferase, catechol *O*-methyltransferase[20], have recently been described. Catechol *O*-methyltransferase has the three sequence motifs, and *Hha*I methylase has a sequence similar to motif I but not sequences corresponding to motifs II and III. The two enzymes, however, are shown

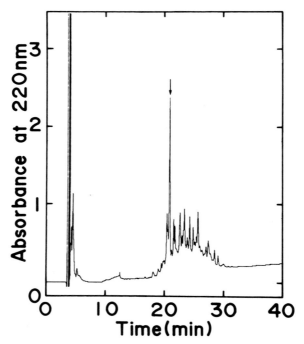

*Fig. 4. Rechromatography of peak b of Fig. 3.*
*The effluent corresponding to peak b was*
*rechromatographed on the same column with a gradient of*
*acetonitrile in 5 mM ammonium acetate (pH 6.8).*
*Acetonitrile gradient: 0–16 per cent between 10 and 15 min;*
*16–80 per cent between 15 and 60 min. The radioactive*
*peptide is indicated by an arrow. Reprinted with permission*
*from Takata and Fujioka[18]. Copyright [1992] American*
*Chemical Society.*

to fold into a typical α/β structure, and have a very similar AdoMet-binding motif that resemble the nucleotide-binding fold found in many dehydrogenases. Despite low degrees of sequence similarity, it is most likely that all methyltransferases have a common structural feature to bind AdoMet. In catechol *O*-methyltransferase Gly66, which is a conserved residue of motif I, participates in AdoMet binding through hydrogen bonding between the main-chain oxygen and the amino group of the methionine moiety of AdoMet. The region comprising motif II is distant from the bound AdoMet, and direct contacts of the residues contained therein with AdoMet seem impossible. In GAMT, however, data described in this article are consistent with the idea that motif II forms part of the AdoMet-binding site. Motifs I and II of catechol methyltransferase are separated by 62 residues, and intervened by several secondary structural elements. Although the two motifs of GAMT are similarly spaced, the amino acid sequence between them is greatly different. It is possible that in GAMT motif II is placed close to AdoMet and contributes to its binding. In addition to Asp134, a preliminary study has shown that Tyr136 also has a role in AdoMet binding. Whereas mutation of the tyrosine to a phenylalanine does not affect the binding of AdoMet (Table 4), conversion to a valine leads to a large decrease in affinity. Motif II is rich in hydrophobic, especially in aromatic residues[8]. AdoMet bound to GAMT may be stabilized through aromatic–aromatic interaction or sulfonium-π interaction with Tyr136[11]. It should be noted that conservative changes of 5 individual invariant residues of motif III of GAMT did not markedly alter $k_{cat}$ and $K_m$ values for both substrates[6].

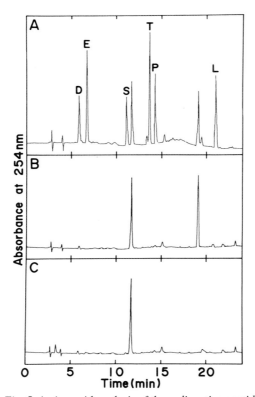

*Fig. 5. Amino acid analysis of the radioactive peptide from GAMT photolabelled with [methyl-$^3$H]AdoMet. (A), The radioactive peptide from the [$^3$H]AdoMet-modified enzyme was hydrolyzed in vacuo in the vapor phase of 5.7 M HCl containing 0.1 per cent phenol and 10 mM dithiothreitol for 24 h at 108 °C. The hydrolysate was treated with phenylisothiocyanate[1] and subjected to reverse phase HPLC on Develosil ODS-7 (0.46 × 25 cm) (Nomura Chemicals, Nagoya, Japan) with a linear gradient from 0.1 M ammonium acetate to 60 per cent acetonitrile (v/v) at a flow rate of 1.0 ml/min. The effluent was monitored by absorbance at 254 nm. AdoMet (B) and adenine (C) were treated similarly with HCl and phenylisothiocyanate and chromatographed under the same conditions. Reprinted with permission from Takata&Fuioka[18]. Copyright [1992] American Chemical Society.*

## References

1.   Bidlingmeyer, B.A., Cohen, S.A. & Tarvin, T.L. (1984): Rapid analysis of amino acids using pre-column derivatization. *J. Chromatogr.* **336**, 93–104.

2.   Cheng, X., Kumar, S., Posfai, J., Pflugrath, J.W. & Roberts, R.J. (1993): Crystal structure of the *Hha*I DNA methyltransferase complexed with *S*-adenosyl-L-methionine. *Cell* **74**, 299–307.

3.  Fujioka, M. (1992): Mammalian small molecule methyltransferases: their structural and functional features. *Int. J. Biochem.* **24,** 1917–1924.

4.  Fujioka, M., Konishi, K. & Takata, Y. (1988): Recombinant rat liver guanidinoacetate methyltransferase: reactivity and function of sulfhydryl groups. *Biochemistry* **27,** 7658–7664.

5.  Fujioka, M., Takata, Y. & Gomi, T. (1991): Recombinant rat guanidinoacetate methyltransferase: structure and function of the $NH_2$-terminal region as deduced by limited proteolysis. *Arch. Biochem. Biophys.* **285,** 181–186.

6.  Gomi, T., Tanihara, K., Date, T. & Fujioka, M. (1992): Rat guanidinoacetate methyltransferase: mutation of amino acids within a common sequence motif of mammalian methyltransferase does not affect catalytic activity but alters proteolytic susceptibility. *Int. J. Biochem.* **24,** 1639–1649.

7.  Ingrosso, D., Fowler, A.V., Bleibaum, J. & Clarke, S. (1989): Sequence of the D-aspartyl/L-isoaspartyl protein methyltransferase from human erythrocytes. Common sequence motifs for protein, DNA, RNA, and small molecule S-adenosylmethionine-dependent methyltransferases. *J. Biol. Chem.* **264,** 20131–20139.

8.  Kagan, R.M. & Clarke, S. (1994): Widespread occurrence of three sequence motifs in diverse S-adenosylmethionine-dependent methyltransferases suggests a common structure for these enzymes. *Arch. Biochem. Biophys.* **310,** 417–427.

9.  Konishi, K. & Fujioka, M.(1991): Reversible inactivation of recombinant rat liver guanidinoacetate methyltransferase by glutathione disulfide. *Arch. Biochem. Biophys.* **289,** 90–96.

10. Kunkel, T.A., Roberts, J.D. & Zakour, R.A. (1987): Rapid and efficient site-directed mutagenesis without phenotypic selection. *Methods Enzymol.* **154,** 367–382.

11. McCurdy, A., Jimenez, L., Stauffer, D.A. & Dougherty, D.A. (1992): Biomimetic catalysis of SN2 reactions through cation-π interactions. The role of polarizability in catalysis. *J. Am. Chem. Soc.* **114,** 10314–10321.

12. Mudd, S.H. & Poole, J.R. (1975): Labile methyl balances for normal humans on various dietary regimens. *Metabolism* **24,** 721–735.

13. Ogawa, H., Date, T., Gomi, T., Konishi, K., Pitot, H.C., Cantoni, G.L. & Fujioka, M. (1988): Molecular cloning, sequence analysis, and expression in *Escherichia coli* of the cDNA for guanidinoacetate methyltransferase from rat liver. *Proc. Natl. Acad. Sci. USA* **85,** 694–698.

14. Ogawa, H., Ishiguro, Y. & Fujioka, M. (1983): Guanidoacetate methyltransferase from rat liver: purification, properties, and evidence for the involvement of sulfhydryl groups for activity. *Arch. Biochem. Biophys.* **226,** 265–275.

15. Penefsky, H.S. (1979): A centrifuged-column procedure for the measurement of ligand binding by beef heart $F_1$. *Methods Enzymol.* **56,** 527–530.

16. Takata, Y., Date, T. & Fujioka, M. (1991): Rat liver guanidinoacetate methyltransferase. Proximity of cysteine residues at positions 15, 90, and 219 as revealed by site-directed mutagenesis and chemical modification. *Biochem. J.* **277,** 399–406.

17. Takata, Y. & Fujioka, M. (1990): Recombinant rat guanidinoacetate methyltransferase: study of the structure by trace labeling lysine residues with acetic anhydride. *Int. J. Biochem.* **22,** 1333–1339.

18. Takata, Y. & Fujioka, M. (1992): Identification of a tyrosine residue in rat guanidinoacetate methyltransferase that is photolabeled with S-adenosyl-L-methionine. *Biochemistry* **31,** 4369–4374.

19. Takata, Y., Konishi, K., Gomi, T. & Fujioka, M. (1994): Rat guanidinoacetate methyltransferase. Effect of site-directed alteration of an aspartic acid residue that is conserved across most mammalian S-adenosylmethionine-dependent methyltransferases. *J. Biol. Chem.* **269,** 5537–5542.

20. Vidgren, J., Svensson, L. A. & Liljas, A. (1994): Crystal structure of catechol O-methyltransferase. *Nature* **368,** 354–358.

*Section IV*

Physiological importance of creatine–creatinine, phosphocreatine and creatine phosphokinase

*Guanidino Compounds : 2*, eds. by P.P. De Deyn, B. Marescau, I.A. Qureshi and A. Mori.
©1997 John Libbey & Company Ltd., pp. 133–142.

# Chapter 16

# The role of creatine kinase in smooth muscle

Joseph F. CLARK

*Department of Biochemistry, University of Oxford, Oxford OX1 3QU, UK*

## Summary

Despite the energy flux being much lower in smooth than in striated muscle, creatine kinase (CK) has been found present and active in all smooth muscles studied to date. Like other tissues with dynamic energy demand, smooth muscle contains all of the components of the phosphocreatine/CK energy circuit. CK is localized to the contractile proteins, mitochondria, membrane pumps and cytoplasm. Brain CK homodimer has been identified as the major isoenzyme in smooth muscle tissue and is specifically bound to the contractile protiens. Mito-CK (found in the intermembrane space in the mitochondria) can produce PCr from nascent ATP and the resulting phosphocreatine (PCr) can supply the contractile protiens with energy. The CK system appears to be acting as an energy transducer at sites of ATP, production and utilization. During development or pathology, the CK system responds by changing its isoenzyme profile and specific activity. The conclusion to be made from these observations is that the CK system is intimately involved with energy metabolism in smooth muscle.

## Introduction

Like other tissues with dynamic energy demand, smooth muscle contains the components of the phosphocreatine/creatine kinase (CK) energy circuit. All smooth muscles studied to date have active mitochondrial CK (mito-CK) and cytosolic CK, along with phosphocreatine (PCr) and creatine (Cr). Brain CK homodimer (BB-CK) has been found to be selectively localized to contractile proteins with mito-CK in the mitochondria. Therefore, mito-CK can produce PCr from nascent ATP. Oxidative PCr is then used to rephosphorylate ADP at the contractile proteins. The resultant Cr diffuses back to the mitochondria and completes the circuit. In Fig. 1, we see how this arrangement can be used to shuttle high-energy phosphates between sites of energy production and energy utilization.

The role or function of the CK/PCr system in smooth muscle is far from clear. Along with functioning as an energy circuit, it may also function as an energetic signal or energy buffer. This review will address each of these three putative functions ascribed to the CK/PCr system as it relates to smooth muscle and the isoenzymes of CK that compose this system.

## CK isoenzymes

The dominant CK isoenzyme found in smooth muscle is BB-CK, so called because it was found in the brain and later in smooth muscle whereas the muscle homodimer (MM-CK) was found in skeletal muscle. BB-CK is sometimes referred to as the foetal form because it dominates in foetal tissues. It is found in all smooth muscles and is seen to increase in the heart and skeletal muscle in various

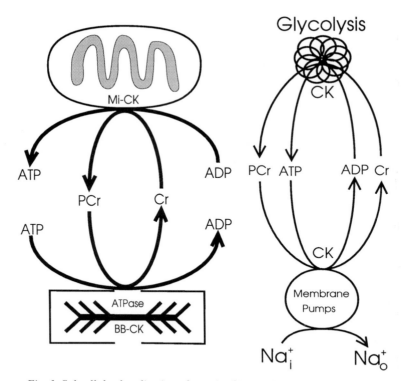

*Fig. 1. Subcellular localization of creatine kinase. Creatine kinase can be localized to discrete locations. These locations are the contractile proteins, mitochondria, glycolysis and membrane pumps. In this representation, we can see that sites of energy production (the mitochondria and glycolysis – at the top of the figure) will produce PCr and ATP. These high-energy phosphates will be used at sites of energy utilization. Because of functional coupling, between the mitochondria and the contractile proteins, and between glycolysis and membrane pumps, mixing between the two pathways is limited, but not excluded.*

pathological conditions. Clark *et al.* found an increase in BB-CK activity in the guinea-pig uterus during gestation[7]. There was also a striking seven-fold increase in mito-CK whereas oxidative metabolism increased by only three-fold. The authors inferred from these data an increase in mito-CK per mitochondrion. There is a two-fold increase in total CK activity during gestation, occurring concomitantly with a large increase in protein synthesis. Originally the drastic increase in synthesis of the mRNA of BB-CK was observed without knowing the function of the resulting protein/enzyme. The protein whose synthesis was induced during gestation was called the induced protein and Reiss & Kaye[25] showed that it was BB-CK.

In mature uterine and vascular smooth muscle,~23 per cent of the total CK activity is tightly bound to the contractile protein[8]. Furthermore it is BB-CK which is specifically bound. These results were obtained in Triton X-100 skinned smooth muscle. (Triton is a detergent which will solubilize all membranes and liberate the cytoplasmic contents.) The CK that remains after 1 h of Triton would have to be very tightly bound. This bound BB-CK, we will find later, is also capable of generating sufficient ATP from PCr and ADP to produce 60 per cent of maximal tension[6,7,9].

BB-CK, purified from porcine carotid artery, has a very low affinity for ATP ($K_m$ ~10 mM)[4]. The low affinity for ATP may be of physiological significance because it might aid in maintaining ATP at the contractile proteins. BB-CK could therefore be functioning as an energy transducer by using PCr to maintain ATP and thus supply (or buffer) the contractile proteins with the energy required for contraction.

So far, we have talked about mito-CK and BB-CK, but there are two other isoenzymes which have been reported in smooth muscle[6,7]. MM-CK and MB-CK may indeed have important functions in smooth muscle, but their subcellular localization or detailed function has not been reported. There are reports of CK activity associated with membranes and membrane pumps[19,28]. This subject will be discussed in greater detail below. These authors did not, however, determine the isoenzymes involved. Therefore it remains to be determined what role the other two isoenzymes play. The function of the soluble CK free in the cytoplasm is probably due to one or both of these isoenzymes.

In the heart, during pathology such as hypertrophy, there is a shift in the isoenzyme profiles to an increased percentage of BB-CK[16]. Despite BB-CK being the dominant isoenzyme in vascular smooth muscle, Boehm et al.[1] demonstrated a significant increase in the percentage of BB-CK in the aortas of hypertensive/hypertrophic rat and guinea-pig vessels (Fig. 2). The authors did not determine whether there was an alteration in the localization of CK, but they did speculate that increased BB-CK at the contractile proteins might be beneficial in maintaining ATP/ADP.

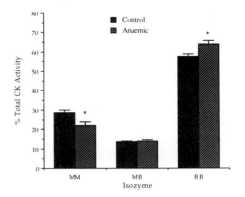

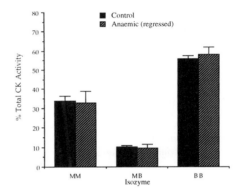

*Fig. 2. Creatine kinase isoenzymes during anaemia. The isoenzymes of creatine kinase in vascular smooth muscle undergo adaptation during anaemic hypoxia. This is seen as a shift in the isoenzyme profile towards the BB isoenzyme of creatine kinase. Interestingly, after regression of the anaemic hypoxia there is a return of the isoenzyme profile to normal.*

## Buffer function

One of the functions ascribed to the CK/PCr system is as a buffering system for ATP/ADP. A fall in ATP (or concomitant rise in ADP) leads to hydrolysis of PCr. A schematic representation of the buffering of ATP is seen in Fig. 3. Rapid hydrolysis of PCr helps to maintain the ATP concentration. This hypothesis requires: (1) the CK reaction to be rapid compared with the sum of the ATPases, (2) a reservoir of PCr to act as a high-energy phosphate supply, and (3) the energy 'stored' as PCr to be readily interconvertible to ATP. These conditions will be addressed below.

## CK rate and equilibrium

There is strong NMR evidence that the CK reaction is indeed at equilibrium in smooth muscle. Yoshizaki et al.[30], Fisher & Dillon[17], Clark & Dillon[5,11]. Magnetization exchange and other NMR

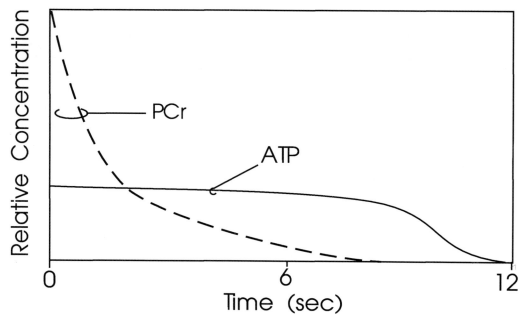

*Fig. 3. A 10 s sprint. The responses of PCr and ATP concentrations in skeletal muscle as seen during a 10 s sprint. The [PCr] falls rapidly to buffer the [ATP]. Therefore, during the first few seconds, there is a negligible fall in ATP. However, as PCr reaches low levels, the [ATP] begins to fall. By the end of a 10 s sprint essentially all of the PCr and ATP can be used up. Without the energy reservoir of PCr, however, all of the ATP could be depleted in the first few seconds. Therefore, PCr is buffering the [ATP] in the sprinting anaerobic muscle.*

experiments are capable of determining the equilibrium status through the CK reaction within functioning smooth muscle tissue (*ex vivo*)[12,14]. NMR experiments are not, however, capable of determining subcellular localization fluxes or specific isoenzymes involved.

Magnetization exchange experiments were performed on bullfrog stomach by Yoshizaki *et al.*[30]. They found that the flux through the CK reaction was 0.77 μmol/g wet weight and that this rate was 100 times faster than the ATPase rate. When they calculated the flux ratio of the forward and reverse reactions they obtained a value of 1.7. This value indicates a non-equilibrium situation because equilibrium would yield a value of 1. Nonetheless, they concluded that the CK reaction was in fact at equilibrium because the CK rate was much faster than the ATPase rate, and they attributed the discrepancy in the flux ratio to subcellular compartmentation of metabolites. Thus, the PCr and ATP that were visible in the NMR experiments were not freely accessible to all other metabolites or enzymes.

Clark & Dillon[5,11] also performed magnetization exchange experiments on porcine carotid artery. They concluded that the CK reaction was also at equilibrium. Unlike Yoshizaki *et al.*[30], though, they found the flux ratio to be consistent with equilibrium in the porcine carotid artery (a flux ratio of 1). Like Yoshizaki *et al.*[30], they concluded that the CK reaction was rapid compared with the ATPase rate, but they did not speculate on compartmentation in vascular smooth muscle. Recent NMR evidence[1a] and unpublished observations have shown that a discrete cytosolic component of CK may be compartmentalized in porcine carotid arteries. The compartmentalized CK was not able to phosphorylate the creatine analogue β-guanidinopropionic acid in the absence of oxidative ATP. Glycolytic metabolism was sufficient to maintain ATP levels similar to control levels as observed by NMR.

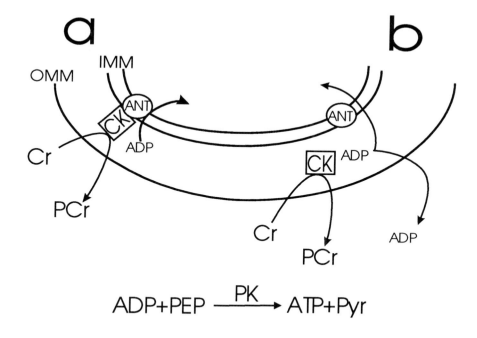

$$ADP+PEP \xrightarrow{\text{PK}} ATP+Pyr$$

*Fig. 4. Coupling of mito-CK to oxidative phosphorylation. This is a schematic representation of how mito-CK can be coupled to oxidative phosphorylation via restricted diffusion of ADP (a) or how ADP might be readily exchangeable with the cytosol (b). Pyruvate kinase (PK) and phosphoenol pyruvate form a competitive enzyme system for the ADP produced by mito-CK. If ADP has preferential access to the adenine nucleotide translocase (ANT) the stimulation of respiration will not be effected by the PK (b). If, however, ADP is freely able to interact with PK or ANT (b) there will be less functional coupling of mito-CK to oxidative phosphorylation.*

Fisher & Dillon[17] used NMR to directly determine free ADP in porcine carotid artery. Knowing total tissue creatine, and using NMR to determine ATP, ADP, PCr and pH, they were able to determine a $K_{eq}$ for CK. They then used this $K_{eq}$ to determine the ADP concentration under various conditions. Their success with this $K_{eq}$ required CK to be at equilibrium. They were indeed able to monitor the changes in [ADP] during periods of changing PCr and concluded that the CK reaction is rapid compared with the sum of the ATPases and that it is at equilibrium.

All these NMR studies support the CK reaction functioning in smooth muscle at or near equilibrium. These studies do not, however, differentiate between equilibrium by a rapid, reversible reaction or two unidirectional fluxes running at approximately the same rate. The scenario shown in Fig. 1. demonstrates how microcompartmented CK could be functioning with separate unidirectional synthesis or hydrolysis of PCr. With the different pools of CK operating in opposite directions, the net result seen in the NMR experiments would be to show CK to be at equilibrium. In fact the individual CK pools may be unidirectional. In either case, one would have a rapid interconversion of PCr to ATP.

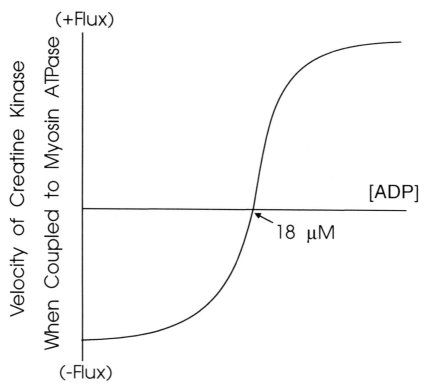

Fig. 5. A schematic representation of the kinetic response of creatine
kinase to increasing [ADP]. At low [ADP], the net flux is negative – or in the direction of
PCr synthesis. At high [ADP], the flux is in the direction of PCr hydrolysis. The x-inter-
cept occurs at about 18 μM and represents equilibrium.

### PCr as a reservoir

The capacity of PCr to act as a reservoir to buffer ATP is directly related to its concentration and synthesis rate. The greater the concentration, the more effective the reservoir. Vascular smooth muscle has a fairly low PCr concentration (compared with striated muscle), in the range of 0.5–0.8 mmol/kg. This is about 60 per cent of the ATP concentration, but in other smooth muscles it can be over twice the ATP concentration[13,30]. By comparison, in striated muscle its concentration is several times the ATP concentration. PCr is therefore a less effective energy reservoir in smooth muscle because of its feeble concentration. Its relative capacity as a reservoir, however, will depend on the smooth muscle type and PCr concentration.

### PCr and ATP interconversion

This is addressed somewhat above in the section CK rate and equilibrium. The interconversion of PCr is 100-fold more rapid than the action of ATPase[30] so it will not be rate limiting when supplying energy to buffer ATP. One of the ways this is accomplished is by substrate channelling and microcompartmentation of CK. The CK reaction is microcompartmented to sites of energy utilization, e.g. contractile proteins. It is also localized in such a way as to form an energetic microcompartment with myosin ATPase, myosin light chain kinase and bound nucleotide[8].

As an energy buffer, CK can help to maintain ATP in the region of the contractile proteins because of its subcellular localization. An equally important function, however, may be to prevent a build-up of ADP. Indeed, Nishiye et al.[23] and Fuglsang et al.[18] have shown that micromolar ADP will significantly inhibit relaxation of smooth muscle. This inhibition was much more pronounced in vascular smooth muscle. Rephosphorylation of ADP by CK is therefore likely to aid in the normal relaxation of vascular smooth muscle.

Butler et al.[2,3] found that there was ADP bound to smooth muscle myosin. They estimated that 60–70 μM of radiolabelled ADP was bound to myosin in their experiments. This bound ADP was released with two time constants. They speculated that the two time constants represented two populations of ADP, being released from two populations of cycling cross-bridges. They did not speculate on the energetic availability of this bound nucleotide.

## Transducer/messenger function

One of the hypothesized roles of the CK system is as an intracellular messenger or energy transducer. The two are inexorably related in that as an energy transducer it can convert PCr to ATP at the contractile proteins. Creatine is produced in this process, and creatine may act as a signal at the mitochondria to increase respiration. One common thread for these two is that CK is localized to the sites of energy production and energy consumption. Below I will present evidence that CK is localized to the contractile proteins, membrane pumps, mitochondria and glycolysis (see Fig. 1).

The ability of mito-CK to act as a signalling mechanism to modulate respiration is well established in the heart[16,26]. Experiments using isolated uterine mitochondria have shown mito-CK to be present and active in the mitochondria but that the CK reaction may not be functionally coupled to oxidative phosphorylation[7,10]. Mito-CK from the guinea-pig uterus is not capable of stimulating respiration under conditions where other enzymes are competing for the ADP produced[7,10]. Mito-CK is active within the uterine mitochondria but the ADP produced does not have preferential access to the adenine nucleotide translocase. ADP will, however, mix readily with the cytoplasmic contents because there is negligible restriction of diffusion (Fig. 4). Restricted diffusion of ADP in heart mitochondria has been described by Saks et al.[26], but it is conspicuously absent in the uterine mitochondria.

At first glance, control and synthesis of PCr appear to be of uncertain origin. It is clear that in heart much control is effected by mito-CK[16,26]. This is done via functional coupling of mitochondrial CK to oxidative phosphorylation with a unidirectional flux through mito-CK for net synthesis of PCr[26] and is represented by scheme a in Fig. 4. We have shown, however, that mito-CK is not functionally coupled to oxidative phosphorylation in uterine smooth muscle (Fig. 4b). This, however, does not prevent mito-CK from playing a part in [PCr] regulation in smooth muscle. Indeed, we have shown a striking decrease in NMR-visible PCr in vascular smooth muscle with cyanide inhibition[1a]. This loss of PCr is maintained for over 10 h with only partial return of PCr and no significant decrease in ATP. A rapid CK reaction, at equilibrium (observable PCr did not return), would allow PCr to return to equilibrium levels. Because this was not observed, the authors concluded that PCr was synthesized by the mitochondria. This is also consistent with mito-CK being the starting point of the PCr circuit[28,29]. Furthermore, the authors concluded that the [ATP] was controlled by glycolysis. Taken together, we believe that these data support a mitochondrial role in controlling porcine carotid artery PCr concentration.

## Compartmentation of creatine kinase

Microcompartmentation of CK, and its isoenzymes, occurs in smooth muscle. Smooth muscle is a tissue that has been well studied to characterize the microcompartmentation of its energetic metabolism. There is limited mixing of the energetic substrates with other energetic enzymes[22,24]. For example, exogenous glucose given to vascular smooth muscle is preferentially converted to lactate even while oxidative metabolism is functioning normally. Oxidative metabolism is supplied with glycosyl units passing through glycolysis but originating from glycogen. Exogenous glucose will pass directly through glycolysis but ends up being converted to lactate. The mixing of these two glycolytic processes appears to be quite limited[22,24]. Membrane-bound CK supplies ATP to $Ca^{+2}$ ATPase. This ATP has preferential access to the $Ca^{+2}$ ATPase over exogenous ATP[19]. This is complemented by glycolytic ATP present on the plasma membrane vesicles. These data are consistent with coupling between CK and the glycolytic enzyme pyruvate kinase[5].

Localization of CK to the contractile proteins of smooth muscle, as well as striated muscle, has been reported by several groups[6,7,9,21,28,29]. In smooth muscle ~23 per cent of the total CK activity is bound to the contractile proteins in the form of BB-CK[7,8]. Bound BB-CK is capable of producing enough ATP to enable the contractile proteins to produce and maintain submaximal tension. In addition, the bound CK can produce sufficient ATP to significantly increase the rate of relaxation from high-tension rigor conditions. The BB-CK, bound to the contractile proteins, had an affinity for PCr such that at concentrations between 0.5 and 12 mM there was no change in the tension generated by the tissue[6]. The importance of PCr in smooth muscle contractility has been addressed by Scott et al.[27].

CK is specifically localized to the contractile proteins, membrane pumps, mitochondria and glycolysis. These energy producing and consuming sites are functionally linked via PCr and creatine. It can be concluded that the CK/PCr system is an essential component of smooth muscle energetics (Fig. 1).

## Creatine kinase in pathology

We have seen that in smooth muscle the isoenzymes of CK are dominated by BB-CK. In pathological conditions such as hypertrophy and hypertension, vascular smooth muscle responds by increasing the percentage of BB-CK. These changes consistently occur when the tissue is becoming abnormal. We have seen, for example, that chronic anaemic hypertrophy leads to a 16 per cent increase in BB-CK.

When anaemic rats are supplied with a normal iron diet, the cardiac hypertrophy reverses (as do many biochemical changes). In the smooth muscle, there is also a regression of the biochemical changes. The CK isoenzymes of the aorta return to normal levels after 6 weeks of a normal iron diet. The ability of the vascular smooth muscle CK system to respond to pathology and eventually return to normal levels bodes well for the clinician (as well as the scientist) who is trying to reverse a disease process in patients. The ability of the smooth muscle energetic system to adapt to metabolic signals should be interpreted as evidence of the important role the CK/PCr system plays in smooth muscle energetics.

## What is the benefit of a BB-CK shift?

That is a question which is very germane to the energetic system. One may speculate that, because BB-CK is localized to the contractile proteins, an increase in BB-CK could aid in maintaining normal contractility. Indeed, BB-CK is strongly coupled to the myosin ATPase[6,7,10]. Vascular smooth muscle relaxation can be drastically slowed by ADP[23]. A functioning CK system could, therefore, permit the vascular smooth muscle to relax normally during the increased contractile work that occurs during hypertrophy. More support for CK working to supply energy to the contractile proteins comes from the observation that BB-CK has a low affinity for ATP. This results in ATP being readily released from CK after it is formed, with the result of a unidirectional flux towards ATP production. CK therefore has a dual role in smooth muscle contractile function. First, it acts as an ATP buffer, producing ATP for the myosin ATPase. The second function is complementary in that it also acts to prevent a build-up of ADP.

Clark et al. (unpublished data) used the kinetic determinations of Jacobs and Kuby[20] to examine the effect of a changing [ADP] on CK flux when coupled to myosin ATPase (Fig. 5). Using literature values for vascular smooth muscle they found that if [ADP] were lowered to 1.3 µM (this value was chosen because it is the $K_d$ reported by Nishiye et al.[23]) there would be a rapid flux in the direction of PCr synthesis. Thus CK would be in competition with myosin ATPase for ATP. These two enzymes would quickly reach a steady state at an elevated [ADP]. Clark *et al.* (unpublished data) concluded that 18 µM ADP is a reasonable concentration for free ADP in vascular smooth muscle where CK and myosin ATPase function effectively. The impact of the relative sensitivity of vascular smooth muscle relaxation to ADP highlights the importance of the CK system in this tissue.

## Conclusion

The smooth muscle CK/PCr system is an important component in smooth muscle energy metabolism. PCr and Cr act as energy donor and energy acceptor because CK is coupled to energy consuming and producing processes. This importance is further exemplified by the changes seen in the CK isoenzyme profile in various patholgical conditions. There exists, therefore, a complete PCr/Cr energy circuit in smooth muscle. PCr is synthesized in the mitochondria and then used to phosphorylate ADP at the contractile proteins. The circuit is completed when Cr returns to the mitochondria and is rephospho-rylated again.

### Acknowledgements

I would like to thank E.A. Boehm for proof-reading this manuscript and for supplying the data on CK isoenzymes. Special thanks to Dr M.L. Field for his unique insight and discussions. Parts of this work have been supported by Searle, The Medical Research Foundation, England and Knoll Pharmaceuticals. The author is a Junior Research Fellow of Linacre College. Lastly, thanks to the members of the Clinical and Biochemical Magnetic Resonance Spectroscopy Unit, Oxford, England, for supporting me during the preparation of this manuscript.

### References

1.   Boehm, E.A., Clark, J.F. & Radda, G.K. (1992): Creatine kinase isoenzyme changes in the vascular smooth muscle during hypertension. *J. Mol. Cell Cardiol.* **24**, S108.

1a.  Boehm, E.A., Clark, J.F. & Radda, G.K. (1995): Metabolite utilization and possible compartmentation in the porcine carotid artery: a study using β-Guanidinopropionic acid. *Am. J. Physiol.* **37**, C628–C635.

2.   Butler, T.M., Pacifico, D.S. & Siegman, M.J. (1989): ADP release from myosin in permeabilized smooth muscle. *Am J. Physiol.* **256**, C59–C66.

3.   Butler, T.M., Siegman, M.J., Moors, S.U. & Narayan, S.R. (1990): Myosin product complex in the resting state and during relaxation of smooth muscle. *Am. J. Physiol.* **258**, C1092–C1099.

4.   Clark, J.F. (1990): Kinetics of porcine carotid artery brain isoform creatine kinase *in situ* and *in vitro*. Thesis, Michigan State University.

5.   Clark, J.F. & Dillon, P.F. (1990): Porcine carotid artery creatine kinase kinetics using NMR saturation transfer. *Biophys. J.* **57**, 156a.

6.   Clark, J.F., Khuchua, Z. & Ventura-Clapier, R. (1992): Creatine kinase binding and possible role in chemically skinned guinea-pig taenia coli. *Biochim. Biophys. Acta* **1100**, 137–145.

7.   Clark, J.F., Khuchua, Z., Kuznetsov, A.V., Saks, V.A. & Ventura-Clapier, R. (1993): Compartmentation of creatine kinase isoenzymes in myometrium of gravid guinea-pig. London: *J. Physiol.* **466**, 553–573.

8.   Clark, J.F. (1994): The creatine kinase system in smooth muscle. *Mol. Cell. Biochem.* **133/134**, 221–232.

9.   Clark, J.F., Khuchua, Z., Boehm, E. & Ventura-Clapier, R. (1994): Creatine kinase activity associated with the contractile proteins of the guinea pig carotid artery. *J. Musc. Res. Cell Motil.* **15**, 432–439.

10.  Clark, J.F., Kuznetsov, A.V., Khuchua, Z., Veksler, V.I., Ventura-Clapier, R. & Saks, V.A. (1994): Creatine kinase function in mitochondria isolated from gravid and non-gravid guinea pig uteri. *FEBS Lett.* **347**, 147–151.

11.  Clark, J.F. & Dillon, P.F. (1995): Phosphocreatine and creatine kinase in energetic metabolism of the porcine carotid artery. *J. Vasc. Res.* **32**, 24–30.

12.  Dawson, M.J. & Wray, S. (1983): [31]P NMR studies of isolated rat uterus. *J. Physiol. (London)* **336**, 19P–20P.

13.  Dawson, M.J. & Wray, S. (1985): The effects of pregnancy and parturition on phosphorus metabolites in rat uterus studied by [31]P NMR. *J. Physiol.* **368**, 19–31.

14.  Degani, H., Shaer, A., Victor, R.A. & Kaye, A.M. (1984): Estrogen induced changes in high energy phosphate metabolism in rat uterus: [31]P NMR studies. *Biochemistry* **23**, 2572–2577.

15. Dillon, P.F. & Clark, J.F. (1990): The theory of diazymes and functional coupling of pyruvate kinase and creatine kinase. *J. Theor. Biol.* **143**, 275–284.

16. Field, M.L., Clark, J.F., Henderson, C., Seymour, A.-M. & Radda, G.K. (1994): Alterations in the myocardial creatine kinase system during chronic anaemic hypoxia. *Cardiovasc. Res.* **28**, 86–91.

17. Fisher, M.J. & Dillon, P.F. (1988): Direct determination of ADP in hypoxic porcine carotid artery using $^{31}$P NMR. *NMR Biomed.* **1**, 121–126.

18. Fuglsang, A., Gromov, A., Torok, K., Somlyo, A.V. & Somlyo, A.P. (1993): Flash photolysis studies of relaxation and cross-bridge detachment: higher sensitivity of tonic than phasic smooth muscle to MgADP. *J. Musc. Res. Cell Motil.* **14**, 666–673.

19. Hardin, C.D., Raeymaekers, L. & Paul, R.J. (1992): Comparison of endogenous and exogenous sources of ATP in fueling $Ca^{2+}$ uptake in smooth muscle plasma membrane vesicles. *J. Gen Physiol.* **99**, 21–40.

20. Jacobs, H.L. & Kuby, S.A. (1980): Studies on muscular dystrophy. *J. Biol. Chem.* **255**, 8477–8482.

21. Ishida, Y., Wyss, M., Hemmer, W. & Wallimann, T. (1991): Identification of creatine kinase isoenzymes in the guinea-pig: presence of mitochondrial creatine kinase in smooth muscle. *FEBS Lett.* **283**, 37–43.

22. Lynch, R.M. & Paul, R.J. (1987): Compartmentation of carbohydrate metabolism in vascular smooth muscle. *Am. J. Physiol.* **252**, C328–C334.

23. Nishiye, E., Somlyo, A.V., Torok, K. & Somlyo, A.P. (1993): The effects of MgATP on cross-bridge kinetics: a laser flash photolysis study of guinea pig smooth muscle. *J. Physiol. (London)* **460**, 247–271.

24. Paul, R.J. (1983): Functional compartmentalization of oxidative and glycolytic metabolism in vascular smooth muscle. *Am. J. Physiol.* **244**, C399–C409.

25. Reiss, N.A. & Kaye, A.M. (1981): Identification of the major component of the estrogen induced protein of rat uterus as the BB isoenzyme of creatine kinase. *J. Biol. Chem.* **256**, 5741–5749.

26. Saks, V.A., Belikova, Y.O. & Kuznetsov, A.V. (1991): *In vivo* regulation of mitochondrial respiration in cardiomyocytes: specific restrictions for intracellular diffusion of ADP. *Biochim. Biophys. Acta* **1074**, 302–311.

27. Scott, D.P., Davidheiser, S. & Coburn, R.F. (1987): Effects of phosphocreatine on force and metabolism in rabbit aorta. *Am. J. Physiol.* **253**, H461–H465.

28. Wallimann, T., Wyss, M., Brdiczka, D., Nicolay, K. & Eppenberger, H.M. (1992): Intracellular compartmentation, structure and function of creatine kinase isoenzymes in tissues with high and fluctuating energy demands: the phosphocreatine circuit for cellular energy homeostasis. *Biochem. J.* **281**, 21–40.

29. Wyss, M., Smeitink, J., Wevers, R.A. & Wallimann, T. (1992): Mitochondrial creatine kinase: a key enzyme of aerobic energy metabolism. *Biochim. Biophys. Acta* **1102**, 119–166.

30. Yoshizaki, K., Radda, G.K., Inubushi, T. & Chance, B. (1987): $^{1}$H and $^{31}$P-NMR studies on smooth muscle of bullfrog stomach. *Biochim. Biophys. Acta* **928**, 26–44.

*Guanidino Compounds : 2*, eds. by P.P. De Deyn, B. Marescau, I.A. Qureshi and A. Mori.
©1997 John Libbey & Company Ltd., pp. 143–155.

# Chapter 17

# Creatine, phosphocreatine and β-guanidinopropionic acid: their uses in biochemical and medical research

Joseph F. CLARK and Mark L. FIELD

*MRC Biochemical and Magnetic Resonance Spectroscopy Unit, Department of Biochemistry, South Parks Road, Oxford, OX1 3QU, UK*

## Summary

Creatine and β-guanidinopropionic acid have been used to manipulate the cytosolic phosphocreatine concentration and thus the kinetic and thermodynamic properties of the cytosolic high-energy phosphate pool in a variety of tissue types. Relevant methods pertaining to the study of creatine and its analogues are described with particular emphasis on the acute and chronic cellular loading protocols. The efficacy of oral supplementation with creatine as a performance enhancing modality is discussed together with recent laboratory work on the cellular control of creatine uptake. Phosphocreatine can be administered intravenously to decrease muscle membrane damage and to decrease loss of cytoplasmic contents. Indeed, phosphocreatine has been used as a therapeutic agent for exercise enhancement and cardioplegia. We present evidence that dietary β-guanidinopropionic acid has effects on the anatomy and biochemistry of the rat heart. Despite a significant elevation in the aerobic capacity of skeletal muscle after dietary β-guanidinopropionic acid the creatine kinase isozyme profile remains unaltered. We conclude that loading tissues with creatine and its analogues can be useful in dissecting the role of the CK system. These compounds can therefore be considered as very useful scientific and clinical tools provided care is taken to document and control the collateral consequences of the loading protocols.

## Introduction

Creatine kinase (CK) is a key enzyme in the spatial and temporal buffering of ATP in many tissues of mesodermal and ectodermal origin. The traditional role of CK as an enzyme that catalyses the transfer of high-energy phosphate groups between phosphocreatine (PCr) and ADP is shown in Equation 1:

$$PCr^{2-} + MgADP^- + H^+ \Longleftrightarrow MgATP^{2-} + Cr \qquad (1)$$

The functional dependence of cellular energy transduction on CK has been a source of many conflicting and controversial publications[4,20,31]. The energy transducing role of CK is more completely discussed in Chapter 16 by J.F. Clark.

In brief, isozymes of CK (MM, MB, BB and mito-CK) have been found localized with a number of subcellular structures, the relative distribution and presence depending on muscle type[39]. Localization of mito-CK on the outside of the inner mitochondrial membrane has led to the suggestion that the enzyme may have a role amplifying the effective local mitochondrial [ADP] and thus reducing the

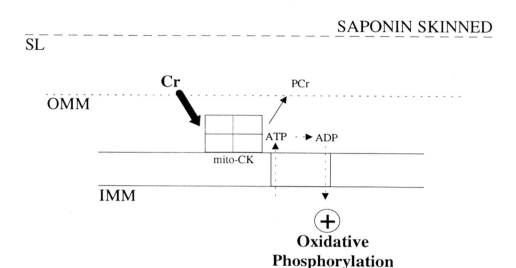

Fig. 1. The mechanism of creatine-stimulated respiration in saponin-skinned cardiac fibres. In this representation, mito-CK is situated near the adenine nucleotide translocase such that the ADP produced by the CK reaction readily returns to the mitochondrial matrix. In this way, oxidative ATP is converted to ADP in the presence of creatine. SL, sarcolemma; OMM, outer mitochondrial membrane; IMM, inner mitochondrial membrane.

$K_m$ of oxidative phosphorylation with respect to [ADP][36]. The suggestion is based on experiments in cardiac fibres where the plasma membrane has been removed by saponin treatment (Fig. 1). These experiments demonstrate an increase in respiratory rate in the presence of creatine (Cr). Because creatine is capable of pushing the CK reaction towards production of ADP (Equation 1), this phenomenon was termed *creatine-stimulated respiration*[12,21,22,36]. The proximity of the mito-CK and the adenine nucleotide translocase allows ADP to drive ATP production via a stimulation of oxidative phosphorylation. Interestingly, Field *et al.*[12] have shown that an increase in mito-CK per mitochondrion is associated with an elevation in creatine-stimulated respiration. Taken together with data from Khuchua *et al.*[21,22] reporting a decrease in mito-CK associated with several cardiac myopathies, this suggests that the enzyme may have a central role in the adaptive and/or maladaptive process of hypertrophy and heart failure.

Localization of CK with the $Ca^{2+}$ ATPases, $Na^+/K^+$ ATPases and myosin ATPases has led to the suggestion that the enzyme regulates the local ATP/ADP ratio and therefore the thermodynamic efficiency of ATP hydrolysis. Figure 2 shows the proposed mechanism with the sarcoplasmicreticular $Ca^{2+}$ ATPase in heart[23]. Functional coupling between the $Ca^{2+}$ ATPase and co-localized CK has been demonstrated by Levitsky *et al.*[23]. CK can also be localized to the contractile proteins of heart as well as smooth muscle. Indeed, Ventura-Clapier *et al.*[37] have shown that Triton X–100 skinned cardiac fibres have CK localized with the myosin ATPase with sufficient functional activity to generate enough ATP from ADP to provide for normal contractile capacity. Although there is widespread agreement on the existence of such mechanisms measured *in vitro*, the exact functional importance

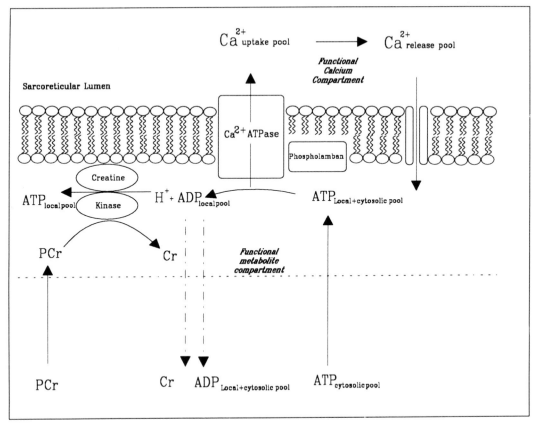

*Fig. 2. The mechanism whereby CK localized at the sarcoplasmic reticulum may regulate the local ATP/ADP ratio. In this way phosphocreatine can become an important energy source to the Ca$^{2+}$ ATPase of the sarcoplasmic reticulum.*

*in vivo* remains unclear. The relative importance of each of the reported roles of CK might therefore be expected to be dependent on ATP turnover, tissue type and the presence of pathology.

Creatine is synthesized *de novo* in pancreas, liver and kidney, the exact mechanisms being species dependent[38]. It is interesting to note that creatine is synthesized in cells which are not able to phosphorylate it (e.g. cells lacking CK). Walker[38] has speculated that the separation of synthesis and utilization allows the process to be regulated more easily. By having the alimentary blood supply passing via the hepatic portal vein to the liver, the liver may act as a regulator of creatine released into the central hepatic vein and hence the systemic blood supply. Walker has suggested that the arrangement also allows further levels of regulation by the Na$^+$- linked creatine transporter and kidney excretion. Creatine is metabolized to creatinine non-enzymatically and released into the blood for excretion by the kidney at a rate of about 2 per cent per day. The $K_m$ of the transporter is in the range of serum creatine concentration (50 μM)[27].

To increase or decrease the efficacy of the phosphocreatine circuit it is possible to manipulate the intracellular activity of CK by either (a) loading the cell with creatine or (b) loading the cell with CK 'antagonists' such as cyclocreatine or β-guanidinopropionic acid (β-GPA). Such manipulations are done to try to alter the cellular energy metabolism in tissues that utilize the CK energy circuit. Manipulation of the phosphocreatine circuit in this manner has been used to examine the functional importance of the CK system in different muscle types (Fig. 3). Because the relative functional importance of CK is tissue dependant, the efficacy of cyclocreatine, creatine and β-GPA as agonists and antagonists of the phosphocreatine circuit may vary with the tissue studied.

The objective of this chapter is to discuss the utility of creatine, phosphocreatine and β-GPA as research tools to manipulate the energetic systems associated with CK. We will also address some of the possible clinical applications of oral creatine and intravenous phosphocreatine.

## Methods and results

### Loading protocols for creatine, cyclocreatine and β-GPA

#### β-GPA loading protocols

The most commonly used method of loading tissue with β-GPA in laboratory rats has been by feeding a 1–2 per cent diet (w/w with laboratory chow) for a period of 5–19 weeks[2,10,13,20,24,31,40]. In one study, pregnant female rats were fed 0.5, 1.0 and 2.0 per cent β-GPA from the 16th day of gestation until their litters were weaned at 21 days[30]. Similar protocols have been adopted with mice[1,35]. Figure 4 shows the [31]P MRS spectrum of an isolated Langendorff perfused rat heart taken from a rat fed a 1 per cent β-GPA diet for 5–6 weeks. Phosphorylated β-GPA appears as a peak 0.44 ppm upfield from the phosphocreatine peak. Phosphocreatine may be depleted by as much as 90 per cent with little change in the cytosolic ATP. A similar spectral profile is seen in skeletal muscle with regard to PCr depletion; however, this is usually accompanied by a 50 per cent depletion in ATP. Pulmonary vascular smooth muscle has also been found to have an almost complete depletion of phosphocreatine as a result of β-GPA feeding[3].

Recently, several studies have shown that it is possible to load certain tissues with β-GPA over an acute time course. Unitt *et al.*[34] have shown that supplementation of the Krebs–Henseleit buffer with 150 mM β-GPA leads to a linear accumulation of phosphorylated β-GPA (Fig. 5) over 2 h. Boehm *et al.*[5] showed that the accumulation of Pβ-GPA into porcine carotid artery was substrate dependent. In this study glucose as substrate produced a more rapid rate of phosphorylation of β-GPA to Pβ-GPA than pyruvate (Fig. 6). More recent work[8] has shown a similar loading of creatine into porcine carotids over a period of 12 h.

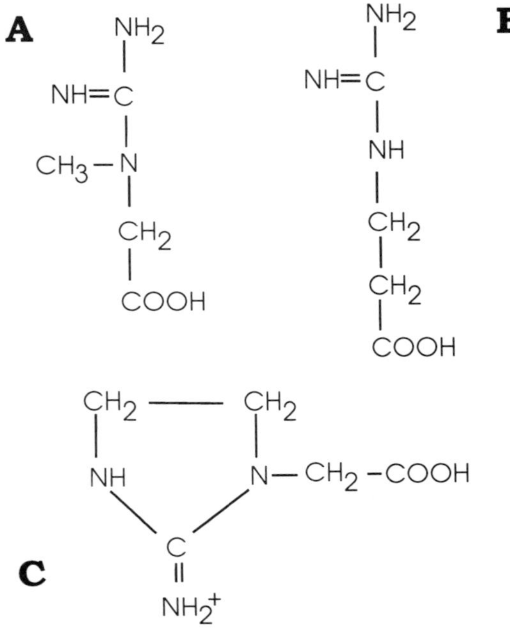

#### Creatine loading protocols

Human dietary creatine supplementation has received much attention recently because of its reported efficacy as a performance-enhancing compound. Harris *et al.*[19] have found that a dosage regimen of 4–5 g/day elevated skeletal muscle creatine and PCr levels in human subjects. They concluded that there is a maximal level of total creatine content in skeletal muscle (150 mmol/kg) and that dietary supplementation could not surpass this level. Indeed, some experimental subjects showed no increase in muscle creatine if their basal creatine levels were close to this apparent maxi-

*Fig. 3. The structures of creatine (A), β-GPA (B) and cyclocreatine (C).*

mum. In the subjects that showed increased creatine levels, there was an enhancement in anaerobic exercise capacity[17]. Acute loading of superfused porcine carotids with creatine results in significant loading over 12 h with an elevation in creatine and phosphocreatine by over 100 per cent with little change in ATP[8].

## Cyclocreatine loading protocols

Walker[38] has reported that feeding rats a 1 per cent (w/w) diet of cyclocreatine results in accumulation of phosphorylated cyclocreatine and a decrease in phosphocreatine in cardiac and skeletal muscle. The percentage of the phosphorylated form is significantly higher than that of phosphocreatine and phosphorylated β-GPA, which is approximately 40–60 per cent depending on substrate. Figure 7 shows the large cyclocreatine phosphate peak just upfield of the phosphocreatine peak. This spectrum was obtained by incubating porcine carotid arteries with 50 mM cyclocreatine for 12 h at 22 °C.

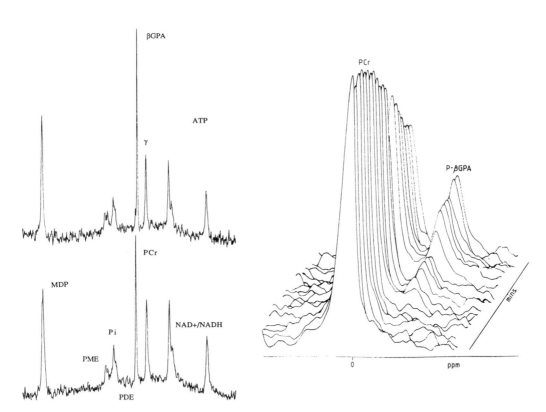

*Fig. 4. A $^{31}P$ MRS spectrum of an isolated Langendorff perfused rat heart loaded with β-GPA (upper) compared with its control (lower). β-GPA, phosphorylated β-GPA; Pi, inorganic phosphate; PME, phosphomonoesters; PDE, phosphodiesters; MDP, methylene diphosphonate. Each spectrum is the result of additions from five hearts, each scanned 64 times at a 20 s receiver delay and 90° pulse angle.*

*Fig. 5. $^{31}P$ NMR spectrum of an isolated Langendorff perfused rat heart perfused with 150 mM β-GPA in Krebs–Henseleit buffer for 2 h. (With kind permission of Unitt et al.)[34].*

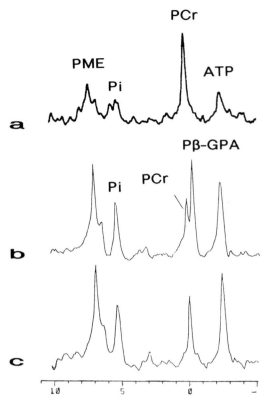

*Fig. 6. A stacked plot of three $^{31}P$ NMR spectra from porcine carotid arteries. In spectrum (a) the phosphocreatine concentration is increased by about four-fold by incubating the carotids with 50 mM creatine for 12 h. In spectrum (b) porcine carotid arteries were exposed to 50 mM β-GPA for 12 h. Note that the Pβ-GPA peak surpasses that of the PCr peak on the up-field side. Spectrum (c) is a control spectrum obtained before β-GPA perfusion.*

## Tissue extraction of guanidino compounds

Tissue (50 mg) is weighed into a frozen tube (–70 °C) onto which is placed 1 ml of 6 per cent perchloric acid (PCA) (4 °C)[29]. The tissue is then homogenized with a polytron (4 °C) and centrifuged for 2 min in a bench Eppendorff microfuge (4 °C). To 400 μl is added 600 μl of 50/50 (v/v) 1,1,2-trichloro-trifluoroethane and tri-n-octylamine (4 °C). After vigorous mixing the solution is left on ice for 30 min. A 2 min centrifugation results in three immiscible layers. The uppermost layer, which is the PCA-free aqueous extract containing the guanidino compounds, was stored at -70 °C until analysis.

## Assays of creatine, cyclocreatine and β-GPA

*Biochemical*

A colorimetric assay for guanidino groups has been described by Bonas *et al.*[6] and based on the Sakaguchi reaction: 0.1 ml of tissue extract is diluted with 0.4 ml of water (4 °C) and added to 0.1 ml 10 per cent NaOH (w/v) in a cuvette (4 °C). This is followed by addition of 0.25 ml of a 50/50 (v/v) solution of thymine (2 per cent w/v in 10 per cent NaOH) and α-napthol solution (40 mg/100 ml 95 per cent ethanol). Addition of 0.1 ml 20 per cent NaOCl (v/v) is followed by an incubation period of 1 min and addition of thiosulphate (2 per cent weight per volume of solution). After a 30 min incubation at 4 °C the samples are read at 515 nm against a standard curve.

Walker[38] has described an assay for cyclo-creatine which involves adding an aged solution of $Na_3[Fe(CN)_5NH_3]$ at alkaline pH (0.03M $Na_2CO_3$) to give a blue complex with a molar extinction coefficient of 4900 M$^{-1}$cm$^{-1}$ at 605 nm.

*High-pressure liquid chromatography*

Phosphocreatine, creatine and β-GPA can be determined by anion-exchange HPLC using a modified method of Sharps and McCarl[29]. A Hichrom Hypersil 3APSI weak anion exchange column is used. The sample application buffer consists of 5 mM $KH_2PO_4$ (pH 2.85) and the gradient buffer of 0.4M $KH_2PO_4$ in 0.4M KCl (pH 4.00). The elution time is set to 13 min with a flow rate of 2 ml/min and is monitored by a variable-wavelength detector set to 210 nm.

*Magnetic resonance spectroscopy*

Creatine, cyclocreatine and β-GPA may be quantified *in vivo* or *in vitro* by [1]H MRS, and their phosphorylated forms by [31]P MRS. Their phosphorylated forms are separated by about 0.4 ppm from each other. Using [31]P NMR, phosphocreatine is at about –2.52 with β-GPA-P at about –2.1 and cyclocreatine phosphate at about –3.0 ppm.

## Applications of creatine, phosphocreatine, cyclocreatine and β-GPA

Walker[38] has defined the criteria expected of synthetic phosphagens which are used to perturb the CK/PCr system. The compound might be expected to: (1) antagonize the synthesis of creatine; (2) antagonize cellular creatine uptake; (3) antagonize CK activity *in vivo*; (4) replace creatine and phosphocreatine with a phosphagen with altered thermodynamic properties. It should be remembered that the above criteria are not mutually exclusive. To this list one may add the properties of compounds that might behave as agonist of the CK/PCr system. It is clear that, by using creatine as an agonists and cyclocreatine and β-GPA as antagonists, the CK circuit can be manipulated for the purposes of dissecting the functional role of the system.

## Properties of CK substrates

*Creatine*

As pointed out earlier, creatine uptake into cardiac cells is through a $Na^+$-linked co-transporter with a $K_m$ of 0.05 mM and a $V_{max}$ of 20 nmol/min/g dry wt[27]. The beneficial consequences of elevating skeletal muscle creatine and phosphocreatine during certain forms of exercise and recovery has been well documented by Harris and co-workers[15,16,18,19]. Greenhaff has reported that oral creatine supplementation allowed a greater peak torque production in the final few contractions of a series. Similarly, they found that phosphocreatine resynthesis was enhanced, which is an important beneficial effect during repeated bouts of exercise. The rapid resynthesis of phosphocreatine may be due to a stimulation of respiration via mitochondrial CK. Such creatine-stimulated respiration has been discussed in chapter 16 of this volume. It is important to note here that creatine will increase the mitochondria's sensitivity to ADP for stimulating respiration. Therefore, with increased creatine in the muscle, there will be increased respiration at any given ADP concentration. One can conclude, therefore, that increasing phosphocreatine with creatine supplementation aids in maintaining muscle ATP and hence increases muscular performance.

*Phosphocreatine*

Phosphocreatine has another use, which does not involve CK and is due to its amphipathic nature. The ionic characteristics of phosphocreatine enable it to bind to the polar phospholipid heads of membranes. Binding the phospholipids decreases membrane fluidity and makes them less leaky (Fig. 8). Stabilization of the membranes prevents some of the membrane damage caused by transient ischaemia and hypoxia. Transient ischaemia can occur in heart damaged as well as in skeletal muscle during strenuous exercise. Phosphocreatine can therefore protect cells from damage by decreasing loss of cytoplasmic contents.

The membrane-stabilizing actions of phosphocreatine have been exploited by giving it to patients following myocardial infarction[28] or to those with ischaemic heart disease[14]. Intravenous treatment with phosphocreatine improves the energetic status of hearts and decreases electrical anomalies, which can lead to lethal arrhythmias. Preservation of hearts with phosphocreatine used as a cardio-plegic additive has been demonstrated as well. The utility of phosphocreatine has not been limited to heart muscle, though. Athletic performance can be enhanced by phosphocreatine. Administration of phosphocreatine improved maximum anaerobic power, potentiating the ability of muscles to produce maximum burst output[9]. There is also a decrease in muscle stiffness and enhanced endurance performance[33]. The decreased muscle stiffness may be due to a decrease in muscle damage that can occur with severe exercise. The membrane-stabilizing capabilities of phosphocreatine may be having a beneficial effect on the athletes as evidenced by the decreased muscle stiffness. Phosphocreatine therefore has the same effects as creatine with the added benefit of membrane stabilization.

## Cyclocreatine

Phosphorylated cyclocreatine has a Gibbs free energy of hydrolysis slightly larger than that of ATP and lower than that of phosphocreatine (8.4 kJ/mol)[38]. Walker reports that creatine is a 5.6-fold better substrate than cyclocreatine, and phosphocreatine is a 160-fold better substrate than phosphorylated cyclocreatine. Interestingly, cyclocreatine is 170-fold more efficacious than β-GPA as a substrate for CK[2]. Phosphorylated cyclocreatine constitutes almost 98 per cent of the total cyclocreatine content in a cyclocreatine-loaded muscle. Walker reports that muscle from rats fed cyclocreatine have a significant delay in the onset of rigor compared with controls and β-GPA-loaded muscle. The authors suggest that the delay is due to the increased reservoir of phosphorylated cyclocreatine, which is utilized at 1 per cent of the rate of phosphocreatine. Therefore there is a large pool of phosphagen available to replenish ATP, although rephosphorylation will occur at a slower rate.

**Table 1. Anatomical alterations in the rat as a result of feeding a diet of 1 per cent w/w β-guanidinopropionic acid**

|  | 5–6 weeks | | | | 10–12 weeks | | | |
|  | 50 g rats | | 200 g rats | | 50 g rats | | 200 g rats | |
|  | C | F | C | F | C | F | C | F |
|---|---|---|---|---|---|---|---|---|
| Body weight (BW) | 225.4 ± 7.9 | 196.3 ± 6.8 | 357.9 ± 15.9 | 351.7 ± 7.8 | 336.6 ± 13.8 | 373.7 ± 11.3 | 420.5 ± 14.0 | 449.0 ± 7.8 |
| Wet heart weight | 1.383 ± 0.082 | 1.399 ± 0.037 | 1.598 ± 0.023 | 1.731 ± 0.050 | 1.664 ± 0.060 | 2.217 ± 0.040 | 2.070 ± 0.049 | 2.310 ± 0.067 |
| Dry heart weight (DHW) | 1.177 ± 0.009 | 0.162 ± 0.007 | 0.212 ± 0.004 | 0.231 ± 0.007 | 0.229 ± 0.007 | 0.283 ± 0.007 | 0.255 ± 0.013 | 0.288 ± 0.004 |
| Wet lung weight | 0.121 ± 0.061 | 0.993 ± 0.014 | 1.279 ± 0.100 | 1.613 ± 0.081 | 1.361 ± 0.126 | 1.599 ± 0.063 | 1.462 ± 0.053 | 1.817 ± 0.092 |
| Dry lung weight (DLW) | 0.214 ± 0.014 | 0.208 ± 0.115 | 0.268 ± 0.030 | 0.341 ± 0.012 | 0.260 ± 0.020 | 0.338 ± 0.017 | 0.307 ± 0.019 | 0.393 ± 0.033 |
| DLW/BW | 0.954 ± 0.061 | 1.072 ± 0.115 | 0.743 ± 0.052 | 0.969 ± 0.020 | 0.783 ± 0.084 | 0.904 ± 0.039 | 0.694 ± 0.036 | 0.894 0.041 |
| DHW/BW | 0.784 ± 0.032 | 0.828 ± 0.032ns | 0.597 ± 0.027 | 0.656 ± 0.014ns | 0.684 ± 0.020 | 0.751 ± 0.026ns | 0.567 ± 0.026 | 0.688 ± 0.026* |

*$P < 0.01$; ns = not significant.  C = control; F = fed.

## β-GPA

β-GPA uptake has similar kinetics to that of creatine but a higher $K_m$ (0.2 mM)[34]. β-GPA is a thousand times less active as a substrate than creatine; however, the phosphorylated compound does accumulate in the tissue[38] with an acute time course[5]. The $K_m$ of CK for phosphorylated β-GPA (Pβ-GPA) is twice that of creatine and the $V_{max}$ three-fold less, thus making it a poor substrate compared with phosphocreatine[1]. It has been suggested that the rate of hydrolysis is not sufficient to buffer ATP during contraction; however, hydrolysis of Pβ-GPA over a relatively fast time course has been observed by Roberts and Walker[25]. Clark et al.[7] has provided evidence that β-GPA is not phosphorylated by mito-CK. This observation has also been supported by van Deursen et al.[35] using CK deficient mice.

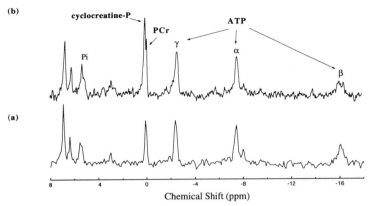

Fig. 7. ³¹P NMR spectra of superfused porcine carotids perfused with 50
mM cyclocreatine (b) compared with a control carotid artery (a).

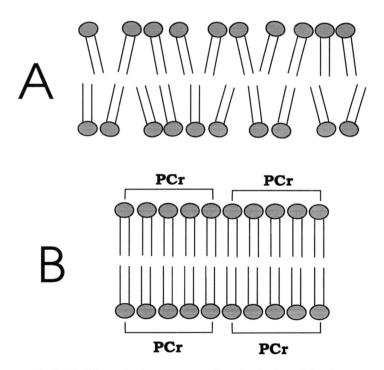

Fig. 8 (A). Schematic representation of randomization of the phospholipids which may occur in the plasma membrane.
(B) Phosphocreatine can bind to the phospholipid heads to decrease the
randomization/fluidity of the membrane by restoring the ordered
structure. In this way phosphocreatine can make membranes less leaky.

**Table 2. Alterations in the myocardial creatine kinase isozyme profile as a result of feeding a diet of 1 per cent w/w β-guanidinopropionic acid**

| | 5–6 weeks | | | | 10–12 weeks | | | |
| | 50 g rats | | 200 g rats | | 50 g rats | | 200 g rats | |
| | C | F | C | F | C | F | C | F |
|---|---|---|---|---|---|---|---|---|
| Total activity (IU/g wet wt) | 225 ± 8 | 196 ± 7 | 358 ± 8 | 352 ± 15 | 321 ± 16 | 333 ± 5 | 335 ± 10 | 276 ± 16** |
| Mito-CK isozyme (%) | 8.86 ± 2.10 | 6.64 0.71 | 14.85 ± 1.40 | 15.27 ± 2.10 | 10.72 ± 2.10 | 8.85 ± 1.40 | 7.85 ± 1.39 | 9.51 ± 1.14 |
| MM isozyme (%) | 60.98 ± 2.16 | 66.18 ± 3.30 | 60.82 ± 1.60 | 60.06 ± 1.50 | 61.18 ± 3.12 | 64.20 ± 0.93 | 69.99 ±1.40 | 64.01 ± 1.70* |
| MB isozyme (%) | 24.69 ± 1.15 | 22.64 ± 2.50 | 20.70 ± 0.42 | 20.20 ± 1.75 | 24.21 ± 1.04 | 23.34 ± 1.15 | 19.75 1.10 | 23.23 ± 0.78* |
| BB isozyme (%) | 5.46 ± 0.56 | 4.50 ± 0.70 | 3.57 ± 0.27 | 3.88 ± 0.54 | 3.88 ± 0.29 | 3.71 ± 0.49 | 2.38 ± 0.21 | 3.96 ± 0.23*** |

* = $P < 0.05$; ** = $P < 0.025$; *** = $P < 0.001$. (For methods see "Field et al.[11]")

**Table 3. Alterations in the skeletal muscle creatine kinase isozyme profile as a result of feeding a diet of 1 per cent w/w β-guanidinopropionic acid**

| | 50 g rats | | 200 g rats | |
| | C | F | C | F |
|---|---|---|---|---|
| Total activity (IU/g wet wt) | 1856 ± 91 | 1826 ± 83 | 2184 ± 116 | 1990 ± 76 |
| Mito-CK isozyme (%) | 2.75 ± 0.87 | 2.28 ± 0.56 | 2.92 ± 0.59 | 1.96 ±0.56 |
| MM isozyme (%) | 96.03 ± 1.61 | 96.65 ± 0.41 | 96.46 ± 0.55 | 97.72 ±0.72 |

(For methods see "Field et al.[11]"). C = control. F = fed.

The use of dietary β-GPA has been complicated by the difficulty in interpreting results because of the variety of dietary regimens used and the varied effects observed. It is clear that many of the contradictory findings reported may be accounted for by the difference in feeding regimens–hence the presence or absence of cardiac hypertrophy[7,20,24,31]. Because cardiac hypertrophy is inevitably associated with a switch in the CK isozyme expression; interpretation of the data can be difficult[11,12]. In order to define the range of feeding regimens cited, we have fed two groups of rats (50 g and 200 g rats) with a 1 per cent diet for 5–6 weeks or 10–12 weeks[10,12,34]. Our results show that cardiac hypertrophy is exclusively associated with the older animals after longer exposure (200 g; 10–12 weeks) (Table 1). This group also has a switch in CK isozyme composition from the M subunit towards the B subunit (Table 2). The β-GPA-fed group also had a decrease in the total CK activity, as well as all the biochemical signs of cardiac hypertrophy. Interestingly, we have found that results may vary depending on the animal batch or genetic strain.

Skeletal muscle also has dramatic alterations in fibre type and aerobic capacity as a result of feeding β-GPA(Table 3)[32]. We found no alteration in the total CK activity as a result of β-GPA feeding; the proportion of neither mito-CK nor MM-CK changed. Apart from the associated pathology associated with the use of β-GPA feeding, an additional problem is incomplete depletion of phosphocreatine and creatine. The apparent $K_m$ of CK measured *in vitro* has been reported as around the 1–2 mM range[26]. If this concentration is physiologically relevant it brings into question the use of this compound, as this degree of depletion is not always achieved.

## Conclusion

We have found that the CK energy shuttle can be manipulated by an agonist action of creatine loading as well as the antagonist action of cyclocreatine and β-GPA. These analogues have been used to study the function and importance of energy metabolism in muscle as it relates to CK. Enhancement of energy metabolism is found when muscle is loaded with creatine but the benefits from the analogues is less, probably due to the differing kinetics. Administration of creatine and phosphocreatine to athletes in training has enhanced athletic performance. This is consistent with an improvement in the intracellular energy metabolism of skeletal muscle. We conclude therefore that these compounds are useful tools for the clinician and scientist because they can alter the high-energy phosphate concentration in muscle.

### Acknowledgements

MLF would like to express his sincere thanks to George K. Radda, Anne-Marie Seymour and J.F. Unitt for their supervision. JFC is a Junior Research Fellow of Linacre College, Oxford. Thanks go to E.A. Boehm for performing the NMR and electrophoresis experiments. We also thank the British Heart Foundation, Zeneca Pharmaceuticals and the Medical Research Council for financial assistance.

## References

1.    Adams, G.R. & Baldwin, K.M. (1995): Age dependence of myosin heavy chain transitions induced by creatine depletion in rat skeletal muscle. *J. Appl. Physiol.* **78**, 368–371.

2.    Annesley, T.M. & Walker, J.B. (1980): Energy metabolism of skeletal muscle containing cyclocreatine phosphate. *J. Biol. Chem.* **255**, 3924–3930.

3.    Archer, S.L., Nelson, D.P., Zimmer, S., From, A.H.L. & Wier, E.K. (1989): Hypoxic pulmonary vasoconstriction is unaltered by creatine depletion induced by dietary β-guanidinopropionic acid. *Life Sci.* **45**, 1081–1088.

4.    Bessman, S. & Gieger, P. (1981): Transport of energy in muscle: the phosphorylcreatine shuttle. *Science* **211**, 448–452.

5.    Boehm, E.A., Clark, J.F. & Radda, G.K. (1995): Metabolite utilization and possible compartmentation in the porcine carotid artery: a study using β-guanidinopropionic acid. *Am. J. Physiol.* **37**, C628–C635.

6.    Bonas, J., Cohen, B. & Natelson, S. (1963): *Microchem. J.* **7**, 421–424.

7.    Clark, J.F., Khuchua, Z., Kuznetsov, A.V., Vassil'eva, E., Boehm, E., Radda, G.K. & Saks, V. (1994): Actions of the creatine analogue β-guanidinopropionic acid on rat heart mitochondria. *Biochem. J.* **300**, 211–216.

8.    Clark, J.F. & Dillon, P.F. (1995): Phosphocreatine and creatine kinase in energetic metabolism of the porcine carotid artery. *J. Vasc. Res.* **32**, 24–30.

9.    Dal Monte, A., Leonardi, L.M., Figura, F., Cappozzo, A., Marchetti, M., De Sanctis, A.M. & Nofroni, I. (1976): Effetti dell'apporto esogeno di fosfocreatina sulla potenza muscolare umana. *Gaz. Med. It.* **135**, 2–11.

10.   Field, M.L., Unitt, J.F., Radda, G.K., Henderson, C. & Seymour, A-M.L. (1991): Age-dependent changes in cardiac muscle metabolism upon replacement of creatine by β- guanidinopropionic acid. *Biochem. Soc. Trans.* **19**, 208S.

11.   Field, M.L., Gree, Y., Radda, G.K., Henderson, C. & Seymour, A.-M.L. (1992): The creatine kinase isozyme profile in myocardial pressure overload, volume overload and idiopathic hypertrophy. *J. Mol. Cell Cardiol.* **24**, 156S.

12.   Field, M.L., Clark, J.F., Henderson, C., Seymour, A.-M.L. & Radda, G.K. (1994): Alterations in the myocardial creatine kinase system during chronic anaemic hypoxia. *Cardiovasc. Res.* **28**, 86–91.

13    Fitch, C.D., Jellinek, M., Fitts, R.H., Baldwin, K.M. & Holloszy, J.O. (1975): Phosphorylated ß-guanidinopropionic acid as a substitute for phosphocreatine in rat muscle. *Am. J. Physiol.* **228,** 1123–1125.

14.   Gelfgan, E.B., Dalili, I.G., Shakhtakhtinskaya, F.N., Yagisarov, N.M. & Shirinova, E.A. (1987): The effect of phosphocreatine on the tolerance of physical exercise in patients with ischemic heart disease. In: *Creatine phosphate: biochemistry, pharmacology, and clinical efficiency*, eds. V.A. Saks, Y.G. Bobkov & E. Strumia, pp. 270–275. Turin: Edizoni Minerva Medica.

15.   Greenhaff, P.L., Bodin, K., Harris, R.C., Hultman, E., Jones, D.A., McIntyre, D.B., Soderlund, K. & Turner, D.L. (1993): The influence of oral creatine supplementation on muscle phosphocreatine resynthesis following intense contraction in man. *J. Physiol.* **467,** 75P.

16.   Greenhaff, P. L., Casey, A. H., Short, R., Harris, K.,Soderlund, K. & Hultman, E. (1993): Influence of oral creatine supplementation on muscle torque during repeated bouts of maximal voluntary exercise in man. *Clin. Sci.* **84,** 565–571.

17.   Greenhaff, P.L., Bodin, K., Soderlund, K. & Hultman, E. (1994): Effect of oral creatine supplementation on skeletal muscle phosphocreatine resynthesis. *Am. J. Physiol.* **266,** E725–E730.

18.   Harris, R.C., Soderlund, K. & Hultman, E. (1992): Elevation of creatine in resting and exercized muscle of normal subjects by creatine supplementation. *Clin. Sci.* **83,** 367–374.

19.   Harris, R.C., Viru, M., Greenhaff, P.L. & Hultman, E. (1993): The effect of oral creatine supplementation on running performance during maximal short term exercise in man. *J. Physiol.* **467,** 74P.

20.   Kapelko, V.I., Saks, V.A., Novikova, N.A., Golikov, M.A., Kupriyanov, V.V. & Popovich, M.I. (1989): Adaptation of cardiac contractile function to conditions of chronic energy deficiency. *J. Mol. Cell Cardiol.* **21,** 79–83.

21.   Khuchua, Z.A., Ventura-Clapier, R., Kuznetsov, A.V., Grishin, M.N. & Saks, V.A. (1989): Alterations in the creatine kinase system in the myocardium of cardiomyopathic hamsters. *Biochem. Biophys. Res. Commun.* **165,** 748–757.

22.   Khuchua, Z., Kuznetsov, A.V., Vasilyeva, E.V., Ventura-Clapier, R., Clark, J.F., Steinshneider, A., Korchazhkina, O.V., Lakomkin, V.L., Branishte, T., Ruuge, E.K., Kapelko, V.I. & Saks, V.A. (1992): The creatine kinase system and cardiomyopathy. *Cardiovasc. Pathol.* **4,** 223–233.

23.   Levitsky, D.O., Levchenko, T.S., Sharov, V.G. & Smirnov, V.N. (1977): The functional coupling between $Ca^{2+}$ ATPase and creatine phosphokinase in heart muscle sarcoplasmic reticulum. *Biokhimiia* **42,** 1766–1773.

24.   Mekhfi, H., Hoeter, J., Lauer, C., Wisnewsky, C., Schwartz, K. & Ventura-Clapier, R. (1990): Myocardial adaptation to creatine deficiency in rats fed with β-guanidinopropionic acid, a creatine analogue. *Am. J. Physiol.* **258,** H1151–H1158.

25.   Roberts, J.J. & Walker, J.B. (1982): Conversion of dietary N-ethylguanidinoacetate by Ehrlich ascites tumour cells and animal tissues to a functionally active analogue of creatine phosphate. *Arch. Biochem. Biophys.* **215,** 564–570.

26.   Saks, V.A., Lipina, N.V., Smirnov, V.N. & Chazov, E.I. (1976): Studies of energy transport in heart cells. The functional coupling between mitochondrial creatine phosphokinase and ATP–ADP translocase: kinetic evidence. *Arch. Biochem. Biophys.* **173,** 34–42.

27.   Seppet, E.K., Adoyaan, A.J., Kallikorm, A.P., Chernousova, G.B., Lyulina, N.V., Sharov, V.G., Severin, V.V., Popovich, M.I. & Saks, V.A. (1985): Hormone regulation of cardiac energy metabolism. I. Creatine transport across cell membranes of euthyroid and hyperthyroid rat heart. *Biochem. Med.* **34,** 267–279.

28.   Sharov, V.G., Saks, V.A., Kupriyanov, V.V., Lakomkin, V.L., Kapelko, V.I., Steinschneider, A.Y. & Javodov, S.A. (1987): Protection of ischemic myocardium by exogenous phosphocreatine. Morphologic and phosphorus [31]P NMR studies. *J. Thorac. Cardiovasc. Surg.* **94,** 749–761.

29.   Sharps, E.S. & McCarl, R.L. (1982): A HPCL method to measure [32]P incorporation into phosphorylated metabolites in cultured cells. *Anal. Biochem.* **124,** 421–424.

30.   Shields, R.P. & Whitehair, C.K. (1973): Muscle creatine: *in vivo* depletion by feeding β-guanidinopropionic acid. *Can. J. Biochem.* **51,** 1046–1049.

31.   Shoubridge, E.A., Challiss, J.R.A., Hayes, D.J. & Radda, G.K. (1985): Biochemical adaptation in the skeletal muscle of rats depleted of creatine with the substrate analogue β-guanidinopropionic acid. *Biochem. J.* **232,** 125–131.

32.   Shoubridge, E.A., Jeffry, F.M.H., Keogh, J.M., Radda, G.K., & Seymour, A.-M.L. (1985): Creatine kinase kinetics, ATP turnover and cardiac performance in hearts depleted of creatine with the substrate analogue β-guanidinopropionic acid. *Biochim. Biophys. Acta* **847,** 25–32.

33. Tegazzin, V., Rossi, M., Schiavon, M., Schiavon, R., Bettini, V., Aragno, R. & Bitozzi, A. (1991): Indagine sulla performance di ciclisti trattati e non tratti con fosfocreatina. *Biol. Med.* **13,** 121–135.

34. Unitt, J.F., Radda, G.K. & Seymour, A.-M.L.(1993): The acute effects of the creatine analogue, β-guanidinopropionic acid on cardiac energy metabolism and function. *Biochim. Biophys. Acta* **1143,** 91–96.

35. van Deursen, J., Jap, P., Heerschap, A., Laak, H., Ruitenbeek, W. & Wieringa, B. (1994): Effects of the creatine analogue β-guanidinopropionic acid on skeletal muscles of mice deficient in muscle creatine kinase. *Biochim. Biophys. Acta* **1185,** 327–335.

36. Veksler, V.I., Kuznetsov, A.V., Sharov, V.G., Kapelko, V.I., & Saks, V.A. (1987): Mitochondrial respiratory parameters in cardiac tissue: novel method of assessment by using saponin-skinned fibres. *Biochim. Biophys. Acta* **892,** 191–196.

37. Ventura-Clapier, R., Mekhfi, H. & Vassort, G. (1987): Role of creatine kinase in forced development in chemically skinned rat cardiac muscle. *J. Gen. Physiol.* **89,** 815–837.

38. Walker, J.B. (1979): Creatine: biosynthesis, regulation and function. *Adv. Enzymol.* **50,** 178–237.

39. Wallimann, T., Wyss, M., Brdiczka, D., Nicolay, K. & Eppenberger, H.M. (1992): Intracellular compartmentation, structure and function of creatine kinase isozymes in tissues with high and fluctuating energy demands: the phosphocreatine circuit for cellular energy homeostasis. *Biochem. J.* **281,** 21–40.

40. Zweier, J.L., Jacobus, W.E., Korecky, B. & Brandejs-Barry, Y. (1991): Bioenergetic consequences of cardiac phosphocreatine depletion induced by creatine analogue feeding. *J. Biol. Chem.* **266,** 20296–20304.

*Guanidino Compounds : 2*, eds. by P.P. De Deyn, B. Marescau, I.A. Qureshi and A. Mori.
©1997 John Libbey & Company Ltd., pp. 157–163.

# Chapter 18

# The comparative significance of adenylate and phosphocreatine pathways of energy transport for cardiac pump function

Valeri I. KAPELKO, Vladimir L. LAKOMKIN and Olga V. KORCHAZHKINA

*Cardiology Research Center, 15A Cherepkovskaya St, 121552, Moscow, Russia*

## Summary

Myocardial functional and metabolic alterations caused by an inhibition of creatine kinase (CK) activity have been compared with those occurring after profound reduction of myocardial ATP content. The perfusion of isolated rat heart with iodoacetamide (IAA, 0.5 mM) was followed by almost complete inhibition of CK activity and loss of cardiac pump function. A partial inhibition of CK, from $295 \pm 7$ to $56 \pm 6$ μmol/g dry weight (81 per cent), was achieved after 0.1 mM IAA treatment and coincided with no significant changes in adenine nucleotides or phosphocreatine content. After such treatment maximal aortic output was lower by 38 per cent and left ventricular systolic pressure by 29 per cent. The cardiac distensibility as judged by the relationship between LV filling volume and LV end-diastolic pressure was significantly decreased. Similar reductions in both maximal aortic output and LV systolic pressure were observed after deoxyglucose (13 mM) treatment that was followed by a reduction of myocardial ATP content to 32 per cent. However, at minimal volume load, both parameters were distinctly lower after deoxyglucose treatment than after IAA. The difference seems to be due to decreased cardiac distensibility in the latter. These data suggest that the CK-related phosphocreatine pathway of energy transport in myocardial cells is essential for the cardiac pump function particularly due to its participation in the maintenance of normal myofibrillar compliance, which is necessary for heart filling.

## Introduction

The energy transport in myocardial cells from mitochondria to sites of its utilization is accomplished both by CK-dependent transformation of ATP into phosphocreatine (PCr) and back (the PCr pathway[1,2,5,12–14]), and by direct diffusion of ATP through the myoplasm, (the adenylate pathway[5]). A reduction of energy flow in cardiomyocytes induced by a deterioration of either pathway has been shown to result in cardiac contractile failure that is more pronounced in conditions requiring intensive energy supply[8–11]. Notably, cardiac pump function ceases after almost complete inhibition of CK[9]. The purpose of this study was to induce partial inhibition of the enzyme and to analyse the pattern of contractile function.

## Materials and methods

### Heart perfusion

Experiments have been performed on rats which were injected with heparin (500 U/kg) and urethane (1.8 mg/kg) before heart isolation. The isolated hearts were perfused according to Neely's working

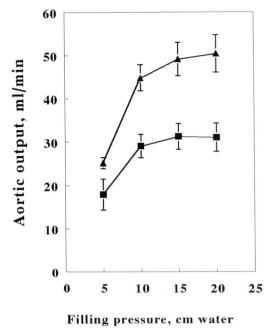

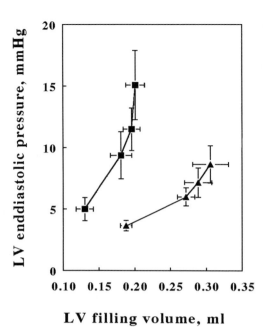

Fig. 1. Effect of increased left atrial filling pressure, ($cmH_2O$), on the aortic output (ml/min), of the isolated rat heart before, (▲), and after iodoacetamide treatment (0.1 mM), (■) (mean ± SEM).

Fig. 2. Relationship between LV filling volume (ml) and LV end-diastolic pressure (mmHg) during volume load in the isolated heart. Groups and symbols as in Fig. 1.

heart model at 37 °C with standard Krebs–Henseleit solution as described previously[9]. The left atrium and aorta were cannulated and initial filling and resistance pressures were set at 15 and 80 $cmH_2O$, respectively.

The cardiac output was continuously recorded with a Caroline Medical Electronics 501D electromagnetic transducer inserted into the inflow line. The coronary flow was measured by counting flow per unit of time and the aortic output was calculated by subtracting the coronary flow from the cardiac output. A needle was inserted into the left ventricular (LV) cavity through a wall, and LV pressure and aortic pressure were continuously recorded with electromanometers Gould Statham P23Gb and Gould 2600S recorder. Cardiac volume–pressure work was calculated as a product of the cardiac output and mean pressure in the aortic chamber representing the real ejection pressure.

The LV diastolic stiffness index was determined as a ratio of the rise in LV diastolic pressure from minimal to maximal values and the rise in LV volume for a diastole, the latter value in stable conditions always being equal to the stroke volume. Elevation of diastolic stiffness corresponds to a decrease in LV distensibility.

## Experimental protocol

After 30 min to stabilize the cardiac function, the functional curve during volume load was determined by gradual elevation of the left atrium filling pressure from 5 to 20–25 $cmH_2O$. The heart was allowed to stabilize under the previous conditions and then the hydrostatic resistance was increased by steps from 60 to 110 $cmH_2O$ to evaluate the cardiac ability to withstand the resistance. After these functional loads the heart was perfused retrogradely at a constant perfusion flow rate of 15 ml/min with perfusate

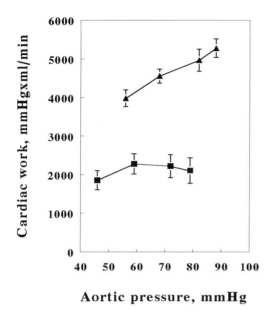

**Aortic pressure, mmHg**

Fig. 3. Relationship between mean aortic pressure (mmHg) and cardiac work (the product of mean aortic pressure and cardiac output) (mmHg × ml/min). Groups and symbols as in F  ig. 1.

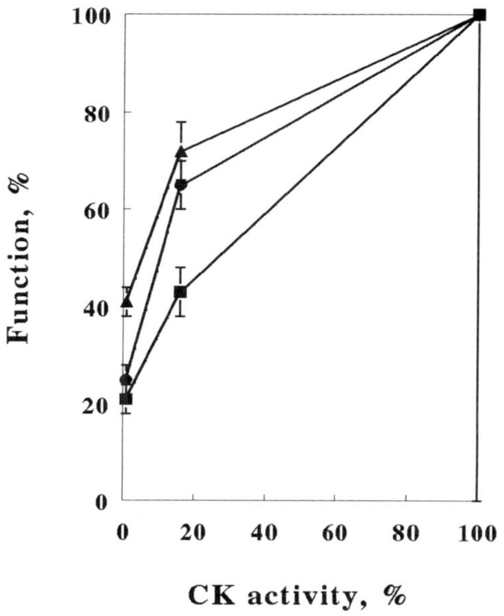

**CK activity, %**

Fig. 4. Relationship between creatine kinase activity remained after iodoacetamide treatment at various doses and relative depression of cardiac functional parameters expressed as a percentage of the corresponding control values before treatment. ▲; maximal LV systolic pressure; ●, maximal cardiac output; ■, maximal cardiac work.

containing iodoacetamide (IAA), 0.1 mM, purchased from Sigma. The perfusion with IAA lasted for 10 min, followed by 5 min period of retrograde perfusion without IAA. Control experiments were performed similarly except drug was not added. In all experiments the retrograde perfusion was followed by transition to working mode and the functional loads were again applied.

At the end of experiment, the hearts were frozen by Wollenberger clamps cooled in liquid nitrogen, and biochemical analyses were performed by routine enzymatic methods[9].

Data are presented as mean ± SEM; $P < 0.05$ Student's $t$-test is considered statistically significant.

## Results

The IAA treatment resulted in 81 per cent inhibition of the enzyme activity, from 295 ± 7 to 56 ± 6 µmol/g dry weight ($P < 0.001$). No significant changes in adenine nucleotides or PCr content were found (Table 1). Lower CK activity was associated with a marked reduction in the cardiac pump function (Table 2). Both cardiac and aortic outputs decreased by 30 per cent and LV systolic pressure decreased similarly. The cardiac work, (that is, the derivative of both pump function and aortic resistance) decreased more, by 40 per cent. Both LV minimal and end-diastolic pressures, LV diastolic stiffness, were slightly but not significantly higher. The heart rate was the only parameter that remained unchanged after IAA treatment.

159

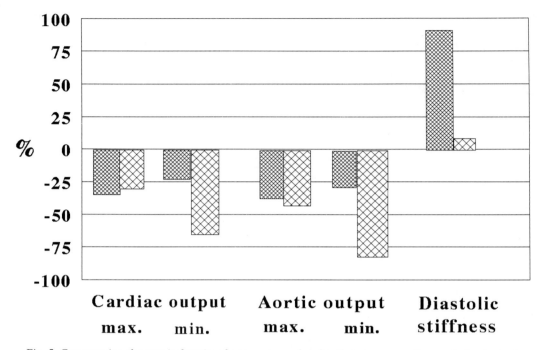

*Fig. 5. Comparative changes in functional parameters related to their corresponding control values after iodoacetamide treatment (0.1 mM), hard cross-hatching, or 2-deoxyglucose treatment, light cross-hatching.*

**Table 1. Effect of iodoacetamide treatment on adenine nucleotides, PCr and Cr content.**

| Indices µmol/g dry weight | Control (n = 7) µmol/g dry weight | IAA-treated (n = 7)µmol/g dry weight |
|---|---|---|
| ATP | 24.6 ± 1.8 | 21.8 ± 1.0 |
| ADP | 4.90 ± 0.38 | 4.78 ± 0.57 |
| AMP | 0.66 ± 0.05 | 0.72 ± 0.06 |
| Phosphocreatine | 29.6 ± 1.4 | 28.6 ± 1.9 |
| Creatine | 28.6 ± 1.9 | 28.6 ± 3.7 |

The cardiac pump failure was observed in IAA-treated hearts at any level of left atrial filling pressure (Fig. 1), maximal aortic output being 38 per cent lower. The LV systolic pressure was also 21–29 per cent lower. The cardiac distensibility as judged by the relationship between LV filling volume and LV end-diastolic pressure was significantly decreased (Fig. 2).

When hydrostatic resistance was elevated at stable left atrial filling pressure, control hearts success-fully increased their work almost linearly with increased resistance (Fig. 3) due to adequate elevation of LV systolic pressure and only slight reduction of cardiac output. The IAA-treated hearts failed to increase the work substantially although they retained the ability to increase LV systolic pressure to match increased resistance.

The results of partial inhibition of CK can be compared with those of almost complete inhibition[9] since both studies have been conducted in similar experimental conditions. Changes in three functional parameters, namely systolic pressure, cardiac output and cardiac work are shown in relation to CK inhibition (Fig. 4). The least affected is LV systolic pressure; cardiac output and cardiac work decreased more profoundly at almost complete inhibition.

The extent of functional alterations induced by partial inhibition of CK is similar to that observed at reduction of adenylate way of energy transport[10] , but some important differences may be noted. During volume load, the extent of reduction of both maximal cardiac and aortic outputs compared with control values is similar after iodoacetamide or deoxyglucose treatment (Fig. 5). However, at minimal volume load both parameters have been affected distinctly more by deoxyglucose treatment. This means that hearts with a disturbed adenylate pathway but with a preserved PCr pathway were able to increase cardiac output while iodoacetamide-treated hearts failed. An apparent difference in the extent of rise in diastolic stiffness is in accordance with this view. Thus less distensible hearts due to partial inhibition of CK lose the adequate response to cardiac filling.

**Table 2. Functional indices of working rat hearts treated with iodoacetamide (IAA) at standard load**

| Indices | Control | IAA | % of control | |
|---|---|---|---|---|
| Cardiac output (ml/min) | $71 \pm 3$ | $50 \pm 3$ | 70 | $P < 0.001$ |
| Aortic flow (ml/min) | $53 \pm 3$ | $37 \pm 3$ | 70 | $P < 0.01$ |
| Coronary flow (ml/min) | $18 \pm 1.1$ | $14 \pm 1.3$ | 78 | $P < 0.05$ |
| Heart rate (min$^{-1}$) | $235 \pm 5$ | $225 \pm 11$ | 96 | |
| Cardiac work (mmHg × ml/min) | $5108 \pm 215$ | $3041 \pm 233$ | 60 | $P < 0.001$ |
| LV systolic pressure (mmHg) | $166 \pm 9$ | $118 \pm 9$ | 71 | $P < 0.01$ |
| LV minimal diastolic pressure (mmHg) | $0.6 \pm 1.8$ | $2.0 \pm 1.6$ | – | |
| LV end-diastolic pressure (mmHg) | $7.8 \pm 0.7$ | $9.9 \pm 1.1$ | – | |
| LV diastolic stiffness (mmHg/ml) | $20 \pm 1.4$ | $30 \pm 7$ | 150 | |

Standard load conditions: left atrium filling pressure 15 cmH$_2$O, resistance pressure 80 cmH$_2$O; all values are represented as mean $\pm$ SEM; n=8.

## Discussion

As expected, the relative depression of cardiac pump function after partial inhibition of CK was less than at almost complete inhibition of the enzyme[9,11]. In both cases, the comparative alterations of parameters associated with pressure development were less pronounced than those associated with ejection. The most affected parameter was maximal cardiac works which requires an extremely high rate of energy supply to myofibrils and oxygen consumption[8]. Similar alterations in cardiac pump functioning were found under influence of various metabolic inhibitors[3,7] and deoxyglucose treatment[10]. In all these studies, a disturbance in energy supply to the myofibrils seems to occur. Our data suggest that cardiac filling rather than force development is the crucial factor leading to cardiac pump failure in these conditions.

The partial CK inhibition in this study (19 per cent activity remained) was combined with almost no changes in high–energy phosphate contents. This finding seems surprising in view of the marked functional deterioration observed. The reason for this discrepancy is not clear. The myofibrillar CK isoenzyme that is involved in PCr–ATP transformation may be more accessible to IAA molecules. An alternative explanation is based on the role of the flux rate through CK: only moderate alterations of PCr stores were found at almost complete CK inhibition[5,9], so the flux rate through CK seems to be a more critical factor than absolute PCr content, as has been demonstrated in several works from Ingwall's laboratory[2-4]. Even very severe depletion of PCr due to chronic creatine deficiency induced by β-guanidinopropionate diet did not show distinct deterioration of the cardiac pump function[9,13] except at high functional loads[9]. However, the flux through CK, in the latter study estimated as being approximately 50 per cent[9,11], was still higher than the rate of ATP hydrolysis[9,11]. On the other hand, complete inhibition of CK function (fluxes) resulted in severe impairment of the cardiac pump function in spite of the high content of ATP or PCr.

Besides similarities in functional deterioration caused by both IAA and deoxyglucose treatment, an important difference has been found in this study. The diastolic stiffness is elevated in accordance with the extent of CK inhibition but remains almost unchanged after a profound fall in ATP content, with the flux through CK presumably remaining unaffected[10,11]. The functional consequence of elevated diastolic stiffness is the disturbance in cardiac filling, which affected both cardiac pump functional response and cardiac ejection at increased resistance.

The reason for this may be the crucial role of this enzyme in myofibrillar energy supply. It has been shown that CK bound to myofibrils can maintain functional compartmentation of adenine nucleotides in myofibrillar space, normal $Ca^{2+}$ sensitivity of myofibrils and low myofibrillar stiffness[15,16]. Thus, CK-produced and PCr-related phosphorylation of ADP in myofibrils seems to play a very important role in maintaining high local phosphorylation potential[6] and adequate supply of ATP for myosine ATPase to elicit dissociation of actomyosin cross-bridges, which is necessary both for a high rate of muscular shortening and for profound relaxation.

These data suggest that the CK-related PCr pathway of energy transport in myocardial cells is essential for the cardiac pump function[3,8,9], particularly due to its participation in the maintenance of normal myofibrillar compliance, which is necessary for heart filling. In fact, changes of cardiac work during graded hypoxia were reported to be linearly related to the changes in the flux through CK and not to absolute values of ATP or PCr content[3].

In conclusion, (1) limited myocardial distensibility after CK inhibition is the crucial factor causing cardiac pump failure; (2) the PCr pathway of energy transport is extremely important for cardiac pump function because it presumably maintains myofibrillar ATP supply and prevents ADP accumulation within myofibrils, permitting myofibrillar distension and cardiac filling.

## Acknowledgement

This work was supported by the Russian Fund of Fundamental Investigations, RFFI 93–04–21868.

## References

1. Bessman, S.P. & Geiger, P.G. (1981): Transport of energy in muscle: the phosphorylcreatine shuttle. *Science* **211,** 448–452.

2. Bittl, J.A. & Ingwall, J.S. (1986): The energetics of myocardial strech. Creatine kinase flux and oxygen consumption in the noncontracting rat heart. *Circul. Res.* **58,** 378–383.

3. Bittl, J.A., Balschi, J. & Ingwall, J.S. (1987): Contractile failure and high energy phosphate turnover during hypoxia: $^{31}$P-NMR surface coil studies in living rat. *Circul. Res.* **60,** 871–878.

4. Bittl, J.A., Balschi, J. & Ingwall, J.S. (1987): Effects of norepinephrine infusion on myocardial high energy phosphate content and turnover in the living rat. *J. Clin. Invest.* **79,** 1852–1859.

5. Fossel, E.T. & Hoefeler, H. (1987): Complete inhibition of creatine kinase in isolated perfused rat hearts. *Am. J. Physiol.* **252** (*Endocrinol. Metab.* **15**), E124–E130.

6. Kammermeier, H., Schmidt, P. & Jungling, E. (1982): Free energy change of ATP hydrolysis: a causal factor of early hypoxic failure of the myocardium. *J. Mol. Cell. Cardiol.* **14,** 267–277.

7. Kapelko, V.I., Veksler, V.I. & Novikova, N.A. (1987): Myocardial contracture caused by disturbances in energy production: mechanism and significance. In: *Myocardial Metabolism.* Eds. A. Katz & V.N. Smirnov, pp. 522–543. London: Harwood Academic.

8. Kapelko, V.I., Kupriyanov, V.V., Novikova, N.A., Lakomkin, V.L., Steinschneider, A.Y., Severina, M.Y., Veksler, V.I. & Saks, V.A. (1988): The cardiac contractile failure induced by chronic creatine and phosphocreatine deficiency. *J. Mol. Cell. Cardiol.* **20,** 465–479.

9. Kapelko, V.I., Saks, V.A., Ruuge, E.K., Kupriyanov, V.V., Novikova, N.A., Lakomkin, V.L., Steinschneider, A.Y. & Veksler, V.I. (1992): The crucial role of creatine kinase for cardiac pump function. In: *Guanidino Compounds In Biology and Medicine.,* Eds. P.P. De Deyn, B. Marescau, V. Stalon & I.A. Qureshi, pp. 249–251. London: John Libbey.

10. Kupriyanov, V.V., Lakomkin, V.L., Kapelko, V.I., Steinschneider, A.Y., Ruuge, E.K. & Saks, V.A. (1987): Dissociation of adenosine triphosphate levels and contractile function in isovolumic hearts perfused with 2-deoxyglucose. *J. Mol. Cell. Cardiol.* **19,** 729–740.

11. Kupriyanov, V.V., Lakomkin, V.L., Korchazhkina, O.V., Steinschneider, A.Y., Kapelko, V.I. & Saks, V.A. (1991): Control of cardiac energy turnover by cytosolic phosphates: $^{31}$P-NMR study. *Am. J. Physiol.* **261** (Suppl. 4), 45–53.

12. McClellan, G., Weisberg, A. & Winegrad, S. (1983): Energy transport from mitochondria to myofibril by a creatine phosphate shuttle in cardiac cells. *Am. J. Physiol.* **245,** C423–C427.

13. Shoubridge, E.A., Jeffry, F.M.H., Keogh, J.M., Radda, G.K & Seymour, A.M.L. (1985): Creatine kinase kinetics, ATP turnover, and cardiac performance in hearts depleted of creatine with the substrate analogue beta-guanidine propionic acid. *Biochim. Biophys. Acta* **847,** 25–32.

14. Veksler, V.I. & Kapelko, V.I. (1984): Creatine kinase in regulation of heart function and metabolism. II. The effect of phosphocreatine on the rigor tension of EGTA-treated rat myocardial fibers. *Biochim. Biophys. Acta* **803,** 265–270.

15. Ventura-Clapier, R., Mekhfi, H. & Vassort, G. (1987): Role of creatine kinase on force development in chemically skinned rat cardiac muscle. *J. Gen. Physiol.* **89,** 815–837.

16. Ventura-Clapier, R., Saks, V.A., Vassort, G., Lauer, C. & Elizarova, G.V. (1987): Reversible MM-creatine kinase binding to cardiac myofibrils. *Am. J. Physiol.* **253** (*Cell. Physiol.* **22**), C444–C455.

*Guanidino Compounds : 2*, eds. by P.P. De Deyn, B. Marescau, I.A. Qureshi and A. Mori.
©1997 John Libbey & Company Ltd., pp. 165–180.

# Chapter 19

---

# Mathematical modelling of the phosphocreatine shuttle in respiring heart mitochondria

---

## Mayis K. ALIEV and Valdur A. SAKS

*Laboratory of Experimental Cardiac Pathology, Cardiology Research Center, Moscow, Russia; Laboratory of Bioenergetics, Institute of Chemical and Biological Physics, Ravala 10, EEOOO1 Tallinn, Estonia*

## Summary

A probability approach was used to describe mitochondrial respiration in the presence of ATP, ADP, creatine (Cr) and phosphocreatine (PCr). Respiring mitochondria were considered as a three-component system, including: (1) oxidative phosphorylation reactions, which provide stable ATP and ADP concentrations in the mitochondrial matrix; (2) adenine nucleotide translocase, which performs exchange transfer of matrix adenine nucleotides for those from outside, supplied from medium and by creatine kinase; (3) CK, starting these reactions when activated by the substrates from medium. The specific feature of this system is the close proximity of the CK and translocase molecules. This results in a high probability of direct activation of translocase by CK-derived ATP or ADP without their leaking into the medium. In turn, the activated translocase with the same high probability directly provides CK with matrix-derived ATP or ADP. This results in strong CK-supported activation of mitochondrial respiration in the medium without added ADP. The main output product of mitochondria in such a system is PCr instead of ATP. In a system with ATP, ADP, Cr and PCr the model satisfactorily describes a drastic respiration-induced shift of CK reaction from equilibrium, providing sustained PCr production in thermodynamically unfavourable conditions. The model predicts extremely high (up to 8) values of control strength of mitochondrial respiration by CK, originating from tight coupling of CK and translocase.

## Introduction

The kinetics of the soluble creatine kinase, have been well known for a long time[17,20,22]. However, in cardiomyocytes, skeletal muscle cells, and also in the brain and other specific types of cell, creatine kinase (CK) is specifically compartmentalized[30]. The behaviour of the structurally bound enzymes is controlled and highly dependent on their microenvironment, mostly by the surrounding enzymes. In mitochondria, CK is controlled by adenine nucleotide translocase (T) and in other structures it is controlled by ATPases[30]. It has been known since 1975 that the behaviour of the CK in mitochondria under conditions of oxidative phosphorylation is not determined by substrate concentration in the medium and soluble enzyme kinetics, but even the direction of the CK reaction may be different depending on the oxidative phosphorylation which very significantly increases the rate of PCr production and decreases the rate of the reverse reaction of ATP formation[22]. Kinetically, the control of oxidative phosphorylation over the mitochondrial CK reaction is manifested as change in dissociation constant for MgATP from the ternary enzyme–substrate complexes[14,25], and thermodynamically it is manifested as a maintenance of the reaction in the direction opposite to that predicted from the mass action ratio in the medium and the equilibrium constant value[22,26]. All that evidence was taken to show functional coupling between mitochondrial CK and adenine nucleotide translocase:

translocase directs ATP molecules to the active site of CK and simultaneously removes the reaction product – ADP[4,5,14,19–23,25,26,30]. The functioning of this complex has not been described quantitatively in sufficient detail: until 1993 only a short and rather general model has been published[27] in 1976. In 1993 we were able to develop a mathematical model[1,2] quantitatively describing the forward reaction of aerobic phosphocreatine synthesis coupled to oxidative phosphorylation in cardiac mitochondria. In this chapter we describe a general mathematical model quantitatively describing the process of aerobic phosphocreatine synthesis (forward and reverse reactions of mitochondrial CK) coupled to oxidative phosphorylation in cardiac mitochondria.

## The model

### General considerations

The model is based on a probability approach to the description of the enzyme–enzyme interaction. In the model, the respiring mitochondria are considered as a three-component system (see Fig. 1), including:

(i) oxidative phosphorylation (OP) reactions, which provide stable ATP and ADP concentrations in the mitochondrial matrix[1,2];

(ii) adenine nucleotide translocase (ANT), which performs the exchange of matrix ATP or ADP for outside CK-supplied ADP or ATP when both substrates are simultaneously bound to translocase[29];

(iii) CK, which starts these reactions when activated by the substrates from medium (shaded areas in Fig. 1).

The specific feature of this system is a close proximity of CK and translocase molecular complexes provided by anchoring them to cardiolipin molecules[21,23]. This results in a high probability of direct activation of translocase by ADP from the CK forward reaction without its leaking into medium (the concept of direct channelling, pathways X and Y in Fig. 1). In turn, the activated translocase with the same high probability directly provides CK with matrix-derived ATP. This 'local' ATP ($A_{loc}$), when forming effective $E.A_{loc}.Cr$ complexes with CK (pathway X in Fig. 1), provides sustained PCr production from mitochondrial ATP in the CK forward reaction. When CK cannot accept or retain locally supplied ATP molecules, they diffuse into solution (pathway Y in Fig.1). In pathway Z, PCr in the CK forward reaction is exclusively produced from ATP in the medium and the ADP formed is released into medium, so this pathway is solely governed by substrate concentrations in medium and well defined soluble CK kinetics[10,15,20,22,27]. In pathway W, translocase directly exchanges medium ADP or ATP for matrix ATP, providing DmA (ADP medium/ATP matrix) or AmA (ATP medium/ATP matrix) exchanges. These exchanges operationally should be distinguished from the similar DlA (ADP local/ATP matrix, pathways X,Y in Fig. 1) and AlA (ATP local/ATP matrix, not shown) exchanges, which originate from activation of translocase (T) by locally supplied adenine nucleotides[3].

The scheme in Fig. 1 can also be used to illustrate the mirror image case of the CK reverse reaction[3], when respiring mitochondria produce, from PCr and CK-supplied local ADP, the creatine instead of PCr (pathway X) or Cr and ADP instead of PCr and ATP (pathway Y). The translocase in this case works in the regimen of matrix ADP export at AlD (ATP local/ADP matrix) or DlD (ADP local/ADP matrix) exchanges. These locally activated exchanges should also be distinguished from similar AmD (ATP medium/ADP matrix) or DmD (ADP medium/ADP matrix) ones, originating from direct activation of translocase by substrates from the medium[3] (pathway W). In the pathway like Z, the CK reverse reaction is directly activated by ADP and PCr from the medium and ATP formed is released into the medium, so this pathway is solely governed by substrate concentrations in the medium and well defined soluble CK kinetics[10,15,20,22,27].

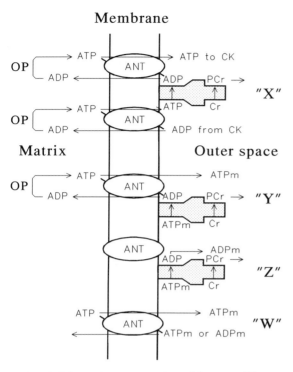

*Fig. 1. Schematic representation of three possible combinations of sources of ATP for mitochondrial PCr production in the CK forward reaction (pathways X,Y,Z) and of medium ATP or ADP induced exchanges of translocase (pathway W). For further explanations see text.*

All combinations of translocase exchanges and CK reactions for all pathways are schematically illustrated in Fig. 2. Figure 2A depicts the forward direction of the CK reaction, the synthesis of PCr and ADP from ATP and Cr. Translocase in this part of the scheme exports ATP from matrix in four species (AmA, AlA, DmA, DlA) of two main kinds (ATP/ATP, ADP/ATP) of exchange. We take this main *in vivo* direction of translocase activity as forward. In our previous paper[1] only one (DlA) of 4 forward-directed species of translocase exchange was considered. In the complete system 'local' ATP ($ATP_{loc}$) for direct CK activation can be derived not only from DlA exchange on translocase, but also from DmA, AmA and AlA exchanges (pathways like X in Fig. 1). When CK cannot accept or retain locally supplied ATP molecules, they can diffuse out into solution (pathways like Y in Fig. 1, $ATP_{loc} \rightarrow ATP_m$ path in Fig. 2A). The unique feature of a complete system is participation of the reverse CK activity in the stimulation of the forward CK reaction through AlA exchange: a CK-derived 'local' ATP from the reverse reaction (arrow 2 in Fig. 2) may directly activate forward AlA exchange of translocase (the concept of direct channelling); activated translocase in turn with the same high probability will provide CK with ATP for the forward reaction (a pathway like X in Fig. 1, AlA → $ATP_{loc}$ path in Fig. 2A).

Figure 2B depicts the reverse direction of the CK reaction, the synthesis of ATP and Cr from ADP and PCr. Translocase here exports ADP from matrix in four species (DmD, DlD, AmD, AlD) of exchange. This direction of translocase functioning is taken as reverse. Figure 2B scheme is a mirror image of Fig. 2A; the system is completely symmetrical. The forward CK reaction participates in reverse CK reaction activation through the reverse DlD exchange on translocase (arrow 1 in Fig. 2); the sources of activation of the reverse CK reaction are also multiple. It is important to note that different ways of activating translocase result in local activation of CK.

### CK activation from medium in pathways 'Z' and 'Y'

The CK reaction mechanism is of rapid equilibrium random binding BiBi type, according to Cleland's classification[15,20,26]. According to this mechanism the reaction rate in the forward direction is determined by interconversion of the central catalytically effective ternary complex E.A.Cr (enzyme–ATP–creatine) into that for the reverse reaction E.D.PCr (enzyme-ADP-phosphocreatine) with the rate constant $kk_1$. In the reverse reaction E.D.PCr is converted to E.A.Cr with the rate constant $kk_{-1}$. The equations for this reaction are given in several earlier papers[20,22,27]. At equilibrium or steady state the distribution of the enzyme between enzyme–substrate complexes can be expressed in terms of probabilities for the purpose of modelling of coupled reactions:

$P$ (E) = [E]/[Etot] = 1 / ( 1 + [Cr]/$K_{icr}$ + [PCr]/$K_{icp}$ + [A]/$K_{ia}$ + [D]/$K_{id}$ + [A]*[Cr]/($K_{ia}$*$K_{cr}$) + [PCr]*[A]/($K_{icp}$*$K_{Ia}$) + [D]*[PCr]/($K_{id}$*$K_{cp}$) + [D]*[Cr]/($K_{id}$*$K_{Icr}$)) = 1/Den          (1)

$P$ (E.A) = [E.A]/[Etot] = [A]/($K_{ia}$*Den)          (2)

$P$ (E.D) = [E.D]/[Etot] = [D]/($K_{id}$*Den)          (3)

$P$ (E.Cr) = [E.Cr]/[Etot] = [Cr]/($K_{icr}$*Den)          (4)

$P$ (E.PCr) = [E.PCr]/[Etot] = [PCr]/($K_{icp}$*Den)          (5)

$P$ (E.A.Cr) = [E.A.Cr]/[Etot] = [A]*[Cr]/($K_{ia}$*$K_{cr}$*Den)          (6)

$P$ (E.PCr.A) = [E.PCr.A]/[Etot] = [PCr]*[A]/($K_{icp}$*$K_{Ia}$*Den)          (7)

$P$ (E.PCr.D) = [E.PCr.D]/[Etot] = [PCr]*[D]/($K_{cp}$*$K_{id}$*Den)          (8)

$P$ (E.D.Cr) = [E.D.Cr]/[Etot] = [D]*[Cr]/($K_{id}$*$K_{Icr}$*Den)          (9)

In these equations $P$ is the probability for enzyme to subsist in the free state, E, or in the indicated enzyme–substrate complexes; the dissociation constants, $K$, of these complexes are given for primary

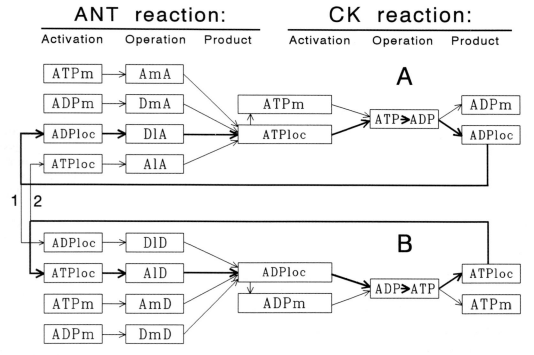

*Fig. 2. Schematic presentation of the interaction of adenine nucleotide translocase and CK in the complete system. (A). Activation of the forward CK reaction of PCr production by ATP from medium (ATPm) and by ATP from local sources (ATPloc) due to translocase forward exchange reactions. (B). Interaction between translocase and CK under conditions of CK reverse reaction of Cr production, which is initiated by ADP from the medium (ADPm) and by local ADP (ADPloc) from the different reverse exchange processes of translocase. Thick arrows mark the main circles of coupled CK activation in the forward (A) and reverse (B) directions. For further explanations see text.*

complexes with an index 'i' (initial), and for ternary complexes only with the symbol for the substrate: 'a' (A) for ATP, 'd' (D) for ADP, 'cr' (Cr) for creatine and 'cp' (PCr) for phosphocreatine. Dissociation constants of substrates from catalytically ineffective E.D.Cr and E.A.PCr dead-end complexes are given with the symbol 'I' (their formation is Inhibitory for the reaction); Etot is the total concentration of enzyme and Den is an abbreviation for denominator.

The rate of reaction product formation, ADP and PCr, in the forward reaction is given by the following equation:

$$V_{fw} = \text{Etot}*P \text{ (E.A.Cr)}*kk_1 \tag{10}$$

The rate of ATP and Cr formation in the reverse reaction is

$$V_r = \text{Etot}*P \text{ (E.D.PCr)}*kk_{-1} \tag{11}$$

The net rate of ADP and PCr production by CK is

$$V_{net} = V_{fw} - V_r \tag{12}$$

## The probabilities of translocase activation from medium in pathway 'W'

Mitochondrial adenine nucleotide translocase binds ATP and ADP in a competitive manner[28,29] both on the outer and on the inner matrix side of the mitochondrial inner membrane (Fig. 1). The equations for the probabilities of ATP and ADP binding on the outer side of translocase ($P$ (T.A)$_o$ and $P$ (T.D)$_o$, respectively) and on the inner side of the translocase ( $P$ (T.A)$_{in}$ and $P$ (T.D)$_{in}$, respectively) are[3,28]:

$$P \text{ (T.D)}_o = D_o/K_{do} / (1 + A_o/K_{ao} + D_o/K_{do}) \tag{13}$$

$$P \text{ (T.A)}_o = A_o/K_{ao} / (1 + A_o/K_{ao} + D_o/K_{do}) \tag{14}$$

$$P \text{ (T.A)}_{in} = A_i/K_{ai} / (1 + A_i/K_{ai} + D_i/K_{di}) \tag{15}$$

$$P \text{ (T.D)}_{in} = D_i/K_{di} / (1 + A_i/K_{ai} + D_i/K_{di}) \tag{16}$$

In these equations the dissociation constants for ATP ($K_a$) and ADP ($K_d$) have symbols to indicate the orientation of the adenine nucleotide binding sites: i for the inner side and o for outer side. The substrates for translocase also are given by these symbols to discriminate between their locations.

When translocase molecular complexes are occupied simultaneously by adenine nucleotides on both side, the translocase provides an exchange transfer of adenine nucleotides. The kind of exchange is determined and signed by the pair of bound adenine nucleotides, DmA, AmD, AmA and DmD (Fig. 2). The probabilities of the formation of these effective exchange complexes can be estimated[1–3] simply as a product of two independent partial probabilities, the probabilities of translocase occupation by substrates from medium, $P$ (T.A)$_o$ or $P$ (T.D)$_o$, and by substrates from matrix, $P$ (T.A)$_{in}$ or $P$ (T.D)$_{in}$:

$$P \text{ (T)}_{DmA} = P \text{ (T.D)}_o * sP \text{ (T.A)}_{in} \tag{f1}$$

$$P \text{ (T)}_{AmA} = P \text{ (T.A)}_o * sP \text{ (T.A)}_{in} \tag{f2}$$

$$F \text{ (T)}_{AmD} = P \text{ (T.A)}_o * sP \text{ (T.D)}_{in} \tag{r1}$$

$$P \text{ (T)}_{DmD} = P \text{ (T.D)}_o * sP \text{ (T.D)}_{in} \tag{r2}$$

These equations and main ones in the following sections will be repeatedly used in presteady state calculations (see pages 171–172). These equations will be distinguished by separate numbering with preceding the symbol 'f' (forward) or 'r' (reverse) to indicate the direction of the reaction. The steady state probability ($sP$) in these equations f,r is introduced to indicate that the probability changes its value in the presteady time calculations due to multiple interferences in the complex set of interacting reactions (Fig. 2, see pages 171–172).

The rates ($v_T$) of adenine nucleotide translocation in different exchanges of translocase can be given by following four Eqns 17–20 for four kinds of exchange:

$$v_{TDmA} = N_T * P \text{ (T)}_{DmA} * kt_1 \tag{17}$$

$$v_{TAmD} = N_T * P \text{ (T)}_{AmD} * kt_2 \tag{18}$$

$$v_{TDmD} = N_T * P \text{ (T)}_{DmD} * kt_3 \tag{19}$$

$$v_{TAmA} = N_T * P\ (T)_{AmA} * kt_4 \tag{20}$$

In these equations $N_T$ is the total concentration of translocase; $kt_1$–$kt_4$ are the rate constants for corresponding kinds of translocase exchange[3,16].

### The probabilities of translocase local activation in pathways 'X' and 'Y'

These probabilities are expressed by the following four equations[3]:

$$P\ (T)_{DIA} = P\ (CK_{ef})_{fw} * P\ (T_{Dloc})_{out} * sP\ (T.A)_{in} \tag{f3}$$
$$P\ (T)_{AIA} = P\ (CK_{ef})_r * P\ (T_{Aloc})_{out} * sP\ (T.A)_{in} \tag{f4}$$
$$P\ (T)_{AID} = P\ (CK_{ef})_r * P\ (T_{Aloc})_{out} * sP\ (T.D)_{in} \tag{r3}$$
$$P\ (T)_{DID} = P\ (CK_{ef})_{fw} * P\ (T_{Dloc})_{out} * sP\ (T.D)_{in} \tag{r4}$$

where $P(CK_{ef})_{fw}$ and $P\ (CK_{ef})_r$ are the total probabilities of CK activation in the forward and reverse directions, respectively; $P(T_{Dloc})_{out}$ and $P(T_{Aloc})_{out}$ designate the general probability of CK-supplied local ADP (Aloc) or local ATP (Aloc) to meet with translocase on its outer side.

### The total probability of translocase activation in pathways 'X', 'Y' and 'W'

The total probability of translocase activation in the forward ($P\ (T_{ef})_{fw}$) or reverse ($P\ (T_{ef})_r$) directions is the sum of exchanges, described by Eqns f1–f4 and r1–r4, respectively:

$$P\ (T_{ef})_{fw} = P\ (T)_{DmA} + P\ (T)_{AmA} + P\ (T)_{DIA} + P\ (T)_{AIA} \tag{f5}$$
$$P\ (T_{ef})_r = P\ (T)_{AmD} + P\ (T)_{DmD} + P\ (T)_{AID} + P\ (T)_{DID} \tag{r5}$$

### The total probability of CK forward reaction activation

The total probability of CK forward reaction activation ($P\ (CK_{ef})_{fw}$) is a sum of CK activations[3] from medium ($sP\ (E.A.Cr)$) and from translocase-supplied local ATP (Aloc) in the pathways X,Y,W (Fig. 2):

$$P\ (CK_{ef})_{fw} = sP\ (E.A.Cr) + P\ (T_{ef})_{fw} * sP\ (CK_{Alact}) \tag{f6}$$

The multiplier $sP\ (CK_{Alact})$ in Eqn f6 is the integral steady–state probability of E.Al.Cr complex formation (Aloc activation of CK) from the donor complexes E.Cr, E and E.PCr. This multiplier can be calculated from a separate set of equations:

$$sP\ (CK_{Alact}) = P\ (CK_{Aloc}) * [sP\ (E.Cr) + sP\ (E)*Pc1 + sP\ (E.PCr)\ *Pc2 * Pc1\ ] \tag{f7}$$
$$Pc1 = [Cr]*K_{Cr+}\ /([Cr]*K_{Cr+} + k_{-1}) \tag{21}$$
$$Pc2 = k_{-4}\ /\ (k_{-4} + k_{-3}) \tag{22}$$

where $P\ (CK_{Aloc})$ is the probability for translocase-supplied local ATP to meet with CK; Pc1 (partition coefficient[9]) is the probability of $CK.A_{loc}$ complex transformation to $E.A_{loc}.Cr$ complex; Pc2 is the probability of $CK.PCr.A_{loc}$ complex transformation to $CK.A_{loc}$ complex; $K_{Cr+}$ is the diffusion-limited association rate constant for Cr; $k_{-1}$ is the rate constant for ATP dissociation from $CK.A_{loc}$ complex; $k_{-3}$ and $k_{-4}$ are the rate constants for ATP and PCr dissociation, respectively, from $E.A_{loc}.PCr$ complex. The details of these equations can be found in our previous papers[1,3].

### The total probability of CK reverse reaction activation

The total probability of CK reverse reaction activation ($P\ (CK_{ef})_r$) is the sum[3] of CK activations from medium ($sP\ (E.D.PCr)$) and from translocase-supplied local ADP (Dloc) in the pathways X,Y,W (Fig. 2):

$$P\ (CK_{ef})_r = sP\ (E.D.PCr) + P\ (T_{ef})_r * sP\ (CK_{Dlact}) \tag{r6}$$

The multiplier $sP\ (CK_{Dlact})$ in Eqn r6 is the integral steady state probability of E.Dl.PCr complex formation (Dloc activation of CK) from the donor complexes E.PCr, E and E.Cr. This multiplier can be calculated from a separate set of equations:

$$sP\ (CK_{Dlact}) = P\ (CK_{Dloc}) * [\ sP\ (E.PCr) + sP\ (E)*Pc3 + sP\ (E.Cr) * Pc4 * Pc3\ ] \tag{r7}$$
$$Pc3 = [PCr]*K_{cp+}\ /\ (\ [PCr]*K_{cp+} + k_{-5}\ ) \tag{23}$$

$$Pc4 = k_{-8} / (k_{-8} + k_{-7}) \tag{24}$$

where $P$ ($CK_{Dloc}$) is the probability for translocase-supplied local ADP to meet with CK; PC3 in Eqn 23 is the probability of $CK.D_{loc}$ complex transformation to $CK.D_{loc}.PCr$ complex; $K_{cp+}$ is the diffusion-limited association rate constant for PCr; $k_{-5}$ is the rate constant for ADP dissociation from $CK.D_{loc}$ complex. Pc4 in Eqn r9 is the probability of $CK.D_{loc}.Cr$ complex transformation to $CK.D_{loc}$ complex; $k_{-7}$ and $k_{-8}$ are the rate constants for ADP and Cr dissociation, respectively, from $CK.D_{loc}.Cr$ complex. The probabilities Pc3 and Pc4 are described in details in our previous paper[3].

### The arrangement of calculations in time

The set of interacting equations f1–f7, r1–r7 can be solved numerically only in the real time scale, as the rate constants of translocase and CK differ considerably. For example, the rate constant for the CK reverse reaction exceeds that for the ATP/ATP exchange of translocase by about 12 times (see 'Choice of the parameters for modeling'). This means that during one persisting cycle of ATP/ATP translocation CK would additionally activate AIA exchange (Fig. 2) about eleven times. It is clear that at each of the eleven activations from CK the population of translocase donor complexes available for AIA activation will be progressively decreased. If we take into account that in the complete system AIA activations from CK take place in parallel with other reactions (Fig. 2) with differing rate constants, the changes in the population of donor molecules available for activation of translocase will be more profound and complex. All such changes ($sP$ probabilities in Eqns f1–f7, r1–r7) are taken into account in the mathematical model on the basis of the assumed time sequence of catalysis and translocation cycles in the coupled system.

According to the reaction mechanism of CK, the binding and dissociation of substrates and products is very rapid and the reaction rate is determined by interconversion of the catalytically effective ternary complexes[10,15,20]. The enzyme molecules that perform the catalysis do not freely exchange substrate with the surrounding medium[6,7]. This means that CK can locally provide translocase with ATP or ADP only very shortly after the end of the rate-limiting catalysis cycle and, vice versa, can be locally supplied by translocase products only very shortly before the catalysis cycle. This topic is discussed in more details in our previous paper[3]. Figure 3 illustrates the accepted sequence of events in the complete coupled system.

### Time-dependent changes in the populations of CK complexes

Figure 3 shows the simulated presteady–state kinetics of CK activation; it can be used to illustrate the mathematical description of the events in time. We have simulated the mitochondrial respiration and CK activation in medium, containing ATP (0.15 mM), ADP (0.02 mM), Cr (26 mM) and PCr (1.6 mM). Figure 3A illustrates the time sequence of forward CK reaction activation ($CK_{fw}$) by different sources of ATP supply: ATP from medium ($ATP_m$) and local ATP from DmA (DmA) and DIA (DIA) exchanges of translocase. Horizontal lines show the probability values and time of activation of CK from different sources. The resulting total activation of the CK forward reaction ($CKef_{fw}$), which is a sum of different activations, and transient levels of E, E.Cr and E.PCr donor complexes of CK (Eqns f7, r7), used in its local activation, are depicted in Fig. 3B. After addition of substrates at zero time CK and translocase are activated according to the classic equations for free enzymes (Eqns 1–9 for CK and f1–f2, r1–r2 for translocase). Because of chosen substrate concentrations (ATP = $K_a$, Cr = $K_{icr}$, PCr = $K_{icp}$) initially the levels of E.A.Cr, E, E.Cr and E.PCr in Fig. 3B coincide at time 0–4. At the time moment 4 CK receives $ATP_{loc}$ from the accomplished first cycle of DmA exchange of T and is activated from DmA source in addition to $ATP_m$ one (Fig. 3A, time 4–8). As a result, the total activation of CK ($CKef_{fw}$) is more than doubled (Fig. 3B, time 4–8) at the expense of significant decrease in levels of E, E.Cr and E.PCr CK donor complexes (Fig. 3B, time 4–8). At time 8, besides local DmA and medium ATP activations CK effective complexes are also formed by local DIA activation (Fig. 3A, time 8–12) from the just completed cycle of DIA exchange, and also from completed AmA and AIA exchange cycles. The latter activations are not shown in Fig. 3A because of their very low value, less than 0.01. Nevertheless, the influence of AIA exchange is clearly seen in the integral curve $CKef_{fw}$ (Fig. 3B) as small steps of activity change within the time interval 9–12. As the result of these local activations, the levels of CK donor complexes E, E.Cr and E.PCr are further decreased at time 8–12 and thereafter (Fig. 3B). AlA activations are started by $CK_r$ at each time-point, and because of this the concentrations of donor complexes are recalculated at each time-point. General equations of these periodical calculations for any time-point are:

$$sP\ (E)_{(t)} = P\ (E)_{(tf)} - P\ (E)_{(tf)} * P\ (CK_{Aloc}) * Sum\ P\ (T_{ef})_{fw(t)} * Pc1 \tag{f8}$$
$$sP\ (E.Cr)_{(t)} = P\ (E.Cr)_{(tf)} - P\ (E.Cr)_{(tf)} * P\ (CK_{Aloc}) * Sum\ P\ (T_{ef})_{fw(t)} \tag{f9}$$
$$sP\ (E.PCr)_{(t)} = P\ (E.PCr)_{(tf)} - P\ (E.PCr)_{(tf)} * P\ (CK_{Aloc}) * Sum\ P\ (T_{ef})_{fw(t)} * Pc1 * Pc2 \tag{f10}$$
$$sP\ (E.A.Cr)_{(t)} = sP\ (E.Cr)_{(t)} * A/K_a \tag{f11}$$
$$sP\ (E.D.PCr)_{(t)} = sP\ (E.PCr)_{(t)} * D/K_d \tag{r8}$$

In this set of equations the symbols (t) and (tf) in assignment of probability mean the considered current time and time of formation of this probability, respectively. Sum $P\ (T_{ef})_{fw(t)}$ in Eqns f8—10 is the sum of all translocase sources for CKfw local activation (see Fig. 2A) at the considered time moment $t$. This sum usually includes the values of AlA, DlA and DmA exchanges all accomplished at time $t$ and at the certain times the accomplished AmA ones. This is because AmA exchange is activated rarely in comparison to DmA exchange and $CK_{fw}$ cycling rate ($kt_4/kk_1$ is assumed 1/3 in our model, see 'Choice of the parameters for modeling'). Eqns f11 and r8 are analogous to Eqns 6 and 8, respectively; they reflect the consequences of changes in concentrations of CK donor complexes.

Eqns f8–f10 can be used to find $sP\ (CK_{Alact})_{(t)}$ and $sP\ (CKDlact)_{(t)}$ from Eqns f7 and r7, respectively.

## Time-dependent changes in the populations of translocase complexes

Now to solve Eqns f6 and r6 we must define $P\ (T_{ef})_{fw}$ and $P\ (T_{ef})_r$ and the co-related presteady state probabilities in Eqns f1–f5, r1–r5. The terms $sP\ (T.A)_{in}$ and $sP\ (T.D)_{in}$ in these equations are the real quantities of translocase complexes with matrix ATP and ADP, respectively, available for activation of translocase from its outer side at the time of consideration. We need to subtract from the initial values of $P\ (T.A)_{in}$ and $P\ (T.D)_{in}$ the corresponding values of effective adenine nucleotide translocating complexes already existing at the time $t$. So for $sP\ (T.A)_{in(t)}$ and $sP\ (T.D)_{in(t)}$ in Eqns f1–f4 and r1–r4, respectively, we have:

$$sP\ (T.A)_{in(t)} = P\ (T.A)_{in} - Sum\ P\ (T_{ef})_{fw(t)} \tag{f12}$$
$$sP\ (T.D)_{in(t)} = P\ (T.D)_{in} - Sum\ P\ (T_{ef})_{r(t)} \tag{r9}$$

where Sum $P\ (T_{ef})_{fw(t)}$ in Eqn. f12 is the sum of the all existing effective complexes of forward reaction of translocase at the considered time moment $t$; Sum $P\ (T_{ef})_{r(t)}$ in Eqn r9 is the sum of all existing effective complexes of the reverse reaction of the translocase at the considered time $t$. The formulated set of equations f1–f12, r1–r9 is the one used for calculations in our Pascal programe. This set of equations is calculated for every relative time unit until the steady-state probabilities will be constant in time. Finally, the steady-state any probabilities of interest are integrated at the time periods like interval 8–16 in Fig. 3 to obtain mean probability values. We display the net rate of PCr production by coupled mitochondrial CK ($CK_{mit}$) as:

$$v_{PCrnet} = N_{CK} * (\text{mean } sP\ (CK_{ef})_{fw} * kk_1 - \text{mean } sP\ (CK_{ef})_r * kk_{-1}) \tag{25}$$

and the net rate for ATP export by the translocase of respiring mitochondria as:

$$v_{Tnet} = N_T * [\ (\text{mean } sP\ (T)_{DmA} + \text{mean } sP\ (T)_{DlA}) *$$
$$kt_1 - (\text{mean } sP\ (T)_{AmD} + \text{mean } sP\ (T)_{AlD}) * kt_2\ ] \tag{26}$$

$N_{CK}$ and $N_T$ in Eqns 25 and 26 are the known values of the contents of $CK_{mit}$ and translocase, respectively, in mitochondria.

All calculations were carried out on an IBM-compatible PC/AT 286/287 computer. The Pascal program we used is available from M.K. Aliev on request.

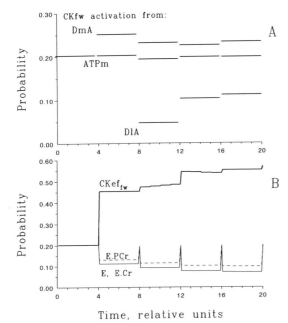

Time, relative units

*Fig. 3. Simulation of the early presteady kinetics of forward CK reaction activation in the complete coupled system. (A). Time dependence of partial activations of CK from DmA and DlA exchanges of translocase and by ATP from medium (ATPm). (B). Time dependence of total activation of CK (CKef_fw) from ATPm and DmA, DlA, AmA, AlA exchanges of translocase and of the levels of E , E.Cr and E.PCr CK donor complexes, used in the process of local activation of CK. The unit of relative time is the duration of the cycle of the reverse CK reaction, which is the fastest in the system. Simulation parameters: as in 'Choice of the parameters for modeling', but $kt_3 = kt_4 = kk_1/2$ to speed the events in time. For further explanations see text.*

## Choice of the parameters for modelling

The following parameters were used. For the mitochondrial forward CK reaction all the data were taken from Jacobus & Saks[14]: $K_{ia} = 0.75$ mM; $K_{icr} = 26$ mM; $K_{icp} = 1.6$ mM; $K_a = 0.15$ mM; $K_{cr} = 5.2$ mM; $K_{Icp} = 24$ mM; $K_{Ia} = 11.25$ mM was calculated from the thermodynamic equation:

$$K_{icp} * K_{Ia} = K_{ia} * K_{Icp}. V_{fwmax} = NCK$$
$$* kk_1 = 1.0 \text{ μmol/(min * mg of protein)}.$$

For the reverse CK reaction we have accepted[3]: $K_{Icr} = K_{cr}$; $K_{cp} = 1.0$ mM; $K_{id} = 0.208$ mM; $K_d = 0.13$ mM; $K_{Id} = 0.042$ mM; $V_{rmax} = NCK * kk_{-1} = V_{fwmax} * 4.0$. The value of the apparent equilibrium constant of the CK reaction has been settled to 0.0133 at pH 7.4.

For calculation of $k_{-1}$ the value of the diffusion-limited association rate constant for ATP, $K_{a+} = 2 \times 10^7$ M $-1g^{-1}$ was used[12]. From that value we found[1] $k_{-1} = 15 \times 10^3$ s$^{-1}$. For calculation of Pc1 the value of the constant was taken to be twice the $K_{a+}$ value, $K_{Cr+} = 4 \times 10^7$ M $-1g^{-1}$, as the Cr and PCr molecules diffuse about two-fold faster than larger ATP ones[18]. Other rate constants were[3]: $k_{-3} = 225 \times 10^3$ s$^{-1}$; $k_{-4} = 960 \times 10^3$ s$^{-1}$; $k_{-5} = 64 \times 10^3$ s$^{-1}$; $k_{-7} = 840$ s$^{-1}$; $k_{-8} = 1040 \times 10^3$ s$^{-1}$.

$P$ (T.A)$_{in} = 0.9$ was found by a method of best approximation to the experimental data[1]. The value of this parameter was taken to be constant in each particular experiment, as the matrix ATP/ADP ratio has been shown to be constant, at about 4, on stimulation of mitochondrial respiration from 0 to 75 per cent of maximum[11]. $P$ (T.D)$_{in}$ was taken to be 1.0 − 0.9 = 0.1.

The probabilities of CK-derived local ADP or ATP meeting with translocase, $P$ (T$_{Dloc}$)$_{out}$ and $P$ (T$_{Aloc}$)$_{out}$, respectively, and of translocase-derived local ATP or ADP meeting with CK, $P$ (CK$_{Aloc}$) and $P$ (CK$_{Dloc}$), respectively, were taken to be equal to 1.0 (using the concept of direct channelling).

$K_{do}$ and $K_{ao}$ for translocase have been taken to be 0.02 mM[19] and 20 mM[13], respectively. $kt_1$, the rate constant for ADP$_o$/ATP$_i$ exchange of translocase, has been taken equal to $kk_1$, the rate constant for the CK forward reaction[1]. Based on our modelling of Kramer & Klingenberg's data[16] on kinetics of reconstituted adenine nucleotide translocase (our unpublished results) we have arbitrarily chosen the following rate constants for the remaining ATP$_o$/ADP$_i$, ADP$_o$/ADP$_i$, ATP$_o$/ATP$_i$ exchanges of translocase, respectively: $kt_2 = kt_1/2$; $kt_3 = kt_1/3$; $kt_4 = kt_1/3$. The differing rate constants for translocase exchanges have been previously considered by Bohnensack in his kinetic model of mitochondrial adenine nucleotide translocase[8]. The number of functioning molecular complexes of CK$_{mit}$ and translocase has been taken equal[1,24].

The concentrations of ATP, ADP, Cr and PCr were the same as those we used experimentally[14,26]. When indicated, the concentration of ADP was taken to be zero.

## Results and discussion

In 1993, using an incomplete model with only one (DlA) exchange on the translocase, we successfully simulated the experimental data of Jacobus & Saks[14], who in 1982 exhaustively explored the mitochondrial CK reaction in the forward direction both in the absence of mitochondrial respiration and coupling of this reaction to mitochondrial oxidative phosphorylation. Now, using a complete system with all eight species of translocase exchange, we have simulated all the experimental data of Jacobus & Saks[14]. In Table 1 the final results of this simulation (column 4) are compared with experimental values (column 3) and also with the results of a previous[1] simulation (column 5).

As can be seen from the data presented in Table 1, in both cases the simulation may be regarded as satisfactory. The most interesting and important feature of the data is the theoretical confirmation that in the case of a mitochondrial CK reaction coupled with oxidative phosphorylation the value of $K_m$ for MgATP ($K_a$) is decreased about ten times, while the apparent affinity of the E.MgATP complex for creatine ($1/K_{cr}$) is not changed. That shows an apparent specific elevation of the affinity of the system for MgATP. Other features of the simulation have been discussed in detail previously[1], but two new peculiarities deserve special consideration.

First, in the complete coupled system the simulated decrease in $K_a$ is more profound (12.5-fold, columns 4 and 2) than that estimated previously (10-fold, columns 5 and 2). The difference is due to a stimulating action of AmA exchange, as this exchange, as well as the AmD one (Fig. 2) occurring in medium with ATP and Cr, was ignored for simplicity in our previous publication[1].

**Table 1. Experimental and simulated kinetic constants for rat heart mitochondrial CK as a function of oxidative phosphorylation (OP)**

| Constants | Without OP (mM) | | With OP (mM) | | |
| | Experimental | Used in model | Experimental | Simulated Now | + Ref1 |
| | 1 | 2 | 3 | 4 | 5 |
| --- | --- | --- | --- | --- | --- |
| $K_{ia}$ | $0.75 \pm 0.06$ | 0.75 | $0.29 \pm 0.04$ | 0.14 | 0.15 |
| $K_a$ | $0.15 \pm 0.01$ | 0.15 | $0.014 \pm 0.005$ | 0.012 | 0.015 |
| $K_{icr}$ | $28.8 \pm 8.45$ | 26.0 | $29.4 \pm 12.0$ | — | — |
| $K_{cr}$ | $5.20 \pm 0.30$ | 5.20 | $5.20 \pm 2.30$ | 5.25 | 5.20 |
| $K_{icp}$ | $1.60 \pm 0.20$ | 1.60 | $1.40 \pm 0.20$ | 1.15 | 0.91 |
| $K_{Icp}$ | $20-50$ | 24.0 | $20-50$ | 9.0–14.6 | 24.0 |
| $V_{fwmax}$ | $1.06 \pm 0.20$ | 1.00 | $0.99 \pm 0.07$ | 0.90 | 0.90 |

Oxygraph measurements were conducted at 30 °C in medium containing 0.25 M sucrose, 10 mM HEPES, pH 7.4, 0.2 mM EDTA, 5 mM potassium phosphate, 5 mM potassium glutamate, 2 mM potassium malate, 3.3 mM Mg(OAc)2, 0.3 mM dithiothreitol, 1.0 mg/ml bovine serum albumin, 0.5–0.1 mg/ml rat heart mitochondrial protein. In respiring mitochondria the ATP/ADP translocation rates were calculated from oxygen consumption rates. In mitochondria, where respiration was completely inhibited by pretreatment with 10 $\mu$M rotenone and 5 $\mu$g/mg oligomycin, the CK reaction rates were determined spectrophotometrically, using a phosphoenolpyruvate (2 mM) – pyruvate kinase (2 IU/ml) system to regenerate exogenous ATP. Mean values and SD are given for five experiments[14]. Simulation parameters are indicated in section 'Choice of the parameters for modeling'. The dissociation constants are given in mM and $V_{fwmax}$ in $\mu$mol ATP*min$^{-1}$*mg$^{-1}$.

Second, in the complete coupled system, $K_{Icp}$, the simulated dissociation constant of PCr from E.A.PCr dead-end complex, is lower (Table 1, column 4) than that in the incomplete system (Table 1, column 5), and has two values, 9.0 and 14.6 mM. These two values were obtained from graphical data analysis, using Eqns 6 and 9, respectively, from the original paper of Jacobus & Saks[14]. These unexpected findings probably reflect the experimental uncertainty in $K_{Icp}$ determination, demonstrated by Jacobus

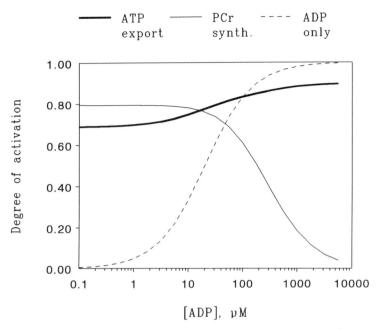

*Fig. 4. Simulated apparent ADP independence of mitochondrial respiration and ATP export (ATP export) under conditions of activation of CK by substrates, 0.75 mM ATP and 25 mM creatine, from medium. The 'ATP export' curve is calculated from Eqn. 26, 'PCr synthesis' from Eqn 25. The 'ADP only' curve is calculated for medium without ATP and Cr from Eqn f1.*

& Saks[14]. They undoubtedly originate from the complexity of the system, as in the presence of ATP, Cr and PCr we have by model an activated CK reverse reaction (from AmD, DlD, AlD – supplied 'local' ADP, see Fig. 2B), which in turn additionally stimulates the CK forward reaction through AlA exchange by the translocase (Fig. 2A). The results of slight activation of the CK reverse reaction predicted by the model remains to be verified by experiment.

All values in Table 1 were obtained for reaction medium without ADP, when respiration was stimulated by ADP, locally supplied by CK in a coupled system. In such a coupled system the classic concept of $K_m$ for ADP becomes meaningless.

To further investigate this topic, we have performed the simulation of adding increased concentrations of ADP to mitochondria respiring in the medium with two substrates of the forward CK reaction, ATP (0.75 mM) and Cr (25 mM) (Fig. 4). In such a system the ADP addition only slightly increases the respiration of mitochondria (curve 'ATP export'), which is already high at ADP = 0 due to local ADP production in the CK reaction. The PCr production curve (curve 'PCr synthesis') shows an initial high rate, which is then reversed due to high ADP: a gradual switching of the system from the regimen of sole CK-supported activation of oxidative phosphorylation (0–1 μM ADP in the medium) to the regimen of $ADP_m$ (medium ADP) - supported one (millimolar ADP concentrations in the medium; see for comparison the 'ADP only' theoretical curve, drawn for the case of no ATP and creatine with sole activation of oxidative phosphorylation by ADP from the medium with $K_{mADP} = 20 \mu M$, Eqn f1).

Thus, the data of Fig. 4 clearly show that in a coupled system with both substrates of the CK forward reaction, ATP and Cr, evaluation of the apparent $K_m$ for ADP for activation of oxidative phosphorylation loses any meaning.

Figure 5 shows the results of simulation of the respiration-induced shift of the mitochondrial CK reaction off the equilibrium experimentally described several years ago[26]. Figures 5A and B reproduce these data[26]. In these experiments the rat heart mitoplasts (mitochondria with disrupted outer membrane) were added to the medium, which initially contained all CK substrates: 0.12 mM ATP, 0.05 mM ADP, 40 mM Cr and 4 mM PCr. The mass action ratio in this medium calculated from the measured concentrations of substrates[26] was $5.6 \times 10^{-2}$ and considerably higher than an experimentally measured apparent equilibrium constant for $CK_{mit}$ ($0.79 \times 10^{-2}$). After the steady state was reached, the samples were periodically withdrawn and analysed for ATP, ADP, Cr and PCr contents. Figure 5A shows changes in the PCr and ADP concentrations, and Fig. 5B the experimentally determined mass action ratio. The dotted line in Fig. 5B shows the experimentally observed value of the apparent equilibrium constant, $K_{eq}= [PCr] \times [MgADP]/[Cr] \times [MgATP]$, which was equal[26] to $0.79 \times 10^{-2}$. In the presence of oxidative phosphorylation the mass action ratio, instead of approaching the apparent $K_{eq}$ value, deviated in the direction of higher values due to aerobic PCr production (Fig. 5B). Figure 5C shows the experimental relationship between the rate of PCr production and the mass action ratio, obtained by nonlinear regression analysis of the PCr production curve in Fig. 5A. The values of substrate concentrations corresponding to different mass action ratios were used for calculations of PCr production rates under aerobic conditions, using the mathematical model of coupled reactions described in the previous section. The results of these calculations are shown in Fig. 5C. This figure shows good approximation of experimental and calculated data: in both cases the rates are close to each other and positive in spite of the fact that the mass action ratio exceeds the value of the equilibrium constant and, consequently, on the basis of thermodynamic predictions for pure solutions, the rate of PCr production should be negative (PCr utilization).

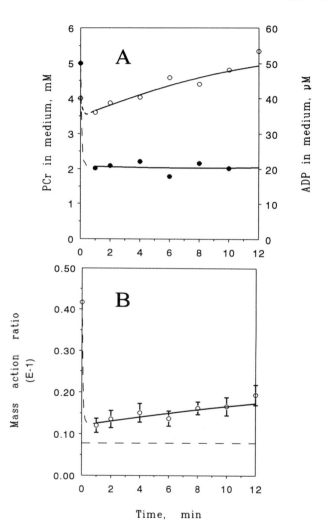

*Fig. 5A & B. Analysis of experimental data from Fig. 9 of Saks et al.[26] on the respiration-induced shift of the CK reaction out of equilibrium. (A) Changes in PCr (O) and MgADP (●) concentrations. (B) Changes in mass action ratio when the mitochondrial CK reaction was coupled to oxidative phosphorylation. Mean values for five experiments and SD (bars) are presented. For further explanations see text.*
*(Data reproduced from Saks et al.26 with permission.)*

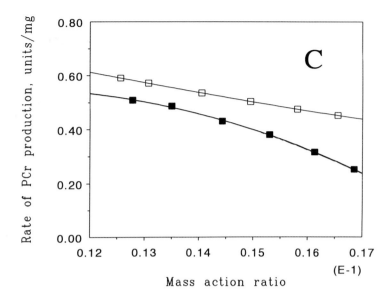

*Fig. 5C: Experimental (■⇨□) and simulated aerobic PCr production rates*
*as a function of mass action ratio in the*
*medium. The data were simulated at* $V_{fwmax}$ = 2.22 *μmol/min per mg*
*mitoplast protein. For further explanations see text.*

All these results show that the mathematical model of the coupled mitochondrial reactions of PCr production both qualitatively and quantitatively describe the experimental data. This allows us to use the model for theoretical analysis of the behaviour of the coupled reactions in mitochondria. One of the most interesting questions in this respect is the efficiency of regulation of the rate of oxidative phosphorylation in heart mitochondria. According to metabolic control analysis, the efficiency of any step of a consecutive reaction in flux regulation is given by the flux control coefficient[31], $C$. We have used the developed model to estimate the flux control coefficient for the CK reaction, $C$ = $(dV_{cp}/V_{cp})/(dV_{max}{}^{ck}/V_{max}{}^{ck})$, calculating the rate of oxygen consumption, ATP transport and aerobic PCr production ($V_{cp}$) when the activity of CK, $V_{max}{}^{ck}$, was changed by 1 per cent (1 per cent inhibition). The non-competitive high-affinity inhibition of CK was simulated by a corresponding decrease of $P$ (E) in Eqn 1. For example, 1 per cent inhibition was attained by multiplying $P$ (E) in Eqn 1 by 0.99. This analysis for alterating ATP concentrations in the presence of 25 mM creatine (PCr = 0, ADP = 0) is given in Fig. 6. At high ATP concentrations the flux control coefficient for CK in regulation of overall rate of oxidative phosphorylation and PCr production is close to 1. However, when the ATP concentration decreases and approximates zero value, the flux control coefficient increases very rapidly and approximates a value as high as 8.39. This is probably the highest value of the flux control coefficient ever reported, and shows the potentially very high efficiency of the coupled mitochondrial CK reaction in regulation of oxidative phosphorylation due to increased turnover of adenine nucleotides in these coupled reactions[2]. Because of this, the coupled CK reaction in mitochondria may be a very effective amplifier of any metabolic signals from cytoplasm[24].

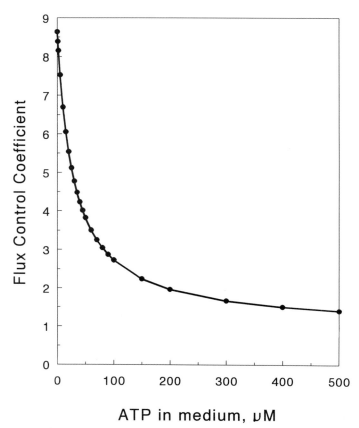

ATP in medium, µM

*Fig. 6. Calculated flux control coefficients for the forward-coupled creatine kinase reaction as a function of ATP concentrations at 25 mM Cr. For further explanations see text.*

## References

1.  Aliev, M.K. & Saks, V.A. (1993): Quantitative analysis of the 'phosphocreatine shuttle'. I. A probability approach to the description of phosphocreatine production in the coupled creatine kinase–ATP/ADP translocase –oxidative phosphorylation reactions in heart mitochondria. *Biochim. Biophys. Acta* **1143**, 291–300.

2.  Aliev, M.K. & Saks, V.A. (1993): Mathematical model of the creatine kinase reaction coupled to adenine nucleotide translocation and oxidative phosphorylation. In: *Modern Trends in Biothermokinetics*, Eds. S. Schuster, M. Rigoulet, R. Ouhabi & J.-P. Mazat, pp. 85–92. New York: Plenum Press.

3.  Aliev, M.K. & Saks, V.A. (1994): Mathematical modelling of intracellular transport processes and the creatine kinase systems: a probability approach. *Mol. Cell. Biochem.* **133/134,** 333–346.

4.  Barbour, R.L., Ribaudo, J. & Chan, S.H.P. (1984): Effect of creatine kinase activity on mitochondrial ADP/ATP transport. Evidence for a functional interaction. *J. Biol. Chem.* **259,** 8246–8251.

5.  Bessman, S.P. & Geiger, P.J. (1981): Transport of energy in muscle: the phosphocreatine shuttle. *Science* **211,** 448–452.

6.  Blumenfeld, L.A. (1983): *Physics of Bioenergetic Processes*. (Springer series. in synergetics, vol. **16**). Berlin: Springer-Verlag.

7.  Blumenfeld, L.A. & Tikhonov, A.N. (1994): *Biophysical Thermodynamics of Intracellular Processes. Molecular Machines of the Living Cell*. New York: Springer-Verlag.

8.   Bohnensack, R. (1982): The role of the adenine nucleotide translocator in oxidative phosphorylation. A theoretical investigation on the basis of a comprehensive rate law of the translocator. *J. Bioenerg. Biomembr.* **14**, 45–61.

9.   Boyer, P.D., de Meis, L., Carvalho, M.G. & Hackney, D.D. (1977): Dynamic reversal of enzyme carboxyl group phosphorylation as the basis of the oxygen exchange catalyzed by sarcoplasmic reticulum adenosine triphosphatase. *Biochemistry* **16**, 136–140.

10.   Cleland, W.W. (1967): Enzyme kinetics. *Ann. Rev. Biochem.* **36**, 77–112.

11.   Davis, E.J. & Lumeng, L. (1975): Relationship between the phosphorylation potentials generated by liver mitochondria and respiratory state under conditions of adenosine diphosphate control. *J. Biol. Chem.* **250**, 2275–2292.

12.   Froehlich, J.P. & Taylor, E.W. (1975): Transient state kinetics studies of sarcoplasmic reticulum adenosine triphosphatase. *J. Biol. Chem.* **250**, 2013–2021.

13.   Jacobus, W.E., Moreadith, R.W. & Vandegaer, K.M. (1982): Mitochondrial respiratory control: evidence against the regulation of respiration by extramitochondrial phosphorylation potentials or by [ATP]/[ADP] ratios. *J. Biol. Chem.* **257**, 2397-2402.

14.   Jacobus, W.E. & Saks, V.A. (1982): Creatine kinase of heart mitochondria: changes in its kinetic properties induced by coupling to oxidative phosphorylation. *Arch. Biochem. Biophys.* **219**, 167–178.

15.   Kenyon, G.L. & Reed, G.H. (1983): Creatine kinase: structure–activity relationships. *Adv. Enzymol.* **54**, 367–426.

16.   Kramer, R. & Klingenberg, M. (1982): Electrophoretic control of reconstituted adenine nucleotide translocation. *Biochemistry*, **21**, 1082-1089.

17.   Kuby, S.A. & Noltmann, E.A. (1962): Adenosine triphosphate-creatine transphosphorylase. In: *The Enzymes*, eds. P.D. Boyer, H. Lardy & K. Myrback, vol. 6, pp.515-603. New York: Academic Press.

18.   Meyer, R.A., Sweeney, H.L. & Kushmerick, M.J. (1984): A simple analysis of the 'phosphocreatine shuttle'. *Am. J. Physiol.* **246**, C365-C377.

19.   Moreadith, R.W. & Jacobus, W.E. (1982): Creatine kinase of heart mitochondria: functional coupling of ADP transfer to the adenosine nucleotide translocase. *J. Biol. Chem.* **257**, 899–905.

20.   Morrison, J.F. & James, E. (1965): The mechanism of the reaction catalyzed by adenosine triphosphate-creatine phosphotransferase. *Biochem. J.* **97**, 37–52.

21.   Muller, M., Moser, R., Cheneval, D. & Carafoli, E. (1985): Cardiolipin is the membrane receptor for mitochondrial creatine phosphokinase. *J. Biol. Chem.* **260**, 3839–3843.

22.   Saks, V.A., Chernousova, G.B., Gukovsky, D.E., Smirnov, V.N. & Chazov, E.I. (1975): Studies of energy transport in heart cells. Mitochondrial isoenzyme of creatine phosphokinase: kinetic properties and regulatory action of Mg$^{2+}$ ions. *Eur. J. Biochem.* **57**, 273–290.

23.   Saks, V.A., Khuchua, Z.A. & Kuznetsov, A.V. (1987): Specific inhibition of ATP–ADP translocase in cardiac mitoplasts by antibodies against mitochondrial creatine kinase. *Biochim. Biophys. Acta* **891**, 138–144.

24.   Saks, V.A., Khuchua, Z.A., Vasilyeva, E.V., Belikova, Y.O. & Kuznetsov, A.V. (1994): Metabolic compartmentation and substrate channeling in muscle cell:. role of coupled creatine kinases in in vivo regulation of cellular respiration – a synthesis. *Mol. Cell. Biochem.* **133/134**, 155–192.

25.   Saks, V.A., Kupriyanov, V.V., Elizarova, G.V. & Jacobus, W.E. (1980): Studies of energy transport in heart cells: the importance of creatine kinase localization for the coupling of mitochondrial phosphorylcreatine production to oxidative phosphorylation. *J. Biol. Chem.* **255**, 755–763.

26.   Saks, V.A., Kuznetsov, A.V., Kupriyanov, V.V., Miceli, M.V. & Jacobus, W.E. (1985): Creatine kinase of rat heart mitochondria: the demonstration of functional coupling to oxidative phosphorylation in an inner membrane-matrix preparation. *J. Biol. Chem.* **260**, 7757–7764.

27.   Saks, V.A., Lipina, N.V., Smirnov, V.N. & Chazov, E.I. (1976): Studies of energy transport in heart cells. The functional coupling between mitochondrial creatine phosphokinase and ATP–ADP translocase: kinetic evidence. *Arch. Biochem. Biophys.* **173**, 34–41.

28.   Souverijn, J.H.M., Huisman, L.A., Rosing, J. & Kemp, A. (1973): Comparison of ADP and ATP as substrates for the adenine nucleotide translocator in rat-liver mitochondria. *Biochim. Biophys. Acta* **305**, 185–198.

29.   Vignais, P.V., Brandolin, G., Boulay, F., Dalbon, P., Block, M.R. & Gauche, I. (1989): Recent developments in the study of the conformational states and the nucleotide binding sites in the ADP/ATP carrier. In: *Anion Carriers of Mitochondrial Membranes*, eds. A. Azzi, K.A. Nalecz, M.J. Nalecz & L. Wojtczak, pp. 133–146. Berlin: Springer-Verlag.

30.   Wallimann, T., Wyss, M., Brdiczka, D., Nicolay, K. & Eppenberger, H.M. (1992): Intracellular compartmentation, structure and function of creatine kinase isoenzymes in tissues with high and fluctuating energy demands: the 'phosphocreatine circuit' for cellular energy homeostasis. *Biochem. J.* **281**, 21–40.

31.   Westerhoff, H.V. & Van Dam, K. (1987): *Thermodynamics and Control of Biological Free-energy Transduction.* Amsterdam: Elsevier.

# Clinical importance of creatine phosphokinase activity and creatinine determination

*Guanidino Compounds : 2*, eds. by P.P. De Deyn, B. Marescau, I.A. Qureshi and A. Mori.
©1997 John Libbey & Company Ltd., pp. 183–195.

# Chapter 20

# Creatine kinase in human spermatozoa: cell biology and clinical significance

## Gabor HUSZAR

*Department of Obstetrics and Gynecology, Yale University School of Medicine,
333 Cedar Street, New Haven, Connecticut 06510, USA*

## Summary

In looking for objective biochemical markers of sperm fertility, we have discovered that the creatine kinase (CK) activity varies both between semen samples from different donors and between sperm subpopulations within a single sample. CK activity predicted male fertility independently of the sperm concentration in the sample. We have established that the increased CK and cytoplasmic protein content result from incomplete extrusion of the cytoplasm in late spermatogenetics and diminished sperm maturity. We have identified two sperm CK isoforms that correlate with the cytoplasmic content and maturity of sperm. CK isoform ratios have also reliably predicted lower fertilization rates, lack of fertilization and occurrence of pregnancies in a blinded study of couples treated with *in vitro* fertilization.

Another important issue is the relationship between biochemical parameters and abnormal sperm morphology. A high CK content in spermatozoa is associated with an increase in the size and roundness of sperm heads and an increased incidence of amorphous sperm forms, along with diminished zona pellcida binding. Thus, incomplete extrusion of the cytoplasm and diminished sperm maturity, abnormal sperm morphology, low CK-M isoform ratios and increased CK activity all correlate with a deficiency in zona binding and fertilizing potential. The increased rates of lipid peroxidation are also associated with diminished cellular maturity and the unextruded cytoplasm in immature sperm. We have now extended our studies to the sperm surface to find markers that are expressed simultaneously with the CK-M isoform. These will facilitate the selection of single, mature and fertile sperm for assisted reproduction – particularly for the currently emerging technique of injection of a single sperm into the oocyte, which provides an opportunity for paternity to men with immotile sperm or with very low sperm concentrations.

## Introduction

The midpiece of the spermatozon contains the mitochondria that generate adenosine triphosphate (ATP) for sperm motility. The ATP energy is channelled to the sperm tail via the creatine–creatine phosphate shuttle: the mitochondrial ATP is utilized to phosphorylate creatine to creatine phosphate, which moves as an energy carrier through the tail (Fig. 1). When an ATP molecule is hydrolysed by the tubulin–dynein interaction during sperm motility, the resulting adenosine diphosphate is rephosphorylated from the creatine phosphate and the creatine is recycled in the mitochondrion. Both in the mitochondrion and in the sperm tail, the phosphate transfer is facilitated by creative phosphokinase (CK). Thus, one may perceive CK as the key enzyme of the generation, transport and utilization of sperm energy[10,18,24,27].

We became interested in developing an objective biochemical method of assessing sperm fertility in 1986 when the question of fertilizing potential differences among sperm of various men or, particu-

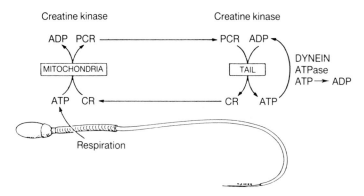

Fig. 1. The sperm creatine phosphate shuttle model. The sperm
mitochondrion and tail are the sites of ATP production and ATP consumption (dynein ATPase). The creatine phosphate–creatine system and
the mitochondrial and muscle type CK isozymes facilitate the
energy transfer between the two compartments. Creatine is phosphorylated by the mitochondrial CK, and the creatine phosphate diffuses
to the sperm tail. The high-energy phosphate is used for the rephosphorylation of ADP to ATP by the flagellar CK. The creatine is then
available for a further phosphorylation cycle in the mitochondia.
PCR/creatine phosphate[10,24]. Reproduced from Huszar et al.[10]

larly, among sperm populations within the same ejaculate had not been explored. The most widely
used tests, the measurements of sperm concentrations, motility, velocity and morphology in the
ejaculate, were shown to be of limited value. Indeed, following intrauterine insemination, a treatment
that compensates for low motile sperm concentrations, no significant differences were found in semen
parameters among those who did or did not achieve pregnancies[11]. Other available assays probing for
selected sperm functions, such as membrane integrity, acrosin activity, bovine cervical mucus
penetration test, zona-free hamster oocyte penetration test and sperm binding to various carbohy-
drates, have all failed thus far to consistently predict fertilizing potential. It has become increasingly
obvious that there was a need to identify cellular markers of sperm quality and fertilizing potential[1,10].

With the aim of finding an objective biochemical marker of sperm function, we decided to study
sperm CK, looking for possible differences in enzymatic activity between fertile and infertile men.
The first line of comparison was between oligospermic and normospermic men because, although
sperm concentrations do not predict sperm fertilization potential, the incidence of infertility is higher
in oligospermia (< 20 million sperm/ml semen).

## Sperm CK activity in oligospermic and normospermic specimens

We determined sperm CK activities (the sperm were washed free of seminal fluid) in 180 specimens
and explored the correlation between CK activities and sperm concentrations. The mean CK activity
of the specimens with > 30 million sperm/ml (expressed in IU/$10^8$ sperm) was $0.106 \pm 0.09$ SEM (n
= 90), whereas the comparable values in oligospermic specimens of < 10 and $10–20 \times 10^6$ sperm/ml
groups were $2.24 \pm 0.46$ and $0.58 \pm 0.12$, respectively (n = 30 and 30). Because sperm motility is
related to CK function, we examined the possible relationship between sperm motility and sperm CK
activity. We divided the 90 normospermic specimens into three motility groups of < 30 per cent (n
= 21), 30–45 per cent (n = 27), and >45 per cent (n = 42). The CK activities of the three groups were
(mean ± SEM): $0.10 \pm 0.03$, $0.19 \pm 0.06$ and $0.13 \pm 0.01$, respectively, with no appreciable differences

($P > 0.15$ in all comparisons). Thus, motility was not a factor in the CK activity differences among the samples[10].

## CK activity in seminal fluid and other possible artefacts

To validate these results, we examined the following four areas for possible artefacts:

i) Sperm CK activities were unrelated to the CK activity of the seminal fluid.

ii) Measurements were carried out in the absence of creatine phosphate and there was no measurable ATP synthesis; thus sperm homogenate did not generate ATP spontaneously.

iii) The presence of the myokinase inhibitor delta-4-guanidine did not affect the CK activity of either the oligospermic or the normospermic specimens.

iv) Spontaneous ATP synthesis without CK.

In another line of experiments, we measured CK activities in the direction of both ATP synthesis and creatine phosphate synthesis in 15 normospermic and 15 oligospermic men. In agreement with previous data, the mean CK values were significantly higher in the oligospermic group in both the ATP and creative phosphate directions (oligospermic vs. hormospermic groups: P<0.001). However, the rates of ATP and creatine phosphate synthesis within each sample were closely related (r > 0.90, Fig. 2.)

### CK activity in the swim-up sperm fractions: CK activity predicts fertility

Although we could not yet explain the phenomenon, it became apparent that the normospermic specimens, which are more likely to be fertile, contained sperm with lower CK activities. We became further convinced that this observation was valid when we discovered that the swim-up sperm fractions, which are improved in sperm motility, velocity and morphology parameters[11], showed significantly lower CK activities than the sperm in semen. The improvement in the swim-up *vs* semen sperm was particularly impressive in the oligospermic groups of < 10 and $10–20 \times 10^6$ sperm/ml ($P < 0.001$ Fig. 3). The differences in sperm CK activities suggested to us that each semen specimen is composed of various sperm populations with various biochemical properties. We made the assumption that sperm CK values in the initial semen specimens and the degree of improvement in the swim-up fractions is proportional to the presence of sperm with biochemical properties and fertilizing potential similar to those of fertile normospermic men[10].

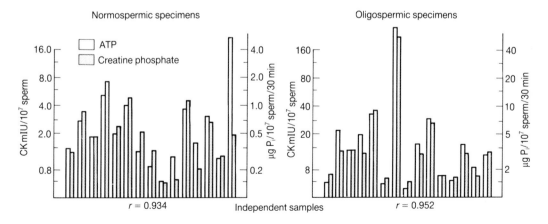

*Fig. 2. Correlation between sperm CK activities and the directions of ATP (open bars) and creatine phosphate (cross-hatched bars) synthesis in 15 normospermic and 15 oligospermic specimens. Reproduced from Huszar et al.[10]..*

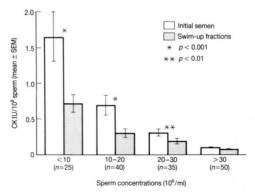

*Fig. 3. Sperm CK activity in the initial semen (open bars) and in the swim-up fractions (cross-hatched bars).*

We tested the possible relationship between sperm concentrations, sperm CK activity parameters and fertilizing potential in 160 samples from 72 fertile and infertile men (who did or did not achieve pregnancies) treated in our intrauterine insemination programme. The study population was composed of three groups: fertile oligospermic men (32 men, 46 samples), infertile oligospermic men (19 men, 82 samples), and normospermic fertile men (21 men, 32 samples). In the oligospermic groups (sperm concentrations, $18.1 \pm 2.0 \times 10^6$ sperm/ml, sperm motility $26.42 \pm 2$ per cent), the couples were treated with intrauterine insemination to compensate for the low motile sperm concentrations of the husbands. The normospermic group (sperm concentrations $49.8 \pm 5.2 \times 10^6$ sperm/ml, motility $41.3 \pm 3.4$ per cent) consisted of husbands of woman receiving treatment for female infertility[15].

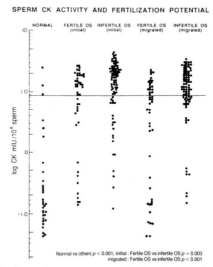

*Fig. 4. The five columns represent the CK activity distribution in the swim-up sperm fractions of the normal group, and in the initial and swim-up-fractions of the fertile and infertile oligospermic groups. The horizontal line represents the log value corresponding to $0.25$ IU CK/$10^8$ sperm. Reproduced from Huszar et al.[15] with permission from JB Lippincott Company.*

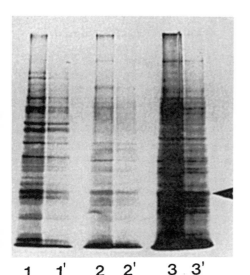

1   1'   2   2'   3   3'

*Fig. 5. Autoradiograph of $^{14}C$-FDNB-labelled sperm extracts (two-thirds and one-third of each sperm extract was applied). Lanes 1,1$^1$: sample with 160 million sperm/ml (48 million sperm extracted). Lanes 2,2$^1$ sample with 9 million sperm/ml million sperm extracted). (4 Lanes 3,3$^1$: sample with 12 million sperm/ml (9 million sperm extracted). Arrowhead designates the localization of CK.*
*Reproduced from Huzar & Vigue$^{13}$ with permission from Wiley-Liss Inc, a division of Wiley & Sons Inc.*

Figure 4, a scattergram, provides a graphic presentation of the CK activity differences among the normospermic and the fertile and infertile oligospermic groups. The horizontal line represents 0.25 CK IU/$10^8$ sperm, the mean ± 2 SD of the CK value of normospermic specimens with sperm concentration of 30 million sperm/ml. The five columns represent the CK activity of fertile normospermic men and the initial semen and migrated sperm fractions of the fertile and infertile oligospermic men. The comparison between normospermin and oligospermic men and the CK activity values of initial and migrated fractions in the fertile and infertile groups is noteworthy. In agreement with our hypothesis, differences between the initial and swim-up sperm fractions, as well as the swim-up sperm fractions of the fertile and infertile oligospermic groups, were significantly different. The swim-up sperm fractions of the fertile oligospermic men were similar to those of the fertile normospermic men. The correlation between CK activity parameters and the occurrence of pregnancies is well discernible whether one considers the CK activities in the initial ($P = 0.003$) or the swim-up ($P < 0.001$) fractions or the proportion of samples below the 0.25 IU CK/100 million sperm value in the swim-up fractions ($P = 0.002$). Thus, the CK activities in the initial semen and the degree of improvement in the swim-up sperm fractions predicted the fertilizing potential of the fertile and infertile oligospermic men.

To further examine the value of sperm CK activities *vs* sperm concentrations in the prediction of fertilizing potential, we utilized two additional approaches. First, we selected all samples with 5–20 × $10^6$ sperm/ml concentrations from the fertile and infertile oligospermic groups. The sperm concentration and motility parameters were practically identical in the two groups (about 11 million sperm/ml and 23 per cent motility). However, the sperm CK activities of the fertile 5–20 group were significantly lower in both the initial samples ($P = 0.02$) and in the migrated fractions ($P = 0.002$) than in those of the infertile 5–20 group. Second, to assess the relative contributions of CK activities and sperm concentrations in predicting fertilization potential in all 160 samples, we performed a logistic regression analysis. The prediction values for sperm concentrations ($-2 \log LR = 14.2$, $P = 0.0002$) *vs* for the CK activities ($-2 \log LR = 1.2$, $P = 0.3$) provided further evidence that the CK activities are the principal contributors in predicting fertilizing potential. The sperm concentrations themselves did not contribute to the predictive power[15]. The predictive value of sperm CK activity for sperm quality and fertility has been recently confirmed by two other laboratories[4,21].

## Relationship between sperm CK activity and sperm maturity

Our studies clearly demonstrated that sperm CK activity reflects sperm fertilizing potential. For a better understanding of the sperm CK activity differences, our working hypothesis included two possible explanations:

i) Sperm in the high-activity samples may contain a CK with higher specific activity, due to iso-formic differences or to post-translational amino acid modifications.

ii) Sperm in the high-CK-activity samples may have higher constituent CK concentrations.

We examined these alternatives in two experimental approaches[13]. To measure sperm CK concentrations, we used covalent modification of the SH-group within the CK active site with [14]C-FDNB and analyzed the sperm extract with sodium dodecyl sulphate (SDS) Tgel electrophoresis and subsequent autoradiography. If our first explanation was true, the sperm CK concentrations detected by autoradiography were expected to be similar in the normal and high-CK-activity samples. However, if the second possibility was true, the intensity of the CK bands on the radiographic should be proportional to the sperm CK activity in the samples. The sperm CK content was also examined by CK immunocytochemistry and direct visualization of the CK in individual spermatozoa.

Figure 5 demonstrates the SDS–acrylamide gradient gel electrophoresis patterns of three sperm extracts after preferential [14]C-FDNB labelling of the high–affinity SH-group within the enzymatic active site of CK. Lanes 1 and $1^1$ are an extract of a normospermic sample (160 million sperm/ml; two-thirds and one-third of each sample were applied to adjacent lanes), lanes 2, 2′; 3 and 3′ are extracts of two semen samples with low sperm concentrations (9 million and 12 million sperm/ml). The total sperm extracted in the three samples were 48 million, 4.0 million and 9.0 million, and the respective CK activities were 0.72, 7.5 and 9.9 IU CK/$10^8$ sperm. The proportions of actual CK activity

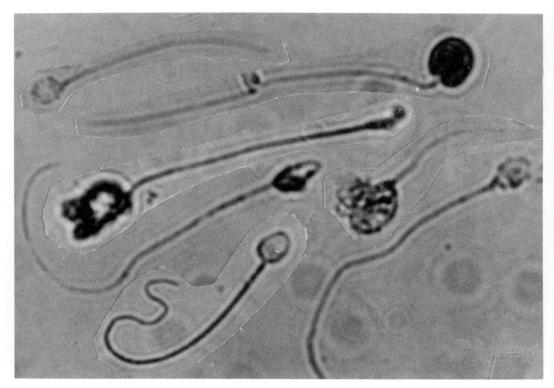

*Fig. 6. Montage of CK immuno-stained sperm with different patterns: (a) normal sperm with lightly stained or clear heads; (b) sperm with different degrees of CK stippling; (c), (d) sperm with different degrees of solid CK staining; (e) amorphoursperm (× 1000). Reproduced from Huszar & Vigue[13] with permission from Wiley-Liss Inc, a division of Wiley & Sons Inc.*

applied to the gels, or CK-equivalents (millions of sperm × CK activity), were 38, 30 and 96 in the three extracts.

Reviewing the X-ray patterns one can make two observations of note. First, the intensity of the CK band (arrow) in the normospermic sample (lanes 1 and 1', extract of 48 million sperm) is comparable to that of the CK bands in lanes 2 and 2' and much less than that of the bands in lanes 3 and 3' (generated by the extraction of only 4.0 million and 9.0 million sperm, respectively). This is about 5–10 times higher–intensity relative to the numbers of sperm than in lanes 2,2' and 3,3', and also corresponds to the differences in CK activity in the three extracts. Thus, the data support the hypothesis that the higher per sperm CK activity is due to increased sperm CK concentrations. A second observation of note is that the intensity of protein bands other than CK were also proportionally increased in the high-CK-activity specimens (2,2' and 3,3'). This suggests that, in addition to the CK, the concentrations of other sperm proteins are also elevated in the high-CK-activity specimens[13].

In the CK immunocytochemical approach, using a first antibody to CK and a second antibody coupled to horseradish peroxidase, we studied the content and distribution of CK in individual spermatozoa. Figure 6 is a montage of sperm representative of semen samples with low and high sperm CK activity. In normal sperm (a) of semen samples with the CK staining is present in the tail and in the sperm mitochondrion (the spiral pattern may be observed) leaving the sperm heads clear or lightly stained. In sperm more characteristic for high-CK-activity specimens, we detected a diffuse light stippling (b), and there were sperm in which the head was partially (c) or fully (d) filled with solid CK staining. Amorphous sperm (e) also showed a random pattern of heavy CK staining. These data confirmed that in normal sperm the CK content is low, whereas the presence of heavy CK staining in sperm arising from high-CK-activity samples is in agreement with the more intense CK autoradiography signal detected in the corresponding sperm extracts[13].

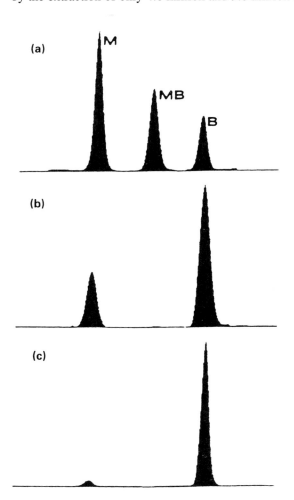

*Fig. 7. Separation of CK isoform: (a) muscle extract; (b) sperm extract from a normospermic man; (c) sperm extract from an oligospermic man. M, CK-MM dimer; MB, CK-MB dimer; B, CK-MB dimer. The electrophoretic and scanning procedures were carried out using the Ciba-Corning (Palo Alto, CA, USA) Cardiotrac-CK and 780 fluorimeter/densitometer system. During electrophoresis, the cathode is on the left side and the anode is on the right side of the gels. Reproduced from Huszar & Vigue[13].*

## Relationship between sperm CK concentrations and head morphology

In addition to sperm concentration and motility, the classic semen analysis parameters

189

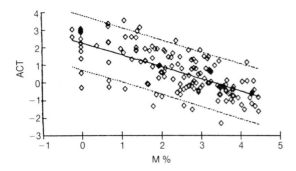

Fig. 8. Correlation between CK activity (ACT) and CK-M ratio [%CK-M/(CK-M+CK-B)] in the 159 normospermic and oligospermic specimens. The dotted lines represent the 95 per cent confidence limits. Reproduced from Huszar & Vigue[12].

also include sperm head morphology. It is quite clear that sperm morphology is related to sperm fertility; also, sperm morphology parameters reflect testicular function and spermatogenesis.

A review of the various sperm forms (a–e) in Fig. 6 demonstrates that there is a relationship among the intensity of CK staining and the size and shape of the sperm heads[13]. We have studied this association by computer assisted morphometry, focusing on the parameters of sperm head perimeter and the ratio of the longitudinal and transverse diameters – which describes the 'roundness' of the sperm head. We have also evaluated the incidence of irregular sperm heads (e), which included asymmetric shapes, elongated and bisected heads, and those with protruding cytoplasmic pockets or similar abnormalities. There was a close relationship between the increase in CK staining and sperm head size; sperm with higher CK content and larger head size were also rounder and demonstrated a higher incidence of abnormal head morphology. All differences in size, roundness or morphology among the groups with various degrees of CK intensity were highly significant ($P < 0.001$)[13].

The biochemical and immunocytochemical evidence led us to the recognition of the biological rationale underlying these phenomena: an incomplete extrusion of the cytoplasm, which during normal spermatogenesis would be retained by the Sertoli cell as residual bodies[7]. Thus, sperm with high concentrations of retained CK and cytoplasmic proteins, which also exhibit increased size and roundness and abnormal morphology of the sperm head, have failed to complete cellular maturation with a consequent impairment of various functions, including fertilization potential. The relationship between abnormal sperm morphology and diminished fertility is further highlighted by the fact that the two major morphological features that classically signify diminished fertility (round sperm heads and amorphous sperm heads) showed the most intense CK staining (Fig. 6). Our data are also in excellent agreement with the recent 'strict criteria' evaluation of abnormal sperm forms developed by Dr Krueger's group. For instance, the roundness of the sperm head and fullness around the neck region were related to lower rates of fertilization *in vitro*[20].

### Sperm development-related changes in the sperm CK isoform concentrations

Although we had established the CK activity differences and the relationships among abnormal sperm morphology, diminished sperm fertility and incomplete cytoplasmic extrusion and spermatogenesis, we still had not identified a biochemical marker that would be specific for fertile sperm populations. In the next phase of the studies, we wished to pursue the biochemical differences related to sperm maturity and immaturity, and this led us to the analysis of CK isoforms in spermatozoa. In another contractile system, skeletal muscle, CK occurs as the B (brain)-type and M (muscle)-type isoforms[25]. There are developmental and differentiation-related changes in the appearance of these two CK isoforms[8,22]. In foetal muscle, or in cells of unfused muscle cultures, only the CK-B isoform is present. However, in postnatal development of animals concurrently with the development use of their muscles, or in muscle cell cultures following fusion and the initiation of contractile activity, the M-type CK isoform is increasingly synthesized. We explored the presence of the CK-B and CK-M isoforms in human sperm and examined whether, in line with sperm maturation as reflected by CK activity differences, a change in the synthesis of sperm CK isoforms may also occur. We determined

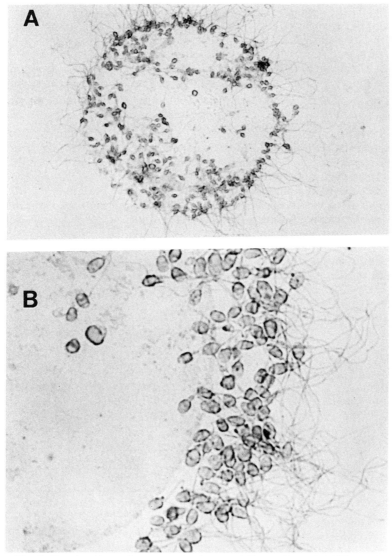

*Fig. 9. Creatine kinase-immunostained sperm–HZ complexes. (a) 400 × and*
*(b) 1000 × overall magnification. Reproduced from Huszar et al.[17] with*
*permission of the publisher, The American Fertility Society.*

the relative concentrations of the CK-M and CK-B isoforms,or the CK-M ratio [ in various sperm extracts using agarose gel electrophoresis (Fig. 7a).

In analyzing 159 normospermic and oligospermic samples, we found that the sperm CK-M ratios in the oligospermic specimens were lower ($P < 0.001$) than those in the normospermic samples. The mean CK-M ratios of the two groups showed a four-fold difference ($P < 0.001$). Indeed, in some oligospermic samples, only trace amounts of CK-M were detectable (Fig. 7c). The comparison of CK activities and CK-M ratios demonstrated that, in samples with diminished CK-M ratios, the  sperm CK activities were also proportionally higher. The close correlation between CK activity and CK-M isoform ratios ($r = -0.69$, $P < 0.001$, Fig. 8) suggests that the loss of cytoplasm and the activation of sperm CK-M synthesis are related events during sperm development[12]. Subsequent studies with density gradients have established that the most mature sperm fraction may contain as much as 60–80 per cent CK-M; thus, the CK-M ratio reflects the incidence of fully mature sperm in the ejaculate rather than the CK-M ratio of each sperm. Thus, the M-type CK isoform provides a cellular marker of sperm differentiation, which may serve as a direct measure of the mature sperm subpopulation in a semen specimen.

## Characterization of the sperm CK-M isoform

We named the sperm CK isoforms 'M' and 'B' according to their mobility in agarose gels, which correspond to the respective CK isoforms in muscle. We have not yet determined the structure of these CK isoforms in sperm; however, we have indications that the sperm CK-isoforms are different from the those of respective CK isoforms in muscle. One indication is the lack of the MB heterodimer in sperm (Fig. 7b). To examine this question, we performed hybridization experiments[22] following urea dissociation of the sperm CK dimers in the presence or absence of added muscle CK-M. The urea-dissociated and reassociated sperm CK-B and CK-M dimers did not form MB hybrids, whereas the muscle CK-M supplied in different ratios to the sperm extract yielded proportional amounts of the sperm–muscle MB CK hybrids. This suggested that the structure of the sperm CK-M isoform is different from that of the muscle CK-M, whereas the CK-B isoforms are homologous in the two sources (Fig. 9)[12].

**Table 1. IVF results**

|  | CK-M fertile group (n = 62 couples) | CK-M infertile group (n = 22 couples) |
|---|---|---|
| No. of oocytes inseminated (per cycle) | 4.9 | 6.2 |
| Oocyte fertilization (%) | 53.4 (2.4/cycle)* | 14.2 (0.6/cycle) |
| Lack of fertilization in couples (%) | 25.8* | 77.3 |
| No. of pregnancies | 14 | 0 |
| Overall PR (n = 84) (%) | 17 | — |
| PR in the CK-M infertile group (n = 22) | 0 | — |
| PR in the CK-M fertile group (n = 62) (%) | 22.5 (14/62) |  |
| PR in the CK-M fertile couples with oocyte fertilization (n = 47) (%) | 30.4 (14/46) | — |

PR = pregnancy rates; *$P < 0.001$.
Reproduced from Huszur *et al.*[16] with permission of the publisher, The American Fertility Society.

## CK-M percentages predict sperm fertility

To ascertain the predictive value of CK-M percentages with respect to fertility, we carried out a blinded study on 84 couples treated with *in vitro* fertilization, a paradigm in which, in addition to the presence or absence of pregnancies, fertilization rates and the lack of any fertilization in a couple may be detected[16]. Based on the correlation between CK-M percentages and CK activities, for the purpose of the study, we chose 10 per cent CK-M as the single dividing value: men with <10 per cent CK-MT were classified as low likelihood for fertilization or 'CK–M-infertile', whereas those with a sperm CK-M of >10 per cent were classified as high likelihood for fertilization or 'CK-M-fertile'. Based on the CK-M percentages of 62 of the 84 men were placed in the >10 per cent CK-M, 'CK-M-fertile group' and 22 men were in the < 10 per cent CK-M, 'CK-M-infertile group' (CK-M: $31.1 \pm 1.8$ per cent *vs* $4.9 \pm 0.6$ per cent).

## CK-M percentages *vs* sperm concentrations

Similar to the CK activity, the CK-M percentage were not related to sperm concentrations[16]. The sperm concentrations and motilities in the samples of the 62 CK-M-fertile and 22 CK-M-infertile men were $78.4 \pm 5.9$ *vs* $31.5 \pm 6.9 \times 10^6$ sperm/ml and $54.0 \pm 2.0$ *vs* $45.6 \pm 5.0$ per cent (n = 88 and 27), all in the normospermic range. There were 20 oligospermic men in the study group: 9 in the CK-M-fertile group and 11 in the CK-M-infertile group. Also, 11 normospermic men were in the CK-M-infertile group, which indicates that the CK-M percentages, for the first time, provide a method to detect unexplained male infertility (men with normal sperm parameters but diminished fertility).

## CK-M percentages and prediction of fertilization rates and pregnancies in *in vitro* fertilizations

The rate of oocyte fertilization showed about a four-fold difference in the CK-M-fertile and CK-M-infertile groups (53.4 per cent *vs* 14.2 per cent, $P < 0.001$). In addition to the differences in gross fertilization rates, it is of note that in 17 of the 22 CK-M-infertile couples there was a lack of any oocyte fertilization, whereas among the 62 CK-M-fertile couples 46 had fertilized oocytes and there were 16 women in whom no fertilization occurred (77.3 *vs* 25.8 per cent, $P < 0.001$). In the latter group, the lack of fertilization may be due to sperm dysfunctions other than immaturity (acrosomal defects, antisperm antibodies, etc.) or to an oocyte defect that prevents fertilization. The measurement of sperm fertilization potential allows a better focus on the female partner and this is yet another benefit of the sperm CK-M percentages.

In the 84 couples receiving *in vitro* fertilization (IVF), there were 14 clinical pregnancies with an overall rate of 17 per cent (Table 1). All 14 pregnancies were in the CK-M-fertile group and none in the CK-M-infertile group. The pregnancy rate in the 62 CK-M-fertile couples was 23 per cent and considering the 47 couples in the CK-M-fertile group in which oocyte fertilization occurred, the rate of prediction was 30 per cent, which corresponds to the maximum presently achievable rate in IVF[16].

## Sperm with abnormally high CK levels fail to bind to the zone pellucida

We have further studied the relationship between increased sperm CK concentration and diminished fertilizing potential in a sperm-binding assay that utilizes unfertilized human oocytes cleaved by microtome into two halves (hemizonae or HZ), which allows the comparison of sperm binding in men in infertile couples with that in fertile donors[17]. Sperm–HZ complexes were treated for immunocytochemistry (similar to the studies in Fig. 6), which demonstrates the CK immunostaining pattern differences that are characteristic for single sperm were normal and diminished fertilizing potential, to determine whether the distribution of mature (clear heads), intermediate (sperm heads with light stippling) and immature (heads with heavy stippling or with solid CK staining) spermatozoa bound to the HZ would follow the incidence of these sperm in the samples tested, or whether there is preferential binding by the mature sperm.

In Fig. 9a and 9b a typical HZ with the bound spermatozoa at lower (400×) and higher (1000×) magnification is presented. As opposed to the sperm samples in which sperm with various degrees of immature dark staining patterns were detected, virtually all sperm in the HZ complexes are mature (96.4–98.1 per cent) with no cytoplasmic retention. This demonstrated the selectivity of zona binding by the mature sperm fractions. The lack of competition by the immature sperm for the zona-binding sites was further demonstrated in two groups of semen samples (five experiments each) in which the concentrations of immature sperm were the lowest and highest (3.5 per cent *vs* 12.2 per cent, $P < 0.001$). In these two groups the incidence of immature sperm (0.6 per cent *vs* 1.0 per cent) bound to the respective HZs was not different. This suggests that spermatozoa with immature CK staining patterns and diminished CK-M percentages are deficient in the site(s) of zona recognition and binding[17].

In considering the present data, one can form an overall hypothesis. Sperm with high retained cytoplasmic and CK content, intense immunostaining and lower CK-M percentages are immature and deficient in various functions, including zona-binding and fertilizing potential. We believe that this selective binding occurs because, in sperm with incomplete cellular maturity similar to the CK isoforms, the oocyte recognition/binding site is not fully developed.

## Sperm CK, cellular maturity and lipid peroxidation

Another aspect of our studies on objective biochemical markers of sperm function focused on lipid peroxidation (LP). The major destructive effects of LP in human sperm, which have low levels of antioxidant enzymes and no mechanism to repair mem-

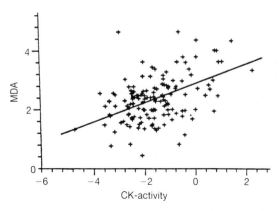

Fig. 10. Correlation between MDA production and CK activity in the 142 study samples. The values were expressed in nmol MDA/1086 sperm, respectively, and were subjected to logarithmic transformation. Reproduced from Huszar & Vigue[14].

brane damage, have been long recognized[3,5,6,9]. Also, it has been demonstrated that in oligospermic semen samples, or in men with decreased sperm motility and/or abnormal sperm morphology, the concentrations of sperm reactive oxygen species were higher than in normospermic samples[2,19]. We wished to study whether the biochemical and developmental heterogeneity characterized by the CK activity differences among sperm subpopulations may be also reflects the rates of LP. We measured malondialdehyde (MDA), which is a stable peroxidation product of polyunsaturated fatty acids, usually cross-linked to proteins. The detectable MDA is proportional to the cumulative LP damage that has occurred within the short lifetime of the spermatozoa[28].

**Table 2. MDA rates and CK activity in oligospermic and normospermic specimens**

| Specimen | n | Sperm concentrations ($10^6$ sperm/ml) | Malondialdelyde (nmol/$10^9$ sperm) | Creatine kinase activity (CK IU/$10^8$ sperm) |
|---|---|---|---|---|
| Oligospermic | 42 | 24.4 ± 0.9 | 24.4 ± 3.3* | 1.17 ± 0.26* |
| Normospermic | 101 | 63.6 ± 4.3 | 9.9 ± 1.0* | 0.21 ± 0.03* |

Values are mean ± SEM. *$P < 0.001$ in both comparisons.
Reproduced from Huszur et al.[16] with permission of the publisher, The American Fertility Society

## Sperm lipid peroxidation and CK activity in oligospermic and normospermic specimens

Based on the simultaneous measurements of sperm MDA production and CK activity, we examined whether there was a relationship between sperm cellular maturity and the rate of LP in 41 oligospermic and 101 normospermic specimens[14]. The rate of lipid peroxidation, as measured by MDA, was higher in the oligospermic vs the normospermic specimens ($P < 0.001$). In line with our previous findings[10], the sperm CK activity was significantly higher in the oligospermic samples ($P < 0.001$, Table ). We tested the relationship between CK activity and MDA by linear regression analysis and found a highly significant correlation ($r = 0.43$, $P < 0.001$, Fig. 10). This indicates that the rate of lipid peroxation and CK activity in spermatozoa are closely related and that both biochemical markers are elevated in oligospermic men, a group which in general, show a higher incidence of infertility[14].

As with the CK activity and CK-M isoform parameters[14,16], LP rates were also independent of the sperm concentrations in the specimens. Among normospermic men 12 per cent of the 101 samples showed MDA and CK activity values well within the oligospermic range. Moreover, about 40 per cent of the oligospermic specimens had MDA and CK activity values similar to those of the normospermic group. This distribution pattern further demonstrates that biochemical markers are more useful than sperm concentrations in the prediction of sperm maturity and fertilizing potential. The correlation between the CK concentrations and LP parameters indicate that the increased rate of LP is related to the increased retained cytoplasmic content and underdeveloped membrane structure, which reflects diminished sperm maturity. Thus, the increased rate of LP, along with the increased CK activity, is an 'inborn' rather than an 'acquired' defect of spermatozoa.

Another important related issue is the relationship among the biochemical parameters and abnormal sperm morphology. The relationship among high CK concentration, the presence of unextruded cytoplasm and the increased circumference and roundness of the sperm head indicates common elements between CK parameters and strict sperm morphology[13,20]. Sperm samples with increased rates of LP have a higher incidence of sperm with midpiece defects[23]. High CK content of spermatozoa is associated with an increase in the size and roundness of the sperm heads and an increased incidence of amorphous sperm forms[13]. Thus, one can make a case for a correlation among the incomplete extrusion of cytoplasm, diminished sperm maturity, abnormal sperm morphology, increased CK activity and increased rate of LP and deficiency in oocyte binding and fertilization potential. In addition to the unextruded extra cytoplasm, we believe that another likely contributing factor to the accelerated rate of LP is the lability of the sperm membrane. In earlier work we have shown that, along with the extrusion of cytoplasm, the synthesis of a new M-type CK isoform also commences in human sperm[12]. Similar maturation steps may also occur in the sperm membrane, and the lack of them is likely to cause diminished membrane integrity. Changes in membrane phospholipid related to LP rates have already been documented[4,6].

## Acknowledgments

I wish to thank Ms Terry Remeika for her expert assistance in the preparation of the manuscript. This work was supported by the NIH (HD–19505).

## References

1.   Aitken, R.J. (1988): Assessment of sperm function for IVF. *Hum. Reprod.* **3**, 89–95.

2.   Aitken, R.J., Buckingham, D., West, K., Wu, C.F., Zikopoulos, K. & Richardson, D.W. (1992): Differential contribution of leukocytes and spermatozoa to the generation of reactive oxygen species in the ejaculates of oligozoospermic patients and fertile donors. *J. Reprod. Fertil.* **94**, 451–462.

3.   Aitken, R.J. & Clarkson, J.S. (1988): Significance of reactive oxygen species and antioxidants in defining the efficacy of sperm preparation techniques. *J. Androl.* **9**, 367–376.

4.   Aitken, R.J., Krausz, C. & Buckingham, D. (1994): Relationships between biochemical markers for residual sperm cytoplasm, reactive oxygen species generation and the presence of leukocytes and precursor germ cells in human sperm suspensions. *Mol. Reprod. Dev.* (in press).

5.   Alvarez, J.G. & Storey, B. (1992): Evidence for increased lipid peroxidative damage and loss of superoxide dismutase activity as a mode of sublethal cryodamage to human sperm during cryopreservation. *J. Androl.* **13**, 232–241.

6.   Alvarez, J.G., Touchstone, J.C., Blasco, L. & Storey, B. (1987): Spontaneous lipid peroxidation and production of hydrogen peroxide and superoxide in human spermatozoa superoxide dismutase as major enzyme protectant against oxygen toxicity. *J. Androl.* **8**, 338–348.

7.   Clermont, Y. (1963): The cycle of the seminiferous epithelium in man. *Am. J. Anat.* **112**, 35–51.

8.   Cohen, A., Buckingham, M. & Gros, F. (1978): A modified assay procedure for revealing the M-form of creatine kinase in cultured muscle cells. *Exp. Cell Res.* **115**, 201-206.

9.  Hooper, C. (1989): Free radicals: research on biochemical bad boys comes of age. *J. NIH Res.* **1**,181–184.

10. Huszar, G., Corrales, M. & Vigue, L. (1988): Correlation between sperm creatine phosphokinase activity and sperm concentrations in normospermic and oligospermic men. *Gamete Res.* **19**, 67–75.

11. Huszar, G. & DeCherney, A.H. (1987): The role of intrauterine insemination in the treatment of infertile couples: the Yale experience. In: *Artificial Insemination: seminars in Reproductive Endocrinology,* Ed. J. Quagliarelo, pp. 11–21. New York: Grune and Stratton.

12. Huszar, G. & Vigue, L. (1990): Spermatogenesis-related change in the synthesis of the creatine kinase B-type and M-type isoforms in human spermatozoa. *Mol. Reprod. Dev.* **25**, 258–262.

13. Huszar, G. & Vigue, L. (1993): Incomplete development of human spermatozoa is associated with increased creatine phosphokinase concentrations and abnormal head morphology. *Mol. Reprod.* **34**, 292–298.

14. Huszar, G. & Vigue, L. (1994): Correlation between the rate of lipid peroxidation and cellular maturity as measured by creatine kinase activity in human spermatozoa. *J. Androl.* **15**, 71–77.

15. Huszar, G., Vigue, L. & Corrales, M. (1990): Sperm creatine kinase activity in fertile and infertile oligospermic men. *J. Androl.* **11**, 40–46.

16. Huszar, G., Vigue, L. & Morshedi, M. (1992): Sperm creatine phosphokinase M-isoform ratios and fertilizing potential of men: a blinded study of 84 couples treated with in vitro fertilization. *Fertil. Steril.* **57**, 882–888.

17. Huszar, G., Vigue, L. & Oehninger, S. (1994): Creatine kinase immunocytochemistry of human sperm-hemizona complexes: selective binding of sperm with mature creatine kinase-staining pattern. *Fertil. Steril.* **61**, 136–142.

18. Huszar, G., Vigue, L., & Wallimann, T. (1985): Creatine kinase in sperm: the presence of mitochondrial and brain-type creatine kinase isozymes. Abstract, American Fertility Society Annual Meeting, Chicago, Illinois.

19. Iwasaki, A. & Gagnon, C. (1992): Formation of reactive oxygen species in spermatozoa of infertile patients. *Fertil. Steril.* **57**,409–416.

20. Menkveld, R., Stander, F.S.H., Kotze, T.J.W., Kruger, T.F. & van Zyl, J.A. (1990): The evaluation of morphological characteristics of human spermatozoa according to stricter criteria. *Hum. Reprod.* **5**, 586–592.

21. Orlando, C., Krausz, C., Forti, G. & Casano, R. (1994) Simultaneous measurement of sperm LDH, LDH-X, CPK activity and ATP content in normospermic and oligozoospermic men. *Int. J. Androl.* (in press).

22. Perriard, J.-C., Caravatti, M., Perriard, E. & Eppenberger, H.M. (1978): Quantitation of CK isoenzyme transition in differentiating chicken embryonic breast muscle and myogenic cell cultures by immunoadsorption. *Arch. Biochem. Biophys.* **191**, 90–100.

23. Rao, B., Soufir, J.C., Martin, M. & David, G. (1989): Lipid peroxidation in human spermatozoa as related to midpiece abnormalities and motility. *Gamete Res.* **24**, 127–134.

24. Tombes, R.M. & Shapiro, B.M. (1985): Metabolite channeling: a phosphorylcreatine shuttle to mediate high energy phosphate transport between sperm mitochondrion and tail. *Cell* **41**, 325–334.

25. Turner, D.C., Maier, V. & Eppenberger, H.M. (1974): CK and aldolase isoenzyme transitions in cultures of chick skeletal muscle cells. *Dev. Biol.* **37**, 63–89.

26. Van Steirteghem, A.C., Liu, J., Joris, H., Nagy, Z., Janssenswillen, C., Tournaye, H., Derde, M.-P., Van Assche, E. & Devroey, P. (1993) Higher success rate by intracytoplasmic sperm injection than by subzonal insemination. Report of a second series of 300 consecutive treatment cycles. *Hum. Reprod.* **8**, 1055–1060.

27. Walliman, T., Moser, H., Zurbriggen, B., Wegman, G. & Eppenberger, H.M. (1986): Creatine kinase isoenzymes in spermatozoa. *J. Muscle Res. & Cell Motil.* **7**, 25–34.

28. Warren, J.S., Johnson, K.J. & Ward, P.A. (1987): Oxygen radicals in cell injury and cell death. *Pathol. Immunopathol. Res.* **6**, 301–315.

*Guanidino Compounds : 2*, eds. by P.P. De Deyn, B. Marescau, I.A. Qureshi and A. Mori.
©1997 John Libbey & Company Ltd., pp. 197–203.

# Chapter 21

# Usefulness of serum creatine kinase-MB determinations for assessing coronary artery reperfusion following thrombolytic therapy

Jean-Paul CHAPELLE

*University of Liège, Department of Clinical Chemistry, University Hospital, CHU B35, B–4000 Liège, Belgium*

## Summary

Thrombolysis has become the standard therapeutic approach in patients with acute myocardial infarction. In this study, we sought to determine the value of total creatine kinase (CK) and of its cardiac isoenzyme (CK-MB) in identifying patients with successful coronary artery reopening after attempted thrombolysis from non-reperfused patients, with special attention to the first 180 min following initiation of therapy. The usefulness of CK-MB was compared with that of myoglobin, one of the earliest myocardial markers. In 27 acute myocardial infarction patients with short hospitalization delays who underwent thrombolysis immediately after their admission to the coronary care units (CCU), blood samples were drawn at 1 h intervals for 6 h and regularly thereafter. CK-MB and myoglobin were determined in each sample using a fluorimetric enzyme immunoassay. Twenty patients were successfully reperfused, as attested by early angiography. Total CK, CK-MB and myoglobin peaks occurred significantly earlier in the reperfused group than in the non-reperfused patients. Immediately after thrombolysis, CK-MB and myoglobin appearance rates were higher in the patients for whom reperfusion occurred than in the patients with persistently occluded artery. For CK-MB, the best discrimination between the two groups was obtained on the basis of increasing rates recorded over the period 60–120 min. For myoglobin, the most important differences between reperfused and non-reperfused patients were recorded during the first 60 min after initiation of therapy. With the use of rapid immunoassays for CK-MB and myoglobin, a useful index for predicting success or failure of reperfusion may be derived over the initial 1–2 h after thrombolysis.

## Introduction

As early thrombolysis has been shown to improve ventricular function and reduce mortality in acute myocardial infarction (AMI) patients[18,19], fibrinolytic agents are now widely used in CCUs. It is important for the clinician to know as rapidly as possible after thrombolysis whether or not the treatment has been successful. The absence of recanalization of the obstructed vessel can be the deciding factor in the decision whether or not to carry out emergency catheterization. To avoid acute angiographic evaluation immediately after intravenous thrombolysis, non-invasive indicators of coronary reperfusion are needed. Because clinical indices – such as reperfusion arrhythmias, electrocardiographic changes and relief of pain – lack sensitivity and specificity[13], alternative solutions have been investigated. One of the most promising approaches is to make use of changes resulting from reperfusion in release parameters and in the evolution kinetics of traditional myocardial markers. The aim of this study was to evaluate the value of serial determinations of the MB isoenzyme of creatine kinase (CK-MB) in serum for assessing the success or failure of thrombolytic therapy in AMI patients. We also compared the performances of CK-MB with those of myoglobin, an earlier tissue marker of myocardial necrosis.

## Materials and methods

### Patient population

Our study population consisted of 27 AMI patients (mean age ± SD 56 ± 8 years) who were admitted to the CCU, University Hospital, Liège, Belgium. AMI was diagnosed on the basis of typical clinical history, electrocardiographic evidence, and a characteristic increase and decrease of total CK activity in serum. To be included in the study, the patients had to reach the CCU within 2 h following the onset of chest pain and initial CK values had to be < 150 U/l. All patients received fibrinolytic therapy (intravenous perfusion of $0.8-1.0 \times 10^6$ units of streptokinase within 30 min following admission). Blood was sampled on admission, at 1 h intervals for 6 h and after 8, 10, 12, 16, 20, 24, 28, 32, 36, 48, 60 and 72 h. For all patients, serum was separated promptly, kept at 4 °C and analysed within 24 h. Angiography of the infarcted coronary artery was performed 90 min after the start of thrombolysis. Coronary angiograms were interpreted by three experienced cardiologists using the criteria of the TIMI-study[20].

## Biochemical measurements

### Total CK

We used an optimized spectrophotometric method[17] (Enzyline CK NAC; Bio Mérieux, Lyon, France) to measure total CK activity (reference limits 0–120 U/l) with an automated analyser (ERIS 6170, Olympus-Eppendorf, Hamburg, Germany) at 37 °C.

### CK-MB

Serum CK-MB (reference limits 0–6 µg/l) was measured in terms of mass by a fluorimetric enzyme immunoassay using a double antibody system and adapted on an automated analyser[4] (Stratus, Baxter Dade, Miami, FL, USA). In this technique, the first monoclonal antibody is complexed onto the surface of glass-fibre paper, which constitutes the solid phase of the system. In a first step, the antigen is allowed to bind to the solid phase-bound antibody during a short incubation period. Then the second antibody, labelled with alkaline phosphatase, is applied, initiating the formation of a 'sandwich' solid phase antibody–CK-MB-conjugate. Finally, the wash buffer containing the fluorogenic substrate is applied, eluting unbound conjugate to the tab periphery. The bound enzyme conjugate is quantified by measuring the rate of increase in fluorescence and is directly proportional to the CK-MB concentration in the sample.

### Myoglobin

Serum myoglobin (reference limits 12–80 µg/l) was also measured by fluorimetric enzyme immunoassay using an automated analyser[5] (Stratus, Baxter Dade). For myoglobin, as for CK-MB, the assay takes 8 min.

## Results

In 20 patients, the infarcted vessel was patent at 90 min after thrombolytic therapy; the vessel remained occluded in the seven other subjects. Serum total CK activities increased faster in the patients for whom reperfusion occurred, leading to an earlier peak than in the non-reperfused patients (9.6 ± 1.9 vs 17.7 ± 5.6 h, $P < 0.001$). As can be seen from Fig. 1, CK-MB concentrations also peaked earlier in the reperfused patients than in the non-reperfused patients (8.1 ± 1.7 vs 14.0 ± 3.4 h, $P < 0.001$).

Myoglobin was the marker exhibiting the fastest changes in the blood (Fig. 2): its peak was observed after $1.8 \pm 0.7$ h in the reperfusion group and after $5.1 \pm 2.5$ h in the non reperfusion group, a significant difference ($P < 0.001$). Peak levels recorded for total CK, CK-MB and myoglobin were lower in the patients for whom recanalization occurred; these differences were not significant, except for CK-MB (Table 1).

**Table 1. Comparison of serum total CK, CK-MB and myoglobin peak levels (mean ± SD) in patients with a patent *vs* an occluded infarcted vessel**

|  | Reperfusion (n = 20) | No reperfusion (n = 7) |  |
|---|---|---|---|
| Myoglobin | $1720 \pm 1440$ µg/l | $1970 \pm 1600$ µg/l | NS |
| CK-MB | $174 \pm 119$ µg/l | $339 \pm 298$ µg/l | $P < 0.05$ |
| Total CK | $1710 \pm 1490$ U/l | $2177 \pm 1205$ U/l | NS |

NS = not significant.

In our study group of 27 patients, ten (37 per cent) had positive myoglobin values at admission, whereas only two (7 per cent) had positive CK-MB values. Initial CK-MB values were not-significantly different in the two groups: $5 \pm 2$ µg/l (mean ± SEM) in the reperfused patients and $5 \pm 3$ µg/l in the others. Sixty minutes after thrombolysis, CK-MB rose significantly ($P < 0.05$) to a mean value of $20 \pm 7$ µg/l in the reperfused patients (Fig. 3). At that time, four patients in this group (20 per cent) still had normal CK-MB values. After 60 min, CK-MB averaged $10 \pm 3$ µg/l in the non-reperfused patients, but the differences with initial values were not significant; at that time, three patients (42 per cent) still demonstrated normal levels. During the next 2 h, CK-MB increased at a higher rate in the reperfused patients (59 µg/l at 120 min and 92 µg/l at 180 min) than in the others (22 and 47 µg/l, respectively). After 120 min, two patients (one in each group) still demonstrated normal CK-MB levels; after 180 min, the test was positive in all patients.

**Table 2. Comparison of the rate of increase in total CK, CK-MB and myoglobin in serum over the first 180 min following thrombolysis in reperfused and non-reperfused patients**

| Time (min) | Myoglobin increase (µg/l/min) | | | CK-MB increase (µg/l/min) | | |
|---|---|---|---|---|---|---|
|  | Reperfusion | No reperfusion | $P^*$ | Reperfusion | No reperfusion | $P$ |
| 0–60 | 11.4 | 1.2 | <0.001 | 0.30 | 0.06 | NS |
| 60–120 | 11.1 | 6.8 | NS | 0.63 | 0.20 | < 0.02 |
| 120–180 | Decrease | 13.1 | – | 0.56 | 0.41 | NS |

NS = not significant.   * Mann-Whitney test.

From low levels at admission (Fig. 3), myoglobin rapidly increased in the reperfused patients to reach a mean value of 793 µg/l after 1 h, approximately 10 times the upper limit of normal ($P < 0.001$). The test was positive in all subjects. However myoglobin increased at a markedly lower rate in the non-reperfused patients (193 µg/l *vs* 123 µg/l before thrombolysis, not significant), with two patients still demonstrating normal myoglobin levels at 60 min. Sixty minutes later – 2 h after thrombolysis was initiated – myoglobin had markedly increased and all patients had positive myoglobin levels: myoglobin concentrations averaged 1460 µg/l in the reperfused patients as compared to 600 µg/l in the non reperfused patients. At 180 min, myoglobin levels already decreased in the reperfused group while they had markedly inceased (1390 µg/l) in the non reperfused group. The comparison of the CK-MB and myoglobin increasing rates, expressed in µg/l/min, during the first, the second and third hours following thrombolysis in the two groups of patients appeared in Table 2. For CK-MB, increasing rates were higher in the reperfused patients during the three periods, but the most significant differences between the two groups were observed between 60 and 120 min following thrombolysis.

In contrast, the greatest differences in myoglobin levels between the two groups were observed during the first 60 min, when increasing rates in the reperfused patients were approximately 10 times those in the non-reperfused group.

## Discussion

The aim of thrombolytic therapy is to restore coronary flow by removal of the thrombus blocking the coronary artery, by activating the natural fibrinolytic system. With the growing use of this treatment early in the course of AMI, there is increasing need to determine in individual patients whether reperfusion has been achieved and to rapidly identify subjects who are in need of additional intervention. Among myocardial markers potentially useful in this application, proteins that appear early in serum after tissue necrosis hold the most promise. For this reason, we studied CK-MB, generally considered the molecular marker of choice at the acute phase of the disease, and myoglobin, whose changes in the blood after the onset of AMI are among the fastest.

### Evolution of myocardial markers following AMI

#### Creatine kinase and its isoenzymes

Cytoplasmic (non-mitochondrial) CK is a dimeric enzyme consisting of either M or B type subunits, which associate to form three isoenzymes called MM, MB and BB[6]. The three CK isoenzymes have different tissue locations. In human skeletal muscle, the enzyme is mainly of the MM type, whereas CK-BB dominates in other tissues, particularly in the brain and in the intestinal tract. The hybrid MB, for its part, is located mainly in the myocardium, where it represents around 20 per cent of the total CK activity[21,22].

Myocardial necrosis is responsible for the release into the blood of CK-MM (± 80 per cent) and CK-MB (± 20 per cent). In non-thrombolyzed AMI patients, total CK, consisting chiefly of CK-MM, reaches its activity peak in serum between 16th and 24 following the onset of symptoms. The point at which the enzyme activity peaks is variable from one patient to another, depending on the release parameters following myocardial injury (quantity of enzyme released, possible extension of the infarcted area, degree of myocardial reperfusion) and on the clearance from the blood (which is specific to the patient)[16]. Because CK-MB is eliminated from the blood at a higher rate than CK-MM, the CK-MB peak is observed earlier, between 12 and 20 h. The return to normal is also more rapid for CK-MB (48–72 h) than for total CK (72–96 h).

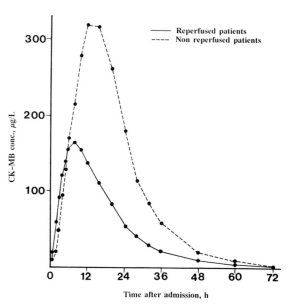

Fig. 1. Mean CK-MB concentration–time curves during the first 60 h following thrombolysis in patients successfully reperfused (n = 20) and in subjects with a permanently occluded coronary artery (n = 7).

#### Myoglobin

Myoglobin is an oxygen-binding heme protein of low relative molecular mass (17 700) found in cardiac and skeletal muscles of vertebrates. As a result of cell necrosis, myoglobin leaks from damaged muscle or myocardial tissue into the circulation. The early elevation of myoglobin in the serum is characteristic. Even patients who ex-

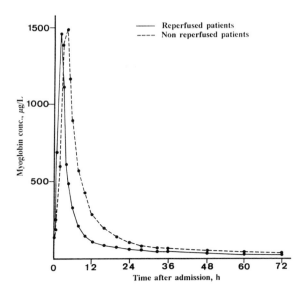

*Fig. 2. Mean myoglobin concentration–time curves in patients successfully reperfused (n = 20) and in subjects with a permanently occluded coronary artery (n =7).*

perience short delays until hospitalization show marked changes in serum myoglobin concentration on admission, whereas they have total CK or CK-MB levels that are still normal or only slightly increased. In this study, 37 per cent had positive initial myoglobin levels (7 per cent for CK-MB). In another study, more than 65 per cent of the patients hospitalized shortly after infarction (on average 2–3 h) demonstrated myoglobin levels higher than 80 µg/l on admission[3]. Therefore, in the absence of fibrinolytic therapy, the time needed for serum myoglobin to reach the upper limit of normal is approximately 1–4 h after onset of symptoms. Maximum concentrations are generally recorded after 8–12 h. Peak values are followed by a regular and rapid decrease in concentration, returning to normal generally within 24–36 h. In some patients, however, normalization is even faster, with values below 50 µg/l being observed as early as 24 h after admission.

### Influence of coronary reperfusion on the plasma concentration curves

Myocardial reperfusion following the recanalization of the occluded coronary artery is thought to cause a 'wash out' phenomenon resulting in the delivery of a large bolus of myocardial proteins to the circulation. As a consequence, serum concentrations of the tissue markers increase at a more rapid rate than in the absence of reperfusion; this results in an earlier occurrence of the peak[10,14].

In a previous study[7], we observed that total CK activities peaked sooner in streptokinase-treated AMI patients than in patients who received intravenous heparin ($15.0 \pm 1.9$ vs $20.4 \pm 2.2$ h, $P < 0.05$). The same was true for CK-MB concentrations, which reached their maximum values approximately 4 h earlier in the thrombolyzed patients ($13.0 \pm 1.8$ vs $16.7 \pm 2.0$ h, $P < 0.05$). The treatment had, however, no influence on the amplitude of the changes of CK-MB concentrations or on the evolution of the CK-MB mass/total CK activity ratios in the course of the disease. In patients with successful coronary reperfusion, myoglobin release into the blood is also accelerated and peaks occur 4– 6 h earlier than in non–thrombolysed patients.

### Differentiating reperfused and non-reperfused patients using serial CK-MB and myoglobin measurements

After thrombolytic therapy, the treatment may be successful or it may fail. In the non-reperfused patients, the time to peak from initiation of therapy is similar to that of non-thrombolyzed patients. Conversely, early peaks are associated with recanalization. In the study of Abendschein *et al.*[1], a time to peak total CK levels of < 4 h was highly predictive of recanalization; a time to peak of > 16 h was a good indication of the absence of recanalization. According to Katus *et al.*[12], the optimal discriminatory limits for infarct reperfusion based on time to peak are 16 h for total CK and 14.4 h for CK-MB. In the present study, total CK activity peaked at 9.6 h in the reperfused patients and at 17.7 h in the non reperfused patients, values in close agreement with those of Zabel *et al.* (Table 3)[23]. For CK-MB, peaks were recorded after 8.1 h and 14.0 h in the two groups respectively. Again, these values are very close to those of Zabel *et al.* [23]. For CK-MB, Grande *et al.*[11] also found time to peak values of 8.5 h in the reperfused group and 17.6 h in the non-reperfused group. So, a shorter time to peak total CK and CK-MB may be highly indicative of coronary reperfusion.

**Table 3. Time to peak from thrombolysis (mean ± SD) for total CK, CK-MB and myoglobin in patients with a patent *vs* an occluded infarcted vessel.:comparison with the study of Zabel *et al.***

| This study | | Zabel *et al.*[23] | |
|---|---|---|---|
| Reperfusion (h) | No reperfusion (h) | Reperfusion (h) | No reperfusion (h) |
| 1.8 ± 0.7 | 5.1 ± 2.5 | 2.1 ± 2.8 | 4.9 ± 4.2 |
| 8.1 ± 1.7 | 14.0 ± 3.4 | 7.5 ± 4.1 | 12.0 ± 5.3 |
| 9.6 ± 1.9 | 17.7 ± 5.6 | 9.0 ± 6.2 | 17.1 ± 6.4 |

However, especially for total CK, but also for CK-MB, measuring the time to peak after thrombolysis has the disadvantage of taking longer to predict artery patency. This study was therefore designed to investigate the usefulness of cardiac markers during the first 180 min following initiation of

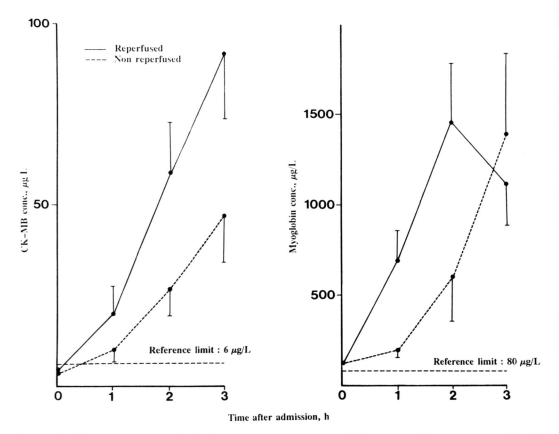

*Fig. 3. CK-MB (left) and myoglobin (right) concentrations (mean ± SEM) at start of thrombolysis and 60, 120 and 180 min later in the reperfused (n = 20) and non-reperfused (n = 7) groups.*

thrombolysis. We found that CK-MB increases at a highter rate in the reperfused patients than in the others, even during the first 60 min. The most significant differences were, however, observed during the second hour following initiation of thrombolysis. During the first 60 min, the mean increase in CK-MB concentrations was approximately four fold the initial value in the reperfused group and twice that in the non-reperfused group. Garabedian *et al.*[9] had previously determined that a 2.5-fold increase (anterior infarct) or a 2.2-fold increase (inferior infarct) at the end of rt-PA infusion (90 min) could distinguish patients with or without reperfusion.

For myoglobin, we found that the peak occurred after 1.8 h in the reperfused group and after 5.1 h in the non-reperfused patients. These values compare favourably with those of Zabel *et al.*[23] (Table 3) but also with those of Ellis *et al.*[8], who reported mean peak myoglobin levels occurring at $111 \pm 8.1$ min mean (SEM) in the patients with known vessel reopening, a shorter time than in the others (360 $\pm$ 61 min). The very fast occurrence of myoglobin peak in the reperfused patient is explained by a higher rate of increase in myoglobin concentrations during the first two hours following thrombolysis. We found that the differences between the two groups were the highest during the first 60 min, probably because some reperfused patients had their peak after 1 h. These values confirm the results of Apple *et al.*[2] who calculated the appearance rate of myoglobin during the first 90 min in patients who underwent early angiography after thrombolytic therapy and found significantly greater values in patients who successfully reperfused than in others ($38 \pm 8$ *vs* $3 \pm 1$ µg/l min, $P < 0.05$). According to Laperche *et al.*[15], the rate of increase of myoglobin during the first 90 min was $19 \pm 20$ µg/l min in the reperfused patients and $6 \pm 7$ µg/l min in the others.

In conclusion, analysis of the early initial rise of CK-MB by measuring this marker before application of thrombolytic therapy and 2 h later may be a valuable approach for predicting the success or failure of thrombolysis. Myoglobin, which exhibits the earliest rise of all markers examined after thrombolytic therapy, may also be extremely useful for prompt evaluation of coronary patency at 60 min. However, due to its early peak, myoglobin may be of poor utility in patients with increased delay in hospitalization, and for this reason, joint evaluation of CK-MB and myoglobin is probably the best strategy.

With the use of rapid immunoassays for CK-MB and myoglobin, the probability of success or failure of attempted thrombolysis could be assessed within 1–2 h. This would allow for measures such as rescue angioplasty that aim to salvage myocardium.

### References

1.  Abendschein, D.R., Ellis, A.K., Eisenberg, F.R., Klocke, F.J., Sobel, B.E. & Jaffe, A.S. (1991): Prompt detection of coronary recanalization by analysis of rate of change of concentrations of macromolecular markers in plasma. *Coronary Artery Dis.* **2**, 201–212.

2.  Apple, F.S., Henry, T.D. & Berger, C.R. (1993): Immunoassay for serum CK-MB: application to the determination of reperfusion following thrombolytic therapy. *Ann. Biol. Clin.* **51**, 349.

3.  Chapelle, J.P., Albert, A., Smeets, J.P., Boland, J., Heusghem, C. & Kulbertus, H.E. (1982): Serum myoglobin determinations in the assessment of acute myocardial infarction. *Eur. Heart J.* **3**, 122–129.

4.  Chapelle, J.P. & El Allaf, M. (1990): Automated quantification of creatine kinase MB isoenzyme in serum by radial partition immunoassay, with use of the Stratus analyser. *Clin. Chem.* **36**, 99–101.

5.  Chapelle, J.P., Lemache, K., El Allaf, M., El Allaf, D. & Piérard, L. (1994): Fast determination of myoglobin in serum using a new radial partition immunoassay. *Clin. Biochem* **27**, 423–428.

6.  Dawson, D.M., Eppenberger, H.M. & Kaplan, N.O. (1975) : Creatine kinase. Evidence for a dimeric structure. *Biochem. Biophys. Res. Commun.* **21**, 346–353.

7.  El Allaf, M., Chapelle, J.P., El Allaf, D., Adam, A., Faymonville, M., Laurent, P. & Heusghem, C. (1986): Differentiating muscle damage from myocardial injury by means of the serum creatine kinase (CK) isoenzyme MB mass measurement/total CK activity ratio. *Clin. Chem.* **32**, 291–295.

8.  Ellis, A.K., Little, T., Masud, A.R.Z., Liberman, H.A., Morris, D.C., Klocke, F.J. (1988): Early noninvasive detection of successful reperfusion in patients with acute myocardial infarction. *Circulation* **78**, 1352–1357.

9. Garabedian, H.D., Gold, H.K., Yasuda, T., Johns, J.A., Finkelstein, D.M., Gaivin, R.J., Cobbaert, C., Leinbach, R.C. & Collen, D. (1988): Detection of coronary artery reperfusion with creatine kinase-MB determinations during thrombolytic therapy: correlation with acute angiography. *J. Am. Coll. Cardiol.* **11,** 729–734.

10. Golf, S.W., Temme, H., Kempf, K.D., Bleyl, H., Brüstle, A., Bödeker, R. & Heinrich, D. (1984): Systemic short-term fibrinolysis with high dose streptokinase in acute myocardial infarction: time course of biochemical parameters. *J. Clin. Chem. Clin. Biochem.* **22,** 723–729.

11. Grande, P., Granborg, J., Clemmensen, P., Sevilla, D.C., Wagner, N.B. & Wagner, G.S. (1991): Indices of reperfusion in patients with acute myocardial infarction using characteristics of the CK-MB time–activity curve. *Am. Heart J.* **122,** 400–408.

12. Katus, H.A., Diederich, K.W., Scheffold, T., Uellner, M., Schwarz, F. & Kübler, W. (1988): Non-invasive assessment of infarct reperfusion: the predictive power of the time to peak value of myoglobin, CK-MB, and CK in serum. *Eur. Heart J.* **9,** 619–624.

13. Kircher, B.J., Topol, E.J., O'Neill, W.W., Pitt, B. (1987): Prediction of infarct coronary artery recanalization after intravenous thrombolytic therapy. *Am. J. Cardiol.* **59,** 513–516.

14. Kwong, T.C., Fitzpatrick, P.G. & Rothbard, R.L. (1984): Activities of some enzymes in serum after therapy with intracoronary streptokinase in acute myocardial infarction. *Clin. Chem.* **30,** 731–734.

15. Laperche, T., Steg, P.G., Benessiano, J., Dehoux, M., Juliard, J.M., Himbert, D. & Gourgon, R. (1992): Patterns of myoglobin and MM creatine kinase isoforms release early after intravenous thrombolysis or direct percutaneous transluminal coronary angioplasty for acute myocardial infarction, and implications for the early noninvasive diagnosis of reperfusion. *Am. J. Cardiol.* **70,** 1129–1134.

16. Roe, R. (1977): Validity of estimating myocardial infarct size from serial measurements of enzyme activity in the serum. *Clin. Chem.* **23,** 1807–1812.

17. Rosalki, S.B. (1967): An improved procedure for serum creatine phosphokinase determination. *J. Lab. Clin. Med.* **69,** 696–705.

18. Sheehan, F.J., Mathey, D.G., Schofer, J., Dodge, H.T. & Bolson, E.L. (1985): Factors that determine recovery of left ventricular function after thrombolysis in patients with acute myocardial infarction. *Circulation* **71,** 1121–1128.

19. Simoons, M.L., van de Brand, M., de Zwaan, C., Verheugt, F.W.A., Remme, W.J., Serruys, P.W., Bär, F., Res, J., Krauss, X.H., Vermeer, F. & Lubsen, J. (1985): Improved survival after early thrombolysis in acute myocardial infarction. *Lancet* **ii,** 578–582.

20. The TIMI study group. (1985): The thrombolysis in myocardial infarction (TIMI) trial. Phase I findings. *N. Engl. J. Med.* **312,** 932–936.

21. Tsung, S.H. (1976): Creatine kinase isoenzyme patterns in human tissue obtained at surgery. *Clin. Chem.* **22,** 173–175.

22. Urdal, P., Urdal, K. & Stromme, J.H. (1983): Cytoplasmic creatine kinase isoenzymes quantitated in tissue specimens obtained at surgery. *Clin. Chem.* **29,** 310–313.

23. Zabel, M., Hohnloser, S.H., Köster, W., Prinz, M., Kasper, W. & Just, H. (1993): Analysis of creatine kinase, CK-MB, myoglobin, and troponin time-activity curves for early assessment of coronary artery reperfusion after intravenous thrombolysis. *Circulation* **87,** 1542–1550.

*Guanidino Compounds : 2*, eds. by P.P. De Deyn, B. Marescau, I.A. Qureshi and A. Mori.
©1997 John Libbey & Company Ltd., pp. 205–213.

# Chapter 22

# Logical conditions under which clinical diagnosis utilizing urinary free cortisol may be enhanced by concomitant measurement of creatinine

Philip V. BERTRAND

*School of Mathematics and Statistics, University of Birmingham, Birmingham B15 2TT, UK*

## Summary

The objective of this paper is to establish where possible the theoretical grounds by which a diagnostic screening test for abnormalities of cortisol production may be enhanced by concomitant measurement of creatinine. In particular to establish the circumstances in which the ratio of urinary free cortisol (UFC) to creatinine may provide a better diagnostic screening test than measurement of UFC alone, and vice versa.

The statistical properties and virtues of these methods are compared and the circumstances in which the UFC to creatinine ratio is preferred to UFC, and vice versa, are investigated. It is found from theoretical considerations of the known values of the coefficients of variation of values of UFC and of creatinine in the normal population, and from the value of their product moment correlation coefficient, that the measurement of UFC in a 24 h sample of urine is a better diagnostic screening tool than is measurement of the ratio of UFC to creatinine in a sample of urine collected over an interval of 24 h or less.

## Introduction

The diagnosis of Cushing's syndrome and other endocrine diseases requires investigation of the ability of the adrenal to produce cortisol. Measurement of urinary free cortisol (UFC) for this purpose was established by Cope & Black[4], Mattingly[9], De Moor *et al.*[6] and Murphy[10]. Although it is now possible to measure cortisol and creatinine in plasma it is still pertinent to consider their measurement in urine samples, which are non-invasive and can be of great value for diagnostic screening tests.

The 24 h collection has been considered essential because of the well established diurnal variation in the secretion of cortisol (Tietz[11]). Difficulties of the method have concerned the collection of 24 h urine samples, which can be inaccurate and inconvenient. Walker[12,13] suggested the use of the ratio of the measurements of UFC to creatinine as a method to avoid 24 h urine collections and as being far more convenient for patients and clinicians. The rationale for this being that total creatinine is very stable over time so that use of the ratio avoids the necessity for 24 h samples.

This clearly has some merit, but Bertrand *et al.*[2] found far greater variability in the ratio than occurred in either substance alone. This contrast stimulated this theoretical study.

This study considers the various sources of variation and of transformation of scales of measurement required to make such a comparison. Sources of variation include measurement errors, the variation of UFC and creatinine values in the population in the absence of measurement errors, and the nature of the relationship between UFC and creatinine. The problem is further complicated by the well established diurnal variation in cortisol excretion (Tietz[11]). It is also necessary to consider the variability associated with errors in collection of volumes of urine over different periods of time and the variability of urine measurements that will result from the known pulsatility of cortisol observed in blood samples (Horrocks *et al.*[8]).

Many other considerations have to be kept in mind in this study. The substances we are measuring are known to vary in the population with genetic make-up, sex, age, body mass, diet, constitution, exercise, prior medical history and events, and no doubt a number of other factors (day to day, height above sea level, carry-over effects from the past, etc.). Also, should we work in terms of concentration per sample, per 24 h, per subject or per kilogram per subject? In order to consider these matters it has been necessary to make certain simplifying assumptions to make the problem tractable. Nevertheless the results obtained are found to provide good insight into the problem and should be of value for the

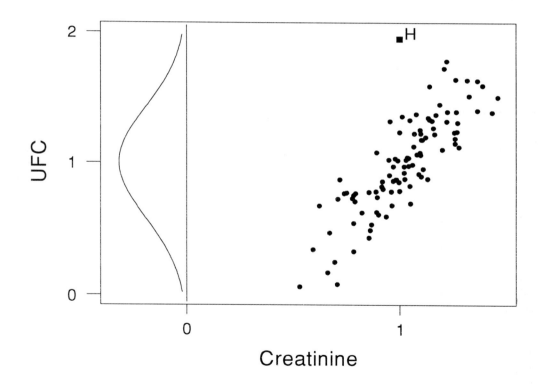

*Fig. 1. The scatter diagram shows a diagrammatic representation of the model for the joint distribution of UFA and creatinine in the normal population based on 24 h urine samples. The marginal distribution of UFC is illustrated as the projection of the joint distribution onto the y-axis. An abnormal value considered for detection is that at the solid square labelled H.*

practitioner in deciding the circumstances where one method may be preferred to the other. Also, the results may often be transferrable to measurements of other substances where a marker, possibly radioactive, is used as a reference in the measurement of some substance.

## Methods

### Background to the problem

Information on creatinine and UFC in the normal population are given in from Tietz11. Creatinine is known to be very stable within any one subject, with a day-to-day coefficient of variation usually less than 15 per cent. Its excretion may be increased by up to 30 per cent by increased dietary intake of meat. The normal reference ranges are M: 124–230, F: 97–177 (μmol/day) /kg. In addition to fluctuations of creatinine over time in a subject there is also a measurement error, but this is small, CV around 2.8 per cent (G. Holder, personal communication).

UFC exhibits diurnal variation within patients. There is a maximum between 0600 h and 0800 h, thereafter a decline until at 2000 h the level is about 66 per cent of that at 0800 h. The level remains low until after midnight when the UFC starts to increase until the early morning maximum. The pattern is, however, stochastic in nature, with trauma and stress leading to increased UFC. Excretion of UFC in the adult is usually 27–280 μmol/day with CV of measurements of around 7–10 per cent.

### Diurnal variation of cortisol

Diurnal variation of cortisol confuses the relevant issues. Although it is important, it detracts from consideration of whether UFC or the ratio of UFC to creatinine is the better diagnostic guide. Early morning samples are likely to have up to 50 per cent more UFC in them than samples over an equivalent time period taken at other times of day, and are thus changes then are more likely to be detected, measurements on them having a slightly lower CV when concentrations are high.

If we standardize our units of reference to be the amount of UFC excreted per subject per day, in the absence of diurnal variation we would convert from a measured concentration of $a$ units per ml to $A$ units per day using $A = Va/t$ where $V$ is the total volume of urine collected over time interval $t$. To allow for diurnal variation we need to appropriately transform the time scale, $t$, to $T$ say, so we could obtain the corrected value of $A$ as $A = Va/T$. In what follows I have ignored this time transformation aspect of the problem because it is a separate issue, which can only be resolved with access to sufficient data.

### Pulsatility of cortisol production

It is well established (Horrocks *et al.*[8]) that cortisol production is pulsatile in the human. Because of this the ratio of UFC to creatinine will fluctuate more and more widely with smaller and smaller values of $V$ (or $t$). They found about 14 cortisol peaks per 24 h on average when 15 min plasma levels were sampled. Consequently it can be inferred that if urine samples are taken in small amounts at, say, hourly intervals, pulsatility in blood would be reflected in more variability (CV) between samples than would occur in a 24 h sample, with overnight samples having some intermediate CV. The simplest statistical model for the pulses is to consider that they occur at random intervals of time according to a Poisson distribution with a mean of 14 pulses per day. The mean number occurring in a time interval of length $t$ will thus be $14t$, where $t$ is (usually) a fraction of a day. Making the (over) simplifying assumption that these amounts would be immediately reflected in the urine samples, the mean amount of UFC measured in urine in time $t$ will be $14t$. For a Poisson distribution the variance equals the mean and will be $14t$ also. Thus, the standard deviation will be the square root of this. Hence the coefficient of variation will be proportional to the inverse of the square root of $t$. Whereas the mean concentration over different samples, all else staying equal, will remain constant, the coefficient of variation of the concentration will increase (according to $t^{-0.5}$) as the time intervals, $t$, over which collections are made shorten.

## Notation

**Table 1**

| Variable (per subject per day) | UFC | Creatine |
|---|---|---|
| *True population values* | | |
| Mean concentration | $\mu_s$ | $\mu_n$ |
| Fluctuation | $\varepsilon_p$ | |
| Standard deviation | $\sigma_p$ | |
| *Concentration measurements* | | |
| Fluctuation | $\varepsilon_s$ | $\varepsilon_n$ |
| Standard deviation | $\sigma_s$ | $\sigma_n$ |
| *Observed population values* | | |
| Mean concentration | $\mu_s$ | $\mu_n$ |
| Fluctuation | $\varepsilon_c$ | |
| Standard deviation | $\sigma_c$ | |

Let $t$ be the time interval recorded for the urine sample with its implicit error having standard deviation $\sigma_\tau$, and $V$ be the volume of urine in a sample collected over time $t$.

### Detection of abnormal UFC

Let us first consider how we would detect an abnormal UFC in the absence of any measurement error. We know that in the normal human adult population the scatter diagram of mean UFC against mean creatinine, where the mean is per subject over a long period of time and is expressed in units per day, will be correlated (Burke & Beardwell[3]). The association is not necessarily linear, but as a first approximation we shall consider it to be so, as shown in Fig. 1. This is a diagrammatic representation, not to scale, of the kind of relationship that exists between UFC and creatinine in a large study population. The population fluctuations will be due to a variety of causes: genetic, sex, body mass, developmental, constitutional, environmental, etc.

A subject with an abnormality of adrenal function influencing cortisol production will be considered to have an UFC value substantially outside the range of the normal population values, such as the value at the point H shown in Fig. 1, which is about two standard deviations of population scatter of UFC above the coordinate position corresponding to the population mean of both creatinine and UFC. Clearly a wide variety of abnormal values are possible, but we shall only consider abnormally high values, the extension to abnormally low values being simple. However, even if there were no measurement error, the detection of an abnormal value would not necessarily be trivial. The detection of such an abnormal value is the basis of a diagnostic test for a UFC abnormality.

### Measurement errors

An assay of creatinine or UFC in a volume of urine will be reported in terms of its concentration in the volume, perhaps in units per ml, whence multiplication by urine volume gives a result in terms of units of substance over the urine collection interval. The error will be similarly amplified.

The error induced by the variation in time over which the urine sample is collected is perhaps a critical factor because the unit of time is ill defined. The liquid consumed by the subject prior to and during the collection interval influences renal function. Thus a 24 h collection of urine is usually preferred, which will still have a collection error. But in principle the collection error is the same no matter how long the interval of time over which the collection is made, so the coefficient of variation of any errors due to time uncertainty will be inversely related to time. This is one of the reasons why the ratio of UFC to creatinine in one small sample may be preferred.

**The model for the process under investigation**

In the absence of measurement errors the regression equation expressing UFC in terms of creatinine (using units of each excreted per day) for a normal subject is:

UFC of subject = $a + b \times$ (creatinine of subject) + (a population fluctuation)
Should the line go through the origin, we would have $a = 0$.

But this is only part of the model. The UFC values of the population of normal subjects will have a marginal distribution, which will be the projection of all the values in the scatter diagram onto the UFC axis. This is shown adjacent to the scatter diagram in Fig. 1. We will for convenience consider this distribution to be a normal (Gaussian) distribution with mean $\mu_s$ and individual subject fluctuations from this mean $\varepsilon_p$ having standard deviation $\sigma_p$.

This model has not so far included measurement errors. Consider a 24 h sample collected on a normal subject. Let us assume we observe concentrations $a_c$ and $a_n$ per ml of UFC and of creatinine, respectively. These will deviate from some population values, which we may assume exist in the absence of measurement error.

In the next two sections we shall consider how we would carry out a diagnostic test using UFC alone on a 24 h sample and using the ratio UFC/creatinine measured from a small sample.

**Diagnostic test using UFC alone**

The UFC concentration $a_c$ obtained in our urine sample of volume $V$ collected over time $t$ will be compared with a reference value. This reference value will usually be the mean excretion of UFC over 24 h in the normal population. This mean will be $\mu_s$ (for all practical purposes), around which the sampled values observed in the normal population will fluctuate. To enable comparisons to be made they must be on the same scale of measurement. So consider the value of

$$d_1 = Va_c/t - \mu_s \tag{1}$$

This is chosen because $Va_c/t$ is our estimate of the subject's excretion of UFC per 24 h, which is the unit of measurement of $\mu_s$. Now, the true population values of UFC per day will fluctuate about $\mu_s$ with variance $\sigma_p^2$. The measured population values will have additional variance added to the $(V/t)^2 \sigma_s^2$ population variance due the measurement error.

Thus

$$\mathrm{Var}\,(a_c) = \sigma_p^2 + (V/t)^2 \sigma_s^2$$

Hence

$$\mathrm{Var}\,(d_1) = \mathrm{Var}\,(Va_c/t)$$

Thus

$$\mathrm{Var}\,(d_1) \approx (V/t)^2 \,\mathrm{Var}\,(a_c)$$

Hence

$$\mathrm{Var}\,(d_1) \approx (V/t)^2 \sigma_p^2 + (V/t)^4 \sigma_s^2 \tag{2}$$

Then the diagnostic test statistic is

$$D_1 = d_1/ \mathrm{SE}\,(d_1)$$

where the the standard error of $d_1$ is

$$\mathrm{SE}\,(d_1) = \sqrt{\mathrm{Var}\,(d_1)}$$

Whence, substituting from (1) and (2) above we obtain

$$D_1 \approx \frac{Va_c/t - \mu_s}{\sqrt{(V/t)^2 \sigma_p^2 + (V/t)^4 \sigma_s^2}} \qquad (3)$$

### Diagnostic test using the ratio of UFC to creatinine

Define

$$r = a_c/a_n \qquad (4)$$

to be the ratio of the two variables measured on a urine sample collected over time $t$. Then the variance of $r$ can be written approximately as:

$$\text{Var}(r) \approx r^2 \text{CV}(a_c)^2 + r^2 \text{CV}(a_n)^2 - 2\,\rho r^2\, \text{CV}(a_c)\, \text{CV}(a_n) \qquad (5)$$

where $p$ is the correlation coefficient between observations of $a_c$ and $a_n$ in the population of subjects and $\text{CV}(a_c)$ and $\text{CV}(a_n)$ are the coefficients of variation of $a_c$ and $a_n$ respectively. This is equivalent to the approximate formula given by Finney[7].

In a small urine sample the two variables will have the same relative dilution, so this should not come at all into the variation of the ratio. The most important factor here will be the increased variability of the UFC estimate due to its pulsatility. It has been shown above that the coefficient of variation of UFC concentration would thus be proportional to $t^{-0.5}$. As a consequence the correlation between UFC and creatinine would be proportional to $\sqrt{t}$. Thus the standard error of $r$ when concentrations are obtained from urine collections made over length of time $t$, measured in some fraction of a day, will be

$$\text{SE}(r) = \sqrt{r^2 \text{CV}(a_c)^2/t + r^2 \text{CV}(a_n)^2 - 2\rho(\sqrt{t})\, r^2 \text{CV}(a_c)\, \text{CV}(a_n)} \qquad (6)$$

Hence the diagnostic test statistic for the ratio is:

$$D_2 = \frac{a_c/a_n - \mu_s/\mu_n}{\text{SE}(a_c/a_n)}$$

i.e.

$$D_2 = \frac{a_c/a_n - \mu_s/\mu_n}{\sqrt{r^2 \text{CV}(a_c)^2/t + r^2 \text{CV}(a_n)^2 - 2\rho(\sqrt{t})\, r^2 \text{CV}(a_c)\, \text{CV}(a_n)}} \qquad (7)$$

### Results: comparison of diagnostic test statistics

To determine which of the test statistics, $D_1$ or $D_2$ is the more powerful to detect a significantly high UFC value, we have to find out which is the larger.

Consider the test criterion, $C$, where

$$C = D_2 - D_1 \qquad (8)$$

Thus the ratio test is preferred if $C > 0$.

Now, substituting into (8) the values of $D_1$ from (3) and $D_2$ from (7) we obtain

$$C = \frac{ac/an - \mu_s/\mu_n}{\sqrt{[r^2 \text{CV}(a_c)^2/t + r^2 \text{CV}(a_n)^2 - 2\rho(\sqrt{t})r^2 \text{CV}(a_c)\text{CV}(a_n)]}} - \frac{Va_c/t - \mu_s}{\sqrt{(V/t)^2 \sigma_p^2 + (V/t)^4 \sigma_s^2}} \qquad (9)$$

This expression is rather complex, so to make a comparison we consider how it may be simplified.

Note that:

1. In $D_2$ we may be using a small value of $t$ of less than 1 day, i.e. $t \leq 1$ *for $D_2$.*

2. In $D_1$ we are using a 24 h sample where $t = 1$ day.

3. Let us measure urine volume in such units that the amount excreted in 1 day is one unit so that $V = 1$.

4. Now choose a central value of $a_n$ such as $\mu_n$.

5. Consider that our observed value of $a_c$ is such that $a_c > \mu_s$ and use units of measurement of UFC and creatinine such that $\mu_s$ and $\mu_s = 1$.

Thus the numerators of $D_1$ and $D_2$ are the same and hence $> 0$ if:

$$\sigma_p^2 + \sigma_s^2 > r^2 \text{ CV } (a_c)^{2}/t + r^2 \text{ CV } (a_n)^{2} - 2\rho(\sqrt{t})r^2 \text{ CV } (a_c) \text{ CV } (a_n) \tag{10}$$

Also

$$\sigma_p^2 + \sigma_s^2 = \sigma_c^2$$

and the standard errors will equal the corresponding coefficients of variation.

Thus we obtain $C > 0$ if:

$$\text{CV } (a_c)^2 > r^2 \text{ CV } (a_c)^{2}/t + r^2 \text{ CV } (a_n)^{2} - 2\rho(\sqrt{t}) r^2 \text{ CV } (a_c) \text{ CV } (a_n) \tag{11}$$

We may observe from this equation that as the time interval $t$ over which both creatinine and UFC are measured decreases from 1 day towards a small fraction of 1 day, a less likely it becomes that the ratio of UFC to creatinine will be preferred to UFC alone. Also, if $a_c > \mu_s$, the first term on the left side of the inequality is less than the first term on the right side of it.

Noting that $a_c = r > 1$ and $t > 1$, we can rearrange Eqn 11 to obtain $C > 0$ if

$$\rho > \frac{1}{2\sqrt{(t)}} \left[ \frac{\text{CV } (a_c)}{\text{CV } (a_n)} \left( \frac{1}{t} - \frac{1}{r^2} \right) + \frac{\text{CV } (a_n)}{\text{CV } (a_c)} \right] \tag{12}$$

From the established reference ranges described above, we may easily infer that $\text{CV}(a_c) \approx 0.41$ and $\text{CV}(a_c) \approx 0.15$. Notice that $r = a_c$ and we can express $a_c$ as $a_c = 1 + m \text{ CV } (a_c)$ where $m$ is a positive multiplier.

Substituting these values into Eqn 12, we obtain that $C > 0$ if

$$\rho > \frac{1}{2\sqrt{(t)}} \left[ 2.73 \left( \frac{1}{t} - \frac{1}{(1 + m \text{ CV } (a_c))^2} \right) + 0.37 \right] \tag{13}$$

If we now consider a value of $a_c$ which is just on the border of the reference range, where $m = 2$, we obtain that the condition for $C > 0$ is

$$\rho > \frac{1}{\sqrt{(t)}} \left( \frac{1.37}{t} - 0.22 \right) \tag{14}$$

The value of $a_c$ we have chosen is just on the boundary beyond which larger values of $a_c$ (and of $m$) may be considered abnormally high (only 2.5 per cent of normal UFC values exceed this amount). Now consider the use of ratio method on a sample collected over 1 day so that $t \sim= \sim 1$. Then, substituting into Eqn 14 for this value of $t$, to obtain $C > 0$ we would require

$$\rho > 1.15 \tag{15}$$

Such a value for the correlation coefficient between UFC and creatinine observed on 24 h samples is statistically impossible, for the correlation coefficient can only be in the range $-1 \leq \rho \leq +1$.

For urine samples collected over a period of less than 24 h, the values of $t$ in Eqn 13 will be less than 1 and it is obvious from consideration of Eqn 14 that the value of $\rho$ required will be even greater than that of 1.15 given by Eqn 15. So this will also be logically impossible.

If larger abnormal values of $a_c$ are considered they will have larger values of $m$ ascribed to them, which from consideration of Eqn 13 will also require the correlation coefficient to have a logically impossible high value.

## Discussion

The results obtained in the previous section from the knowledge of the international reference ranges of UFC and creatinine and from theoretical considerations show that on 24 h urine samples the ratio of UFC to creatinine must be inferior to the use of UFC alone. This is because a logically impossible high value for the correlation coefficient between UFC and creatinine would be required in the normal population. It was also shown that urine samples collected over less than 24 h would also require a logically impossible, even higher, value for the correlation coefficient.

Some of the equations upon which these conclusions are based are well established. That concerning the variance of a ratio, Eqn 5, is given by Finney[7], who gives a good discussion of the assumptions. The validity of the approximation depends on certain assumptions that have been made. It is important that the coefficient of variation of the denominator, in our application creatinine, be small enough, and that small values of creatinine do not occur. This is indeed the case, as it is well established that creatinine is excreted at a fairly constant rate in the normal population and it has a population CV of around 15 per cent.

The requirement that $\rho$ exceeds a function of time, as on the right side of Eqn 12, is a consequence of the assumption that the sample collected over time $t$ is used to obtain the data on which the ratio is computed and that the distribution of UFC will follow a Poisson distribution with a mean proportional to the collection interval. Although this latter assumption is unlikely to be exactly true, it is likely to be a close approximation. Moreover, whatever the true distribution is, it is likely to be equivalent to carrying out a monotonic transformation of the time scale, so that the right side of Eqn 12 will always increase as time decreases.

The condition on the correlation coefficient, Eqn 12, will be unlikely to ever be satisfied by increasing the precision of measurements. This is because the major source of fluctuations in the normal population is not the error of measurement but the natural variation between persons in the normal population.

Ratio methods have been useful in other areas of statistical investigation and are widely used in sample survey methodology (see Barnett[1] and Cochran[5]). These areas of application also have logical limits on the correlation between variables which are necessary for the validity of use of the ratio method for estimation in preference to direct estimation.

In the situation considered in this paper a ratio method has been compared with a direct method to determine which of them may be preferable as a diagnostic screening test. The conclusion is that the direct method using UFC alone to detect abnormal high values in urine must be preferable to the use of the ratio of UFC to creatinine when measurements are made on urine samples collected over 24 h or less.

## Acknowledgements

I am indebted to the British Council who provided financial assistance to visit Canada, to the staff in the Mathematics and Statistics Department of McGill University, particularly George Styan and Keith Worsley who have provided useful comments on the preparation of this paper, and to the organizers of the Fourth International Guanidino Conference held in Montreal in 1994.

## References

1.  Barnett, V. (1991): *Sample Survey Principles and Methods,* 2nd ed., pp. 77–91. London: Edward Arnold.
2.  Bertrand, P.V., Rudd, B.T., Weller, P.H. & Day, A.J. (1987): Free cortisol and creatinine in urine of healthy children. *Clin. Chem.* **33,** 2047–2051.

3. Burke, C.W. & Beardwell, C.G. (1971): Urinary free cortisol and compounds. In: *Cushing's Syndrome, .* eds. E.C. Binder & P.E., Hall. p. 53: London: Heinmann Medical Ltd.

4. Cope, C.L. & Black, E.G. (1959): The reliability of some adrenal function tests. *BMJ* **2,** 1117.

5. Cochran, W.G. (1963): *Sampling Techniques,* 2nd ed.pp. 165–166. New York: John Wiley.

6. De Moor, P., Raskin, M. & Steeno, O. (1960): Fluorimetric determination of free plasmatic and urinary 4,3-keto-11-hydroxycorticosteroids.*Ann. Endocrinol.* **21,** 479.

7. Finney, D.J. (1964): *Statistical Method in Biological Assay,* 2nd ed. p. 29. London: Griffin.

8. Horrocks, P.M., Jones, A.F., Ratcliffe, W.A., Holder, G., White, A., Holder, R., Ratcliffe, J.G. & London, D.R. (1990): Patterns of ACTH and cortisol pulsatility over twenty four hours in normal males and females. *Clin. Endocrinol.* **32,** 127–134.

9. Mattingly, D. (1964): A simple fluorimetric method for the estimation of free 11-hydroxycorticoids in human plasma. *J. Clin. Pathol.* **15,** 374–379.

10. Murphy, B.E.P. (1967): Some studies of the protein binding of steroids and their application to the routine micro and ultra-micro measurement of various steroids in body fluids by competitive protein radio-assay. *J. Clin. Endocrinol.* **27,** 973.

11. Tietz, N.W. (1986): *Textbook of Clinical Chemistry,* p.1061. Philadelphia: W.E. Saunders.

12. Walker, M.S. (1977): Urinary free 11-hydroxycorticosteroid/creatinine ratios in early morning urine samples as an index of adrenal function. *Ann. Clin. Biochem.* **14,** 203–206.

13. Walker, M.S. (1979): Screening for Cushing's syndrome using early morning urine samples. *Ann. Clin. Biochem.* **16,** 86–88.

# Section VI
## Metabolism of guanidino compounds

*Guanidino Compounds : 2*, eds. by P.P. De Deyn, B. Marescau, I.A. Qureshi and A. Mori.
©1997 John Libbey & Company Ltd., pp. 217–241.

# Chapter 23

# Biosynthesis of the guanidino compounds in health and disease: the guanidine cycle

Samuel NATELSON

*University of Tennessee, College of Veterinary Medicine, Department of Comparative Medicine, Neyland Drive, Knoxville, TN 37901, USA*

## Summary

Evidence is presented, adduced by numerous experiments over the past 35 years, to indicate that alongside the Urea Cycle there exists a Guanidine Cycle, which comes into play when a shortage of nitrogen exists in the diet (as in concentration camp inmates, during hibernation of bears and probably other animals) to reutilize nitrogen from urea.

The urea is oxidized by the P-450 system to form hydroxyurea. This hydrolyzes to generate hydroxylamine and carbamoyl phosphate. The hydroxylamine reacts with homoserine lactone, enzymatically, with ATP, to form canaline which now assumes the role of ornithine in the urea cycle.

The carbamoyl phosphate reacts with canaline to form ureidohomoserine, analogous to citrulline. Ureidohomoserine reacts with aspartate to form canavanino-succinate, analogous to argininosuccinate, which then is acted on by a lyase to form canavanine. Canavanine transamidinates to glycine which on methylation forms creatine. Thus, one of the functions of both the Urea Cycle and Guanidine Cycle is to form creatine phosphate for storing energy. The guanidino cycle indicates its activation by reduction of canavanino-succinate to form homoserine and guanidino-succinic acid.

In this review we will also discuss the spontaneous reaction of hydroxylamine with L-homoserine lactone to form L-homoserine hydroxamic acid (L-HHA). This can invade various types of tumour cells, which are acidic in nature, releasing hydroxylamine which is toxic to cells in general.

## Introduction and discussion

About 35 years ago I was approached by a student at Brooklyn College who wanted a project leading to a master's degree. At that time I was lecturing in advanced biochemistry in the graduate school. I asked him where he worked and what was his interest. He told me that he was employed by Dr. Burton D. Cohen at the N.Y. Veteran Administration Hospital and that he was studying the distribution of the various guanidino compounds in blood by means of structure specific colorimetric reactions. I told him that I had just finished a project that required the building of an amino acid analyzer, which he was welcome to use for the purpose of analysing the guanidino compounds in urine. This would be a positive contribution, because to my knowledge no one had used a column for that purpose. This should complement his regular activities.

Early in this century, F.G. Benedict and then his student Victor C. Myers carried out a series of studies extending over about 35 years.

They identified the guanidino compounds in human body fluids, and developed methodology for the study of changes that took place in disease[1,2,14–16].

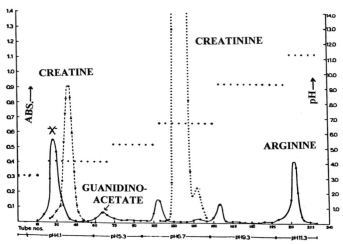

*Fig. 1. Tracing of an eluate from a column, of urine from a uremic patient, for guanidino compound determination. The guanidinosuccinate peak is labelled X. Dotted horizontal lines refer to the pH and duration of buffer being administered.*

These procedures were modified by Van Pilsum *et al.* who described reagents for assaying creatine, creatinine, guanidine, methylguanidine, arginine and guanidinoacetic acid[48].

Using similar methodology, Cohen and Bonas, as I understood it, were exploring blood changes in these parameters in disease. Adding a fractionating column could only improve their studies.

The reagents utilized were the *Sakaguchi reagent* to locate arginine and guanidinoacetic acid, *alkaline picrate* to locate creatinine, the *diacetyl reagent* to locate creatine and the *ninhydrin reagent* to locate compounds with an α-amino acid grouping. To these should be added the *pentacyanoferrate reagent* and $Fe^{3/2+}$ *reagent* to locate certain guanidino compounds such as hydroxy guanidine, canavanine, canaline, hydroxamic acids and other substituted hydroxyl amines. Also to be considered is the less specific *phenanthrene-quinone reagent*, which reacts with any guanidino-containing compound, producing a fluorophor (see Reagents and procedures).

After adding the specimen of urine to the Dowex 50 column, Bonas used a changing pH, from 4 to 11, to free creatine, guanidinoacetic acid, creatinine, and arginine in sequence[3].

When the procedure was applied to urine from a uremic patient, an unexpected peak appeared at the acid end of the tracing (Fig. 1). This substance gave a strong Sakaguchi reaction but did not give the ninhydrin reaction, indicating that it was not an α-amino carboxylic acid. It appeared in the highest concentration (about 25 mg/L) only in the urine of the uremic patient[3,40].

With the appearance of this compound, the guanidinoacetic acid concentration was lowered substantially[3]. This was confirmed, subsequently, by an exhaustive study with Dr Masahide Sasaki of Japan[40] (Table 1, Fig. 1).

**Table 1. The inverse relationship between the concentration of guanidinosuccinate (GSA) and guanidinoacetate (GAA) in adult human urine of healthy controls and those who were hospitalized with uremia. Results listed are mean values in mg/24 h/kg body weight[40]**

| No. of subjects | (GAA) | (GSA) | No. of subjects | (GAA) | (GSA) |
|---|---|---|---|---|---|
| Healthy males 17 | 0.71 | 0.18 | Uremic males 13 | 0.16 | 0.80 |
| Healthy females 10 | 1.23 | 0.25 | Uremic females 10 | 0.16 | 1.28 |

This so-called *X* peak, in the original paper[3] was soon identified by isolation and infrared analysis as being identical with synthetic guanidinosuccinic acid[11,32].

An obvious explanation of guanidinosuccinic acid formation was by transamidination from arginine to aspartic acid. However, Borsook, who first discovered the transamidination reaction, pointed out that he was successful in transaminating from arginine to a basic amino acid such as ornithine or

## TRANSAMIDINATION OF ARGININE TO GLYCINE

H NH
CH$_2$-N-C-NH$_2$    NH$_2$-CH$_2$                    NH                    H$_2$N-CH$_2$
|                        |                    H$_2$N-C-NH
(CH$_2$)$_2$        +    COOH     $\longrightarrow$    |                    (CH$_2$)
|                    GLYCINE                CH$_2$                    |
CH-NH$_2$                                    |                    CHNH$_2$
|                                        COOH    +            |
COOH                                    GUANIDINO-            COOH
ARGININE                                 ACETIC ACID          ORNITHINE

*Fig. 2. Transamidination from arginine to glycine.*

canaline, or to a neutral amino acid, such as glycine, to form guanidinoacetate (Fig. 2). But, he was unsuccessful in transaminating to an acidic amino acid such as aspartic or glutamic acids[4]. This was confirmed by Ratner & Rochovansky[37], who first completed Krebs' urea cycle.

Borsook also noted that canavanine was effective as the donor in the transamidination reaction[4] but not to aspartate. Claims were made by some that they had evidence that guanidinosuccinic acid was made by transamidination[33,34,42]. Careful studies under various conditions, with the assistance of Dr. E.J. Lauber of Switzerland, including the use of isolated viable liver cells with Hsiu-Yu Tseng of Taiwan, could not confirm these claims[11,19,25].

Guanidinosuccinate and guanidinoglutarate, like guanidinoamino acids in general could be readily synthesized in the laboratory by the reaction of an amino acid with methyl isothiourea[22–24,30] (Fig. 3).

In further support of the fact that guanidinosuccinic acid is not formed by transamidination from arginine, *patients with hyperargininemia do not have elevated guanidinosuccinic acid levels in the urine*[13,46].

If transamidination did take place between arginine and aspartate or glutamate in a living organism, it would be a metabolic disaster, because it would interfere with the support of the oxidative cycle by aspartate and glutamate at two points where replacement of losses of oxaloacetate and ketoglutarate are replenished to keep the oxidative cycle going. Thus, the supply of ATP would be compromised.

## GUANIDINO COMPOUNDS FROM AMINO ACIDS

NH$_2$
|
C-S-CH$_3$    $\xrightarrow{\text{OH}^-}$    NH$_2$ -C≡N  + ↑CH$_3$SH
||
NH                    CYANAMIDE

METHYL ISO-            (CYANAMINE)
THIO-UREA

NH
||
R-NH$_2$  +  H$_2$N-C≡N    $\xrightarrow{\text{OH}^-}$    R-NH-C-NH$_2$
AMINE    CYANAMIDE                    GUANIDINO-
COMPOUND

*Fig. 3. General method for preparing guanidino compounds synthetically from methyl isothiourea and amino acids.*

For more than 30 years I have searched for the origin of guanidinosuccinic acid in the uremic patient and what relationship it has to normal metabolism.

I searched for a compound that contained guanidinosuccinic acid in its structure. I examined argininosuccinic acid (Fig. 4), which had been prepared by addition of arginine to fumaric acid, by Ratner *et al.*[36], but could not see how it could be readily

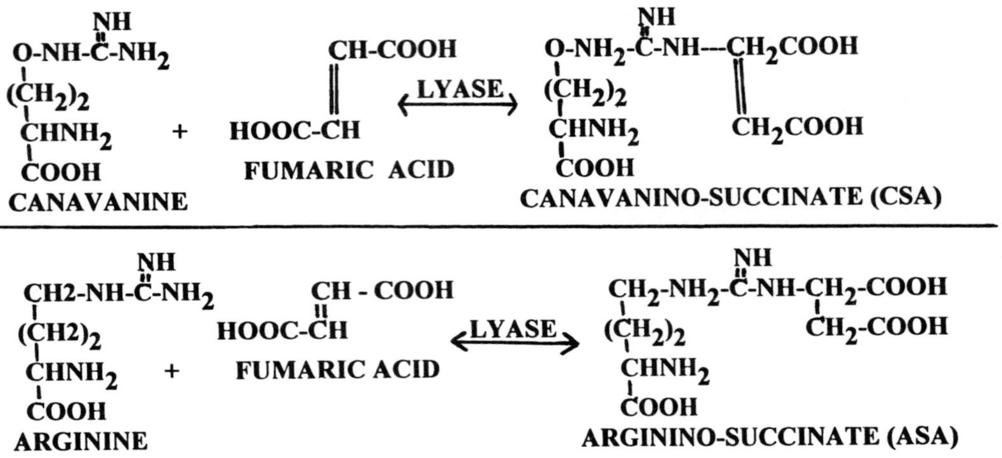

$$
\begin{array}{ll}
\text{NH} & \\
\text{O-NH-}\overset{\|}{\text{C}}\text{-NH}_2 & \\
(\text{CH}_2)_2 & \\
\text{CHNH}_2 \qquad + & \\
\text{COOH} & \\
\text{CANAVANINE} &
\end{array}
\quad
\begin{array}{l}
\text{CH-COOH} \\
\qquad \| \qquad \overset{\text{LYASE}}{\longleftrightarrow} \\
\text{HOOC-CH} \\
\text{FUMARIC ACID}
\end{array}
\quad
\begin{array}{ll}
\text{NH} & \\
\text{O-NH}_2\text{-}\overset{\|}{\text{C}}\text{-NH---CH}_2\text{COOH} & \\
(\text{CH}_2)_2 & \quad \| \\
\text{CHNH}_2 & \quad \text{CH}_2\text{COOH} \\
\text{COOH} & \\
\multicolumn{2}{l}{\text{CANAVANINO-SUCCINATE (CSA)}}
\end{array}
$$

$$
\begin{array}{ll}
\text{NH} & \\
\text{CH2-NH-}\overset{\|}{\text{C}}\text{-NH}_2 & \\
(\text{CH2})_2 & \\
\text{CHNH}_2 \qquad + & \\
\text{COOH} & \\
\text{ARGININE} &
\end{array}
\quad
\begin{array}{l}
\text{CH - COOH} \\
\qquad \| \\
\text{HOOC-CH} \\
\text{FUMARIC ACID}
\end{array}
\quad \overset{\text{LYASE}}{\longleftrightarrow} \quad
\begin{array}{ll}
\text{NH} & \\
\text{CH}_2\text{-NH}_2\text{-}\overset{\|}{\text{C}}\text{-NH-CH}_2\text{-COOH} & \\
(\text{CH}_2)_2 & \quad \text{CH}_2\text{-COOH} \\
\text{CHNH}_2 & \\
\text{COOH} & \\
\multicolumn{2}{l}{\text{ARGININO-SUCCINATE (ASA)}}
\end{array}
$$

Fig. 4. Addition of arginine or canavanine to fumarate to form argininosuccinate (ASA) or canavanino-succinate (CSA) respectively.

converted to guanidinosuccinic acid.

This enzyme was first extracted from liver where it is predominantly a lyase. In the kidney it is predominantly a synthetase. For this reason the urea cycle occurs mainly in the liver, where an alternate method of synthesizing argininosuccinate exists (Fig. 5). Argininosuccinic acid lyase (ASAL) levels in the blood correlate with liver biopsies for measuring liver function in humans[5,41,45].

At that time a paper appeared on the effect of addition of canavanine to fumarate in the presence of kidney homogenate (Fig. 4). Like argininosuccinate, canavaninosuccinate was isolated as its barium salt, we had to find a biochemical pathway for this reaction.

For these studies, I had the cooperation of Dr. Kihachiro Takahara from Japan as a postdoctoral fellow. We prepared an extract of acetone-dried human liver powder as the enzyme. We discovered soon that the only coenzyme that was effective in the reduction was thioctic acid (reduced lipoate). NADH, NADPH and reduced glutathione were ineffective[43]. This also required a small amount of ferrous ion in addition to the liver extract for the best results (Fig. 6).

With Dr. Shigeko Nakanishi, we discovered further that canavanine was an inhibitor of the ribose reduction to deoxyribose system, in the conversion of RNA to DNA[44]. This was expected because the RNA reduction to DNA system requires a protein carrier of a disulfide bond called, *thioredoxin*.

If larger amounts of ferrous ion were added, in the presence of reduced lipoate, canavanine could be reduced to homoserine and guanidine in the absence of enzyme[44](Fig. 7).

*Up to this time there had been no explanation for the presence of guanidine in human urine.*

Once we were convinced that guanidinosuccinate originated by the reduction of canavaninosuccinate, we had to explain how canavaninosuccinate was synthesized in the human.

Argininosuccinate is synthesized in the liver by the condensation of aspartic acid with citrulline (delta ureido-norvaline). This reaction is mediated by an argininosuccinate synthetase, and requires ATP[36] (Fig. 5). We felt that canavaninosuccinate was synthesized similarly by a *canavaninosuccinate synthetase* system present in animal liver. We had to show that ureido-homoserine could also react with aspartate to form *canavanino-succinate*, enzymatically.

With Dr. Anthony Koller, a post doctorate student from the Ill. Inst. of Technol., we prepared ureido-homoserine from canaline and KCNO[9,10] and set up an experiment to demonstrate the condensation reaction between ureido-homoserine and aspartate, using an enzyme prepared from bovine liver. (Fig. 5). The progress of the reaction was followed with paper electrophoresis. (Fig. 8)

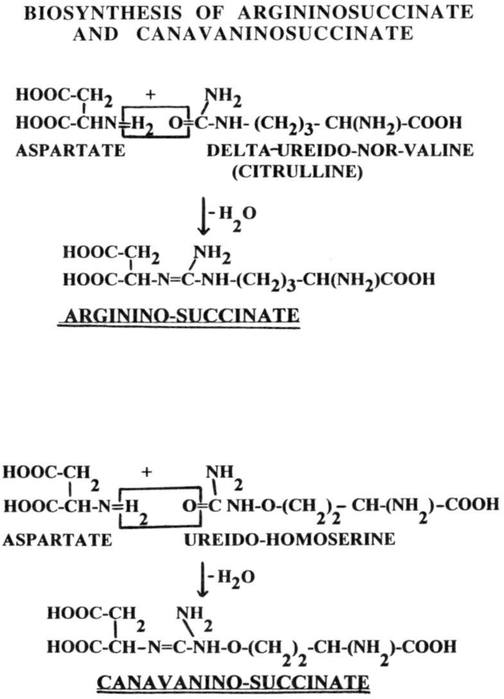

## BIOSYNTHESIS OF ARGININOSUCCINATE AND CANAVANINOSUCCINATE

HOOC-CH$_2$    +    NH$_2$
HOOC-CHN$=$H$_2$   O$=$C-NH- (CH$_2$)$_3$- CH(NH$_2$)-COOH
ASPARTATE       DELTA-UREIDO-NOR-VALINE
                  (CITRULLINE)

$\downarrow$-H$_2$O

HOOC-CH$_2$    NH$_2$
HOOC-CH-N=C-NH-(CH$_2$)$_3$-CH(NH$_2$)COOH

**ARGININO-SUCCINATE**

HOOC-CH$_2$    +    NH$_2$
HOOC-CH-N$=$H$_2$    O$=$C NH-O-(CH$_2$)$_2$- CH-(NH$_2$)-COOH
ASPARTATE       UREIDO-HOMOSERINE

$\downarrow$-H$_2$O

HOOC-CH$_2$    NH$_2$
HOOC-CH-N=C-NH-O-(CH$_2$)$_2$-CH-(NH$_2$)-COOH

**CANAVANINO-SUCCINATE**

*Fig. 5. Condensation of Aspartate with citrulline or ureidohomoserine to form argininosuccinate(ASA) or canavaninosuccinate(CSA), respectively.*

This reaction was carried out successfully (Fig. 8) and is illustrated. (Figs. 5 & 9)[8]. We now saw an opportunity to find an explanation for the inverse relationship between the concentration of guanidino-succinate and guanidino-acetate in the urine.

In the urea cycle an *argininosuccinate lyase* acts on arginino-succinate to form arginine and fumarate. The arginine goes on to transamidinate to glycine to form guanidinoacetate. The argininosuccinate is synthesized by a synthetase for the reaction between aspartate and citrulline. Thus there is one directed pathway for the dissolution of arginino-succinate.

In the guanidine cycle there exists two pathways for the dissolution of canavanino-succinate. One results in the formation of canavanine and by transamidination guanidinoacetate. The other was by reduction to form guanidinosuccinate and homoserine. Thus the formation of guanidino-acetate was at the expense of the formation of guanidino- succinate and *Visa Versa*[40].

At this point, it was felt that the evidence we had accumulated pointed to the fact that a sequence of reactions could be assembled resembling the urea cycle, which could explain the significance of the appearance of guanidinosuccinate in the urine of the uremic patient.

We proposed first, that *the objective of the urea cycle was to form creatine phosphate, needed for muscular contraction*. The formation of urea was a by-product to be excreted in order to allow the cycle to proceed.

It was apparent also that we were dealing with an *alternate pathway to form creatine phosphate*. Since canavanino-succinate was an intermediate, this would readily explain the formation of guanidinosuccinate in the uraemic patient.

Ornithine (from the action of arginase on arginine) is an essential amino acid which needs to be supplied by the diet for the preparation of citrulline for the urea cycle to function. In our case, the essential amino acid supplied from the diet had to be canaline. If this would react with carbamoyl phosphate we would have the enzymatic method for the formation of ureidohomoserine postulated above.

We showed then, with Richard F. Dods a post doctoral fellow, that canaline reacts readily with carbamoyl phosphate, enzymatically, to form ureido-homoserine[27]. (Fig. 10)

The origin of the carbamoyl phosphate for the guanidine cycle had to be the urea in the blood. If urea is placed in sterile water, it is stable indefinitely, this is a useful property of urea since it protects the tissues, such as the bladder, from the caustic action of ammonia. Therefore, we had to demonstrate how this effect was overcome in the uremic patient, since urease does not occur in the human.

For this purpose, with Dr. John Sherwin, a post doctoral fellow, we discovered that urea in blood could be readily oxidized using the p-450 system present in the microsomes of the liver to form hydroxy urea[31] (Table 2).

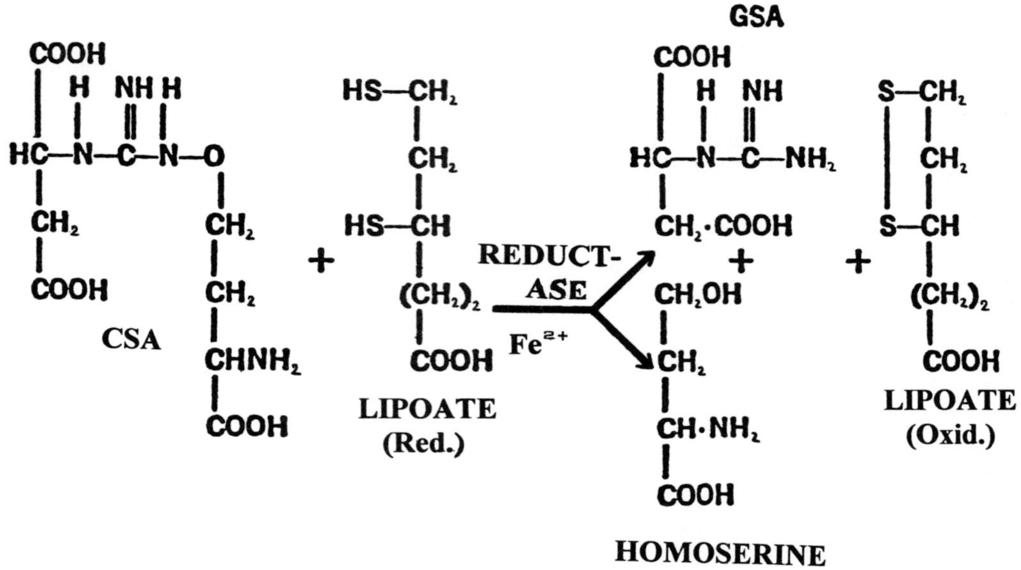

GSA

```
COOH                                              COOH
 |                                                 |
 |  H NH H              HS—CH₂                      |  H NH           S—CH₂
 |  | | |               |                          |  | ‖            |
HC—N—C—N—O              CH₂                        HC—N—C—NH₂        CH₂
 |                      |                           |                |
CH₂          CH₂       HS—CH         REDUCT-       CH₂·COOH    +    S—CH
 |           |    +     |            ASE            +                |
COOH        CH₂        (CH₂)₂                      CH₂OH            (CH₂)₂
            |                 Fe²⁺                  |                |
 CSA       CHNH₂        COOH                        CH₂             COOH
            |           LIPOATE                     |              LIPOATE
           COOH         (Red.)                     CH·NH₂          (Oxid.)
                                                    |
                                                   COOH
```

HOMOSERINE

Fig. 6. Reduction of canavanino-succinate(CSA) to form guanidinosuccinate (GSA) and homoserine (HS).

Table 2. Comparison of the rate of hydroxylation ($\mu$mol/min) of hexobarbital (control), *urea* and *guanidine*, by a rat liver microsome preparations[31]

| Substrate ($\mu$mol/L) | Rate of hydroxylation ($\mu$mol/min/mg protein). (Mean of 4 replicates) |
|---|---|
| Hexobarbital: 40 | $9.8 \pm 1.76$ |
| Urea: 40 | $18.3 \pm 6.01$ |
| Guanidine: 40 | $12.0 \pm 2.34$ |

The hydroxy urea formed hydrolyzes in water to form carbamic acid and hydroxyl amine. The carbamic acid is phosphorylated with ATP, mediated by a kinase. This had been demonstrated before by numerous others.

Thus, the *Guanidine Cycle* opened up an alternate pathway for the synthesis of creatine phosphate when the urea cycle was blocked (as in uraemia).

The hydroxyl amine, generated by the hydrolysis of hydroxy urea, reacts with the lactam of homoserine, to form homoserine hydroxamic acid spontaneously[18], or enzymatically to form canaline (Fig. 9). Excess hydroxyl amine, which is toxic, may be detoxified by reaction with the lactam of homoserine (Fig. 11).

A model for this reaction has been explored[7] to demonstrate that this reaction can take place between an oxime and lactam of homoserine (see Reagents and Procedure Section) (Fig. 12).

All of the above is shown graphically as a Bi-Cycle in Fig. 13[31.]

The *Guanidine Cycle* provides also a method by which urea, which is normally excreted in the urine, may be re-utilized when nitrogen is in short supply[49].

It has been shown that hibernating bears must be reutilizing at least 20 per cent of the urea nitrogen generated by the urea cycle. The presence of a system which provides for the reutilization of urea also explains the survival of individuals on long hunger strikes.

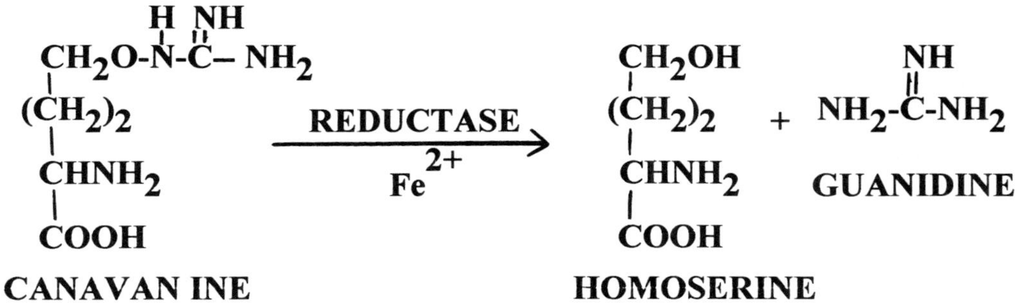

$$
\begin{array}{c}
\overset{H}{\underset{|}{\phantom{}}}\overset{NH}{\underset{\|}{\phantom{}}} \\
CH_2O\text{-}N\text{-}C\text{-}NH_2 \\
| \\
(CH_2)_2 \\
| \\
CHNH_2 \\
| \\
COOH
\end{array}
\quad
\xrightarrow[Fe^{2+}]{\textbf{REDUCTASE}}
\quad
\begin{array}{c}
CH_2OH \\
| \\
(CH_2)_2 \\
| \\
CHNH_2 \\
| \\
COOH
\end{array}
\quad + \quad
\begin{array}{c}
\overset{NH}{\underset{\|}{\phantom{}}} \\
NH_2\text{-}C\text{-}NH_2
\end{array}
$$

**CANAVAN INE**                          **HOMOSERINE**                    **GUANIDINE**

Fig. 7. Reduction of canavanine to form guanidine and homoserine.

This also explains, probably, how some inmates of concentration camps, could survive for as much as four years on an intake of only one gram of nitrogen a day, or less.

It must be stressed that, 'The function of both the Urea and Guanidine Cycles is to form creatine phosphate which is needed for muscular contraction. When the Urea cycle is blocked, or when nitrogen is in short supply, the Guanidine Cycle can take over this function'.

In this case guanidinosuccinic acid levels will appear in the urine in increased amounts. It must be kept in mind that loss of guanidinosuccinic acid refers to a loss of only milligram quantities of nitrogen, while the urea cycle puts out gram quantities of nitrogen.

**Table 3. Guanidinoacetate, guanidinosuccinate and guanidines found in the urine of rats after injection of 0.05 mmoles of saline (controls), canavanine, canavaninosuccinate, guanidinosuccinate, arginine or urea[19]. Guanidine and methylguanidine are not separated cleanly, under the conditions used. When the arginine solution was mixed with 0.05 mmol of glycine, a tenfold increase in GAA in the urine was noted. When 0.05 mmol aspartate was added to the arginine solution, GSA excretion did not increase significantly. Values are μmols excreted per 24 h**

| Added substance | Guanidinoacetate (GAA) | Guanidinosuccinate (GSA) | Guanidines |
|---|---|---|---|
| Saline (controls) | 3.50 | 0.40 | 2.66 |
| Canavanine | 19.4 | 4.27 | 239.0 |
| Canavaninosuccinate | 14.1 | 78.7 | 15.8 |
| Guanidinosuccinate | 1.90 | 316.0 | 2.98 |
| Arginine | 3.45 | 3.15 | 4.37 |
| Urea | 2.34 | 1.01* | 2.64 |

*The 'Y Peak' is not included.

With the introduction of phenanthrene-quinone as a general reagent for the detection of all the different guanidino compounds, fluorometrically[39,51], I decided to re-study the Guanidine Cycle *In Vivo*[19]. Rats held in metabolic cages were inoculated intraperitoneally with normal saline (controls), urea, canavanine, canavaninosuccinate, guanidinosuccinate or arginine. Analysis was done by an Autoanalyzer system. (Fig. 14). The products found supported the idea of a Guanidino Cycle and are listed in Table 3.

When urea was injected into rats, intraperitoneally, and the urine was analysed by column chromatography a prominent 'Y peak' appeared before the guanidinosuccinate peak. This was apparent also in the urine of the human uremic patient and was swamped under the guanidinosuccinic acid peak when guanidinosuccinic acid excretion was very high. Its identity is in doubt (Fig. 15). However, it obviously originated from urea. The compound under this peak is distinct from guanidinosuccinate and locates at a site where carbamoyl-phosphate appears. It is distinct from taurocyamine.

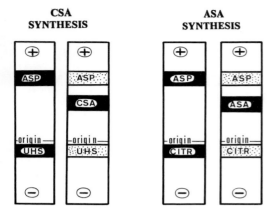

**CSA SYNTHESIS**   **ASA SYNTHESIS**

*Fig. 8. Paper electrophoretic pattern showing the reaction between ureido-homoserine and aspartate to form canavaninesuccinate (CSA), or citrulline with aspartate to form argininosuccinate (ASA). Stained with Ninhydrin.*

It had been postulated by others that certain guanidino compounds derive from hydroxy guanidine[37]. Our experiments did not support that idea[19].

In preparing canavaninosuccinic acid I required substantial amounts of canavanine. For this purpose I extracted the 'Jack Bean' which can contain as high as 25 g of canavanine per kilo[25].

This bean also known as the 'Chickasaw Lima Bean', was a major source of food for the American Indians in Tennessee. The Indians would soak these beans in water and throw the water away. In this way the toxic canavanine was removed. Europeans did not take this precaution, as a result the beans were found to be toxic, and removed from the food market, only in recent times.

**Table 4. Canavanine and arginine content of some commercially available varieties of alfalfa seed and their canavanine to arginine ratio. They are listed in the order of their decreasing canavanine to arginine ratio. A popular variety in use in Tennessee is Classic**

| Variety | Canavanine g/kilo | Arginine g/kilo | Canavanine/Arginine Ratio |
|---------|-------------------|-----------------|---------------------------|
| Team | 10.7 | 0.920 | 11.6 |
| Classic | 9.90 | 0.929 | 10.7 |
| Weevlchek | 13.6 | 1.39 | 9.78 |
| Saranac | 10.9 | 1.13 | 9.65 |
| Arc | 10.8 | 1.13 | 9.55 |
| Buffalo | 8.33 | 0.962 | 8.65 |

The natives used similar techniques to wash the toxins out of other foods, such as acorns, in order to make them edible. Tequila is made from a toxic Agave plant (Maguey). The fermented root (Pulque) is freed of toxins by distillation after fermentation. The toxic residue is discarded. Pulque is toxic.

At this time I was asked to analyse Kentucky 31 grass to see whether I could identify a toxin which was present in aged grass. This produced a disease known as 'Blackleg'. Naturally I looked for canavanine. This was a blind alley because the toxic substance was made by the spore forming mold 'Clostridium chasuoei'. However, when I used different fodders as controls, I confirmed the fact that there was substantial amounts of canavanine in alfalfa and clover[20,25]. The alfalfa with the highest canavanine content contained about 14 g/kilo of seed[25]. I then noted that the various varieties of alfalfa were resistant to the destructive alfalfa weevil (Hypera postica, Gyllenhal) in accordance to their canavanine to arginine ratio[21]. This supported the idea that arginine was a competitive inhibitor of the toxic action of canavanine (Table 4).

The alfalfa weevil's depredations had reduced the amount of alfalfa grown in Tennessee from 300 000 acres per year to 50 000 acres per year. Alfalfa's were developed with a high canavanine to arginine ratio and today over 200 000 acres of alfalfa per year are grown in Tennessee[47].

In working with columns for the analysis of amino acids I was impressed by the ubiquitous presence of homoserine and its lactone, whether the sample was from animal or plant fluids. Yet, homoserine

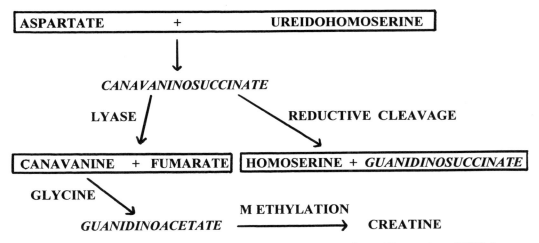

*Fig. 9. The inverse relationship between guanidinoacetate (GAA) and guanidinosuccinate (GSA) due to alternate pathways for the dissolution of canavaninosuccinate (CSA).*

does not occur commonly in plant or human proteins. I felt that homoserine must be a precursor of the formation of canaline, and thus of canavanine.

**Table 5. Homoserine and its lactone in pea seedlings. Mean ± SD**

| *Weight (mg) of root + shoot | Homoserine ( μmol/g) | Lactone (μmmol/g) | Lactone (%) |
|---|---|---|---|
| 190 ± 21 | 87.4± 10.5 | 12.8± 2.1 | 14.7 % |

*Values are for wet weight. Grown in vermiculite, in a pH 5, 0.05 mM acetate buffer containing 0.05 mMol/l potasssium nitrate.

To explore these ideas I grew pea seedlings in vermiculite containing, in a 0.01 M pH 5 acetate buffer, one of urea, nitrate or hydroxylamine[18]. I found that the amount of homoserine formed in the root and shoot was highest in the presence of nitrate (874 μmol/g of root, wet weight of which 14.7 per cent was in the form of the lactone). Thus, the formation of homoserine is stimulated by the presence of nitrate.

In this study, it was also demonstrated that the lactone reacts spontaneously with hydroxylamine to form homoserine hydroxamic acid, thus detoxifying the hydroxylamine[18] (Fig. 11, Table 5). It has been reported by others that in some roots the homoserine content is as high as 12 per cent of the dry weight, of which 20 per cent is in the form of the lactone[12].

At about this time Mori *et al.* reported that they had found α-guanidinoglutaric acid in the central nervous system of animals in whom they had induced epileptic seizures[14]. I knew that glutamic acid cyclized to form a lactam readily to form a pyrollidone derivative (Pyroglutamic acid, pGlu) which was the end group of several low molecular weight neuropeptides from the hypothalamus, such as *thyrotropin releasing hormone* (TRH), *luteinizing releasing hormone* (LRH), *bombesin* and others. For comparison, I prepared the lactam of α-guanidinoglutaric acid (amidinopyroglutamate, ApGlu)[22]. I used this to form the tripeptide, ApGlu–Gly–ProNH$_2$. The corresponding pGlu–Gly–ProNH$_2$ was very active in releasing the thyrotropic hormone. In contrast, ApGlu–Gly–ProNH$_2$ strongly inhibited the release of TRH[23].

With Professor Fred Schell of the University of Tennessee I studied the structure of the lactam of α-guanidinoglutaric acid and explained why it reacted with two molecules of acyl compounds. Thus,

CH₂ONH₂
|
(CH₂)₂          +
|
CHNH₂
|
COOH
**CANALINE**

$$NH_2\overset{O}{\overset{\|}{C}}\text{-}OPO_3H_2$$
**CARBAMOYL**   **TRANS-**
**PHOSPHATE**   **FERASE**  →

$$CH_2ONH\text{-}\overset{O}{\overset{\|}{C}}\text{-}NH_2$$
|
(CH₂)₂
|
CHNH₂
|
COOH
**UREIDO--**
**HOMOSERINE**

*Fig. 10. Reaction of carbamoyl phosphate with canaline to form ureidohomoserine.*

it was dibasic and had a structure resembling that of creatinine. Its formula is 2-imido-4-oxo-5(3'-propane acid) -tetrahydroimidazole[30] (Fig. 16).

I then went on to dehydrate the lactam of α-guanidinoglutamate to form a condensed ring system of an imidazolidine fused to a pyrrol ring[24]. This compound was basic in nature and thus an alkaloid, but I did not explore its biological properties. Its formula is 2-imino-4-oxoimidazolidine-1,5-pyrollidone-8 (Fig. 17).

Feeding an alfalfa with a high canavanine content to the larvae of the alfalfa weevil resulted in never reaching the adult stage. I felt that this toxicity might reflect the fact that canavanine would inhibit

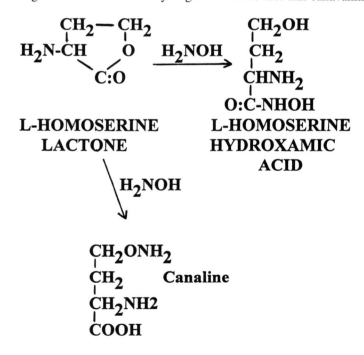

*Fig. 11. Reaction of hydroxylamine with the lactam of homoserine to form
either homoserine hydroxamic acid spontaneously or
canaline enzymatically (with ATP).*

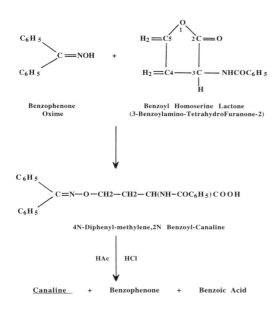

**Condensation of an Oxime with Homoserine Lactone to Form Canaline**

Benzophenone
Oxime

Benzoyl Homoserine Lactone
(3-Benzoylamino-TetrahydroFuranone-2)

4N-Diphenyl-methylene,2N Benzoyl-Canaline

HAc | HCl

Canaline  +  Benzophenone  +  Benzoic Acid

*Fig. 12. Reaction of benzophenone with benzoyl homoserine lactone as a model for the synthesis of canaline.*

the ribose reductase system[44], which would explain its action as an antitumor agent, and decided to explore this.

For this purpose, Desmond Doyle, Professor at the University of Tennessee Veterinary College, and I chose the model of the Novikoff hepatoma cells. These cells can be grown intraperitoneally in rats, because they are deficient in the histocompatibility antigens, HLA, at their surface. They were obtained from the M.D. Anderson Hospital in Houston.

We demonstrated that the canavanine toxicity could be enhanced if the rat was injected with a mitogen such as concanavalin A, to accelerate the metabolic rate of the tumor cell. It was noted that the cells that were subjected to the action of canavanine developed vacuoles, which became larger and coalesced, the cells expanding in size and finally bursting, dumping their contents into the peritoneal cavity[6](Fig. 18).

It was also noted that, when the cells were incubated *in vitro* with canavanine, canaline and urea were formed due to the action of arginase, present in these liver cells, on canavanine (Fig. 19).

Figure 20 shows that canaline was released. This is an O- substituted hydroxylamine. Hydroxyl amine is a well known mutagenic substance, which reacts with pyrimidines (Fig. 21)[8,35]. It was therefore concluded that canaline was the toxic substance, canavanine was acting as the carrier of canaline.

This suggested a possible treatment for the terminally ill patient with metastasizing tumor cells: namely, to use a substance too toxic to be administered to the patient, but masked by an inert carrier until it is absorbed by the cancer cell and the toxic moiety is released inside the cell. I felt that I had such a substance in L-homoserine hydroxamic acid (L-HHA): Warburg noted in the late 1920s that tumor cells tend to be very acid (as low as pH 5), and this should hydrolyze the homoserine hydroxamic acid to release hydroxylamine[18]. Normal cells from which the tumor cells were derived should not hydrolyze the HHA readily.

For this study I enlisted the aid of Ethan A. Natelson MD, of the University of Texas Houston Health Center, and Dr P. Pantazis of the Stehlin Foundation for Cancer Research at St. Joseph Hospital in Houston. The model studied was the 'nude' mouse, which has no thymus. Human tumors can be implanted in this mouse.

Body fluids are generally mildly alkaline, in which medium the homoserine hydroxamic acid is stable. However, inside the rapidly growing tumor cell the pH drops dramatically as $CO_2$ is formed. This releases $H_2NOH$[18,28,29]. Results of these ongoing studies are shown in Fig. 22, which shows, with myo-leukaemia cells in tissue culture, that L-HHA is more toxic to the cells than the D-isomer. Figure 23 shows, by examining changes in the cell cycle (Fig. 24), that malignant melanoma cells respond more readily to therapy than the normal melanocytes from which they were derived. Figures 25 and 26 show that human breast tumor and human peritoneal sarcoma cells respond to L-HHA treatment[29].

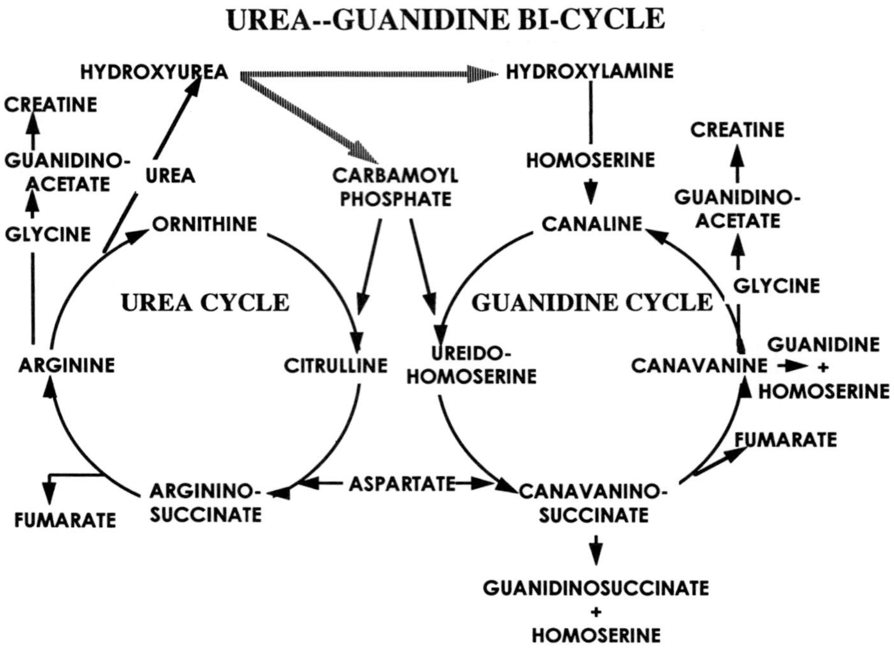

## UREA--GUANIDINE BI-CYCLE

*Fig. 13. The guanidine cycle as it interacts with the urea cycle to form a 'bi-cycle' to assure the continued supply of creatine phosphate for muscular contraction.*

Studies designed to elucidate the origin of guanidinosuccinate in urine of the uremic patients have uncovered the following new ideas. There is a guanidine cycle in animals which assures the continued synthesis of creatine phosphate when the urea cycle is blocked or when nitrogen is in short supply. The appearance of guanidinosuccinate in the urine signals that this guanidine cycle is operative. This explains the following:

a. The guanidinoacetate concentration in the urine is in inverse proportion to the guanidinosuccinate concentration.

b. Guanidine appears in the urine in increased amounts in uremia.

c. Animals can survive for substantial periods of time on a markedly reduced nitrogen intake.

d. Urea and guanidine are readily oxidized by the *P*-450 system (hepatic microsomes) to form hydroxy urea and hydroxyguanidine.

e. High concentrations of homoserine and its lactone in plant roots act as a defense against the toxicity of hydroxylamine formed by the reduction of nitrate. A highly specific test has been developed to monitor the function of the liver, which correlates with liver biopsies[5,41,45].

Some other facts have been uncovered:

a. The canavanine to arginine ratio in alfalfa is related to its resistance to the alfalfa weevil. This permits a rapid evaluation of the potential resistance of a new variety of alfalfa to the alfalfa weevil.

b. If the end amino acid (pGlu, the lactone of glutamic acid) of certain low molecular weight neuropeptides is amidinated to form the lactam of amidinopyroglutamic acid (pAGlu) the biological activity of the neuropeptide is altered significantly.

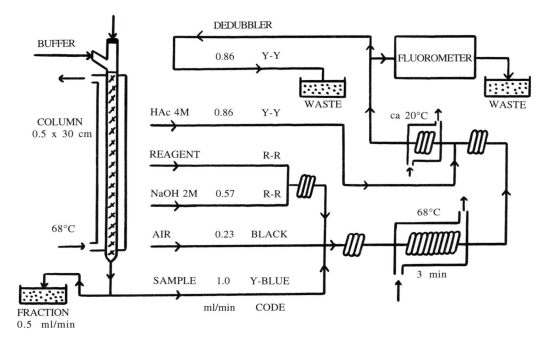

*Fig. 14. Automated system for guanidino compound assay with phenanthrene-quinone, fluorimetrically.*

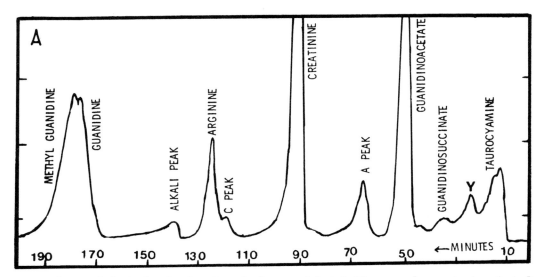

*Fig. 15. The y-peak obtained on fractionating urine from rats injected with urea, or from azotemic patients. It is visualized here with the phenanthrene quinone reagent. The peak occupies a position similar to that of car-bamoyl phosphate and obviously derives from urea. Its composition is in doubt.*

229

COOH (CH)2 CH NH-C:(NH)-NH2 COOH
GUANIDINO-GLUTARATE (AMIDINO-GLUTAMATE)

LACTAM (ApGlu) 2-IMIDO-4-OXO-5-(3' PROPANE ACID)-TETRAHYDRO-IMIDAZOLE

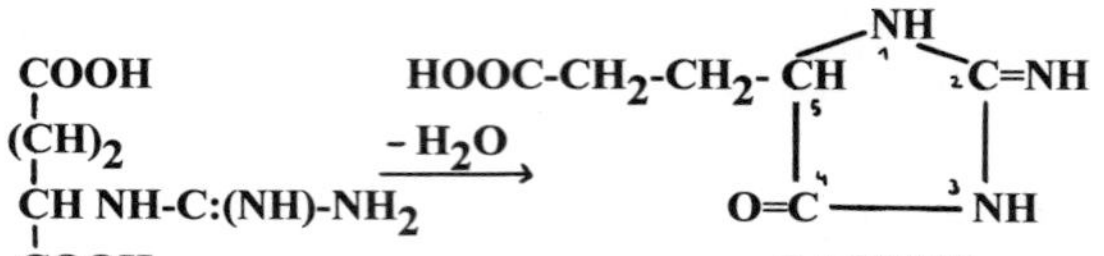

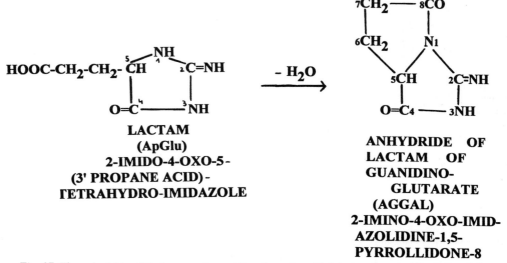

*Fig. 16. The lactam of guanidinoglutaric acid (amidino glutamic acid lactam, ApGlu) 2-imido-4-oxo-5(3'-propane acid)-tetrahydroimidazolidine.*

LACTAM (ApGlu) 2-IMIDO-4-OXO-5-(3' PROPANE ACID)-TETRAHYDRO-IMIDAZOLE

ANHYDRIDE OF LACTAM OF GUANIDINO-GLUTARATE (AGGAL) 2-IMINO-4-OXO-IMID-AZOLIDINE-1,5-PYRROLLIDONE-8

*Fig. 17. The anhydride of the lactam of guanidinoglutamic acid, 2-imino-4-oxoimidazolidine-1,5-pyrollidone-8.*

c. A system for the destruction of tumor cells is under study in which hydroxylamine activity is used to alter the DNA and RNA of the tumor cells, thus destroying them. The hydroxylamine activity is masked by a carrier, which makes the compound inert until it reaches its target where it releases the hydroxylamine.

d. L-Homoserine hydroxamic acid is shown to invade a variety of tumor cells, destroying them. It is proposed that the toxicity of canavanine to certain organisms also results from the release of canaline, which is a substituted hydroxylamine, active in reacting with C = O and C = NH groups of pyrimidines of RNA and DNA[6,29], (Fig. 21).

Practicable reagents, methods and instrumentation have been improved, developed and described in these studies. Numerous new compounds have been formed and methods for their preparation have been described in detail.

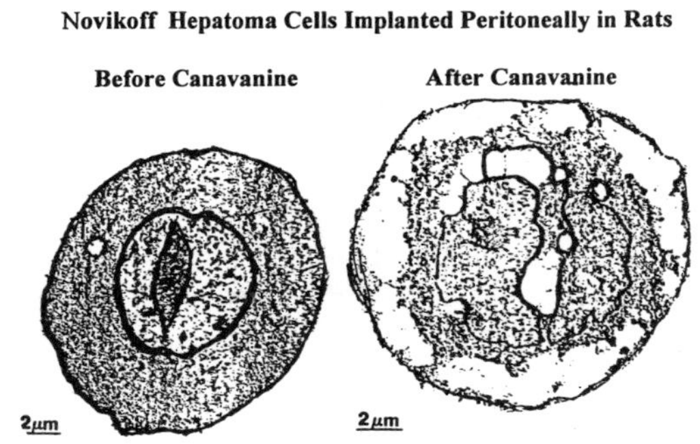

**Novikoff Hepatoma Cells Implanted Peritoneally in Rats**

Before Canavanine     After Canavanine

2μm                    2μm

*Fig. 18. Effect of canavanine on Novikoff hepatoma cells under the electron microscope.*
*The hepatoma cell supplies arginase to convert the canavanine to canaline, which is an O-substituted hydroxylamine (see Figs. 19 and 20).*

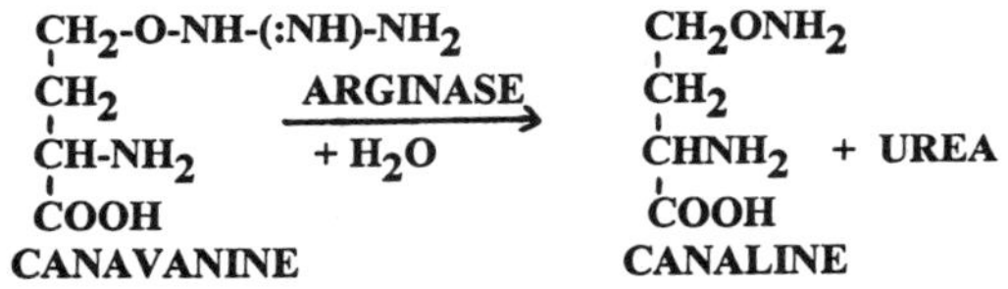

*Fig. 19. Action of arginase on canavanine to form canaline and urea. Note that! canaline is an O-substituted hydroxylamine and is thus toxic to cells in general.*

## Reagents and procedures

### Reagents[3,26]

*Ninhydrin spray*

Dissolve 500 mg ninhydrin in a mixture of 300 ml of isopropanol, 100 ml n-butanol, 15 ml pyridine and 100 ml methyl cellosolve.

*Ninhydrin reaction*

Prepare a 3 per cent solution of ninhydrin in methyl cellosolve. Dissolve sodium acetate trihydrate (36 g, 0.1 mol) in 50 ml $H_2O$, add 10 ml gl. acetic acid to adjust the pH to 5.4, and dilute to 100 ml with $H_2O$.

Dissolve 49 mg NaCN and dilute to 100 ml with $H_2O$ (molar).

On the day of the test, dilute 2 ml of the cyanide solution to 100 ml with the acetate buffer to make the buffered cyanide solution. To 0.5 ml of eluate add 0.2 ml of buffered cyanide solution followed by 0.1 ml of the ninhydrin solution. Heat at 100 °C for 15 min, and dilute with 2 ml of 50 per cent isopropanol in $H_2O$.

### Sakaguchi reagent[43,44]

Solution A: 300 mg of 2,4 dichloronaphthol and 10 g thymine per liter of 95 per cent ethanol. Refrigerate.

Solution B: 5 per cent NaOCl diluted five-fold to make 1 per cent NaOCl. To develop the color, alkalinize 0.5 ml of the eluate with 0.1 ml of 2.5 M NaOH, cool in an ice bath, and add 0.5 ml of solution A. Follow with 0.1 ml of the NaOCl solution B. Read at 10 min at 515 nm.

To use as a spray, spray spot with 0.01 M NaOH, air dry, spray with A, air dry, and spray with B. Keep in deep freezer to preserve the colour.

### Diacetyl reagent[3,32,26]

A: Dissolve 1 g of α- or β- naphthol in 100 ml of an aqueous solution containing 6 g of NaOH and 16 g $Na_2CO_3$. B: Dissolve 50 μl diacetyl(butanedione-2,3) in 100 ml $H_2O$.

Add 0.5 ml solution A to 0.5 ml eluate in an ice bath. Follow with 0.5 ml of solution B. Cool in ice bath for 1 h. Read at 520 nm.

### Alkaline picrate (Jaffée reaction)

Dissolve 1 g of picric acid in 100 ml of $H_2O$ at room temperature. This solution is stable in a dark brown bottle indefinitely at room temperature. On the day of the test add 1 ml of the picric acid solution to 0.5 ml of eluate. Now add 0.2 ml of 2 Molar NaOH. Read the reddish color at 520 nm.

Can be used as a spray:  spray the spots with the picric acid solution. Air dry and spray with 0.1 M NaOH.

231

**CONVERSION OF CANAVANINE TO CANALINE BY
LIVER CELL ARGINASE. PAPER ELECTROPHORESIS
OF THE PICRATE PRECIPITATE**

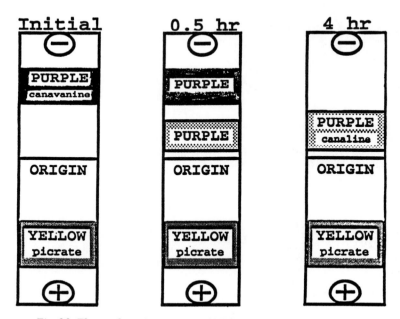

Fig. 20. Electrophoretic pattern on 3MM paper, showing the effect of
arginase on canavanine to convert it to canaline. Both canavanine and
canaline precipitate as picrates. They are distinguished here by paper electrophoresis. Canavanine, being more basic, travels to the cathode more rapidly.
Yellow picric acid travels to the anode. Buffer is pH 5, 0.05 M lithium
acetate. Ninhydrin spray.

## REACTION OF HYDROXYLAMINE WITH A PYRIMIDINE

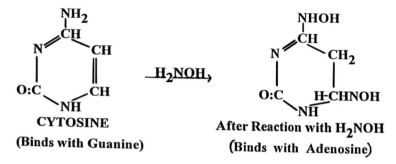

Fig. 21. Reaction of hydroxylamine with a pyrimidine to alter its properties.

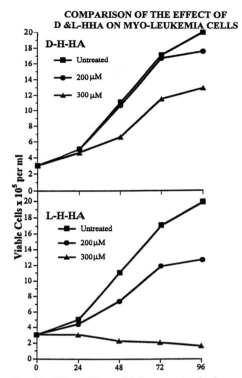

### COMPARISON OF THE EFFECT OF D &L-HHA ON MYO-LEUKEMIA CELLS

Fig. 22. Effect of D- and L-homoserine hydroxamic acids on myelogenous leukemia cells in tissue culture. Note! L-homoserine hydroxamic acid (L-HHA ) is more toxic to the cells than the D- form

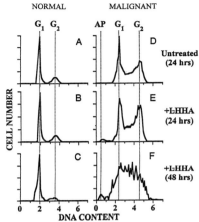

**L-HOMOSERINE HYDROXAMIC ACID DISRUPTS THE CELL CYCLE OF MALIGNANT MELANOMA CELLS BUT NOT THAT OF THE MELANOCYTES FROM WHICH THEY WERE DERIVED**

Fig. 23. Effect of L-homoserine hydroxamic acid (L-HHA) on the cell cycles of normal melanocytes and malignant melanoma cells, measured by flow cytometry[29]. AP refers to non-viable cells (apoptosis). In panel F, $G_1$ is the sum of $G_0 + G_1$ and $G_2$ is the sum of $G_2$ and M (see Fig. 24).

## Pentacyanoferrate reagent[18]

Dissolve 25 g potassium ferricyanide and 25 g of sodium nitroprusside each in 500 ml $H_2O$. Mix and add 500 ml of 2 M NaOH. The solution darkens and then turns lighter. Allow to stand in the refrigerator overnight before use. Keep in the refrigerator.

The pentacyanoferrate reagent can be used directly as a spray. If used with an eluate, treat 0.5 ml of the eluate with 0.2 ml of 2 M NaOH and add 0.5 ml of the reagent plus 1 ml of $H_2O$. Read at 480 nm.

## $Fe^{3/2+}$ reagent

Stir 1 g of ferric ammonium sulfate ($FeNH_4(SO_4)_2.12H_2O$) and 200 mg ferrous sulfate ($FeSO_4.7H_2O$) in a glass-stoppered bottle at room temperature for several hours with 100 ml 0.25 M HCl. Not all of the material dissolves. Leave it to stand overnight at room temperature. In the morning it should be clear with no sediment. If it is not completely dissolved let it stand with occasional shaking. This is a convenient reagent for assaying hydroxylamine derivatives, such as homoserine hydroxamic acid or canaline[18]. It gives no color with hydroxylamine itself. It can be used as a spray or for the analysis of a solution. In a typical application 0.5 ml of an eluate is acidified with 0.1 ml of 2 M HCl and 2 ml of the reagent is added. The maroon color is read at 525 nm. In contrast with the reagents above this color is stable for several days at room temperature.

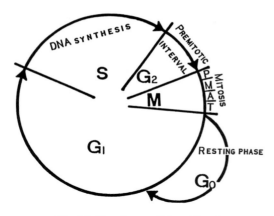

*Fig. 24. The phases of the cell cycle.*

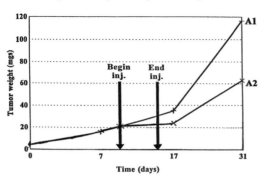

*Fig. 25. Inhibition of the growth of human breast tumour cells implanted subcutaneously in 'nude' mice. Vertical bars show treatment for 1 week. The upper curve represents the typical logarithmic growth of the tumor in the untreated mouse($A_1$). Inhibition of the growth of the tumour with L-HHA was about 40 per cent ($A_2$). 'Nude Mice' are also called 'Thymus-less Mice'. They accept xenographs from humans.*

### Phenanthrene-quinone reagent[21]

Dissolve 42 mg (0.2 mol) of phenanthrene-quinone (analytical grade) in 10 ml of formamide and dilute to 500 ml with isopropanol (0.4 mol/l). Keep in the refrigerator in a brown glass-stoppered bottle. On the day of the test dilute the solution with an equal volume of isopropanol/water (1:1). This solution is 0.2 M. Use this reagent in the automated procedure described below. If analyzing the effluent from the column manually, add 0.1 ml 4 M NaOH to 0.5 ml of eluate to alkalinize and 1 ml of the diluted phenanthrene-quinone reagent (0.2 mol/l). Heat for 15 min at 70 °C, cool, and acidify with 0.1 ml gl. acetic acid. Read the fluorescence. Excitation is at 307 nm and emission is 447 nm in the spectrofluorimeter.

### Chlorine reagent

This reagent is used to locate amino compounds such as carbamate on thin layer chromatography or electrophoresis strips. Add 3 M HCl solution to cover the bottom of a thin layer chromatography jar.

Add about 15 g of $KMnO_4$ crystals to the jar and let stand. The jar fills with chlorine gas. (If a tank of chlorine gas is available use it to fill the jar with chlorine.) Place a glass or plastic support in the jar and expose the strip on the dry support for 1 h. Remove the paper strip from the jar and expose it to the air to allow excess chlorine to evaporate. Spray with a solution of 250 mg of O-tolidine dihydrochloride hydrate (Aldrich), and 1 g of KI, in 100 ml of 2 per cent acetic acid in $H_2O$. The Cl bound to the amine releases iodine. This serves to oxidize the tolidine to enhance the color.

## Procedures

*D-, L- or DL-homoserine lactone hydrobromide[28]*

In a liter flask fitted with a ground glass stopper and a reflux condenser (under a hood) add, with stirring, 200 ml $H_2O$, 200 ml isopropanol, 100 g bromoacetic acid, 100 g of D-, L- or DL- methionine (Aldrich) and 50 ml gl. acetic acid. Heat at 60 °C until the methionine dissolves and reflux for 2 h. Cool and evaporate the solution to dryness on the water bath, with the water aspirator. Add 100 ml of toluene to which has been added 10 ml of isopropanol (to reduce foaming) and evaporate to dryness again to remove the last traces of water. The residue is brown and viscous.

Prepare a solution of anhydrous HCl in cooled anhydrous dioxane by passing the HCl into the dioxane until it absorbs enough HCl to be 2 M with respect to HCl. *This solution can be purchased from Aldrich.*

**PERITONEAL SARCOMA CELLS**
**(BEFORE TREATMENT)**

**PERITONEAL SARCOMA CELLS**
**(AFTER TREATMENT)**

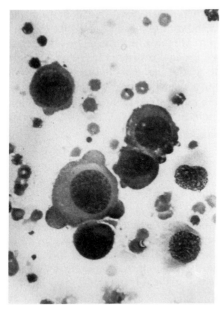 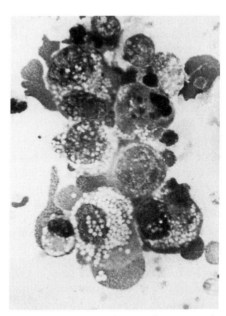

*Fig. 26. Effect of treatment of human peritoneal sarcoma cells implanted in 'nude' mice with*
*L-homoserine hydroxamic acid.*

While the residue in the flask is still warm add 100 ml of the 2 M HCl in dioxane. This is to ensure that all of the homoserine is in the form of the lactone. Stir with a glass rod and let stand overnight in the refrigerator to crystallize. Filter on a Buchner funnel and wash well with anhydrous ethanol. Dry at 70 °C in the vacuum oven. The yield is about 85 g of the D- and L-homoserine lactone hydrobromides, which melt at 151–153 °C. DL-homoserine hydrobromide melts 10 °C lower.

*d-, l-, and dl- homoserine hydroxamic acids*[18,29]

Dissolve 70 g of NH$_2$OH.HCl (ca. 1 mol) in 500 ml anhydrous methanol. Add 56 g (1 mol) of KOH pellets and 200 ml anhydrous ethanol and stir for 1 h. Place in the refrigerator for 1 h and filter off the precipitated KCl with the Buchner funnel. Wash the KCl with an additional 200 ml of ethanol. Refilter the filtrate (NH$_2$OH solution), and combine with the washings through a fluted filter to clarify. Now add 50 g of the homoserine lactone hydrobromide (ca. 0.28 mol) and stir at room temperature. The lactam hydrobromide dissolves and the solution clears. Crystals of homoserine hydroxamic acid then begin to precipitate. Stir for an additional hour and let stand overnight in the refrigerator. Filter off the homoserine hydroxamic acid and wash with ethanol. Dry at 70 °C. This yield is about 30 g. If the methanol is distilled off an additional 5 g will precipitate. This is not recommended because the product obtained is impure.

d- and l-Homoserine hydroxamic acids melt at 140–142 °C. dl-homoserine hydroxamic acid melts at 129–132 °C.

*N-benzoyl homoserine lactone*

Cool 100 ml anhydrous pyridine in a 500 ml flask in an ice and salt bath. Add 40 g of homoserine lactone hydrobromide (ca 200 mmol) and stir to dissolve. Add, drop by drop, 26 ml (excess of 200

235

mmol) of benzoyl chloride. Add at a rate so that the temperature does not rise significantly above 35 °C. When addition is complete stir for an additional hour at room temperature.

Evaporate the pyridine by heating to 60 °C on a water bath at reduced pressure with the water aspirator pump. Cool and add water to dissolve the residue and add to 500 ml of crushed ice. Wash out the flask with water and add to the ice mixture. Now add 4 M HCl to adjust the pH to between 3 and 4, to neutralize the residual pyridine. Stir for 1 h and allow to stand overnight in the refrigerator. Filter off the shiny crystalline needles of benzoyl homoserine lactone and recrystallize from ethyl acetate.

Additional yield can be obtained by evaporating the ethyl acetate to about 50 ml, and adding an equal volume of benzene. The yield is about 32 g. The melting point of the D- and L-benzoyl homoserine lactams is 152–154 °C. The DL-benzoyl homoserine melts at 139–142 °C. The yield is about 32 g.

L-benzoyl homoserine lactone, m.p. 152–154 °C, MW = 205.21.

Calculated: % C = 64.38, % H = 5.40, % N = 6.83

Found: % C = 64.16, % H = 5.43, % N = 6.48

$[\alpha]_D$ 25 °C = –26.7°C (c = 0.1, acetone)

## N-benzoyl homoserine hydroxamic acid

100 ml of a solution of hydroxylamine in methanol (50 mmol/100 ml) is prepared as described above under homoserine hydroxamic acid. To this add 10 g of benzoyl homoserine lactone (48 mmol) dissolved in 100 ml methanol and stir until the color with the $Fe^{3/2}+$ reagent (0.1 ml with 2 ml reagent) stops increasing (about 1 h). Let it stand overnight at room temperature. Evaporate under vacuum to about 50 ml, add 50 ml ethylacetate and allow to crystallize in the refrigerator.

The yield is about 8.5 g benzoyl homoserine hydroxamic acid. The D and L forms melt at 142–143 °C. The DL form melts at 133–135 °C.

## Synthesis of canaline picrate

In this procedure the oxime of benzophenone is converted to its sodium salt and allowed to react with L-benzoyl homoserine lactone to form *2N-benzoyl-4N-Diphenylmethylidine-canaline* (Fig. 12). Benzoic acid and benzophenone are split off with an HCl-HAc solution to form canaline dihydrochloride, this is dissolved in water and picric acid is added to precipitate the picrate. Ureidohomoserine was made from the picrate of canaline[9].

Prepare *benzophenone oxime* as described elsewhere (Org. Synth. Coll. Vol. 1, pp. 70–71). Recrystallize the oxime from benzene to remove adsorbed water and dry at 70 °C in the vacuum oven for 2 h.

Benzophenone oxime is unstable and is preserved in the deep freeze section of the refrigerator in a tightly sealed container.

## 2-benzoyl-4-diphenylmethylene-canaline

The success of the condensation reaction depends on keeping the mixture anhydrous.

Add 100 ml of N-ethylmorpholine (Aldrich) into a heat dried 500 ml flask with a stirring bar, in a silicone oil bath, attached to a condenser, which has been heat dried in the oven also. Distill off, and discard, about 20 ml of the N-ethylmorpholine (b.p. 130–139 °C at ambient pressure). This is to ascertain that the ethylmorpholine is anhydrous and the condenser is dry on the inside. Now, set the condenser up in reflux position, fitted with a drying tube, to keep moisture out. Cool to 60–70 °C, and add 40 g of dried benzophenone oxime (ca 0.2 mol), stirring to dissolve.

Weigh out 5.5 g of metallic sodium (excess of 0.2 mol) and mince finely under toluene. Alternatively (Hood), heat the toluene just to boiling in an Erlenmeyer flask and stir for a few minutes. The sodium fragments into fine globules.

Transfer the sodium to the flask containing the benzophenone oxime and stir for 2 h. The sodium dissolves over a period of 2 h to form the sodium salt of the benzophenone oxime. Keep the temperature between 60 and 70 °C. A small amount of sodium will remain undissolved. This is desirable, as it serves to dehydrate the benzoyl homoserine lactone to be added, if it contains traces of water. Now add 0.2 mol (41 g) of the dried benzoyl homoserine lactone. Raise the temperature of the oil bath to 100 °C. Continue heating and stirring for 2 h.

Cool the homogeneous solution to room temperature and pour into a liter of crushed ice, *cautiously* because of the possible presence of traces of metallic sodium. The pH of the solution after melting is about 8.

Filter the solution through a large Buchner funnel and *save the filtrate*. It contains the sodium salt of the product sought. Refilter the filtrate through a large fluted filter and acidify with gl. acetic acid to a pH of 2.5–3. A copious precipitate of the product sought forms. Allow it to stand overnight in the refrigerator to harden.

Decant the water and dissolve the wet precipitate in acetone and add an equal volume of water. The solution should remain clear but light brown in color. Clarify by adding activated charcoal and filter through a fluted filter.

On evaporation of the acetone, under reduced pressure (water aspirator), an aqueous suspension of crystals in water forms. This is filtered off on a Buchner funnel, washed with water and dried at 70 °C in a vacuum oven. It is recrystallized from benzene. Yield after crystallization is about 41 g 2-benzoyl-4N-diphenylmethylene-canaline, m.p. 122–125°C, MW = 402.43.

Calculated: % C = 71.63, % H = 5.51, % N = 6.96

Found: % C = 71.29, % H = 5.59, % N = 6.74

*Hydrolysis of benzoyl diphenylmethylene-canaline*

Prepare a solution by mixing 12 M HCl (conc. HCl) with an equal volume of 6 M HAc (360 g HAc to 1 λ with H2O). This solution is 6 M with respect to HCl and 3 M with respect to acetic acid. Suspend 25 g of benzoyl diphenylmethylene-canaline in 200 ml of this solution and reflux for 2 h.

Cool to room temperature and allow to stand in the refrigerator overnight. Filter off the benzoic acid and benzoquinone and evaporate the solution to dryness, under vacuum, on a water bath. Add 200 ml water to dissolve the residue, add acid-activated charcoal and filter carefully to remove the color. Evaporate to dryness again under vacuum. Cool, add 100 ml absolute ethanol, stir and evaporate to dryness under vacuum to remove the last traces of HCl and water.

Repeat this once more. Now cool the crystalline residue (canaline dihydrochloride) and stir with 100 ml absolute ethanol to remove the last traces of benzoic acid centrifuge and decant. (This cannot be filtered because it is very hygroscopic.) Dry under vacuum at 70 °C. This yield is about 10 g.

*Canaline dipicrate*

If canaline dihydrochloride is to be stored for an extended time, convert it to the more stable dipicrate by dissolving it in 100 ml of a 1:1 ethanol–water solution, adding 25 g of picric acid, warming until the picrate dissolves. On cooling the dipicrate will precipitate (see below).

The hydrochloride need not be isolated. After hydrolysis of the benzoyl diphenylmethylene-canaline solution and evaporation to dryness, the residue is dissolved in 200 ml $H_2O$ and the solution is filtered. Acid-activated charcoal is added to remove the color and the solution is carefully filtered again.

At this point, add picric acid (25 g) dissolved in warm 100 ml of warm 95 per cent ethanol. Heat to boiling and filter. On cooling, the canaline dipicrate precipitates. This is recrystallized from a water – ethanol solution (1:1). The needle-shaped crystals are air dried at 70 °C in the vacuum oven. The yield is about 25 g, m.p. 192–193 °C with decomposition.

### *In vivo* experiments with rats

Six male rats weighing from 300 to 325 g each were placed in each of six metabolic cages. One cage was used as the control. In the cage labeled control, the rats were injected intraperitoneally with 1 ml of a sterile normal saline solution. The rats in the other cages were injected intraperitoneally with 1 ml of one of the solutions containing 0.05 mmol/ml of arginine, canavanine, canavaninosuccinate, guanidinosuccinate or urea. The urines were collected for the next 24 h and lyophilized. On the day of the test the urines were dissolved in water, filtered through a Buchner funnel and then a Micropore filter, and diluted so that 100 µl contained about 80 nmol of creatinine. This amount was injected in the continuous flow system with phenanthrene-quinone as the reagent (Fig. 14).

In this system, the compounds sought are reacted with phenanthrene-quinone at 68 °C at alkaline pH, cooled, and then acidified. This was analysed then in the continuous flow system to read the fluorescence[19].

### Electrophoresis[7,43,44]

In the experiments using electrophoresis the instrument used was the Beckman Model R paper electrophoresis apparatus with Whatman #3 strips 30 cm in length. The buffer was 0.05 M lithium acetate buffer, pH 5. At this pH guanidinosuccinate moved to the anode while canavanine moved to the cathode. The voltage was 500 V and the current was about 3 mA per strip. This was carried out, usually, for 60 min.

### Guanidinoglutarate lactone

S-Methylisothiourea reacts with amino acids to add an amidine group to the amine, releasing MeSH (Fig. 3). Therefore, this reaction has to be carried out in an efficient hood. The reaction is carried out in a strong alkaline solution. Acidification precipitates the guanidino acid. This is true for all amino acids except glutamic acid. Guanidinoglutamic acid precipitates as a complex salt with S-methylisothiourea in alkaline solution. This may be mistaken for NaCl and filtered off and discarded. If the complex is filtered off and refluxed with acetic acid solution at acid pH, the methylisothiourea moiety of the complex is destroyed, precipitating the lactam of guanidinoglutaric acid (amidinopyroglutamic acid lactone, ApGlu)[22].

Stir 100 g of L-glutamic acid (0.68 mol) and 190 g of S-methylisothiourea (Sigma, excess of 1.36 mol) in 400 ml $H_2O$. Cool a solution of 80 g of NaOH (2 mol) by adding 100 ml crushed ice and add this solution slowly, over 15 min, to the suspension of the l-glutamic acid and the methylisothiourea.

Copious amounts of MeSH are given off and the solution warms and becomes clear. The pH should be between 9 and 10. On cooling to room temperature, the salt of guanidinoglutaric acid and methylisothiourea begin to precipitate. Stir for 2 h, and allow to stand overnight in the refrigerator.

Filter off the precipitate and wash well with ice-water. Destroy the MeSH in the filtrate with crystalline potassium permanganate before pouring it down the drain. Transfer the crystals to a liter flask containing 500 ml of a molar acetic acid (60 ml gl. acetic acid/l.). Reflux for 3 h (hood). Allow to cool overnight in the refrigerator and filter off the lactam of guanidinoglutamate (GGA) or amidino-pyroglutamate (ApGlu). The yield is about 40 g, m.p. 249–252 °C.

### Anhydride of guanidinoglutarate lactone

Add 10 g of guanidinoglutarate lactone (GGAL, ca. 55 mmol) to a 250 ml flask fitted with a stirrer and ground glass stopper. Add 50 ml benzene and 15 ml of trifluoroacetic anhydride (Aldrich, ca. 107 mmol). Stir at room temperature. Warm to 50 °C on a water bath, and the GGAL dissolves. After 2 h evaporate the solution to dryness under vacuum, add 50 ml of anhydrous toluene and evaporate to dryness again to remove the benzene and trifluoroacetic acid and its anhydride. While still warm add 30 ml anhydrous ethanol. The residue dissolves and the anhydride precipitates as it cools. Allow to stand in the refrigerator overnight and filter off the anhydride (AGGAL). Recrystallize it from anhydrous ethanol. On analysis, the anhydride contained no significant amount of fluoride. The yield is about 7 g, m.p. 230–232 °C. Calculated for $C_6H_7N_3O_2$ MW = 153.14. : % C = 47.06, % H = 4.61, % N = 27.44, % O = 21.00, % F = 0.00

Found: % C = 46.49, % H = 27.47, % N = 27.47, % O = 21.20, % F = 0.14

## References

1. Andes, J.E. & Myers, V.C. (1937): Guanidine like substances in the blood. IV. Blood guanidine in patients with parathyroid deficiency and in patients with idiopathic tetany. *J. Lab. Med.* **23**, 123–126.

2. Benedict, F.G. & Myers, V.C. (1907): The determination of creatine and creatinine. *Am. J. Physiol.* **18**, 397–405.

3. Bonas, J.E., Cohen, B.D. & Natelson, S. (1973): Separation and estimation of certain guanidine compounds. Application to human urine. *Microchem. J.* **7**, 63–77.

4. Borsook, H. & Dubroff, J.W. (1941): The formation of glycocyamine in animal tissues. *J. Biol. Chem.* **138**, 389–403.

5. Campanini, R.Z., Tapia, R.A., Sarnat, N. & Natelson, S. (1970): Evaluation of serum arginino-succinate lyase (ASAL) concentrations as an index of parenchymal liver diseases. *Clin. Chem.* **16**, 44–53.

6. Doyle, D.G., Bratton, G.R. & Natelson, S. (1985): Effect of canavanine and 2,3 dimercapto-1-propanol (British Antilewisite) on the proliferation of Novikoff heptoma cells in the presence of concanavalin A. *Clin. Physiol. Biochem.* **1**, 305–317.

7. Gilon, C., Knobler, Y. & Sheradsky, T. (1967): Synthesis of omega-aminoxy acids by oxygen alkyl fission of lactones. Improved synthesis of DL-canaline. *Tetrahedron* **23**, 4441–4447.

8. Gross, P. (1985): Biologic activity of hydroxylamine. *CRC Handbook, Crit. Rev. Toxicol.* **14**, 87–89.

9. Koller, A., Aldwin, L. & Natelson, S. (1975): Hepatic synthesis of canavaninosuccinate from ureidohomoserine and aspartate and its conversion to guanidinosuccinate. *Clin. Chem.* **21**, 1777–1782.

10. Koller, A., Comess, J. & Natelson, S. (1975): Evidence supporting a proposed mechanism explaining the inverse relationship between guanidino-acetate and guanidinosuccinate in human urine. *Clin. Chem.* **21**, 235–242.

11. Lauber, E.J. & Natelson, S. (1966): Separation of guanidino compounds by a combination of electrophoresis and gel filtration. Application to human urine. *Microchem. J.* **11**, 498–507.

12. Lawrence, J.M. (1973): Homoserine in seedlings of the tribe Vicieae of the leguminosae. *Phytochemistry* **12**, 2207–2209.

13. Marescau, B. , De Deyn, P.P., Lowenthal, A., Qureshi, I.A., Antonozzi, I., Bachman, C., Cederbaum, S.D., Lerone, R., Chamoles, N., Colombo, J.P., Hyland, K., Gatti, R., Kang, S.S., Letarte, J., Lamberet, M., Mizutani, N., Possemieres, I., Rezvani, I., Snyderman, S.E., Terheggen, H.G. & Yoshino, M. (1990): Guanidino Compound analysis as a complementary diagnostic parameter for hyperarginnemia: follow-up of guanidino compound levels during therapy. *Ped. Research*, 27, 297 - 303

14. Mori, A., Watanabe, Y., Shindo, S., Agaki, M. & Hiramatsu, M. (1982): Alpha guanidino glutaric acid and epilepsy. *Adv. Exp. Med. Biol.* **153**, 419–436.

15. Myers, V.C. (1910): The physiology and pathology of creatinine and creatine. *Am. J. Med. Sci.* **139**, 256–264.

16. Myers, V.C. & Fine, M.S. (1913): The metabolism of creatine and creatinine. The fate of creatine when administered to man. *J. Biol. Chem.* **21**, 377–381.

17. Myers, V.C. & Lough, W.G. (1915): The creatinine in the blood in nephritis. Its diagnostic value. *Arch. Intern. Med.* **16**, 536–546.

18. Natelson, S. (1982): Specific assay for homoserine and its lactone in *Pisum sativum*. Preparation of homoserine hydroxamic acid. *Microchem. J.* **27**, 466–483.

19. Natelson, S. (1984): Metabolic relationship between urea and guanidino compounds as studied by automated fluorometry of guanidino compounds in urine. *Clin. Chem.* **30**, 252–258.

20. Natelson, S. (1985): Canavanine in alfalfa (*Medicago sativa*). *Experientia* **41**, 257–259.

21. Natelson, S. (1985): Canavanine to arginine ratio in alfalfa (*Medicago sativa*), clover (*Trifolium*), and the jack bean (*Can avanalia ensiformis*). *J. Agric. Food Chem.* **33**, 413–419.

22. Natelson, S. (1986): The lactam of alpha guanidino glutamic acid (1-amidino-2-pyrrolidone-5 carboxylic acid). *Anal. Biochem.* **56**, 31–37.

23.    Natelson, S. (1988): Derivative of the lactam of alpha guanidinoglutaric acid (N-amidinopyroglutamic acid, ApGlu) for peptide formation. *Microchem. J.* **37,** 132–140.

24.    Natelson, S. (1991): On the nature of the anhydride of the lactam of guanidino glutamic acid. *Microchem. J.* **44,** 356–366.

25.    Natelson, S. & Bratton, G.R. (1984): Canavanine assay of some alfalfa varieties (*Medicago sativa*) by fluorescence: practical procedure for canavanine preparation. *Microchem. J.* **29,** 26–43.

26.    Natelson, S., Hsiu-Yu Tseng & Sherwin, J.E. (1978):  On the biosynthesis of guanidinosuccinate. *Clin. Chem.* **24,** 2108–3114.

27.    Natelson, S., Koller, A., Hsiu-Yu Tseng & Dods, R.F. (1977): Canaline carbamoyl transferase in human liver as part of a metabolic cycle in which guanidino compounds are formed. *Clin. Chem.* **23,** 960–966.

28.    Natelson, S. & Natelson, E. (1989): Preparation of D-, DL- and L-homoserine lactone from methionine. *Microchem. J.* **40,** 226–232.

29.    Natelson, S., Pantazis P. & Natelson, E.A. (1994): L-Homoserine hydroxamic acid as an anti-tumor agent. *Clin. Chim. Acta* **229,** 133–149.

30.    Natelson, S. & Shell, F.M. (1990): Structure of the lactam of guanidino glutaric acid (N-amidinoglutamic acid). *Microchem. J.* **42,** 161–169.

31.    Natelson, S. & Sherwin, J.F. (1979): Proposed mechanism for urea nitrogen reutilization: Relationship between urea and proposed guanidine cycle. *Clin. Chem.* **25,** 1342–1343.

32.    Natelson, S., Stein, I. & Bonas, J.E. (1964): Improvements in the method of separation of guanidino organic acids by column chromatography. Isolation and identification of guanidinosuccinic acid from human urine. *Microchem. J.* **8,** 371–382.

33.    Perez, G., Rey, A. & Schiff, E. (1976): Biosynthesis of guanidinosuccinate by perfused rat liver. *J. Clin. Invest.* **57,** 807–809.

34.    Perez, G.O., Rietberg, B., Owens, B. & Schiff, E.R. (1977): Effect of acute uremia on arginine metabolism and urea and guanidino acid production by perfused rat liver. *Pfluegers Arch.* **372,** 275–278.

35.    Philips, T.H. & Brown, D.H. (1967): The mutagenic action of hydroxylamine. *Progr. Nucl. Acid Res. Mol. Biol.* 349–368.

36.    Ratner, S., Petrack, B. & Rochovansky, O. (1953): Biosynthesis of urea.  Isolation and  properties of arginino-succinic acid. *J. Biol. Chem.* **204,** 95–113.

37.    Ratner, S. & Rochovansky, O. (1956): Biosynthesis of guanidino acetic acid.  Purification and properties of transamidinase. *Arch. Biochem. Biophys.* **63,** 277–315.

38.    Reiter, A.J. & Horner, W.H. (1979): Studies on the origin of guanidine in mammals. Formation of guanidine and hydroxyguanidine in the rat. *Arch. Biochem. Biophys.* **10,** 125–131.

39.    Sakaguchi, T., Tanaka, S., Yagi, H. *et al.* (1977): Reaction of guanidines with alpha diketones. IV. The reactivity of mono substituted guanidines and biguanides and their fluorimetric determination with fluorescent reagent. *Yakugaku Zasshi* **97,** 1053–1057.

40.    Sasaki, M., Takahara, K. & Natelson, S. (1973): Urinary guanidinoacetate, guanidinosuccinate ratio. An indicator of urinary dysfunction. *Clin. Chem.* **19,** 315–321.

41.    Sherwin, J.E. & Natelson, S. (1975): Serum and erythrocyte arginino-succinate lyase (ASAL) assay by NADH fluorescence generated from formed fumarate. *Clin. Chem.* **24,** 230–234.

42.    Stein, I.M., Cohen, B.D. & Kornhauser, R.S. (1969): Guanidinosuccinic acid in renal failure, experimental azotemia, and inborn errors of the urea cycle. *N. Engl. J. Med.* **280,** 921–930.

43.    Takahara, K. Nakanishi, S. & Natelson, S. (1969): Cleavage of canavaninosuccinic acid by human liver to form guanidinosuccinic acid, a substance found in the urine of uremic patients. *Clin. Chem.* **15,** 397–418.

44.    Takahara, K., Nakanishi, S. & Natelson, S. (1971): Studies on the reductive cleavage of canavanine and canavaninosuccinic acid. *Arch. Biochem. Biophys.* **145,** 85–95.

45.    Takahara, K. & Natelson, S. (1967): Arginino-succinic acid lyase in human erythrocytes in health and disease. *Am. J. Clin. Pathol.* **47,** 693–700.

46.    Terheggen, H.G., Lowenthal, A., Lavinha, F. & Columbo, J.P. (1975): Familial hyperarginaemia. *Arch. Dis. Child.* **50,** 57–62.

47.  U.S. Dept. Agr. Tech Bull. (1977): #1, 571. *U.S. Dept. Agr., Tenn. Crop Report*, Ser. (1984): 84, #1, 6 (USPHS) 368–470.

48.  Van Pilsum, J.F., Martin, R.A., Kito, E. & Hess, J. (1956): Determination of creatine, creatinine, arginine, guanidinoacetic acid, guanidine and methylguanidine in biological fluids. *J. Biol. Chem.* **222,** 225–236.

49.  Varcoe, R., Halliday, D., Carson, E.R. *et al.* (1975): Efficiency of utilization of urea nitrogen for albumin synthesis by chronically uraemic and normal man. *Clin. Sci. Mol. Med.* **48,** 379–390.

50.  Walker, J.B. (1955): Biosynthesis of canavaninosuccinic acid from canavanine and fumarate in kidney. *Arch. Biochem. Biophys.* **59,** 233–239

51.  Yamamoto, Y., Marui, T., Saito, A. *et al.* (1979): Ion exchange chromatography and fluorimetric detection of guanidino compounds in physiologic fluids. *J. Chromatogr.* **162,** 327–340.

*Guanidino Compounds : 2*, eds. by P.P. De Deyn, B. Marescau, I.A. Qureshi and A. Mori.
©1997 John Libbey & Company Ltd., pp. 243–249.

# Chapter 24

# Role for reactive oxygen and argininosuccinate in guanidinosuccinic acid synthesis

Kazumasa AOYAGI, Sohji NAGASE, Katumi TAKEMURA, Shoji OHBA, Kayo AKIYAMA, Tie TOMITA and Akio KOYAMA

*Department of Internal Medicine, Institute of Clinical Medicine, University of Tsukuba, Tsukuba City, Ibaraki-ken, 305, Japan*

## Summary

Hydroxyl radicals which were generated by hydrogen peroxide (1 mM) and $FeSO_4$ (0.2 mM) reacted with ASA (1 mM) in 50 mM phosphate buffer pH 7.4 for 5 min. GSA was measured by high-performance liquid chromatography by a post-column derivatization method using 9,10-phenanthrenequinone. Hydroxyl radicals formed a large amount of GSA (0.8 per cent conversion), and 0.1 mM dimethylsulphoxide (DMSO), which is a hydroxyl radical scavenger, inhibited GSA synthesis by 56 per cent. Hydrogen peroxide alone or $FeSO_4$ alone did not form GSA. Superoxide radicals generated by xanthine and xanthine oxidase also formed GSA from ASA and this was inhibited by superoxide dismutase and catalase. GSA synthesis in isolated rat hepatocytes (6 h incubation) in the presence of 66 mM urea was inhibited by DMSO dose dependently, and almost the same amount of GSA was formed by the addition of 5 mM ASA without urea. And 0.1 mM $FeCl_2$, which increases hydroxyl radicals, increased GSA synthesis in hepatocytes about two to four times in various conditions. Our *in vitro* data and the data obtained on isolated rat heptocytes suggest that ASA can be the prercursor of GSA. Reactive oxygen can play a role in the biosynthesis of GSA from ASA.

## Introduction

Guanidinosuccinic acid (GSA), a guanidine derivative, was first detected in the urine of patients with uremia[4] and implicated as a uremic toxin[6,7]. GSA is increased in serum and urine of patients with renal insufficiency, which suggests increased biosynthesis. The mechanism of GSA synthesis has been thought to be related to the urea cycle[5], since urinary excretion of GSA was increased depending on the protein intake and related to urinary urea excretions[9]. Addition of urea to rats with renal failure increased GSA synthesis[12]. We have also reported that urea increased GSA synthesis in isolated rat hepatocytes depending on the urea concentration[1]. It has been reported that urea inhibits the urea cycle enzyme argininosuccinic acid lyase activity[11]. From this report, we made the hypothesis that argininosuccinate would be involved in GSA synthesis.

On the other hand, we reported that methylguanidine, known as a uremic toxin is formed from creatinine by reactive oxygen[2]. In addition it has been reported that GSA and methylguanidine synthesis in patients with renal failure was increased by catabolic stress in the same manner[8,9]. Therefore we expected that GSA synthesis also might be dependent on reactive oxygen synthesis. In this report we tested ASA as a precursor of GSA synthesis and we also investigated the effect of reactive oxygen scavengers on GSA synthesis *in vitro* and in isolated rat hepatocytes.

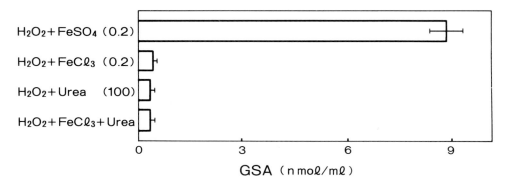

Fig. 1. GSA production from ASA by hydroxyl radicals in vitro. Each column represents the mean of duplicate incubations. Urea (100 mM) had no effect on GSA synthesis.

## Methods

### GSA synthesis from ASA by hydroxyl radicals or superoxide radicals *in vitro*

The reaction mixture consisted of 1 ml of 50 mM potassium phosphate buffer (pH 7.4), 1 mM hydrogen peroxide, 0.2 mM $FeSO_4$, 0.2 mM ethylene diamine tetra-acetic acid (EDTA) and 1 mM ASA, and was kept at room temperature for 5 min. Dimethylsulphoxide (DMSO), a hydroxyl radical scavenger, was added to this mixture, when its effect on GSA synthesis was tested.

Superoxide radicals ($O_2^-$) were generated by the reaction of xanthine oxidase (0.09 U/ml) and 0.22 mM xanthine in 50 mM potassium phosphate buffer (pH 7.4), and 1 mM ASA was added to this mixture and kept at 37 °C for 60 min. Superoxide dismutase (SOD) (40 μg /ml) or catalase (33 μg/ml) was added to scavenge superoxide radicals, and 0.2 mM $FeCl_2$ was added to generate superoxide-driven Fenton-type oxidation.

### GSA biosynthesis by isolated rat hepatocytes

*Preparation of isolated rat hepatocytes*

Male Wistar rats weighing 300–350 g were used in all experiments. The rats were allowed free access to water and laboratory chow containing 25 per cent protein. Rats were anesthetized by intraperitoneal (i.p.) injection of sodium pentobarbital, 60 mg/kg. Isolated hepatocytes were prepared according to the method of Berry & Friend[3] modified by Zahlten et al[16]. However, we added 2.5 mM calcium chloride and 3 per cent bovine serum albumin to replace 1.5 per cent gelatin in the medium for cell-washing. The viability of the cells, judged from the trypan blue exclusion test, was around 90 per cent. Using the method of Zahlten et al.[16], we calculated that $9.8 \times 10^7$ cells corresponded to 1 g of wet liver[16].

*Incubation of cells*

Cells were incubated in 6 ml of Krebs–Henseleit bicarbonate buffer containing 3 per cent bovine serum albumin, 10 mM sodium lactate, and indicated substances with shaking at 60 cycles/min in a 30 ml conical flask with a rubber cap under an atmosphere of mixed gases (95 per cent oxygen and 5 per cent carbon dioxide) at 37 °C. The equilibration of the buffer was repeated every hour. The incubation was stopped by the addition of 0.6 ml of 100 per cent (wt/V) trichloroacetic acid. After sonication, the supernatant was obtained by centrifugation at $1700 \times \gamma$ for 15 min at 0 °C; 0.2 ml of the extract was used for GSA determination.

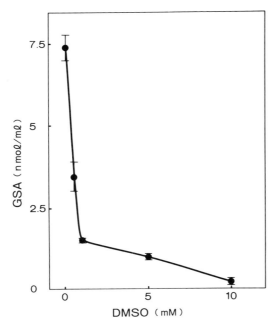

*Fig. 2. Inhibition by DMSO of GSA synthesis by*
*hydrogen peroxide plus FeSO4 in vitro. Each point*
*represents the mean of duplicate incubations*

## Assay of GSA

GSA was determined by high-performance liquid chromatograpic analysis using 9,10-phenanthrene-quinone for the post-column derivatizaton according to Yamamoto et al.[15] as descrived previously[1].

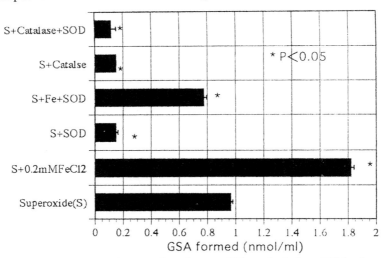

*Fig. 3. Effect of iron, SOD, and catalase on GSA synthesis from ASA by the*
*xanthine and xanthine oxidase reaction. Each column represents the mean of*
*five incubations. Bars indicate the standard error. * P <0.05 vs.*
*control value.*

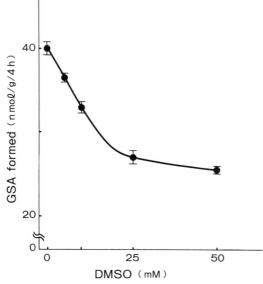

*Fig. 4. Effect of DMSO on GSA synthesis in isolated rats hepatocytes in the presence of 66 mM urea. Each point represents the mean of duplicate incubations.*

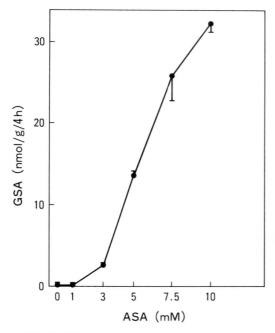

*Fig. 5. GSA synthesis from ASA in isolated rat hepatocytes. Each point represents the mean of duplicate incubations.*

## Results

### *In vitro* experiments

#### *Effect of hydroxyl radicals on in vitro GSA synthesis from ASA*

From 1 µmol ASA, 0.009 µmol GSA was formed by 1 mM hydrogen peroxide and 0.2 mM $FeSO_4$ in a 1 ml system. However, in the presence of 1 mM hydrogen peroxide alone or hydrogen peroxide and $FeCl_2$, GSA synthesis from ASA was not detected. Urea at a concentration of 100 mM had no effect on the *in vitro* GSA synthesis from ASA as shown in Fig.1. DMSO, which is a hydroxyl radical scavenger, inhibited GSA synthesis from ASA almost completely. DMSO (1 mM) inhibited GSA formed by 1 mM hydrogen peroxide and 0.2 mM $FeSO_4$ by 84 per cent as shown in Fig. 2.

#### *Effect of superoxide radicals on GSA synthesis from ASA*

GSA was formed from ASA by superoxide radicals as well as hydroxyl radicals. SOD inhibited this synthesis. Addition of 0.2 mM $FeCl_2$ to this system, which generates hydroxyl radicals, enhanced GSA synthesis 1.8 times. Catalase and SOD inhibited GSA synthesis as shown in Fig. 3.

These results indicate that superoxide radicals as well as hydroxyl radicals produce GSA from ASA.

### Experiments using isolated rat hepatocytes

#### *Effect of reactive oxygen scavengers on GSA synthesis in isolated rat hepatocytes*

In the presence of 66 mM urea, 40 nmol/g hepatocytes/4h GSA was formed. DMSO, which is known as a hydroxyl radical scavenger, decreased GSA synthesis as shown in Fig. 4. DMSO at 50 mM inhibited GSA synthesis by 35 per cent.

#### *GSA synthesis from ASA in isolated rat hepatocytes*

We investigated whether ASA is a precursor of GSA synthesis using isolated rat hepatocytes. GSA was formed by isolated rat hepatocytes depending on the added concentration of ASA.

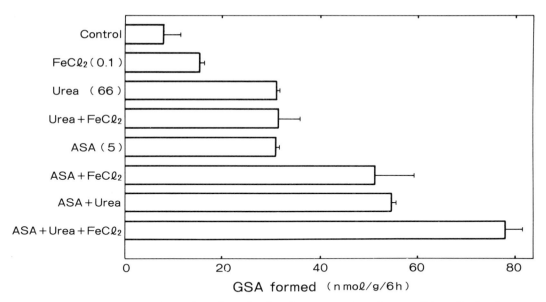

*Fig. 6. Effect of FeCl2 on GSA synthesis by isolated rat hepatocytes in various conditions. Each column represents the mean of duplicate incubations.*

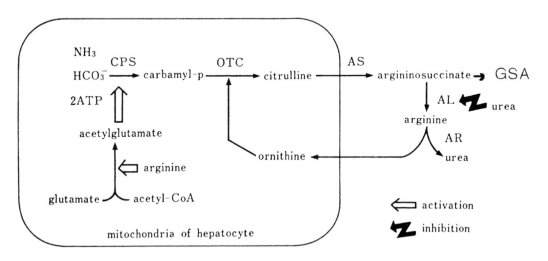

## UREA CYCLE

CPS: carbamyl phosphate synthetase,  OTC: ornithine transcarbamylase,
AS: argininosuccinate synthetase,  AL: argininosuccinase,  AR: arginase

*Fig. 7. Proposed mechanism of GSA biosynthesis .*

The rate of GSA synthesis at 10 mM ASA is 32 nmol/g per 4 h as shown in Fig. 5. This is 75 per cent of the GSA synthesis rate in the presence of 66 mM urea (Fig. 4).

*Effect of FeCl2 on GSA synthesis by isolated rat hepatocytes in various conditions*

Fig. 6 shows that the rate of GSA biosynthesis in the presence of both urea and ASA is increased in an additive fashion. Addition of $FeCl_2$ to the isolated hepatocytes may increase hydroxyl radical generation. $FeCl_2$ at a concentration of 0.1 mM increased GSA biosynthesis from ASA or ASA plus urea. These results support the fact that hydroxyl radicals generated by iron contributed to GSA synthesis in isolated rat hepatocytes. However, GSA synthesis in the presence of urea was not increased by $FeCl_2$. The reason $FeCl_2$ had no effect on the GSA synthesis in the presence of urea is not clear. The amount of increased GSA in the presence of urea plus ASA was the same as the amount of increased GSA in the presence of 5mM ASA alone as shown in Fig.6.

## Discussion

GSA was found in uremics by J.E. Bonas *et al.*[4] and has been investigated as a candidate uremic toxin[6,7]. GSA synthesis was suggested to be related to urea metabolism and the urea cycle[5,9]. We also reported that GSA synthesis in isolated rat hepatocytes depends on the urea concentration[1]. However, the mechanism of GSA synthesis was unclear, including its precursor. In renal failure, the increased urea could inhibit argininosuccinase activity[11] resulting in an increase of ASA, as a candidate GSA precursor. On the other hand, we have discovered that hydroxyl radicals, reactive oxygen species, formed methylguanidine (MG), a uremic toxin, from creatinine[2]. It has also been shown that the biosynthesis of GSA varied as did that of MG in patients with renal insufficiency.

This study demonstrated that ASA is a precursor of GSA *in vitro* and in isolated rat hepatocytes. Furthermore, in this chapter, we demonstrated that reactive oxygen species such as hydroxyl radicals

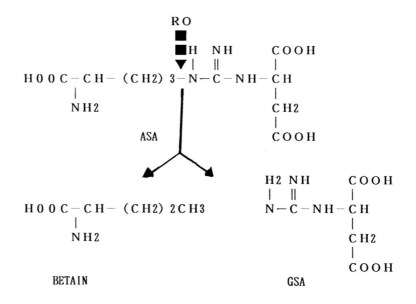

*Fig. 8. Proposed cleavage of ASA into betaine and GSA.*

and superoxide radicals react with ASA to form GSA, as shown in Fig. 7. We propose that ASA is cleaved into GSA and betaine, as shown in Fig. 8. These results suggest that the rate of GSA synthesis will be dependent on the concentration of ASA and the generation of reactive oxygen species in the liver.

Earlier we reported that arginine and ornithine plus 10 mM ammonium chloride, which increased urea synthesis, inhibited GSA synthesis in isolated rat hepatocytes[1]. Moreover, we reported that DL-norvaline, which inhibits ornithine transcarbamylase, the second enzyme of the urea cycle, also inhibits GSA synthesis in isolated rat hepatocytes[1]. These conflicting results may explain the decrease of ASA concentration or reactive oxygen generation in the liver. Also, the reported decrease in urinary excretion levels of GSA in patients with argininosuccinic aciduria[13] and hyperargininemia[10,14] may be explained by decreased reactive oxygen generation, or perhaps GSA is further metabolized in these pathophysiological conditions.

## Acknowledgements

This study was supported in part by the Scientific Research Funds of the Ministry of Education, Science and Culture of Japan (C–05670942, C–06671126), Grant for Scientific Research Expenses for Health and Welfare Programs, Ministry of Health and Welfare, Japan and Grant for Research Project, University of Tsukuba, Japan.

## References

1.  Aoyagi, K., Ohba S., Narit, M. & Tojo, S. (1983): Regulation of biosynthesis of guanidinosuccinic acid in isolated rat hepatocytes and *in vivo*. *Kidney Int.* **24** (S16), S224–S228.

2.  Aoyagi, K., Nagase, S., Narit, M. & Tojo, S. (1987): Role of active oxygen on methylguanidine synthesis in isolated rat hepatocytes. *Kidney Int.* **32** (S22), S29–S233.

3.  Berry, M.N. & Friend, D.S. (1969): High-yield preparation of isolated liver cells. *J. Cell Biol.* **43**, 506–520.

4.  Bonas, J.E., Cohen B.D. & Natelson, S. (1963): Separation and estimation of certain guanidino compounds: application to human urine. *Microchem. J.* **7**, 63–77.

5.  Cohen, B.D., Stein. I.M. & Bonas, J.E. (1968): Guanidinosuccinic aciduria in uremia. A possible alternate pathway for urea synthesis. *Amer. J. Med.* **45**, 63–68.

6.  Giovanetti, S., Cioni, L., Balestri & P.L, Biagini, M. (1968): Evidence that guanidines and some related compounds cause hemolysis in chronic uremia. *Clin. Sci* .**34**, 141–148.

7.  Horowitz, H.I., Stein, I.M. & Cohen, B.D. (1971): Further studies on the platelet-inhibitory effect of guanidinosuccinic acid and its role in uremic bleeding. *Amer. J. Med.* **49**, 336–345.

8.  Ishizaki, M., Kitamoto, H., Sugai, H., Susuki, K., Kurosawa, K., Futaki, G., Sohn, T., Taguma, Y., Takahasi, H. & Nakajima, M. (1989): Changes in urinary methylguanidine in cases with end-stage renal disease. In: *Guanidines 2.* Eds. A. Mori, B.D. Cohen & H. Koide, pp. 289–297. New York: Plenum Press.

9.  Kopple, J.D., Gordon, S.I., Wang, M. & Swendseide, M.E. (1968): Factors affecting serum and urinary guanidinosuccinic acid level in normal and uremic subjects. *J. Lab. Clin. Med.* **90**, 303–311.

10.  Marescau, B., Lowenthal, A., Terheggen, H.G., Esmans, E. & Alderweireldt, F. (1982): Guanidino compounds in hyperargininemia. In: *Urea Cycle Diseases.* Eds. A. Lowenthal, A. Mori & B. Marescau, pp. 427–434. New York: Plenum Press.

11.  Menyhart, J. & Grof, J. (1977): Urea as a selective inhibitor of argininosuccininate lyase. *Eur. J. Biochem.* **75**, 405–409.

12.  Mikami, H., Orita, Y., Ando, A., Fujii, M., Kikuchi, T., Yoshihara, K., Okada, A. & Abe, H. (1982): Metabolic pathway of guanidino compounds in chronic renal failure. In: *Urea Cycle Diseases,* Eds. A. Lowenthal, A. Mori & B. Marescau. pp. 449–458. New York: Plenum Press.

13.  Stein, I.M & Cohen, B.D. Kornhauser, P.S. (1969): Guanidinosuccinic acid renal failure, experimental azotemia and inborn error of the urea cycle. *N. Engl. J. Med.* **280**, 920--930.

14.  Terheggen, H.G., Lowenthal, A., Lavinha, F. & Colombo, J.P. (1975): Familial hyperargininaemia. *Arch. Dis. Child.* **50**, 57–62.

15.    Yamamoto, Y., Manji, T., Saito, A., Maeda, K. & Ohta, K. (1979): Ion-exchange chromatographic separation and fluorometric determination of guanidino compounds in physiological fluids. *J. Chromatogr.* **162,** 327–340.

16.    Zahlten, R.N., Frederick, W.S. & Lady, H.A. (1973): Regulation of glucose synthesis in hormone-sensitive isolated hepatocytes. *Proc. Natl Acad. Sci. USA* **70,** 3213–3218.

*Guanidino Compounds : 2*, eds. by P.P. De Deyn, B. Marescau, I.A. Qureshi and A. Mori.
©1997 John Libbey & Company Ltd., pp. 251–257.

# Chapter 25

## Effect of endotoxemia on guanidino compounds and nitric oxide metabolism

Devendra R. DESHMUKH, Vikram S. GHOLE, Bart MARESCAU and
Peter P. DE DEYN

*Department of Surgery, UMDNJ–Robert Wood Johnson Medical School, New Brunswick, New Jersey, U.S.A. and
the Department of Medicine, Born-Bunge Foundation, Laboratory of Neurochemistry and Behavior, University of
Antwerp, Antwerp, Belgium*

### Summary

The effect of endotoxemia on the levels of amino acids, nitrates, nitrites and guanidino compounds was investigated. Plasma levels of nitrate and nitrite were significantly increased, indicating increased production of nitric oxide during endotoxemia. Plasma concentrations of alanine, glutamine, leucine, methionine, phenylalanine, proline and taurine were also significantly elevated. These results indicate that endotoxin produces a hypercatabolic state. The plasma concentration of arginine was significantly decreased whereas the concentrations of ornithine and urea, the catabolites of arginine, were increased. Decreased plasma arginine coupled with increased plasma ornithine and urea indicate that arginine catabolism is increased and arginine synthesis is decreased during endotoxemia. Plasma levels of creatine, creatinine, guanidine and guanidinosuccinic acid were significantly elevated whereas homoarginine levels were significantly decreased. Nitric oxide synthase utilizes arginine and homoarginine as substrates. The decreased concentration of both substrates may be related to alterations in nitric oxide synthase activity during endotoxemia. These results suggest that, in addition to nitric oxide, other catabolites of arginine such as guanidino compounds may be important in the pathophysiology of endotoxemia. Because of the marked increase in guanidinosuccinic acid, a known uremic toxin, we speculate that guanidinosuccinic acid may be important in the pathophysiology of endotoxemia.

### Introduction

Sepsis is a complex metabolic disorder resulting from invasive infection and the release of microbial toxins into the blood stream. Sepsis occurs in approximately 100 000 to 300 000 patients annually in the United States. Shock is a complication of sepsis in almost half of these patients, with a mortality rate ranging from 40 to 60 per cent[11]. Among the multiple pathophysiologic events occurring during sepsis, hypotension, altered systemic vascular resistance and reduced sensitivity to vasoconstrictors are key factors leading to the development of septic shock.

Gram-negative bacteria are among the most common causes of sepsis. Endotoxin, a lipopolysaccharide component of gram-negative bacterial cell walls, initiates a cascade of endogenous mediators, which leads to the development of septic shock, multiple organ system failure and death. Nitric oxide has recently been implicated as an important biologic mediator of the hemodynamic response to endotoxemia[15,19,28]. Although nitric oxide is synthesized from l-arginine, the relationship between arginine metabolism and endotoxemia has not been investigated. Our first objective was to investigate the effect of endotoxemia on the metabolism of arginine.

Guanidino compounds are catabolites of arginine that have in common a strongly basic guanidinium group. These compounds are present in many tissues including liver, kidney and brain[22]. Guanidino compounds such as arginine and creatine have known physiologic roles. Other guanidino compounds such as guanidinosuccinic acid are toxic at high concentrations[6-8,20]. Guanidinosuccinic acid and other guanidino compounds are altered in tissues and body fluids in many pathologic conditions, especially those involving arginine metabolism[10,17,18]. Our second objective was to investigate the relationship between guanidino compounds and endotoxemia.

## Methods

Adult, male, albino rats (Sprague-Dawley) weighing 400–450 were purchased from the Charles River Laboratories (Wilmington, DE). Rats were injected intraperitoneally (i.p.) with either saline or endotoxin (lipopolysaccharide from *Salmonella typhimurium*, 5 mg/kg). Sixteen hours after injection, blood was collected by heart puncture into heparinized tubes. After rats were anesthetized with pentobarbital (40 mg/kg, i.p.), tissues were removed and frozen in liquid nitrogen. All animal experimentation was performed with Institutional Animal Care and Use Committee approval.

### Amino acid analysis

Plasma amino acids were determined by HPLC (= high performance liquid chromatography) using the Pico-Tag system[4] (Maxima 820, Waters, Millipore Corp., Milford, MA). The samples were passed through a filter with a 5 kDa exclusion limit, dried under vacuum and derivatized with phenyl isothiocyanate. The derivatized amino acids were loaded on a reverse phase column with an autoinjector (Wisp, Waters) and eluted with a gradient of 0.07 M sodium acetate pH 6.5 and a mixture of acetonitrile:methanol:water (45:40:40) at a flow rate of 1 ml/min. The absorbance at 254 nm (Waters, 481 UV detector) was used to calculate amino acid concentrations.

### Guanidino compound analysis

Plasma and tissue homogenates (10 per cent in water) were deproteinized with 30 per cent trichloroacetic acid and centrifuged at 100 000 x $g$ for 30 min. The guanidino compounds in supernatants were separated over a cation exchange column using a Biotronic LC 6001 amino acid analyzer with a stepwise gradient of sodium citrate buffers (0.175 M) of pH ranging from 3.0 to 11.0. The guanidino compounds were derivatized with ninhydrin and detected by a fluorescence detector[16].

### Nitrite and nitrate analysis

The nitrate and nitrite in plasma and tissue homogenates (10 per cent in water) were determined by HPLC[9]. The samples were passed through a filter with a 5 kDa exclusion limit and the filtrate was loaded on an anion exchange column. Nitrate and nitrite were eluted with phosphate buffer (10 mM, pH 6.0) and detected at 214 nm (Waters, UV detector).

### Urea and protein analysis

Urea levels in plasma and tissue homogenates were determined by the diacetyl monoxime–antipyrene method[5]. Protein concentration was determined using the Bradford[1] method.

### Nitric oxide synthase assay

The activity of nitric oxide synthase was determined by measuring the l-[2,3-³H]citrulline formed from l-[2,3-³H]arginine[2,3,13].

## Statistical analysis

Student's *t*-test was used to calculate statistical differences. *P* values < 0.05 were considered statistically significant.

## Results and discussion

### Effects of endotoxin on mortality rate

We carried out preliminary studies to investigate the dose-dependent effect of a bacterial lipopoly-saccharide from *Salmonella typhimurium* on mortality rate in a murine model of endotoxemia. Eighty per cent of rats died within 8 h after the intraperitoneal injection of 15 mg/kg of endotoxin. Rats injected with 5 and 10 mg/kg of endotoxin became lethargic within 4–8 h, exhibited characteristic responses such as raised fur, dysentery and lack of movement, and recovered within 24 h.

### Effect of endotoxin on nitrate and nitrite levels

A previous report[14] suggests that, during endotoxemia, tissue-specific changes occur in nitric oxide synthase activity that lead to an increase in nitric oxide production. We determined the concentration of nitric oxide and its breakdown products such as nitrate and nitrite in plasma and tissues of rats challenged with bacterial endotoxin. Plasma levels of nitrate and nitrite and liver nitrate were significantly elevated in endotoxemic animals, indicating increased production of nitric oxide during endotoxemia (Table 1). Increased nitric oxide may contribute to the hypotension and multiple organ system failure during endotoxemia.

**Table 1. Effect of endotoxin on nitrate and nitrite levels**

|  | Nitrate (μmol) | |
|---|---|---|
|  | Saline (n=5) | Endotoxin (n=7) |
| Plasma | 136 ± 16 | 2504 ± 508* |
| Liver | 0.4 ± 0.1 | 0.97 ± 0.2* |
| Lung | 1.2 ± 0.4 | 1.80 ± 0.3 |
|  | Nitrite (μmol) | |
| Plasma | 83 ± 13 | 135 ± 11* |
| Liver | 26.9 ± 7 | 25.8 ±11 |
| Lung | 0.5 ± 0.1 | 0.4 ± 0.1 |

Results are means ± SEM.
*$P < 0.05$ compared with the saline group.

### Effect of endotoxin on nitric oxide synthase activity

Compared with saline-treated animals, rats challenged with endotoxin had significantly higher levels of nitric oxide synthase activity in liver and lung (Table 2). Increased nitric oxide production during endotoxemia appears to be due to increased nitric oxide synthase activity. Further studies on the constitutive and inducible nitric oxide synthase during endotoxemia will be useful in understanding the mechanism by which nitric oxide mediates the biologic effects of endotoxemia.

### Effect of endotoxin on amino acid concentrations in plasma

The plasma concentration of alanine, glutamine, leucine, methionine, phenylalanine, proline and taurine were significantly higher in endotoxemic animals than in control animals. These results indicate that endotoxin produces a hypercatabolic state. However, the plasma concentration of arginine was significantly decreased whereas the concentrations of ornithine and urea, the catabolites of arginine, were increased in endotoxemic rats (Table 3). Reduced plasma arginine was coupled with raised plasma ornithine in endotoxemic rats (Table 3). Reduced plasma arginine, coupled with raised plasma ornithine and urea, indicates decreased synthesis and increased catabolism of arginine during endotoxemia.

**Table 2. Effect of endotoxin on nitric oxide synthase activity**

| Tissue | Enzyme activity (pmol/h per g) | |
| --- | --- | --- |
| | Saline (n = 5) | Endotoxin (n = 7) |
| Liver | $113 \pm 13$ | $227 \pm 31$* |
| Kidney | $74 \pm 14$ | $136 \pm 28$ |
| Lung | $4 \pm 0.9$ | $28 \pm 6$* |

Results are means ± SEM.
*$P < 0.05$ compared with the saline group.

**Table 3. Effect of endotoxin on amino acid in plasma**

| Amino acid ( μmol/l) | Saline (n= 5) | Endotoxin (n = 7) |
| --- | --- | --- |
| Alanine | $365 \pm 29$ | $1108 \pm 150$* |
| Arginine | $147 \pm 18$ | $50 \pm 5$* |
| Glutamine | $497 \pm 34$ | $810 \pm 88$* |
| Methionine | $46 \pm 4$ | $96 \pm 10$* |
| Ornithine | $128 \pm 14$ | $604 \pm 101$* |
| Phenylalanine | $42 \pm 3$ | $136 \pm 13$* |
| Proline | $179 \pm 13$ | $318 \pm 40$* |
| Taurine | $88 \pm 13$ | $455 \pm 57$* |

Results are means ± SEM.
*$P < 0.05$ compared with the saline group.
The plasma concentration of other guanidino compounds was not significantly altered in rats challenged with endotoxin.

### Effect of endotoxin on guanidino compounds in plasma

Compared with controls, endotoxemic rats exhibited significantly higher levels of plasma creatine, creatinine, guanidine, guanidinosuccinic acid and urea. However, plasma levels of arginine (Table 3) and homoarginine (Table 4) were significantly decreased. These results support the observation that arginine catabolism is increased during endotoxemia. Nitric oxide synthase utilizes arginine as well as homoarginine as a substrates[14]. The decreased concentration of both substrates may be related to alterations in nitric oxide synthase activity during endotoxemia.

**Table 4. Effect of endotoxin on guanidino compounds in plasma**

| Guanidino compound (μmol/l) | Saline (n = 5) | Endotoxin (n = 7) |
| --- | --- | --- |
| Creatine | $310 \pm 16$ | †$699 \pm 47$* |
| Creatinine | $29.8 \pm 2.5$ | $45.6 \pm 2.3$* |
| Guanidine | $0.12 \pm 0.02$ | $0.33 \pm 0.06$* |
| Guanidinosuccinic acid | $0.05 \pm 0.01$ | $0.47 \pm 0.15$* |
| Homoarginine | $1.21 \pm 0.03$ | $0.27 \pm 0.03$* |
| Urea (mmol/l) | $4.7 \pm 0.24$ | $12.1 \pm 1.5$* |

Results are means ± SEM.
*$P < 0.05$ compared with the saline group.
The concentration of other guanidino compounds was not significantly altered in rats challenged with endotoxin.

## Effect of endotoxin on guanidino compounds in kidney

Endotoxemic rats exhibited significantly lower levels of guanidinoacetic acid in kidney (Table 5). Arginine is the only amino acid that provides the amidino group for the synthesis of guanidinoacetic acid. Therefore, decreased guanidinoacetic acid during endotoxemia may be due to the decreased level of arginine. Guanidinoacetic acid is synthesized in the kidney by the action of arginine:glycine transamidinase[25]. Whether the activity of arginine:glycine transamidinase is altered during endotoxemia remains to be investigated.

Although arginine is synthesized in large amounts by the liver, the high activity of hepatic arginase normally prevents release of arginine into the blood. Among extrahepatic tissues, the kidney is the main organ of arginine production[24,27]. The citrulline required for this process is synthesized in the small intestine, released in the blood and taken up by the kidney[27]. However, citrulline is also a by-product of the nitric oxide synthase reaction. Whether nitric oxide synthase in kidney regulates the arginine production remains to be investigated.

### Table 5. Effect of endotoxin on guanidino compounds in kidney

| Guanidino compound ( $\mu$mol/l) | Saline (n = 5) | Endotoxin (n = 7) |
|---|---|---|
| Arginine | $284 \pm 63$ | $374 \pm 45$ |
| Creatine | $2767 \pm 198$ | $4888 \pm 369$* |
| Creatinine | $240 \pm 38$ | $306 \pm 36$ |
| Guanidinoacetic acid | $226 \pm 11$ | $132 \pm 18$* |
| Guanidinosuccinic acid | $0.9 \pm 0.1$ | $2.3 \pm 0.6$* |
| Homoarginine | $3.0 \pm 0.2$ | $0.8 \pm 0.07$* |

Values are means $\pm$ SEM.
*$P < 0.05$ compared with the saline group.
The concentration of other guanidino compounds was not significantly altered in rats challenged with endotoxin.

In liver, guanidinoacetic acid is converted into creatine, a major source of high-energy phosphate in muscle. Creatine levels in kidney, plasma (Table 4) and liver (data not shown) were significantly elevated in endotoxemic animals. These results indicate that increased catabolism of guanidinoacetic acid coupled with decreased arginine may be responsible for the decreased guanidinoacetic acid during endotoxemia.

Endotoxemia has diverse effects on vasoactivity in different tissues. Systemic vasodilation is a primary feature of endotoxemia and has been attributed to the increased production of nitric oxide. This is supported by the observation that the administration of $N^G$-monomethylarginine, an antagonist of l-arginine and inhibitor of nitric oxide synthase, prevents endotoxin-induced shock[12,21]. Pulmonary vasoconstriction is also observed during endotoxemia. Although increased nitric oxide synthase activity has been reported in the lungs of endotoxemic rats[14], the production of nitric oxide appears to be inadequate because the administration of exogenous nitric oxide reverses the vasoconstriction[26]. The diverse effects of endotoxemia in different tissues may be due to differences in the expression of nitric oxide synthase.

Variations in the concentration of arginine metabolites may also be responsible for the diverse effects of endotoxemia in different tissues. For example, in tissues containing arginine, the addition of l-arginine does not potentiate the endothelium-dependent relaxation to acetylcholine whereas, in tissues depleted of l-arginine, the addition of arginine causes an endothelium-dependent relaxation[23]. Thus, the response to l-arginine depends on the initial concentration of l-arginine in a specific tissue. The concentration of arginine is regulated by several enzymes including arginase, arginine:glycine transamidinase, argininosuccinase and nitric oxide synthase. The effects of endotoxemia on the activities of these enzymes remains to be investigated.

Because of its short half-life, direct measurement of nitric oxide is difficult. The hypothesis that nitric oxide is responsible for the development of endotoxic shock is primarily based on a reversal of hypotension by nitric oxide synthase inhibitors[12,21]. These inhibitors are structural analogs of arginine, which not only affect nitric oxide synthesis but are also likely to influence other arginine metabolism and transport. Our results suggest that, in addition to nitric oxide, other catabolites of arginine such as guanidino compounds may be important in the pathophysiology of endotoxemia. Guanidinosuccinic acid has been identified as a uremic toxin[6–8,20] contributing to the bleeding diathesis, increased hemolysis and immunologic disturbances of patients with renal failure. Because of the marked increase in guanidinosuccinic acid, we speculate that guanidinosuccinic acid may be important in the pathophysiology of endotoxemia.

## Acknowledgements

We thank the Flemish Ministry of Education, the Baron Bogaert-Scheid Fund, the Born-Bunge Foundation, the Medical Research Foundation OCMW Antwerp, the Universitaire instelling Antwerpen, the United Fund of Belgium and NFWO grants nos. 3.0044.92 and 3.0064.93.

# References

1.  Bradford, M.M. (1976): A rapid and sensitive method for the quantitation of microgram quantities of protein utilizing the principle of protein-dye binding. *Anal. Biochem.* **72**, 248–254.

2.  Bredt, D.S. & Snyder, S.H. (1989): Nitric oxide mediates glutamate-linked enhancement of cGMP levels in the cerebellum. *Proc. Natl. Acad. Sci. USA* **86**, 9030–9033.

3.  Bredt, D.S. & Snyder, S.H. (1990): Isolation of nitric oxide synthase, a calmodulin-requiring enzyme. *Proc. Natl. Acad. Sci. USA* **87**, 682–685.

4.  Buzzigole, G., Lanzone, L., Ciociaro, D., Reascerra, S., Cerri, M., Scandroglio, A., Coldani, R. & Ferrannini, E. (1990): Characterization of a reverse phase high performance liquid chromatography system for the determination of blood amino acids. *J. Chromatogr.* **507**, 85–93.

5.  Cerriotti, G. (1971): Ultramicro determination of plasma urea by reaction with diacetyl monoxime-antipyrene without deproteinization. *Clin. Chem.* **17**, 400–402.

6.  D'Hooge, R., Pei, Y.-Q., Marescau, B. & De Deyn, P.P. (1992): Convulsive action and toxicity of uremic guanidino compounds: behavioral assessment and relation to brain concentration in adult mice. *J. Neurol. Sci.* **112**, 96–105.

7.  De Deyn, P.P. & Macdonald, R.L. (1990): Guanidino compounds that are increased in uremia inhibit GABA- and glycine responses on mouse neurons in cell culture. *Ann. Neurol.* **28**, 627–633.

8.  De Deyn, P.P., Marescau, B., Cuykans, J.J., Van Gorp, L., Lowenthal, A. & De Potter, W.P. (1987): Guanidino compounds in serum and cerebrospinal fluid of non-dialyzed patients with renal insufficiency. *Clin. Chim. Acta* **167**, 81–88.

9.  Dennis, M.J., Key, P.E., Papworth, T., Pointer, M. & Massey, R.C. (1990): The determination of nitrate and nitrite in cured meat by HPLC/UV. *Food Addit. Contam.* **7**, 455–461.

10.  Deshmukh, D.R., Meert, K., Sarnaik, A.P., Marescau, B. & De Deyn, P.P. (1991): Guanidino compound metabolism in arginine-free diet induced hyperammonemia. *Enzyme* **45**, 128–136.

11.  Groeneveld, J., Bronsveld, W. & Thijs, L. (1986): Hemodynamic determinants of mortality in human septic shock. *Surgery* **99**, 140–152.

12.  Kilbourn, R.G., Jubran, A., Gross, S.S., Griffith, O.W., Levi, R., Adams, J. & Lodato, R.F. (1990): Reversal of endotoxin-mediated shock by $N^G$-methyl-l-arginine, an inhibitor of nitric oxide synthesis. *Biochem. Biophys. Res. Commun.* **172**, 1132–1138.

13.  Klatt, P., Schmidt, K. & Mayer, B. (1992): Brain nitric oxide synthase in a haemoprotein. *Biochem. J.* **288**, 15–17.

14.  Knowles, R.G., Merrett, M., Salter, M., & Moncada, S. (1990): Differential induction of brain, lung and liver nitric oxide synthase by endotoxin in the rat. *Biochem. J.* **270**, 833–836.

15.  Lowenstein, C.J. & Snyder, S.H. (1992): Nitric oxide, a novel biologic messenger. *Cell* **70**, 705–707.

16.  Marescau, B., De Deyn, P.P., Lowenthal, A., Qureshi, I.A., Antonozzi, I., Bachmann, C., Cederbaum, S.D., Cerone, R., Chamoles, N., Colombo, J.P., Hyland, K., Gatti, R., Kang, S.S., Letarte, J., Lambert, M., Mizutani, N., Possemiers, I., Rezvani, I., Snyderman, S.E., Terheggen, H.G. & Yoshino, M. (1990): Guanidino compound analysis as a complementary diagnostic parameter for hyperargininemia: follow up of guanidino compound levels during therapy. *Pediatr. Res.* **27,** 297–303.

17.  Marescau, B., Deshmukh, D.R., Kockx, M., Possemiers, I., Quereshi, I., Weichert, P. & De Deyn, P. (1992): Guanidino compounds in serum, urine, liver, kidney and brain of man and some ureotelic species. *Metabolism* **41,** 526–532.

18.  Meert, K.L., Deshmukh, D.R., Marescau, B., De Deyn, P.P. & Sarnaik, A.P. (1992): Guanidino compound metabolism in hyperammonemic rats. In: *Guanidino Compounds in Biology and Medicine. Eds. P.P. De Deyn, B. Marescau, V. Stalon & I.A. Qureshi, pp. 139–144. London: John Libbey.*

19.  Moncada, S., Palmer, R.M.J. & Higgs, E.A. (1991): Nitric oxide: physiology, pathophysiology, and pharmacology. *Pharm. Rev.* **43,** 109–142.

20.  Mori, A. (1987): Biochemistry and neurotoxicology of guanidino compounds. *Pavlovian J. Biol. Sci.* **22,** 85–94.

21.  Nava, E., Palmer, R.M.J. & Moncada, S. (1991): Inhibition of nitric oxide synthesis in septic shock: how much is beneficial? *Lancet* **338,** 1555–1557.

22.  Robin, Y. & Marescau, B. (1985): Natural guanidino compounds. In: *Guanidines. Eds. A. Mori, B.D. Cohen & A. Lowenthal, pp. 383–438. New York: Plenum Press.*

23.  Schini, V.B. & Vanhoutte, P.M. (1991): l-Arginine evokes both endothelium-dependent and -independent relaxation in l-arginine depleted aortas of the rat. *Circ. Res.* **68,** 209–216.

24.  Shindo, S. & Mori, A. (1985): Metabolism of l-[amidino-[15]N]-arginine of guanidino compounds. In: *Guanidines. Eds. A. Mori, B.D. Cohen & A. Lowenthal, pp. 71–81. New York: Plenum Press.*

25.  Van Pilsum, J.F. (1971): Evidence for a dual role of creatine in the regulation of kidney transamidinase activities in the rat. *J. Nutr.* **101,** 1085–1091.

26.  Weitzberg, E., Rudehill, A., Alving, K. & Lundberg, J.M. (1991): Nitric oxide inhalation selectively attenuates pulmonary hypertension and arterial hypoxia in porcine endotoxin shock. *Acta Physiol. Scand.* **143,** 451–452.

27.  Windmueller, H.G. & Spaeth, A.E. (1981): Source and fate of circulating citrulline. *Am. J. Physiol.* **241,** E473–E480.

28.  Wright, C.E., Rees, D.D. & Moncada, S. (1992): Protective and pathological roles of nitric oxide in endotoxin induced shock. *Cardiovasc. Res.* **26,** 48–57.

*Guanidino Compounds : 2*, eds. by P.P. De Deyn, B. Marescau, I.A. Qureshi and A. Mori.
©1997 John Libbey & Company Ltd., pp. 259–269.

# Chapter 26

# Protein-arginine methylation: enzymology and biological significance

Sangduk KIM and Woon Ki PAIK

*Fels Institute for Cancer Research and Molecular Biology, Temple University School of Medicine, 3420 North Broad Street, Philadelphia, Pennsylvania 19140, USA*

## Summary

Enzymatic methylation on the guanidino nitrogen of arginine residues in protein is catalyzed by protein methylase I (S-adenosylmethionine:protein-arginine N-methyltransferase; EC. 2.1.1.23) utilizing S-adenosyl-L-methionine as the methyl donor, yeilding $N^G$-mono-, $N^G$, $N^G$-di- and $N^G$, $N'^G$-dimethylarginines. Subclasses of protein methylase I, myelin basic protein-specific and nuclear-protein/histone-specific, have been purified to homogeneity from bovine brain and rat liver. Differences and/or similarities in their molecular, catalytic and immunologic properties are presented. The unmethylated recombinant heterogeneous nuclear (hn)RNP protein A1 (one of the most highly *in vivo* methylated proteins) was methylated *in vitro* by the nuclear-protein/histone-specific protein methylase I to the extent of 1.45 mol methyl-groups per mol of protein A1, and one of the methylation sites was identified as the residue-194 arginine, which was identical with the *in vivo* methylation site. The relative binding affinity of the methylated and the unmethylated protein A1 to nucleic acid was compared by their elution behavior on a single-stranded DNA-cellulose; concentration of NaCl to release the bound protein A1 was 0.59 M for the methylated and 0.63 M for the unmethylated, respectively. Employing an isoelectrofocusing technique, pI values of the methylated and the unmethylated proteins were found to be 9.41 and 9.48, respectively. When both species of protein A1 were subjected to controlled trypsin digestion, $T_{1/2}$ of the methylated protein was 1.31 min and of the unmethylated 1.63 min. The difference in their $T_{1/2}$ values was much greater in the presence of coliphage MS2-RNA: 2.41 min for the former and 4.3 min for the latter, indicating that the methylated species was less stabilized by RNA than the unmethylated. All of these results strongly indicate that the binding property of hnRNP protein A1 to single-stranded nucleic acid has been significantly reduced subsequent to its arginine- methylation.

## Introduction

A large number of proteins biosynthesized at the polyribosomes are post-translationally modified to yield functionally active and/or inactive proteins to express their regulatory roles[35] while preserving their nascent structure. One of such modification reactions is the methylation of protein; this occurs on several side chains of amino acids, catalyzed *in vivo* by a group of highly protein-specific methyltransferases[26,27]. In the protein-arginine residue, the guanidino nitrogen is methylated to form three different methylated derivatives; $N^G$-monomethyl-, $N^G$,$N^G$-dimethyl- and $N^G$,$N'^G$-dimethylarginines[17,28]. Indeed, these unusual methylated amino acids are found in a wide variety of proteins, such as myelin basic protein[1,6] (MBP), hnRNP protein[3,24], nucleolin[22], fibrillarin[23], HMG chromosomal protein[4], heat shock protein[34], tooth matrix protein[16] and several as yet unclearly characterized proteins in PC12 cells[25]. This diversity and wide occurrence of the methylated amino

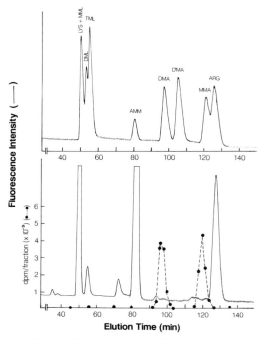

Fig. 1. High-performance liquid chromato-
graphy(HPLC)/analysis of methylated arginine
derivatives in [methyl-³H]protein A1. The acid
hydrolysate of enzymatically methylated [methyl-³H]
protein A1 was analyzed by HPLC using an Amino
Acid Analysis column (Waters). ARG, arginine; MMA,
$N^G$-mono-methylarginine; D' MA, $N^G$,$N^G$-di-
methyl(symmetric) arginine; DMA, $N^G$,$N^G$-di-
methyl(asymmetric) arginine; TML,
ε-N-trimethyllysine; DML, ε-N-dimethyllysine; MML,
ε-N-monomethyllysine; LYS, lysine.

acids, together with the ubiquitous distribution of the protein-arginine methylating enzyme, led us to uncover a multiplicity of methyltransferases which are specific to each methyl-acceptor protein. At present, three subclasses of protein methylase I have been identified and purified to apparent homogeneity: MBP-specific and nuclear-protein/histone-specific protein methylase I from mammalian organs[13,30,32], and the cytochrome *c*-specific protein methylase I from *Euglena gracilis*[18]. The molecular, catalytic and immunologic properties of these enzymes were shown to be quite different from each other. Interestingly, these enzymes recognize not only specific arginine residues in the –Gly–**Arg**–Gly–motifs, but also the total protein molecule to be methylated.

The protein A1 is the major core protein in the hnRNP complex[18]. The complex is known to be a pre-mRNA and also to serve as the platform for processing mRNA after a series of reactions involving splicing, packaging and transport[9,12]. Protein A1 is a basic protein with a molecular mass of 34-kDa, consisting of 320 amino acid residues and containing 3.1 mol of $N^G$,$N^G$-dimethylarginines per mol of the protein[2,18]. The notable property of protein A1 is that it binds single-stranded (ss) RNA and ssDNA[10,19], and stimulates α DNA polymerase activity *in vitro*[15]. The protein has been cloned and overexpressed in *Escherichia coli*[10].

The major difficulty in the past for the investigation of the biochemical significance of protein-arginine methylation *in vitro* has been an unavailability of completely unmethylated methyl-acceptor protein, because the proteins isolated from normal cells are already methylated at the physiologic level. To overcome this difficulty, we used the recombinant hnRNP protein A1, which is yet to be post-translationally methylated and is also available in milligram quantities. Thus, the recombinant protein A1 was methylated with purified nuclear-protein/histone-protein methylase I up to 1.45 mol methyl-groups per mol of protein *in vitro*, and several biochemical properties between the methylated and the unmethylated A1 were compared, specifically: (a) the difference in their binding properties to ss nucleic acids, (b) the overall charge difference, and (c) whether the arginine methyl-modification influences sensitivity to protease activity.

## Methods

### Purification of protein methylase I

All the purification procedures were essentially as described elsewhere[13,30]. Briefly, calf brain was homogenized in 4 volumes of 5 mM ethylene diamene tetraacetic acid (EDTA) sodium phosphate

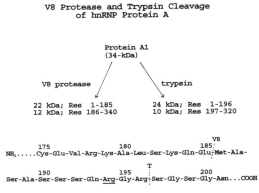

V8 Protease and Trypsin Cleavage
of hnRNP Protein A

Protein A1
(34-kDa)

V8 protease          trypsin

22 kDa; Res  1-185          24 kDa; Res  1-196
12 kDa; Res 186-340          10 kDa; Res 197-320

V8
175          180          185
NH₂.....Cys-Glu-Val-Arg-Lys-Ala-Leu-Ser-Lys-Gln-Glu┊Met-Ala-
T
190          195          200
Ser-Ala-Ser-Ser-Ser-Gln-Arg-Gly-Arg┊Ser-Gly-Ser-Gly-Asn...COOH

*Fig.2. Schematic illustration of V8 endoprotease and trypsin cleavage of protein A1.*

(pH 7.2) containing 5 mM EDTA and 0.32 M sucrose, and the homogenate was centrifuged at 78 500 x $g$ for 60 min. The supernatant was then adjusted to 40 per cent $(NH_4)_2SO_4$ saturation and centrifuged at 39 000 x $g$ for 30 min. The obtained precipitate was used for purification of the MBP-specific protein methylase I, and the remaining supernatant was further adjusted to 70 per cent $(NH_4)_2SO_4$. The second precipitate was used for fractionation of the nuclear-protein/histone-specific protein methylase I.

To purify MBP-specific enzyme, the 0–40 per cent $(NH_4)_2SO_4$ precipitate was further subjected to 60 per cent acetone precipitation (at –20 °C), followed by batchwise treatment on DE52-cellulose first and then column chromatography on DE52. The enzymatically active fractions were pooled and further subjected successively to Sephadex G-200 and hydroxyapatite column chromatographies[13].

Nuclear-protein/histone-specific protein methylase I was purified from the above 40–70 per cent $(NH_4)_2SO_4$ precipitate obtained from calf brain cytosol by successive chromatographies on DE52-cellulose, Sephadex G-200 and hydroxyapatite, similar to the purification of MBP-specific enzyme. However, the elution profiles of these two methylases were quite different in all the chromatography steps[13]. The nuclear-protein/histone-specific protein methylase I from the rat liver cytosol was also purified as described[32], employing DE52-cellulose, Sephadex G-200 and fast protein liquid chromatography (FPLC).

### Assay for protein methylase I

Protein methylase I activity was determined as described previously[13,32], using Ado[*methyl*-[14]C]Met as the methyl donor. MBP was used as the methyl-acceptor substrate for MBP-specific enzyme, and either unmethylated hnRNP protein A1 or histone type II-AS (Sigma) for nuclear-protein/histone-specific enzyme. Specific activity of the enyzme is defined as pmol of [*methyl*-[14]C] group transferred/min/mg enzyme protein. Unit activity of the enzyme is defined as pmol of [*methyl*-[14]C]group transferred/min.

### Purification of MBP

MBP was purified from bovine brain according to the methods described elsewhere[8,11]. Briefly, the brain homogenate in methanol – chloroform was subjected to successive treatments of delipidation, acid extraction, DE52 treatment, CM52 chromatography and finally gel filtration on Sephadex G-25. The purified MBP migrated as a single band on sodium dodecyl sulfate polyacrylamide gel electrophoresis (SDS-PAGE) at an approximate molecular weight of 18 000.

### Isolation and purification of hnRNP protein A1

*Escherichia coli,* which harbored the expression vector plasmid pEX11 to express recombinant hnRNP protein A1, was grown as described by Cobianchi et al.[10]. The protein A1 was then isolated and purified by columns of DEAE-cellulose and ssDNA-cellulose, which were connected in tandem as described by Cobianchi et al.[10]. The eluate from the column was concentrated by polyethylene glycol  20 000.

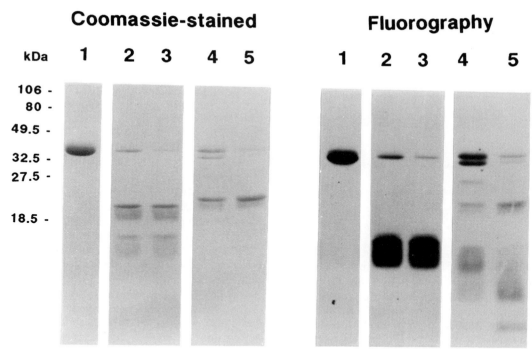

*Fig. 3. SDS-PAGE and fluorography of [methyl-³H]protein A1 treated with V8 endoprotease and trypsin. Lane 1, untreated; 2 and 3, treated with V8 protease (1:10 and 1:50); 4 and 5, treated with trypsin (1:1000 and 1:500) at 0°C for 5 min.*

Preparation of [**methyl**-³H]-labelled hnRNP protein A1

Unmethylated recombinant protein A1 (1 mg) was incubated in 0.1 M phosphate buffer (pH 7.6) in the presence of 120 μM Ado[*methyl*-³H]Met and purified calf brain nuclear-protein/histone-specific protein methylase I (123 units) in a total volume of 10 ml for 2 h at 37°C. The total reaction mixture was directly loaded onto an ssDNA-cellulose column (1 × 5 cm). After extensive washing with the buffer containing 0.4 M NaCl[30] to remove all the unreacted Ado[*methyl*-³H]Met , the [*methyl*-³H]protein A1 was finally eluted with buffer containing 1.0 M NaCl. The methylated protein A1 thus obtained contained 1.45 mol [*methyl*-³H]groups per mol protein.

**Controlled digestion of protein A1 with trypsin**

Methylated or unmethylated protein A1 was digested with N - tosyl - L - phenylalanine chloromethyl ketone (TPCK) - treated trypsin (trypsin:protein, 1:2500) in the presence or absence of MS2-RNA. The reaction was stopped by the addition of three volumes of Laemmli's buffer, and the mixture was subjected to SDS-PAGE with 15 per cent acrylamide as described by Laemmli[20]. Protein bands were visualized by Coomassie Blue staining and quantified by tracing the bands with a LKB Ultroscan X Laser densitometer.

**Other analytic methods**

Protein concentration was estimated using Bradford's Coomassie Blue protein reagent[5]. Isoelectro-focusing of the methylated and the unmethylated protein A1 was performed on a linear gradient of sorbitol[29] according to the method of Vesterberg. Amino acid analysis was carried out by HPLC equipped with a strong cation exchange Amino Acid Analysis Column (0.4 × 25 cm; Waters ) according to the method described by Rawal *et al.*[31]

## Results and discussion

### Purification and characterization of protein methylase I

Mammalian organs contain several subclasses of protein methylase I. We have purified MBP-specific protein methylase I from calf brain and two subtypes of nuclear-protein/histone-specific protein methylase I from calf brain and rat liver. MBP-specific and nuclear-protein/histone-specific protein

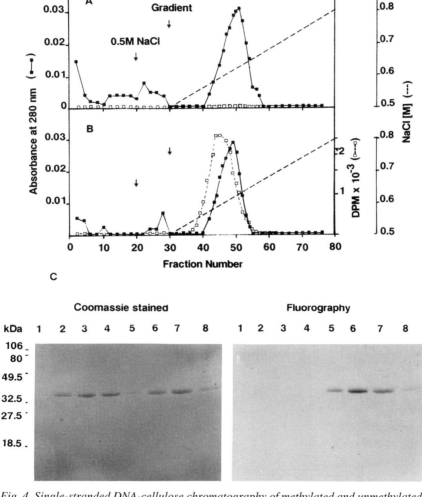

*Fig. 4. Single-stranded DNA-cellulose chromatography of methylated and unmethylated protein A1. The recombinant protein A1, either unmethylated (A) or enzymatically radio methylated (B) was chromatographed on an ssDNA-cellulose column. C: SDS-PAGE of the corresponding protein A1 fractions eluted from the ssDNA-cellulose column.*

methylase I are both present in the cytosolic fraction of calf brain, so an initial fractionation to concentrate each subclass of the enzyme is very important. It could be achieved by $(NH_4)_2SO_4$ precipitation: 0–40 per cent for MBP-specific and 40–70 per cent for nuclear-protein/histone-specific enzyme. Subsequently, both enzymes were purified to homogeneity by a chromatographic procedure on DE52-cellulose, Sephadex G-200 and hydroxyapatite columns. The native molecular mass of the MBP-specific protein methylase I was approximately 500 kDa, consisting of 100 and 72 kDa hetero-subunits, whereas that of the nuclear-protein/histone-specific enzyme was 275 kDa with 110 and 75 kDa hetero-subunits[13,30].

In contrast to the brain enzymes, the purification of nuclear-protein/histone-specific protein methylase I from rat liver was relatively simple. Over 100-fold purification from the cytosol fraction could be achieved by a single chromatographic stage on a DE52-cellulose column[31]. The following two molecular sieve procedures, on a Sephadex G-200 column and FPLC with a Superose 6 column, resulted in a homogeneous enzyme with a molecular mass of 450 kDa consisting of 110 kDa homo-subunits.

It should be noted that, in our earlier investigations, histone was the most commonly used substrate for protein methylase I, even though it was not an ideal substrate in terms of its purity and had a low methyl-accepting efficiency. Recently, the unmethylated recombinant hnRNP protein A1 has become available, so that we have reinvestigated and compared the substrate activity of protein A1 and histone[30]. We found that the $K_m$ value for the unmethylated protein A1 was two orders of magnitude lower than that of histone (0.19 μM $vs$ 21 μM). Furthermore, the maximal extent of methylation between these two substrates differed greatly: whereas 1.08 mol of methyl groups incorporated into the protein A1, only 0.04 mol incorporated into the histone[30]. The greater capacity of the protein A1 for methylation with its higher affinity constant makes it more relevant as a substrate for this enzyme than the histone. Consequently, 'histone-specific' protein methylase I has been renamed 'nuclear-protein/histone-specific' protein methylase I[30].

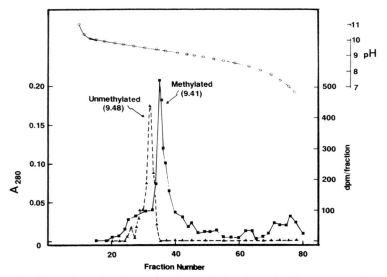

Fig. 5. pI values of methylated and unmethylated protein A1. A mixture of unmethylated and [methyl-³H]methylated protein A1 was applied on an isoelectrofocusing column.

**Table 1. Comparative properties of MBP and nuclear-protein/histone-specific protein methylase I**

| Characteristics | MBP-specific | Nuclear-protein/histone-specific | |
| --- | --- | --- | --- |
| | Calf brain | Calf brain | Rat liver |
| $M_r$ (native) | 500 kDa | 275 kDa | 450 kDa |
| Subunit (SDS-PAGE) | 100 kDa; 72 kDa | 110 kDa; 75 kDa | 110 kDa |
| $p$ I value | 5.09 | 5.68 | – |
| $K_m$ value (M) | | | |
| Protein A1 | – | $0.19 \times 10^{-6}$ | $0.54 \times 10^{-6}$ |
| Histone | $1.0 \times 10^{-4}$ | $21.0 \times 10^{-6}$ | – |
| MBP | $0.23 \times 10^{-6}$ | inhibitor | – |
| AdoMet | $4.4 \times 10^{-6}$ | $8.0 \times 10^{-6}$ | $6.3 \times 10^{-6}$ |
| $K_i$ value (M) | | | |
| AdoHcy | $1.8 \times 10^{-6}$ | $22.3 \times 10^{-6}$ | $8.4 \times 10^{-6}$ |
| Sinefungin | $7.0 \times 10^{-6}$ | $6.6 \times 10^{-6}$ | $0.65 \times 10^{-6}$ |
| MBP | – | $3.42 \times 10^{-6}$ | non- inhibitor |
| 50% inactivation | | | |
| p-chloromercuribenzoate | 0.46 mM | 0.15 mM | 0.12 mM |
| Guanidine-HCl | 3.1 mM | 0.3 mM | 100 mM |
| At 50ÉC for 5 min | 99% remained | 60% remained | 17% remained |

Table 1 shows the overall molecular and catalytic properties of the three purified enzymes. It shows that not only the native molecular sizes of the three methylases but also their subunit structures are quite different. An immunologic difference between MBP-specific and nuclear-protein/histone-specific enzyme has been verified by Western immunoblot analyses using the respective antibodies raised against the purified enzymes[13].

Sensitivities toward p-chloromercuribenzoate and guanidine-HCl as well as heat inactivation profiles also differ markedly among the enzymes. Although the $K_m$ and $K_i$ values for the methyl donor (AdoMet) and its analogs are not much different between the enzyme subclasses, those for the methyl acceptor substrate protein show marked differences as discussed above. Interestingly, MBP, the high-affinity substrate of the MBP-specific protein methylase I, acted as an inhibitor for the nuclear-protein/histone-specific enzyme with a $K_i$ value of $3.42 \times 10^{-6}$ M. These results further confirm that the subclasses of methylases are indeed different molecular species, and further suggest that there must exist in nature many more as yet unknown subclasses of protein-arginine methyltransferases which are specific for each species of protein substrate.

Recently, we have studied the structural specificity of several synthetic peptides as substrates for protein methylase I, and found that the hexapeptide containing the –Gly–**Arg**–Gly– motif is the minimum essential structure required to be methylated[14]. Substitution of the Gly on either the N- or C-flanking side of the methylatable **Arg** with another amino acid completely abolished its ability to accept a methyl group. In addition to this proximal amino acid sequence, the overall three-dimensional protein structure was crucial in the formation of the $N^G$-dimethylarginine isomers, as well as in the determination of the specificity for the protein substrate.

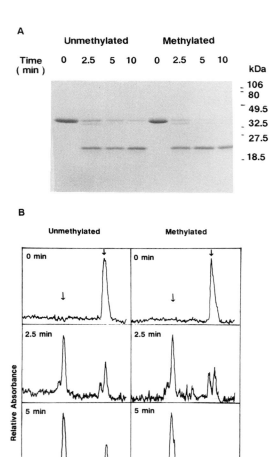

**A**

| Time (min) | Unmethylated | | | | Methylated | | | |
|---|---|---|---|---|---|---|---|---|

Unmethylated: 0  2.5  5  10
Methylated: 0  2.5  5  10

kDa
106
80
49.5
32.5
27.5
18.5

**B**

Unmethylated                    Methylated

0 min                           0 min

2.5 min                         2.5 min

5 min                           5 min

10 min                          10 min

Relative Absorbance

Relative Distance

*Fig. 6. SDS-PAGE and fluorography of controlled trypsin digest of protein A1. Either methylated or unmethylated protein A1 was digested with trypsin (trypsin:protein ratio 1:2500) at 22 °C for the indicated period. The digestion mixtures were then analyzed by SDS-PAGE (A). The Coomassie blue stained gel was quantified by densitometric tracing (B).*

## Enzymatic methylation of the recombinant hnRNP protein A1

*Identification of methylated amino acids and the site of methylation*

Recombinant protein A1 was methylated with purified calf brain nuclear-protein/histone-specific protein methylase I in the presence of Ado[*methyl*-³H]Met, and the methylated amino acids in the [*methyl*-³H] protein A1 have been identified by HPLC (Fig. 1): the radiomethyl groups are exclusively associated with N$^G$-monomethylarginine (52 per cent) and N$^G$,N$^G$-dimethyl(asymmetric)arginine (48 per cent). Neither N$^G$,N$^{'G}$-dimethyl(symmetric)arginine nor methylated lysine derivatives are present, demonstrating the high degree of enzyme specificity. Based on the specific radioactivity, it was calculated that the [*methyl*-³H]protein A1 contained 1.45 mol methyl-groups per mol of protein, which is equivalent to 47 per cent of the *in vivo* methylation level (3.1 mol N$^G$,N$^G$-dimethylarginine per mol of the *HeLa* cell protein A1[18]).

The site of methylation has been identified by specific cleavage of [*methyl*-³H]protein A1 by *Staphylococcus aureus* V8 endoprotease and trypsin. These proteases are known to cleave protein A1 at different sites, selectively, to yield two similar-sized fragments under mild conditions: the larger NH₂-terminal (22 or 24 kDa) and the smaller COOH-terminal (10 or 12 kDa) fragments[19,30] (see Fig. 2). Trypsin cleaves at Arg-196, whereas V8 cuts at Glu-185. These digestion products on SDS-PAGE and autoradiography indicated that the Arg-194 of the NH₂-terminal fragment and unidentified sites at the COOH-terminal fragment were methylated (Fig. 3).

*Effect of arginine methylation on protein A1*

Protein A1 can effectively be purified by ssDNA-cellulose, because the protein possesses a high affinity for ss nucleic acid. Thus, the adsorbed protein A1 on ssDNA is often eluted with 1 M NaCl buffer. To compare the relative binding of the methylated and the unmethylated protein A1 in a microenvironment, an NaCl gradient (0.5–0.8 M) was applied to elute the bound A1 from ssDNA- cellulose. As shown in Fig. 4, the methylated protein A1 (indicated by radiomethyl incorporation) eluted at 0.59 M NaCl, whereas the unmethylated (indicated by A280) eluted at 0.63 M NaCl, indicating that the former is less tightly bound to ssDNA. There were no molecular alterations caused by the methylation reaction, as evidenced by SDS-PAGE (Fig. 4C); proteins from several pooled fractions showed identical mobilities at approximately 34 kDa. In addition, the ultraviolet absorption maxima of the methylated and unmethylated protein A1 were identical at 278.5 nm.

Differences in the pI values between the methylated and the unmethylated protein A1 were also studied by an isoelectrofocusing technique. The methylated protein A1 was shown to have a pI of 9.41 and the unmethylated 9.48 (Fig. 5). The lowering of the pI of protein A1 by methylation may have contributed to the weaker binding of the protein to the highly anionic ssDNA-cellulose column.

Protein A1 is a two-domain molecule, which can be easily cleaved by limited tryptic digestion[19,30]. The relative rate of trypsin digestibility of the methylated and the unmethylated A1 was studied under controlled conditions. As shown in Fig. 6, the methylated A1 was completely digested within 5 min, whereas the undigested unmethylated A1 still remained after 10 min digestion. The difference in trypsin sensitivity of the two species of the protein was much more pronounced in the presence of ssRNA. In the presence of coliphage MS2-RNA, the $T_{1/2}$ was 2.41 min for the methylated and 4.3 min for the unmethylated protein (Table 2). It was therefore concluded that protein A1 became more susceptible to trypsin on arginine methylation and that the sensitivity was further enhanced by the presence of ssRNA, which complexed with the methylated protein A1 less tightly than with the unmethylated.

**Table 2. Differences in properties between recombinant unmethylated and arginine-methylated hnRNP protein A1**

| Properties | Methylated | Unmethylated |
|---|---|---|
| [NaCl] to release from ssDNA-cellulose (M) | 0.59 | 0.63 |
| $p$ I value | 9.41 | 9.48 |
| Trypsin sensitivity (50% digestion) (min): | | |
| no addition | 1.31 | 1.63 |
| + ssMS2-RNA | 2.41 | 4.30 |

In this paper, we have investigated a possible biologic significance for protein-arginine methylation, utilizing hnRNP protein A1 as a model compound. As summarized in Table 2, differential properties between the methylated and the unmethylated protein A1, such as binding to ssDNA-cellulose, pI value and trypsin sensitivity in the presence and absence of MS2-RNA, all indicate that the methylation of arginine residues decreases the binding capacity of protein A1 to ss nucleic acids. These results suggest that arginine methylation of hnRNP protein may modulate the nucleic acid binding *in vivo* as well. In support of this idea, the importance of a single arginine residue in the HIV-I Tat protein as a specific binding site to interact with TAR RNA target has recently been proposed[7,21]; the ethylation (methylation) would block hydrogen bonding between the arginine fork and the phosphate groups of the RNA bulge[7]. Therefore, it is tempting to speculate that the selective enzymatic methylation of arginine residues in hnRNP protein A1 may have a specific effect rather than a non-specific one on the interaction with ss nucleic acids.

### Acknowledgements

This work was supported in part by grants from the National Cancer Institute (5–P30 CA 12227), the National Institute of Diabetes, Digestive and Kidney Diseases (AM09602), and the Research Incentive Fund, Temple University.

# References

1.  Baldwin, G.S. & Carnegie, P.R. (1971): Specific enzymic methylation of an arginine in the experimental allergic encephalomyelitis protein from human myelin. *Science* **171**, 579–581.

2.  Beyer, A., Christensen, M.E., Walker, B.W. & LeStourgeon, W.M. (1977): Identification and characterization of the packaging proteins of core 40S hnRNP particles. *Cell* **11**, 127–138.

3.  Boffa, L.C., Karn, J., Vidali, G. & Allfrey, V.G. (1977): Distribution of $N^G,N^G$-dimethylarginine in nuclear protein fractions. *Biochem. Biophys. Res. Commun.* **74**, 969–976.

4.  Boffa, L.C., Sterner, R., Vidali, G. & Allfrey, V.G. (1979): Post-synthetic modification of nuclear proteins: high mobility group proteins are methylated. *Biochem. Biophys. Res. Commun.* **89**, 1322–1327.

5.  Bradford, M. (1976): A rapid and sensitive method for the quantitation of microgram quantities of protein utilizing the principle of protein-dye binding. *Anal. Biochem.* **72**, 240–254.

6.  Brostoff, S. & Eylar, E.H. (1970): Localization of the methylated arginine in the A1 protein from myelin. *Proc. Natl. Acad. Sci. USA* **68**, 765–769.

7.  Calnam, B.J., Tidor, B., Biancalana, S., Hudson, D. & Frankel, A.D. (1991): Arginine-mediated RNA recognition: the arginine fork. *Science* **24**, 1167–1171.

8.  Chanderkar, L.P., Paik, W.K. & Kim, S. (1986): Studies on myelin basic protein methylation during mouse brain development. *Biochem. J.* **240**, 471–479.

9.  Choi, R.D., Grabowski, P.J., Sharp, P.D. & Dreyfuss, G. (1986): Heterogeneous nuclear ribonucleoproteins: role in RNA splicing. *Science* **231**, 1534–1539.

10. Cobianchi, F., Karpel, R.L., Williams, K.R., Notario, V. & Wilson, S. (1988): Mammalian heterogeneous nuclear ribonucleoprotein complex protien A1: large-scale overproduction in *Escherichia coli* and cooperative binding to single-stranded nucleic acids. *J. Biol. Chem.* **263**, 1063–1071.

11. Deibler, G.E., Boyd, L.F. & Kies, M.W. (1984): Proteolytic activity associated with purified myelin basic protein. In: *Experimental Allergic Encephalomyelitis*. Eds. E.C. Alvord Jr, M.W. Kies & A.J. Suckling, pp. 249–256. New York: Alan Liss, Inc.

12. Dreyfuss, G. (1986): Structure and function of nuclear and cytoplasmic ribonucleoprotein particles. *Ann. Rev. Cell Biol.* **2**, 459–498.

13. Ghosh, S.K., Paik, W.K. & Kim, S. (1988): Purification and molecular identification of two protein methylases I from calf brain: myelin basic protein- and histone-specific enzyme. *J. Biol. Chem.* **269**, 1075–1082.

14. Ghosh, S.K., Syed, S.K., Jung, S., Paik, W.K. & Kim, S. (1990): Substrate specificity for myelin basic protein-specific protein methylase I. *Biochim. Biophys. Acta* **1039**, 142–148.

15. Herrick, G., Delius, H. & Alberts, B. (1976): Single-stranded DNA structure and DNA polymerase activity in the presence of nucleic acid helix-unwinding proteins from calf thymus. *J. Biol. Chem.* **251**, 2142–2146.

16. Kalasz, H., Kovacs, G.H., Nagy, J., Tyhak, E. & Barnes, W.T. (1978): Identification of N-methylated basic amino acids from human adult teeth. *J. Dent. Res.* **57**, 128–132.

17. Kim, S., Ghosh, S.K. & Chanderkar, L.P. (1990): Protein-arginine methylation. In: *Protein Methylation*. Eds. W.K. Paik & S. Kim, pp. 75–95. Boca Raton, FL: CRC Press.

18. Kumar, A., Williams, K.R. & Szer, W. (1986): Purification and domain structure of core hnRNP protein A1 and A2 and their relationship to single-stranded DNA-binding proteins. *J. Biol. Chem.* **261**, 11266–11273.

19. Kumar, A., Casa-Finet, J.R., Luneau, D.J., Karpel, R.L., Merrill, B.M., Williams, K.R. & Wilson, S. (1990): Mammalian heterogeneous nuclear ribonucleoprotein A1: nucleic acid binding properties of the COOH-terminal domain. *J. Biol. Chem.* **265**, 17094–17100.

20. Laemmli, U.K. (1970): Cleavage of structural proteins during the assembly of the head bacteriophage T4. *Nature* **227**, 680–685.

21. Lazinski, D., Grzadzielska, E. & Das, A. (1989): Sequence-specific recognition of RNA hairpins by bacteriophage antiterminators requires a conserved arginine-rich motif. *Cell* **59**, 207–218.

22. Lischwe, M.A., Cook, R.G., Ahn, Y.S., Yeoman, L.C. & Busch, H. (1985): Clustering of glycine and $N^G,N^G$-dimethylarginine in nucleolar protein C23. *Biochemistry* **24**, 6025–6028.

23. Lischwe, M.A., Ochs, R.L., Reddy, R., Cook, R.G., Yeoman, L.C., Tan, E.M., Reichlin, M. & Busch, H. (1985): Purification and partial characterization of a nucleolar scleroderma antigen rich in $N^G,N^G$-dimethylarginine. *J. Biol. Chem.* **260**, 14304–14310.

24. Lischwe, M.A. (1990): Amino acid sequence of arginine methylation sites. In: *Protein Methylation.* Eds. W.K. Paik & S. Kim, pp. 97–123. Boca Raton, FL: CRC Press.

25. Najbauer, J. Johnson, B.A., Young, A.L. & Aswad, D.W. (1993): Peptides with sequences similar to glycine, arginine-rich motifs in proteins interacting with RNA are efficiently recognized by methyltransferase(s) modifying arginine in numerous proteins. *J. Biol. Chem.* **268,** 10501–10509.

26. Paik, W.K. & Kim, S. (1980): Enzymology of protein Methylation. In: *Protein methylation.* Ed. A. Meister, pp. 112–141. New York: John Wiley.

27. Paik, W.K. & Kim, S. (1985): Protein methylation. In: *The Enzymology of Post-translational Modification of Proteins,* Vol. 2. Eds. R.B. Freeman & H.C Hawkins, pp. 187–228. London: Academic Press.

28. Paik, W.K. & Kim, S. (1993): $N^G$-Methylarginines: biosynthesis, biochemical function and metabolism. *Amino Acids* **4,** 267–286.

29. Park, S., Frost, B.F., Lee, W.H., Kim, S. & Paik, W.K. (1989): Effect of enzymatic methylation of proteins on their iso-electric points. *Arch. Pharm. Res. Seoul* **12,** 79–87.

30. Rajpurohit, R., Lee, S.O., Park, J.O., Paik, W.K. & Kim, S. (1994): Enzymatic methylation of recombinant heterogeneous nuclear RNP protein A1: dual substrate specificity for S-adenosylmethionine:histone-arginine N-methyltransferase. *J. Biol. Chem.* **269,** 1075–1082.

31. Rawal, N., Lee, Y.-J., Paik, W.K. & Kim, S. (1992): Studies on $N^G$-methylarginine derivatives in myelin basic protein from developing and mutant mouse brain. *Biochem. J.* **287,** 929–935.

32. Rawal, N., Rajpurohit, R., Paik, W.K. & Kim, S. (1994): Purification and characterization of S-adenosylmethionine-protein-arginine N-methyltransferase from rat liver. *Biochem. J.* **300,** 383–389.

33. Riva, S., Morandi, C., Tsoulfas, P., Pandolfo, M., Biamonti, G., Merrill, B., Williams, K., Multhaup, G., Byrenther, K., Werr, H., Heinrich, B. & Schafer, K.P. (1986): Mammalian single-stranded DNA binding protein UP1 is derived from the hnRNP core protein A1. *EMBO J.* **5,** 2267–2273.

34. Wang, C., Lazarides, R., O'Connor, C.M. & Clarke, S. (1982): Methylation of chicken fibroblast heat shock proteins at lysyl and arginyl residues. *J. Biol. Chem.* **257,** 8356–8362.

35. Wold, F. (1981): Chemical modification of protein. *Ann. Rev. Biochem.* **50,** 783–814.

# Section VII
## Guanidino compounds in renal failure

*Guanidino Compounds : 2*, eds. by P.P. De Deyn, B. Marescau, I.A. Qureshi and A. Mori.
©1997 John Libbey & Company Ltd., pp. 273–277.

# Chapter 27

# Glycine in acute renal tubular necrosis

Burton D. COHEN and Harini PATEL

*Bronx-Lebanon Hospital, 1276 Fulton Avenue, Bronx, NY 10476, USA*

## Summary

A renal cytoprotective effect of glycine, greater than that noted with glutathione, was first reported in 1987 following observations made on cells in culture. This study tests this effect *in vivo* and proposes a dual role for nitric oxide in the kidney.

Eighty patients on potentially nephrotoxic intravenous antibiotics were randomly divided into two groups; one of which received glycine 4.5 g (60 mmol) daily in the form of Knox Drinking Gelatine concurrent with the nephrotoxic agent. Acute tubular necrosis, defined as a 50 per cent increase in serum urea and creatinine, occurred in 14 of 45 untreated subjects and only three of 35 receiving glycine.

Three additional groups of subjects served as their own controls in experiments testing the short term (2 h to 2 day) effect and the longer (2 week) action of gelatine on the renal production of nitric oxide and guanidinoacetic acid, both generated by renal cells from arginine. Constitutive nitric oxide is widely reported to increase following administration of amino acids including glycine. In this study we show that inducible nitric oxide, measured as the urinary ratio of nitrite to nitrate, fell as the output of guanidinoacetic acid increased. We conclude that the protection conferred by glycine may be an effect of its dual role in both stimulating and suppressing renal nitric oxide generation.

## Introduction

A unique protection against lytic cell injury was first described for glycine in incubated renal tubular cells by Weinberg *et al.*[15]. Although this cytoprotective effect is not shared by other amino acids, renal vasodilatation with increased filtration is an effect of a variety of amino acids including glycine and is particularly facilitated by arginine, suggesting a role for nitric oxide in this vasomotor response to protein[2,3]. Paradoxically, nitric oxide, while potentially shielding from hypoxia through this hyperaemic effect, is itself cytotoxic by a direct effect as well as contributing to the generation of free radicals that enhance toxicity[10].

This study tests the tubular protective effect of glycine *in vivo* and offers a hypothesis invoking this dual and apparently antagonistic action of nitric oxide.

## Methods

The source of glycine used throughout these studies was Knox Drinking Gelatine (T. Lipton Co., Englewood Cliffs, NJ). The product, which is orange flavoured, is sweetened with aspartame and provides 40 calories per serving entirely from protein. It contains primarily non-essential amino acids (83.4 per cent) of which glycine contributes 24 per cent by weight. (The approximate amino acid composition is shown in Table 1). Each serving, dissolved in 4 oz (120 ml) of water, is 6 g – making the glycine intake approximately 1.5 g (20 mmol) per serving in an eminently palatable drink. Except in the acute study, which tested the effect of a single serving, all patients received three servings daily for a total of 4.5 g (60 mmol) of glycine. The protocol was submitted to the Institutional Review Board of Bronx-Lebanon Hospital for ethical approval. Patients and volunteers gave informed consent.

Four studies are reported. The first involved 80 hospitalized patients, all of whom were receiving intravenous aminoglycoside, glycopeptide or polyene antibiotics (gentamicin, kanamycin, amikacin, tobramycin, vancomycin, amphotericin) in appropriate doses for periods ranging from 5 to 85 days with a mean of 20 days. Thirty-five of these subjects, randomly selected, received 4.5 g (60 mmol) of glycine daily in the form of 4 oz of Knox Drinking Gelatine three times a day. The mean duration of treatment and dose of antibiotic was not significantly different when the two groups were compared. None of the patients in either group had pre-existing renal disease.

The second study involved 26 hospitalized patients, none of whom had evidence of renal disease or were receiving antibiotics or other nephrotoxic agents. They were given 4.5 g (60 mmol) of glycine daily for 2 days following a control period of 2 days without dietary supplement.

The third study involved five normal controls, who took glycine 4.5 g (60 mmol) daily for 2 weeks bracketed by a pretreatment and a post-treatment control period of 2 weeks, and the fourth was a study of normal controls who received a single dose of Knox Gelatine (1.5 g or 20 mmol of glycine) following an overnight fast. There were 34 nurses and physicians who volunteered for this portion of the study.

All subjects in the first three groups were followed with daily randomly collected urine specimens, which were analyzed for guanidinoacetic acid (GAA) and the stable oxidative products of nitric oxide, $NO_2$ and $NO_3$. GAA was measured using high-performance liquid chromatography[5,6]. Nitrate and nitrite were determined using a procedure and reagents previously described from Hack, Loveland, Co[3]. The method was modified for use in diluted urine samples and consisted of two reactions. In the first step a low range cadmium metal (NitraVer 6 Reagent Powder Pillows) is used for the reduction of nitrate to nitrite. In the second step nitrite ions react with sulphanilic acid to produce a diazonium salt which then reacts with low range chromotropic acid (NitriVer 3 Reagent Powder Pillows) to form a red-orange colour complex. The colour produced is proportional to the total nitrite present in the sample plus that reduced from nitrate and is measured spectrophotometrically at 507 nm. The concentration of nitric oxide is expected to vary with the ratio of nitrite to nitrate as suggested by Marletta et al.[9].

## Results

For the purposes of this study, acute tubular necrosis (ATN) during exposure to nephrotoxic antibiotics is defined as an increase in the blood urea nitrogen and serum creatinine of 50 per cent or greater[14]. The incidence of nephrotoxic ATN in the group receiving concurrent Knox Gelatine was 3 in 35 (9 per cent) compared with 14 in 45 (31 per cent) in the untreated group. This difference is significant at the 95 per cent confidence level by $\chi^2$ analysis. The mean level of urinary GAA varied widely and, although the number achieving an increased output and the percentage increase were greater in those receiving the gelatine, these changes were not statistically significant. Glycine, the principal amino acid constituent of Knox Gelatine, is a reactant in the renal tubular cell synthesis of GAA and an inducer of the enzyme glycine-arginine transamidinase (GAT). We have shown, in a larger series followed over a longer duration, that Knox Gelatine at this dosage will significantly increase urinary GAA[1]. The duration of antibiotic therapy and, therefore, exposure to the nephrotoxin was $18 \pm 15$ days for those receiving gelatine and $20 \pm 17$ days for the controls. These differences are also not statistically significant.

Since the other major contributor to the GAT reaction in renal tubular cells is arginine, it was suggested that the effect of glycine is a reduction in nitric oxide production through competition for available arginine. Values of the ratio of nitrite to nitrate ($NO_2/NO_3$) in spot urines in a random sampling of control and Knox-treated patients are shown in Table 2 (Experiment I). Mean levels reflecting nitric oxide are lower in those receiving supplemental glycine although the differences are not significant by Student's *t*-test. Levels of $NO_2$ / $NO_3$ in spot urines in the 26 subjects given glycine for 2 days following a 2 day control period (Experiment II) showed a statistically significant decrease. Those of GAA are significantly increased. Table 3 shows data for the five subjects receiving Knox Gelatine for a week (Experiment III). There was a fall in the mean urinary $NO_2/NO_3$ ratio, followed by a rise in the mean value during the succeeding week of therapy accompanied by a complementary rise and fall in GAA.

**Table 1. The relative amino acid distribution in gelatine**

| Essential | % | Non-essential | % |
|---|---|---|---|
| Isoleucine | 1.4 | Alanine | 9.0 |
| Leucine | 3.1 | Arginine | 8.3 |
| Lysine | 4.5 | Aspartate | 6.1 |
| Methionine | 0.8 | Cystine | – |
| Phenylalanine | 2.2 | Glutamate | 10.0 |
| Threonine | 2.0 | Glycine | 24.1 |
| Tryptophan | – | Histidine | 0.8 |
| Valine | 2.6 | Hydroxyproline | 6.6 |
| | | Proline | 14.6 |
| | | Serine | 3.5 |
| | | Tyrosine | 0.4 |
| Total | 16.6 | Total | 83.4 |

The change in GAA correlated inversely with the ratio of $NO_2/NO_3$

The correlation coefficient for the pretreatment vs treatment phase (week 1 *vs* week 2) is -0.6598 and that for the treatment *vs* post-treatment phase (week 2 *vs* week 3) is –0.5283. Finally, in the acute study (Experiment IV), there was a statistically insignificant decline in the mean urinary $NO_2/NO_3$ ratio despite an impressive rise in GAA. The effect of glycine on nitric oxide and GAA is apparently neither immediate nor simultaneous.

**Table 2. The effect of gelatine on the urinary output of the derivatives of arginine**

| Product | n | Control | n | Knox |
|---|---|---|---|---|
| | | Output (µmol/g creatinine or 9 mmol creatinine) | | |
| I NO2/NO3 | 32 | 0.3858 ±0.1032 | 16 | 16.1587± .0524 |
| Urinary GAA | 32 | 301.9 ±60.4 | 16 | 16,200.1± 65.5 |
| II NO2/NO3 | 26 | 26.0505 ±.0099 | 26 | 26.0231 ±.0071* |
| Urinary GAA | 26 | 26 228.1 ±42.8 | 26 | 26,337.3± 62.9 *** |

Values are means ±. SEM.
*$P < 0.025$;** $P < 0.005$;*** $P < 0.0005$.

## Discussion

Glycine as a cytopreservative presents several paradoxes. Although the effect *in vitro* is consistent and reproducible in the presence of a variety of cytotoxic materials, it is sadly disappointing *in vivo* when tested in normal animals. Moreover, in the few experiments done in intact animals, glycine can be shown to intensify toxicity in a number of settings, including induced hypoxia[4].

**Table 3. The effect of gelatine on the urinary output of the derivatives of arginine**

| Product | n | Output ($\mu$mol/$g$ creatinine or 9 mmol creatinine) | | | | |
|---|---|---|---|---|---|---|
| | | Control | n | Knox | n | Control |
| III $NO_2/NO_3$ | 5 | $5.0028 \pm 0.0006$ | 5 | $5.0011 \pm 0.0004$ | 5 | $5.0030 \pm 0.0004$** |
| Urinary GAA | 5 | $5612.3 \pm 137.6$ | 5 | $5756.9 \pm 179.1$ | 5 | $5508.7 \pm 120.7$* |
| IV $NO_2/NO_3$ | 34 | $34.0066 \pm .0018$ | 34 | $34.0058 \pm .0022$ | | |
| Urinary GAA | 34 | $34,501.1 \pm 48.9$ | 34 | $34,624.2 \pm 53.0$*** | | |

*$P < 0.025$; **$P < 0.005$; ***$P < 0.0005$. Values are means $\pm$ SEM

Studies performed *in vitro*, however, are carried out in tissues depleted of glycine by design whereas observations in intact normal animals involve tissues where stores of glycine are presumably adequate. The effect of the addition of new glycine is, therefore, blunted. Cellular levels of glycine in our multi-infected and protein-malnourished subjects are undoubtedly reduced as evidenced by the low levels of urinary GAA and the response to the protein supplement[1]. The effect of added glycine should better mimic that seen in protein-depleted tissue preparations.

The sequence of events in ATN is probably as follows: nephrotoxins, which circulate at concentrations which are tolerated by the host tissue, are concentrated in the proximal nephron where they interfere with the function of the tubular lining epithelium. Salt and water entering the distal nephron signal afferent arteriolar vasoconstriction via the process known as tubuloglomerular feedback (TGF). Normally, TGF serves to protect against overwhelming the distal nephron, which would result in polyuria and volume loss. It serves the additional function in ATN of minimizing filtration and, hence, exposure of the epithelium to further toxin damage. The price, however, of this vasoconstriction is 'renal angina' particularly in the medullary regions distal to the vasoactive segment where vigorous salt transfer occurs in an atmosphere of low oxygen availability.

Nitric oxide is generated in endothelial tissue through a continuous, basal, 'constitutive' process where it serves vasodilatation and antithrombosis. Among the many substances, such as acetylcholine and bradykinin, that stimulate this constitutive pathway are a number of amino acids (including glycine). They produce renal vasodilatation and increased filtration, which can be blocked by nitric oxide inhibitors such as N-monomethyl arginine and restored by L-arginine[13].

Nitric oxide is also produced through an inducible pathway available in renal tubular cells and stimulated by various cytokines. Originating in the lining epithelium, it would appear in urine as $NO_2$ as well as $NO_3$ and would, therefore, be reflected in the ratio of nitrite to nitrate ($NO_2/NO_3$). Its function is less clear but it would, by its presence, contribute both to direct cytotoxicity and to the toxic effects of free radicals generated by hypoxia and reperfusion.

In the presence of GAT, maximally concentrated in the renal tubules, arginine transamidinates glycine producing ornithine and GAA. This is obviously not a major tubular cell routing of arginine, which is generally present intracellularly in excess, and the bulk of which is directed via arginase to urea. It shares, however, equal importance with arginine utilization for protein synthesis and nitric oxide generation. Both glycine-arginine transamidinase and nitric oxide synthase (NOS) are substrate -dependent reactions[7,11,12,16].

We propose, therefore, that glycine is the ideal agent to favourably modify the renal response to nephrotoxins. It stimulates the constitutive production of nitric oxide through some direct action shared with other amino acids and, thus, reduces somewhat the vasoconstriction that generates 'renal angina'. It reduces the inducible production of nitric oxide further along the nephron through substrate competition generated by GAA synthesis and, thus, relieves the pressures leading to cytotoxicity.

We suggest, therefore, a Jeckyl and Hyde role for nitric oxide in ATN in which glycine is the ideal prophylactic, reducing its cytotoxic action while preserving its vasoactive function.

## Acknowledgements

This work was supported by a grant from the Thomas Lipton Foundation.

## References

1. Cohen, B.D. & Patel, H. (1992): Urinary guanidinoacetic acid. In: *Guanidino compounds in biology and medicine*. Eds. P. De Deyn, B. Marescau, V. Stalon & I.A. Qureshi, pp. 255–260. London: John Libbey.

2. De Nicola, L., Blantz, R.C. & Gabbai, F.B. (1992): Nitric oxide and angiotensin II: glomerular and tubular interaction in the rat. *J. Clin. Invest.* **89**, 1248–1256.

3. Hack (1992): *Water Analysis Handbook*, 2nd edn, pp. 400–431. Loveland: Hack.

4. Heyman, S.N. & Epstein, F.H. (1993): Glycine and cyto-preservation: disappointments and promises. *J. Lab. Clin. Med.* **121**, 199–200.

5. Hiraga, Y. & Kinoshita, T. (1981): Post-column derivatization of guanidino compounds in high performance liquid chromatography using ninhydrin. *J. Chromatogra.* **226**, 43–51.

6. Hiraga, Y. & Kinoshita, T. (1985): High performance liquid chromatographic analysis of guanidino compounds using ninhydrin reagent II: guanidino compounds in blood of patients on hemodialysis therapy. *J. Chromatogra.* **342**, 269–275.

7. Jansen, A., Lewis, S., Cattell, V. & Cook, H.T. (1992): Arginase is a major pathway of L-arginine in nephritic glomeruli. *Kidney Int.* **42**, 1107–1112.

8. King, A.J., Troy, J.L., Anderson, S., Neuringer, J.R., Gunning, M. & Brenner, B.M. (1991): Nitric oxide a potential mediator of amino acid-induced renal hyperemia and hyperfiltration. *J. Am. Soc. Nephrol.* **1**, 1271–1277.

9. Marletta, M.A., Yoon, P.S., Iyengar, R., Leaf, C.D. & Wishnok, J.S. (1988): Macrophage oxidation of L-arginine to nitrite and nitrate: nitric oxide is an intermediate. *Biochemistry* **27**, 8706–8711.

10. Moncada, S. & Higgs, A. (1993): The L-arginine-nitric oxide pathway. *N. Engl. J. Med.* **329**, 2002–2012.

11. Palmer, R.M.J. (1993): The L-arginine nitric oxide pathway. *Curr. Opin. Nephrol. Hypert.* **2**, 122–128.

12. Pfelschifter, J., Kunz, D. & Muhl, H. (1993): Nitric oxide: an inflammatory mediator of glomerular mesangial cells. *Nephron* **64**, 518–525.

13. Romero, J.C., Lahera, V., Salom, M.G. & Biondi, M.L. (1992): Role of the endothelium-dependent relaxing factor nitric oxide on renal function. *J. Am. Soc. Nephrol.* **2**, 1371–1378.

14. Swartz, R.D., Rubin, J.E., Leeming, B.W. & Silva, P. (1978): Renal failure following major angiography. *Am. J. Med.* **65**, 31–37.

15. Weinberg, J.M., Davis, J.A., Abarzua, M. & Rajan, T. (1987): Cytoprotective effects of glycine and glutathionine against hypoxic injury to renal tubules. *J. Clin. Invest.* **80**, 1446–1454.

16. Wu, G. & Brosnan, J.T. (1992): Macrophages can convert citrulline to arginine. *Biochem. J.* **281**, 45–48.

*Guanidino Compounds : 2*, eds. by P.P. De Deyn, B. Marescau, I.A. Qureshi and A. Mori.
©1997 John Libbey & Company Ltd., pp. 279–288.

# Chapter 28

# Guanidino compound patterns in uraemic brain

P.P. DE DEYN[1,2], B. MARESCAU[1], R. D'HOOGE[1,2], I. POSSEMIERS[1],
J. NAGLER[3] and Ch. MAHLER[3]

[1]*Laboratory of Neurochemistry and Behaviour and* [2]*Department of Neurology, A.Z. Middelheim, Born-Bunge
Foundation, University of Antwerp, Antwerp, Belgium, and* [3]*Department of Internal Medicine, A.Z. Middelheim,
Antwerp, Belgium*

## Summary

Guanidino compounds have been suggested to contribute to the complex neurological complications associated with uraemia. Several of them have previously been reported to accumulate in physiological fluids of renal insufficient subjects. We report on guanidino compound levels in 28 brain regions in control and uraemic brains.

In all brain regions studied, in controls as well as in uraemic patients, concentrations of α-keto-δ-guanidinovaleric acid, α-N-acetylarginine and β-guanidinopropionic acid remained below detection limits. Creatine, guanidinoacetic acid, argininic acid, γ-guanidinobutyric acid, arginine and homoarginine were not increased in uraemic patients. Argininic acid and homoarginine were detectable in some brain regions only. Creatine concentrations varied from 2500 ± 2100 nmol/g tissue in hypophysis to 10 500 ± 1200 nmol/g tissue in cerebellar cortex. Even more pronounced regional differences were found for γ-guanidinobutyric acid with the lowest concentration in the caudate nucleus (0.6 ± 0.3 nmol/g tissue) and the highest in substantia nigra, pallidum and cerebellar dentate nucleus (8.3 ± 2.8 nmol/g tissue). Although guanidinosuccinic acid levels were below the detection limit in controls in the majority of brain regions, important increases (some > 100 fold) were observed in all brain regions of uraemic patients. Accumulation of guanidinosuccinic acid increased with increasing degree of renal failure, with levels up to 65 nmol/g tissue in the hypophysis. Creatinine concentrations were also found to be increased in uraemic brain regions, but increases seemed to be less strictly related to serum urea levels. Guanidine and methylguanidine were found only occasionally in brain regions of controls whereas respectively 100- and 30-fold increases were found in brain regions of uraemic subjects. Levels of guanidinosuccinic acid and creatinine in uraemic brain were comparable to those previously observed in brain of experimental animals displaying convulsions following intraperitoneal injection of the respective compounds. Our findings further establish guanidino compounds as probable uraemic toxins contributing to the neurological complications in uraemia.

## Introduction

The pathophysiology of the uraemic syndrome consists of a variety of neurological and haematological complications, which remain poorly understood[11]. The haematological complications include anaemia and bleeding diathesis, and the neurological syndrome may consist of encephalopathy, the 'uraemic twitch-convulsive syndrome' and polyneuropathy. Among the candidate uraemic toxins are several guanidino compounds that have been previously reported to be increased in uraemic biological fluids and were shown to be experimental toxins[1,3,4,6,–8,16,38,–40,42].

Although there is some disagreement[15], guanidinosuccinic acid is believed to be related to the uraemic bleeding diathesis[23,40]. Methylguanidine could be related to polyneuropathy in uraemic dogs[18] and has been shown experimentally to induce a catabolic state, haemolysis [18] and epilepsy[30]. γ-Guanidinobutyric acid, methylguanidine, homoarginine, creatine and creatinine were found to have a convulsive effect in animals when injected intracisternally[24,25,34,43]. Guanidinosuccinic acid, methylguanidine, guanidine and creatinine were suggested to be chloride channel blockers and were shown to induce myoclonic and generalized seizures after systemic and intracerebroventricular administration in mice[5,13,14]. Moreover, guanidinosuccinic acid was shown to inhibit excitatory neurotransmission in rat hippocampal brain slices: an effect that hypothetically could contribute to the uraemic encephalopathy[12]. Furthermore, *in vitro* studies have shown that guanidinosuccinic acid, creatine, guanidinoacetic acid, guanidine and β-guanidinopropionic acid might be factors responsible for the increased haemolysis[16]. Methylguanidine and creatinine were shown to induce haemolysis *in vitro* and *in vivo*[18]. Finally, guanidinosuccinic acid decreased erythrocyte transketolase activity[26], and methylguanidine inhibited brain sodium-potassium ATPase[32]. Reynolds and Rothermund[37] showed that guanidine, methylguanidine, guanidinosuccinic acid and creatinine interact with the NMDA receptor *in vitro*, as indicated by their ability to inhibit [$^3$H]dizocilpine binding, accelerate the dissociation of [$^3$H]dizocilpine, and inhibit NMDA-induced increases in $Ca^{2+}$ in cultured central neurons. Finally, guanidine, methylguanidine and guanidinosuccinic acid decreased the synaptosomal membrane fluidity of rat cerebral cortex[22].

In an attempt to investigate whether proposed toxic guanidino compounds accumulate in brain and consequently may contribute to the central nervous system complications associated with the uraemic syndrome, we determined their concentrations in uraemic brain. The results of guanidino compound determination in autopsy brain tissue samples of control subjects and patients with various degrees of renal failure are presented and their possible significance discussed.

## Materials and methods

The study considered nine uraemic patients and five control subjects. The control group consisted of two women and three men (average age 62 years). Serum urea concentration in the control subjects was $4.6 \pm 2.3$ mM (mean $\pm$ SD). None of the control subjects suffered from neurological, renal or hepatic disorders. Causes of death were terminal respiratory insufficiency and cardiac arrest. The uraemic patient group consisted of five women and four men (average age 75 years). Serum urea concentration (sampled on the day before or within the hours before death) ranged from 13.3 to 93.8 mM. We will discuss the guanidino compound brain levels in the patients with moderately increased serum urea concentration separate from those in the patients with highly increased urea levels. Four uraemic patients had moderately increased serum urea concentrations of $17.5 \pm 2.9$ mM (mean $\pm$ SD). Three of these patients suffered from discrete renal insufficiency, probably of extrarenal origin, and died of respiratory and cardiac failure; the fourth patient had impaired renal function resulting from rhabdomyolysis. Five patients had highly increased serum urea levels: patient VP (serum urea 37.2 mM) suffered from chronic interstitial nephritis; PH (serum urea 56.9 mM) had impaired renal function due to adenocarcinoma without cerebral metastases; DA (serum urea 62.4 mM) presented with acute renal failure of unknown origin and died of cardiopulmonary arrest; DJ (serum urea 68.9 mM) suffered from chronic interstitial nephritis and colon cancer without cerebral metastases, and died of cardiac arrhythmia; HM (serum urea 93.8 mM) had impaired renal function resulting from rhabdomyolysis and died of cardiac failure. These patients with highly increased serum urea levels displayed severe signs of encephalopathy (intractable status epilepticus, coma).

All autopsy material was taken between 2 and 8 h after death. Until autopsy, the bodies were kept at 2 °C. Brain tissue specimens were taken from the left hemisphere only. Dissections were performed at 0 °C. Specimens from 28 brain regions, weighing 100–400 mg (wet weight), were used for further work-up according to Marescau et al.[27]. Values reported for the cerebral cortex are concentrations measured in specimens from superior frontal gyrus; there was no difference between samples taken from the superior frontal gyrus, cingulate gyrus, superior temporal gyrus, precentral gyrus, postcentral gyrus, occipital cortex or parahippocampal gyrus. Additional brain regions studied are listed in the tables.

Brain samples were homogenized with a Potter homogenizer in 1 ml water at 0 °C. The Teflon pestle was washed twice with 1 ml water. For deproteinization, 1 ml of a 30 per cent trichloroacetic acid solution was added to the homogenate before centrifugation at 100 000 × $g$ for 30 min at 4 °C. The supernatant was removed, and the pellet Vortex-mixed in 1 ml of a 10 per cent trichloroacetic acid solution. After a second centrifugation, the two supernatants were pooled, and aliquots were taken for guanidino compound analysis.

The concentration of 13 guanidino compounds was determined by cation exchange chromatography and fluorescence ninhydrin detection. Determinations were performed according to a method described earlier[29], modified from Hiraga and Kinoshita[21], with the Biotronik LC 5001 amino acid analyser adapted for guanidino compound determination. The concentration of 13 guanidino compounds was determined. Cation exchange chromatography for separation of the guanidino compounds was followed by fluorescence ninhydrin detection using a Spectroflow 980 fluorimeter (ABI Analytical Kratos Division, Ramsey, NJ). Measurements were performed at an excitation wavelength of 301 nm, and with an emission filter of 500 nm. Guanidino compounds used as standards for guanidino compound determination (creatine, creatinine, guanidinosuccinic acid, guanidine, methylguanidine, guanidinoacetic acid, α-N-acetylarginine, argininic acid, β-guanidinopropionic acid, γ- guanidinobutyric acid, arginine and homoarginine) were purchased from Sigma (St Louis, USA), creatine and creatinine from Merck (Darmstadt, Germany). α-Keto-δ-guanidinovaleric acid was synthesized enzymatically according to the method of Meister[31], which was modified as described earlier[29]. All other chemicals used were obtained from Merck and of analytical grade.

Serum urea levels were determined using an automated urease-glutamate dehydrogenase method[19].

## Results

In all brain regions studied, concentrations of α-keto-δ-guanidinovaleric acid, α-N-acetylarginine and β-guanidinopropionic acid remained below detection limits in controls as well as in uraemic patients. Creatine, guanidinoacetic acid, γ-guanidinobutyric acid and arginine were detectable in all brain regions of controls and uraemics. Argininic acid and homoarginine were detectable in some brain regions only. The levels of these compounds were no higher in uraemics than in controls. The concentrations of these guanidino compounds in the studied brain structures in controls are given in Table 1. Creatine concentrations ranged from 2500 ± 2100 nmol/g tissue in hypophysis to 10 500 ± 1200 nmol/g tissue in cerebellar cortex. High concentrations of creatine were also found in putamen and pallidum. Guanidinoacetic acid concentrations ranged between 1.9 ± 0.5 nmol/g tissue in the internal capsule and 4.6 ± 2.0 nmol/g tissue in the optic chiasm. High concentrations were also found in hypophysis and cerebellar cortex. Concentrations of γ-guanidinobutyric acid ranged from below detection limits to 1.51 nmol/g tissue in hypophysis and 8.3 ± 2.8 nmol/g tissue in cerebellar dentate nucleus. High levels of this compound were also found in substantia nigra, pallidum, thalamus, red nucleus and cerebellar subcortical tissue. Arginine levels ranged between 340 ± 80 nmol/g tissue in subcortical tissue of the forebrain and 750 ± 140 nmol/g tissue in medulla oblongata. Homoarginine displayed a low and equally spread level over all brain areas, with a maximum of 3.0 ± 1.8 nmol/g tissue in cerebellar dentate nucleus.

Levels of guanidinosuccinic acid in controls were below detection limits in the majority of brain regions, and trace amounts were found in some others (Table 2). In parallel with the increased serum urea levels, guanidinosuccinic acid was increased in all brain areas in uraemics. Guanidinosuccinic acid levels in the extremely uraemic patients DJ and HM were found to be more than 100 times higher than the control levels. The highest value was found in patient HM with a guanidinosuccinic acid concentration as high as 65.0 nmol/g tissue in the hypophysis. The level of guanidinosuccinic acid in this patient's hippocampus (not shown in Table 2) was 16.9 nmol/g tissue, somewhat lower than the 23.7 nmol/g tissue found in forebrain cortex, putamen and claustrum.

**Table 1. Brain levels of guanidino compounds that are not increased in uraemic patients**

| Brain region | Guanidino compound | | | | | |
|---|---|---|---|---|---|---|
| | Creatine | Guanidinoacetic acid | Argininic acid | γ-Guanidinobutyric acid | Arginine | Homoarginine |
| *Forebrain* | | | | | | |
| Cortex | 7600 ± 700 | 2.3 ± 0.3 | <dl–0.12 | 1.6 ± 0.3 | 390 ± 80 | 1.3 ± 0.1 |
| Subcortical tissue | 7300 ± 1100 | 2.0 ± 0.5 | <dl–1.08 | 1.1 ± 0.5 | 340 ± 80 | 1.7 ± 0.8 |
| Corpus callosum | 6200 ± 1800 | 2.0 ± 0.6 | <dl–0.82 | 0.8 ± 0.5 | 380 ± 60 | 1.7 ± 0.5 |
| Internal capsule | 7100 ± 1200 | 1.9 ± 0.5 | <dl–0.64 | 1.2 ± 0.8 | 420 ± 80 | 1.9 ± 0.8 |
| Caudate nucleus | 6600 ± 2200 | 2.9 ± 0.8 | <dl | 0.6 ± 0.3 | 360 ± 220 | 1.5 ± 1.0 |
| Putamen | 9500 ± 1000 | 3.0 ± 1.2 | <dl–0.87 | 1.1 ± 0.7 | 450 ± 170 | 2.1 ± 1.8 |
| Pallidum | 9400 ± 400 | 3.0 ± 1.1 | <dl–1.22 | 4.2 ± 1.6 | 500 ± 160 | 2.2 ± 1.0 |
| Claustrum | 7100 ± 1300 | 2.0 ± 0.5 | <dl | 1.4 ± 1.1 | 420 ± 190 | <dl –2.65 |
| Mammillary bodies | 4300 ± 1100 | 3.3 ± 0.8 | <dl | 2.1 ± 0.5 | 550 ± 220 | <dl –3.13 |
| Thalamus | 7500 ± 700 | 3.0 ± 1.1 | <dl | 3.9 ± 2.0 | 610 ± 210 | 1.9 ± 1.4 |
| Optic chiasm | 5800 ± 3300 | 4.6 ± 2.0 | <dl | 2.0 ± 1.1 | 520 ± 110 | <dl–3.95 |
| Optic nerve | 4200 ± 500 | 3.2 ± 1.1 | <dl | 1.5 ± 0.2 | 440 ± 150 | <dl |
| Hypophysis | 2500 ± 2100 | 4.2 ± 2.1 | <dl–0.77 | <dl-1.51 | 440 ± 210 | 2.0 ± 0.9 |
| *Midbrain* | | | | | | |
| Cerebral peduncle | 5100 ± 700 | 2.6 ± 0.7 | <dl | 3.3 ± 0.5 | 460 ± 180 | <dl–2.3 |
| Red nucleus | 7000 ± 700 | 3.4 ± 0.9 | <dl | 3.9 ± 1.4 | 520 ± 110 | <dl–3.55 |
| Substantia nigra | 7800 ± 1000 | 3.0 ± 0.9 | <dl–2.24 | 4.8 ± 2.1 | 640 ± 180 | <dl–3.45 |
| *Hindbrain* | | | | | | |
| Pons | 5300 ± 800 | 3.0 ± 1.1 | <dl–1.58 | 2.3 ± 0.7 | 450 ± 160 | 1.9 ± 0.6 |
| Pyramid | 4800 ± 900 | 3.6 ± 0.7 | <dl | 2.0 ± 0.4 | 590 ± 200 | <dl–2.98 |
| Medulla oblongata | 6400 ± 500 | 3.7 ± 1.1 | <dl–0.95 | 3.0 ± 0.6 | 750 ± 140 | <dl–3.40 |
| Cerebellum: | | | | | | |
|    cortex | 10500 ± 1200 | 3.9 ± 0.4 | <dl–0.85 | 2.7 ± 1.2 | 710 ± 370 | 1.8 ± 1.1 |
|    subcortical tissue | 6100 ± 800 | 2.2 ± 0.2 | <dl–1.39 | 3.6 ± 1.9 | 420 ± 170 | 1.8 ± 1.0 |
|    dentate nucleus | 6500 ± 800 | 3.3 ± 0.9 | <dl–1.40 | 8.3 ± 2.8 | 490 ± 80 | 3.0 ± 1.8 |

The levels of six guanidino compounds in 22 brain regions in control subjects are shown. These compounds are not increased in brain of uraemic subjects. Values are means ± SD of five controls, expressed in nmol/g tissue (<dl: below detection limit).

Creatinine concentrations ranged in controls from $130 \pm 70$ nmol/g tissue in hypophysis to $560 \pm 290$ nmol/g tissue in putamen (Table 3). High concentrations were also found in pallidum and caudate nucleus. Although the levels were much higher in the patients with highly increased serum urea levels than in controls and in the four subjects with moderately increased serum urea, brain creatinine levels may not be strictly related to serum urea levels. In most brain regions in moderately uraemic patients, creatinine levels were actually somewhat below the mean levels in controls. The highest level (1698 nmol/g tissue) was found in the cerebellar cortex of patient HM, a value 4.7 times higher than the mean value found in controls. In this patient, the levels varied strongly over the different brain structures. Creatinine reached levels in highly uraemic patients 2.5–5 times higher than in controls.

Guanidine concentrations remained below detection limits in all brain regions in four of five controls (Table 4). Guanidine was present in trace amounts in four brain structures in the remaining control subject. The highest values were found in patient DJ with values up to 100 times higher than in controls.

Methylguanidine levels were below detection limits in most brain regions in controls and were increased in uraemics (Table 5). Concentrations 20–30 times higher than those found in some of the controls were measured in the highly uraemic patients. The extremely uraemic patient HM did not have the highest levels of methylguanidine. Highly increased levels of this compound were found in hypophysis. In patient PH's hippocampus (not shown in Table 5), a methylguanidine concentration of 5.14 nmol/g tissue was measured.

**Table 2. Brain levels of guanidinosuccinic acid in controls and uraemics**

| Brain region | Controls | Uraemics | | | | | |
|---|---|---|---|---|---|---|---|
| | | 17.5 ±2.9  mM* (n = 4) | 37 mM* (Patient VP) | 57 mM* (Patient PH) | 62 mM* (Patient DA) | 69 mM* (Patient DJ) | 94 mM* (Patient HM) |
| *Forebrain* | | | | | | | |
| Cortex | <dl–0.37 | 0.3 ± 0.1 | 4.19 | 10.5 | 13.5 | 21.0 | 23.7 |
| Subcortical tissue | <dl–0.30 | 0.30 ± 0.04 | 2.05 | 9.02 | 3.23 | 12.0 | 18.1 |
| Corpus callosum | <dl–0.19 | <dl–0.33 | 2.06 | 6.16 | 3.87 | – | 12.4 |
| Internal capsule | <dl | <dl–0.17 | 1.92 | 8.52 | 1.45 | – | 15.5 |
| Caudate nucleus | <dl | <dl–0.42 | 3.72 | 10.8 | 1.98 | 13.0 | 19.0 |
| Putamen | <dl | 0.3 ± 0.1 | 3.13 | 13.0 | 2.85 | – | 23.7 |
| Pallidum | <dl | <dl–2.36 | 3.36 | 10.8 | 6.08 | – | 19.0 |
| Claustrum | <dl | <dl–0.42 | 2.90 | 11.2 | 9.34 | – | 23.7 |
| Mammillary bodies | <dl | <dl | 3.44 | 9.87 | 8.19 | – | 18.6 |
| Thalamus | <dl–0.24 | <dl–0.33 | 2.90 | 8.58 | 6.06 | 14.0 | 19.8 |
| Optic chiasm | <dl | <dl–2.52 | 2.52 | 5.75 | 13.8 | – | 24.3 |
| Optic nerve | <dl | <dl–0.69 | 2.36 | 7.59 | 14.1 | – | 20.8 |
| Hypophysis | <dl | 0.9 ± 0.4 | – | 32.7 | 31.3 | – | 65.0 |
| *Midbrain* | | | | | | | |
| Cerebral peduncle | <dl | <dl | 1.63 | 7.19 | 5.28 | 13.0 | 12.8 |
| Red nucleus | <dl–0.54 | <dl–0.39 | 2.17 | 10.4 | 2.92 | – | 16.8 |
| Substantia nigra | <dl | <dl–0.29 | 2.42 | 11.8 | 4.05 | – | 23.3 |
| *Hindbrain* | | | | | | | |
| Pons | <dl–0.11 | <dl–0.33 | 1.84 | 10.3 | 6.43 | 12.2 | 18.9 |
| Pyramid | <dl | <dl–0.37 | 2.18 | 7.69 | 15.6 | – | 11.9 |
| Medulla oblongata | <dl | 0.3 ± 0.1 | 2.12 | 7.29 | 17.2 | 17.3 | 12.7 |
| Cerebellum: | | | | | | | |
| cortex | <dl | <dl–0.34 | 3.68 | 12.1 | 3.01 | 16.5 | 30.2 |
| subcortical tissue | <dl | <dl–0.34 | 1.23 | 6.66 | 2.96 | 11.7 | 13.7 |
| dentate nucleus | <dl | <dl–0.34 | 2.12 | 10.0 | 2.25 | – | 21.6 |

Guanidinosuccinic acid levels in 22 brain regions in five control subjects, and in four moderately and five highly uraemic patients. The results from the moderately uraemic patients are grouped; those of the highly uraemic patients are listed individually. The levels in the highly uraemic patients are shown for each subject separately. Data are expressed in nmol/g tissue (< dl: below detection limit, –: not determined). *Serum urea levels. Where applicable, mean values ± SD were calculated.

## Discussion

Earlier reports demonstrated that the levels of several guanidino compounds are markedly increased in serum and cerebrospinal fluid of non-dialysed uraemic patients. More specifically, it was shown that, along with high urea levels, levels of guanidinosuccinic acid, guanidine, methylguanidine and creatinine are clearly increased[8]. In patients with serum urea levels ten times higher than in controls, guanidinosuccinic acid levels were about 100 times higher. Moreover, the same compounds (guanidinosuccinic acid, guanidine, methylguanidine and creatinine) were markedly (≥ 10×) increased in uraemic cerebrospinal fluid. Levels of γ-guanidinobutyric acid and argininic acid were moderately (< 10×) increased, whereas those of the other guanidino compounds were close to normal. Haemodialysis and peritoneal dialysis were shown to remove guanidino compounds in uraemic patients, but failed to restore their normal levels entirely[9,10].

We have determined guanidino compound concentrations in brain structures of uraemic patients. More than 20 different brain regions were examined in control subjects and in patients with various degrees of uraemia. Concentrations of α-keto-δ-guanidinovaleric acid, α-N-acetylarginine and β-guanidinopropionic acid remained below detection limits in controls as well as in uraemic patients. The same compounds were undetectable or present only in trace amounts in brain of other ureotelic animals as well[28]. All other compounds were detectable in brain tissue samples from control subjects. Argininic acid, homoarginine, guanidinosuccinic acid, guanidine and methylguanidine were present in detectable amounts in some brain regions only.

**Table 3. Brain levels of creatinine in controls and uraemics**

| Brain region | Controls | Uraemics | | | | | |
|---|---|---|---|---|---|---|---|
| | | 17.5 ± 2.9 mM* (n = 4) | 37 mM* (Patient VP) | 57 mM* (Patient PH) | 62 mM* (Patient DA) | 69 mM* (Patient DJ) | 94 mM* (Patient HM) |
| *Forebrain* | | | | | | | |
| Cortex | 290 ± 30 | 220 ± 40 | 566 | 566 | 677 | 733 | 389 |
| Subcortical tissue | 370 ± 120 | 340 ± 40 | 617 | 742 | 791 | 644 | 547 |
| Corpus callosum | 250 ± 80 | 210 ± 30 | 439 | 513 | 973 | – | 438 |
| Internal capsule | 380 ± 150 | 320 ± 60 | 650 | 696 | 1221 | – | 924 |
| Caudate nucleus | 490 ± 150 | 470 ± 180 | 853 | 874 | 1122 | 749 | 1101 |
| Putamen | 560 ± 290 | 370 ± 30 | 724 | 1045 | 1082 | – | 1070 |
| Pallidum | 510 ± 220 | 480 ± 90 | 731 | 1006 | 755 | – | 949 |
| Claustrum | 370 ± 260 | 450 ± 50 | 795 | 929 | 852 | – | 768 |
| Mammillary bodies | 270 ± 70 | 220 ± 20 | 470 | 720 | 613 | – | 429 |
| Thalamus | 400 ± 160 | 330 ± 60 | 712 | 792 | 686 | – | 1202 |
| Optic chiasm | 200 ± 20 | 180 ± 10 | 508 | 464 | 634 | – | 591 |
| Optic nerve | 180 ± 40 | 230 ± 90 | 323 | 585 | 535 | – | 532 |
| Hypophysis | 130 ± 70 | 150 ± 30 | – | 935 | 643 | – | 501 |
| *Midbrain* | | | | | | | |
| Cerebral peduncle | 240 ± 90 | 230 ± 30 | 482 | 669 | 525 | 625 | 426 |
| Red nucleus | 240 ± 60 | 340 ± 80 | 454 | 850 | 679 | – | 747 |
| Substantia nigra | 400 ± 260 | 390 ± 60 | 642 | 1007 | 781 | – | 1086 |
| *Hindbrain* | | | | | | | |
| Pons | 270 ± 150 | 270 ± 60 | 568 | 574 | 664 | 514 | 787 |
| Pyramid | 190 ± 60 | 210 ± 30 | 510 | 590 | 527 | – | 503 |
| Medulla oblongata | 300 ± 120 | 260 ± 30 | 550 | 575 | 505 | 856 | 540 |
| Cerebellum: | | | | | | | |
| cortex | 360 ± 80 | 460 ± 70 | 824 | 973 | 799 | 815 | 1698 |
| subcortical tissue | 320 ± 100 | 360 ± 50 | 536 | 654 | 700 | 502 | 462 |
| dentate nucleus | 310 ± 80 | 330 ± 70 | 673 | 657 | 737 | – | 974 |

Creatinine levels in 22 brain regions in five control subjects, and in four moderately and five highly uraemic patients. The results from the moderately uraemic patients are grouped; those of the highly uraemic patients are listed individually. Data are expressed in nmol/g tissue (–: not determined). *Serum urea levels.

Brain distribution of creatine appeared to be similar to that of its anhydrized form, creatinine. The lowest concentrations were found in hypophysis, optic nerve, pyramid, cerebral peduncles and pons, the highest concentrations in putamen, pallidum and cerebellar cortex. Niklasson & Agren[35] suggested that creatine and creatinine concentrations in cerebrospinal fluid can serve as clinical indices of energy metabolism in brain. Similarly, high creatine and creatinine levels in certain brain structures might be indicative of high energy consumption in these structures. Heiss *et al.*[20] measured brain glucose consumption in healthy volunteers by positron emission tomography. They found low metabolic activity in white matter, pons and cerebral peduncles. High activity was found in several cortical areas in caudate nucleus, putamen, pallidum and especially visual cortex. The general pattern observed by Heiss *et al.*[20] appears to parallel the regional distribution of creatine and creatinine reported here.

An even more pronounced regional distribution was found in the case of γ-guanidinobutyric acid. The compound was most abundant in cerebellar dentate nucleus, substantia nigra and pallidum. The lowest concentration was measured in hypophysis. The regional distribution of this compound corresponds well with the distribution of GABAergic neurons[2]. This might be explained by the metabolic relationship between GABA and γ-guanidinobutyric acid. Pisano *et al.*[36] demonstrated that γ-guanidinobutyric acid can be synthesized in mammalian brain preparations as the result of a transamidination from arginine to GABA.

Levels of guanidinosuccinic acid, creatinine, guanidine and methylguanidine were increased in all studied brain structures of renal insufficient patients, more or less in relation to the increased serum urea levels. Small increases were observed in moderately uraemic patients. More pronounced increases were found in the group of highly uraemic patients. In the most extreme cases, patients, we found highly increased levels of guanidinosuccinic acid (> 100-fold), creatinine (five-fold), guanidine (> 100-fold) and methylguanidine (20-fold). The levels of guanidinosuccinic acid closely followed the increases in serum urea concentration, again demonstrating the metabolic relationship between urea and guanidinosuccinic acid (see also Marescau et al.[28]).

**Table 4. Brain levels of guanidine in controls and uraemics**

| Brain region | Controls | Uraemics | | | | | |
|---|---|---|---|---|---|---|---|
| | | 17.5 ± 2.9mM* (n=4) | 37mM* (Patient VP) | 57mM* (Patient PH) | 62 mM* (Patient DA) | 69 mM* (Patient DJ) | 94 mM* (Patient HM) |
| *Forebrain* | | | | | | | |
| Cortex | <dl | <dl | <dl | 5.03 | 1.39 | 12.0 | 2.86 |
| Subcortical tissue | <dl–0.82 | <dl–1.56 | <dl | 6.34 | 1.55 | 11.3 | 1.99 |
| Corpus callosum | <dl–0.96 | <dl–1.28 | <dl | 5.83 | 1.53 | – | 1.74 |
| Internal capsule | <dl–0.67 | <dl–0.84 | <dl | 6.28 | 0.93 | – | 2.90 |
| Caudate nucleus | <dl | <dl–0.90 | 0.71 | 5.44 | 1.10 | 3.15 | 3.08 |
| Putamen | <dl–0.43 | <dl–0.65 | 0.78 | 6.24 | 0.89 | – | 2.74 |
| Pallidum | <dl | <dl–0.90 | <dl | 5.55 | 1.07 | – | 3.20 |
| Claustrum | <dl | <dl–1.82 | <dl | 7.15 | 2.91 | – | 3.58 |
| Mammillary bodies | <dl | <dl | <dl | 6.39 | <dl | – | 2.60 |
| Thalamus | <dl | <dl–1.34 | <dl | 5.91 | 1.06 | 45.2 | 3.35 |
| Optic chiasm | <dl | <dl | <dl | 4.50 | 2.45 | – | 4.81 |
| Optic nerve | <dl | <dl | <dl | 5.47 | 1.58 | – | 3.81 |
| Hypophysis | <dl | <dl–1.51 | – | 11.5 | 2.43 | – | 9.36 |
| *Midbrain* | | | | | | | |
| Cerebral peduncle | <dl | <dl | <dl | 6.24 | 2.19 | 54.6 | 2.99 |
| Red nucleus | <dl | <dl–2.25 | <dl | 9.06 | <dl | – | 2.72 |
| Substantia nigra | <dl | <dl–1.72 | <dl | 7.28 | 1.17 | – | 4.90 |
| *Hindbrain* | | | | | | | |
| Pons | <dl | <dl–2.03 | <dl | 6.07 | 1.20 | 11.3 | 3.99 |
| Pyramid | <dl | <dl–1.96 | <dl | 3.66 | 1.74 | – | 3.92 |
| Medulla oblongata | <dl | <dl–4.75 | <dl | 4.32 | 1.92 | 44.5 | <dl |
| Cerebellum: | | | <dl | | | | |
| cortex | <dl | <dl–1.58 | <dl | 4.79 | 1.85 | 31.7 | – |
| subcortical tissue | <dl | <dl–0.73 | <dl | 5.92 | <dl | 16.8 | 2.92 |
| dentate nucleus | <dl | <dl–1.49 | <dl | 6.90 | <dl | – | 4.40 |

Guanidine levels are shown in 22 brain regions in five controls, and in four moderately and five highly uraemic patients. The results from the moderately uraemic patients are grouped; those of the highly uraemic patients are listed individually. Data are expressed in nmol/g tissue (< dl: below detection limit, –: not determined). * Serum urea levels.

These same four guanidino compounds were previously shown to be highly increased in serum and cerebrospinal fluid of uraemic patients[7]. The term uraemic guanidino compounds, which we have coined on the basis of their levels in uraemics, also appears to be highly appropriate to their levels in uraemic brain.

In the light of their putative or proven neuro- and haematotoxicity (see Mori[33,34]), increased levels of these compounds might contribute to uraemia-associated complications. Especially the increased brain levels reported here suggest the involvement of these compounds in the neurological and neuropsychological complications of this acquired metabolic disease.

Indeed, levels of guanidinosuccinic acid and creatinine in uraemic brain were comparable to those previously observed in brain of experimental animals displaying convulsions following intraperitoneal injection of the respective compounds[13]. After intraperitoneal administration of $CD_{50}$ doses of creatinine and guanidinosuccinic acid in mice, corresponding brain concentrations of 1328 nmol/g tissue and 56 nmol/g tissue, respectively, were found[13]. These concentrations are similar to those here found in brain regions of uraemic subjects. Moreover, significant inhibitory effects of guanidinosuccinic acid were observed on excitatory neurotransmission in rat hippocampal brain slices and on GABA-responses on mouse neurons in primary dissociated cell culture at presumably comparable concentrations of $\geq 100\ \mu M$[5,12]. Therefore, these experimentally proven toxic guanidino compounds, shown here to reach potentially pathophysiologically relevant levels in uraemic brain, should certainly be considered as possible uraemic toxins contributing to the central neurological complications in uraemia.

**Table 5. Brain levels of methylguanidine in controls and uraemics**

| Brain region | Controls | Uraemics | | | | | |
|---|---|---|---|---|---|---|---|
| | 17.5 ± 2.9 mM* (n =4) | | 37 μM* (Patient VP) | 57 mM* (Patient PH) | 62 mM* (Patient DA) | 69 mM* (Patient DJ) | 94 mM* (Patient HM) |
| *Forebrain* | | | | | | | |
| Cortex | <dl–0.30 | <dl | 1.95 | 4.10 | 4.71 | 6.04 | 1.30 |
| Subcortical tissue | <dl | <dl–0.59 | 2.19 | 4.29 | 2.85 | 4.42 | 1.84 |
| Corpus callosum | <dl | <dl | 2.01 | 2.93 | 2.79 | – | 1.25 |
| Internal capsule | <dl | <dl | 2.78 | 4.06 | 1.62 | – | 3.25 |
| Caudate nucleus | <dl–0.42 | <dl–0.53 | 3.23 | 4.87 | 1.61 | – | 4.12 |
| Putamen | <dl–0.32 | <dl–0.47 | 2.95 | 5.99 | 2.24 | – | 3.56 |
| Pallidum | <dl | <dl–0.55 | 2.57 | 5.42 | 3.46 | – | 3.16 |
| Claustrum | <dl | <dl–0.63 | 3.09 | 5.38 | 4.71 | – | 3.30 |
| Mammillary bodies | <dl | <dl | 3.13 | 3.36 | 3.98 | – | 1.57 |
| Thalamus | <dl | <dl–0.37 | 3.37 | 4.94 | 2.98 | 4.52 | 2.93 |
| Optic chiasm | <dl | <dl | 1.95 | 2.03 | 4.50 | – | 1.90 |
| Optic nerve | <dl | <dl | 1.71 | 2.97 | 5.21 | – | 1.69 |
| Hypophysis | <dl | <dl | – | 9.72 | 9.06 | – | 7.00 |
| *Midbrain* | | | | | | | |
| Cerebral peduncle | <dl | <dl | 2.08 | 4.69 | 3.32 | 3.78 | 1.79 |
| Red nucleus | <dl | <dl | 3.05 | 5.41 | 2.45 | – | 2.58 |
| Substantia nigra | <dl | <dl–0.57 | 3.01 | 6.42 | 2.94 | – | 3.97 |
| *Hindbrain* | | | | | | | |
| Pons | <dl–0.44 | <dl | 3.18 | 5.67 | 4.38 | 4.20 | 2.42 |
| Pyramid | <dl | <dl | 2.18 | 4.47 | 6.78 | – | 4.40 |
| Medulla oblongata | <dl | <dl–0.45 | 2.39 | 3.35 | 6.52 | 5.83 | 2.21 |
| Cerebellum: | | | | | | | |
| cortex | <dl–0.33 | <dl–1.52 | 2.74 | 4.92 | 2.07 | 5.39 | 2.97 |
| subcortical tissue | <dl | <dl–0.73 | 2.06 | 3.48 | 2.38 | 4.11 | 2.21 |
| denate nucleus | <dl | <dl–0.39 | 2.49 | 4.09 | 1.72 | – | 3.48 |

Methylguanidine levels are shown in 22 brain regions in five controls, and in four moderately and five highly uraemic patients. The results from the moderately uraemic patients are grouped; those of the highly uraemic patients are listed individually. Data are expressed in nmol/g tissue (<dl: below detection limit, –: not determined). * Serum urea levels.

## Acknowledgements

This work was supported by the Flemish Ministry of Education, the Baron Bogaert-Scheid Fund, Born-Bunge Founda tion, Medical Research Foundation OCMW Antwerp, the Universitaire Instelling Antwerpen, the United Fund of Belgium and NFWO grants No. 3.0044.92 and 3.0064.93.

## References

1.   Baker, L.R.I. & Marshall, R.D. (1971): A reinvestigation of methylguanidine concentrations in sera from normal and uraemic subjects. *Clin. Sci.* **41**, 563–568.

2.   Björklund, A. & Hökfelt, T., eds. (1985): *Handbook of Chemical Neuroanatomy,* Vol. 4: *GABA and Neuropeptides in the CNS.* Amsterdam: Elsevier.

3.   Cohen, B.D. (1970): Guanidinosuccinic acid in uremia. *Arch. Intern. Med.* **126**, 846–850.

4.   De Deyn, P.P. (1989): Analytical studies and pathophysiological importance of guanidino compounds in uremia and hyperargininemia. Thesis submitted to obtain the degree of 'Geaggregeerde van het hoger onderwijs', Belgium: University of Antwerp.

5.   De Deyn, P.P. & Macdonald, R.L. (1990): Guanidino compounds that are increased in cerebrospinal fluid and brain of uremic patients inhibit GABA and glycine responses on mouse neurons in cell culture. *Ann. Neurol.* **28**, 627–633.

6.   De Deyn, P.P., Marescau, B., Lornoy, W., Becaus I. & Lowenthal, A. (1986): Guanidino compounds in uraemic dialyzed patients. *Clin. Chim. Acta* **157**, 143–150.

7.   De Deyn, P.P., Marescau, B., Lornoy, W., Becaus, I., Van Leuven, I., Van Gorp, L. & Lowenthal, A. (1987): Serum guanidino compound levels and the influence of a single hemodialysis in uremic patients undergoing maintenance hemodialysis. *Nephron* **45**, 291–295.

8.   De Deyn, P.P., Marescau, B., Cuykens, J.J., Van Gorp, L. & Lowenthal, A. (1987): Guanidino compounds in serum and cerebrospinal fluid of nondialyzed patients with renal insufficiency. *Clin. Chim. Acta* **167**, 81–88.

9.  De Deyn, P.P., Marescau, B., Lornoy, W., Becaus, I., Van Leuven, I., Van Gorp, L. & Lowenthal, A. (1987): Serum guanidino compound levels and the influence of a single hemodialysis in uremic patients undergoing maintenance hemodialysis. *Nephron.* **45,** 291–295.

10. De Deyn, P.P., Marescau, B., Swartz, R.D., Hogaerth, R., Possemiers, I. & Lowenthal, A. (1990): Serum guanidino compound levels and clearances in uremic patients treated with continuous ambulatory peritoneal dialysis. *Nephron.* **54,** 307–312.

11. De Deyn, P.P., Saxena, V.A., Abts, H., Borggreve, F., Marescau, B. & Crols, R. (1992): Clinical and pathophysiological aspects of neurological complications in renal failure. *Acta Neurol. Belg.* **92,** 191–206.

12. D'Hooge, R., Manil, J., Colin, F. & De Deyn, P.P. (1991): Guanidinosuccinic acid inhibits excitatory synaptic transmission in Ca1 region of rat hippocampal slices. *Ann. Neurol.* **30,** 622–623.

13. D'Hooge, R., Pei, Y.-Q., Marescau, B. & De Deyn, P.P. (1992): Convulsive action and toxicity of uremic guanidino compounds: behavioral assessment and relation to brain concentration in adult mice. *J. Neurol. Sci.* **112,** 96–105.

14. D'Hooge, R., Pei, Y.-Q., Manil, J. & De Deyn, P.P. (1992): The uremic guanidino compound guanidinosuccinic acid induces behavioral convulsions and concomitant epileptiform electrocorticographic discharges in mice. *Brain Res.* **598,** 316–320.

15. Dobbelstein, H., Edel, H.H., Schmidt, M., Schubert, G. & Weinzierl, M. (1971): Guanidinbernsteinsäure und Urämie. I. Mitteilung: Klinische Untersuchungen. *Klin. Wochenschr.* **49,** 348–357.

16. Giovannetti, S., Balestri, P.L. & Barsotti, G. (1973): Methylguanidine in uremia. *Arch. Intern. Med.* **131,** 709–717.

17. Giovannetti, S., Cioni, L., Balestri, P.L. & Biaginin, M. (1968): Evidence that guanidines and some related compounds cause haemolysis in chronic uraemia. *Clin. Sci.* **34,** 141–148.

18. Giovannetti, S., Biagini, M., Balestri, P.L., Navalesi, R., Giagnoni, P., De Matteis, A., Ferro-Milone, P. & Perfetti, C. (1969): Uremia-like syndrome in dogs chronically intoxicated with methylguanidine and creatinine. *Clin. Sci.* **36,** 445–452.

19. Gutmann, I. & Bergmeyer, H.U. (1974): Urea. In: *Methods of Enzymatic Analysis.* Ed. H.U. Bergmeyer. New York: Academic Press.

20. Heiss, W.B., Pawlik, G., Herholz, K., Wagner, R., Göldner, H. & Wienhard, K. (1984): Regional kinetic constants and cerebral metabolic rate for glucose in normal human volunteers determined by dynamic positron emission tomography of [$^{18}$F]-2-fluoro-2-deoxy-D-glucose. *J. Cereb. Blood Flow Metab.* **4,** 212–223.

21. Hiraga, Y. & Kinoshita, T. (1981): Post-column derivatization of guanidino compounds in high-performance liquid chromatography using ninhydrin. *J. Chromatogr.* **226,** 43–51.

22. Hiramatsu, M., Ohba, S., Edamatsu, R., Kadowaki, D. & Mori, A. (1992): Effect of guanidino compounds on membrane fluidity of rat synaptosomes. In: *Guanidino Compounds in Biology and Medicine.* Eds. P.P. De Deyn, B. Marescau, V. Stalon & I.A. Qureshi, pp. 387–393. London: John Libbey.

23. Horowitz, H.I., Cohen, B.D., Martinez, P. & Papayoanou, M.F. (1967): Defective ADP-induced platelet factor 3 activation in uremia. *Blood* **30,** 331–340.

24. Jinnai, D., Sawai, A. & Mori, A. (1966): Guanidinobutyric acid as a convulsive substance. *Nature* **212,** 617.

25. Jinnai, D., Mori, A., Mukawa, J., Ohkusu, H., Hosotani, M., Mizuno, A. & Tye, L. (1969): Biochemical and physiological studies on guanidino compounds induced convulsions. *Jap. J. Brain Physiol.* **106,** 3668–3673.

26. Lonergan, E.T., Semar, M., Terzel, R.B., Treser, G., Needle, M.A., Voyles, L. & Lange, K. (1971): Erythrocyte transketolase activity in dialyzed patients: a reversible metabolic lesion of uremia. *N. Engl. J. Med.* **284,** 1399–1403.

27. Marescau, B., De Deyn, P.P., Wiechert, P., Van Gorp, L. & Lowenthal, A. (1986): Comparative study of guanidino compounds in serum and brain of mouse, rat, rabbit, and man. *J. Neurochem.* **46,** 717–720.

28. Marescau, B., De Deyn, P.P., Qureshi, I.A., De Broe, M.E., Antonozzi, I., Cederbaum, S.D., Cerone, R., Chamoles, N., Gatti, R., Kang, S.-S., Lambert, M., Possemiers, I., Snyderman, S.E. & Yoshino, M. (1992): The pathobiochemistry of uremia and hyperargininemia further demonstrates a metabolic relationship between urea and guanidinosuccinic acid. *Metabolism* **41,** 1021–1024.

29. Marescau, B., Deshmukh, D.R., Kockx, M., Possemiers, I., Qureshi, I.A., Wiechert, P. & De Deyn, P.P. (1992): Guanidino compounds in serum, urine, liver, kidney and brain of man and some ureotelic animals. *Metabolism* **41,** 526–532.

30.  Matsumoto, M. & Mori, A. (1976): Convulsive activity of methylguanidine in cat and rabbits. *IRCS Med. Sci.* **4,** 65.

31.  Meister, A. (1954): The α-keto analogs of arginine, ornithine and lysine. *J. Biol. Chem.* **206,** 577–583.

32.  Minkoff, L., Gaertner, G., Darab, M., Mercier, C. & Levin, M.L. (1972): Inhibition of brain sodium-potassium ATPase in uremic rats. *J. Lab. Clin. Med.* **80,** 71–78.

33.  Mori, A. (1983): Guanidino compounds and neurological disorders. *Neurosciences* **9,** 149–157.

34.  Mori, A. (1987): Biochemistry and neurotoxicology of guanidino compounds. History and recent advances. *Pavlov. J. Biol. Sci.* **22,** 85–94.

35.  Niklasson, F. & Agren, H. (1984): Brain energy metabolism and blood–brain barrier permeability in depressive patients: analysis of creatine, creatinine, urate, and albumin in CSF and blood. *Biol. Psychiatry* **19,** 1183–1206.

36.  Pisano, J.J., Abraham, D. & Udenfriend, S. (1963): Biosynthesis and disposition of guanidinobutyric acid in mammalian tissues. *Arch. Biochem. Biophys.* **100,** 323–329.

37.  Reynolds, I.J. & Rothermund, K. (1992): Multiple modes of NMDA receptor regulation by guanidines. In: *Guanidino Compounds in Biology and Medicine.* Eds. P.P. De Deyn, B. Marescau, V. Stalon & I.A. Qureshi, pp. 441–448. London: John Libbey.

38.  Awynok, J. & Dawborn, J. (1975): Plasma concentration and urinary excretion of guanidine derivatives in normal subjects and patients with renal failure. *Clin. Exp. Pharmacol.* **2,** 1–15.

39.  Shainkin, R., Giatt, Y. & Berlyne, G.M. (1975): The presence and the toxicity of guanidinopropionic acid in uremia. *Kidney Int.* **7,** S302–S305.

40.  Stein, I.M., Cohen, B.D. & Horowitz, H.I. (1968): Guanidinosuccinic acid: the 'X' factor in uremic bleeding? *Clin. Res.* **16,** 397.

41.  Stein, I.M. & Cohen, B.D. (1969): Guanidinosuccinic acid in renal failure, experimental azotemia and inborn errors of the urea cycle. *N. Engl. J. Med.* **280,** 926–930.

42.  Yamamoto, Y., Saito, A., Manji, T., Nishi, H., Ito, K., Maeda, K., Ohta, K. & Kobayashi, K. (1978): A new automated analytical method for guanidino compounds and their cerebrospinal fluid levels in uremia. *Trans. Am. Soc. Artif. Intern. Organs* **24,** 61–68.

43.  Yokoi, I., Toma, J. & Mori, A. (1984): The effects of homoarginine on the EEG of rats. *Neurochem. Pathol.* **2,** 295–300.

*Guanidino Compounds : 2*, eds. by P.P. De Deyn, B. Marescau, I.A. Qureshi and A. Mori.
©1997 John Libbey & Company Ltd., pp. 289–297.

# Chapter 29

# Guanidino compound metabolism in uremic rats: the guanidine cycle

Olivier LEVILLAIN, Bart MARESCAU[1] and Peter P. DE DEYN[1]

*Collège de France, Laboratoire de Physiologie Cellulaire, 11 place Marcelin Berthelot, 75231 Paris Cedex 05, France; [1]Department of Medicine UIA, Laboratory of Neurochemistry and Behavior-BBS, Universiteitsplein 1, B-2610 Antwerpen (Wilrijk), Belgium*

## Summary

Several guanidino compounds (GCs) synthesized from arginine (Arg) are formed in various tissues including liver, kidney and brain. In mammalian kidney, arginine synthesis takes place in the cortex, mainly in the proximal convoluted tubule. In renal failure, the functional renal mass is reduced. Consequently, synthesis of both might be disturbed. We studied GC metabolism in rats subjected to nephrectomy by ligating branches of the renal arteries. Three weeks later, blood and urine were collected to measure GC.

The results show the following. (1) In rats with 42 per cent nephrectomy (NX), plasma β-guanidinopropionic acid (β-GPA) was increased 5-fold and levels of the other GCs were higher than in controls but not significantly different. Urinary excretion of guanidinoacetic acid (GAA) decreased by 32 per cent whereas urinary excretion of homoarginine was increased two- fold. (2) In rats with 80 per cent nephrectomy, the plasma concentration of all GCs was increased, but creatine was lowered. Urinary excretion of GCs was reduced by 2.5-fold in spite of the higher excretion of guanidinosuccinic acid (GSA), methylguanidine (MG) and guanidine (G).

We conclude that: (1) GC metabolism alterations start in moderate uremia because plasma GC levels are increased (however,) not significantly; (2) in severe uremia, GC metabolism was strongly disturbed. GAA and creatine synthesis were sharply decreased in spite of the renal hypertrophy. Synthesis of GSA, G and MG was strongly increased. These compounds may be related to the pathophysiology as discussed elsewhere in this volume.

## Introduction

Among more than one hundred natural guanidino compounds that have been identified in plant and animal tissues[20], we focused our attention on those derived from arginine and which might be synthesized in the 'guanidine cycle'[16,17]: guanidinosuccinic acid (GSA), guanidine (G), guanidinoacetic acid (GAA), creatinine (CTN), creatine (CT) and methylguanidine (MG) whose precursor is creatinine[13,14].

It has been reported that mammalian kidneys synthesize arginine from the circulating citrulline in order to supply the needs of the body[2,3,19]. Recently, arginine synthesis has been demonstrated to be restricted to the proximal tubule, mainly in the proximal convoluted tubule[8,9]. This nephron segment is also involved in the synthesis of guanidinoacetic acid, the precursors of which are arginine and glycine[12,23,24]. Methylation of GAA occurs in the liver and leads to synthesis of creatine, the precursor of creatinine.

The kidney is indirectly involved in the synthesis of several guanidino compounds because it produces large amounts of arginine. Consequently, renal pathologies, nephrotoxic drugs and nephrec-

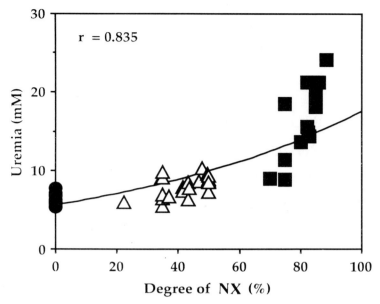

*Fig. 1. Uremia as a function of degree of nephrectomy (NX).*
*Results are expressed in mmol/l;/45 rats were used://10 controls (•), 21 with*
42 per cent NX (open triangles) and 14 with 80 per cent NX (closed squares).
The heavy line represents the correlation.

tomy might disturb the metabolism of several guanidino compounds by reducing the functional mass of the kidney[15,18]. Indeed, GSA, G and MG are increased in uremic animals and patients[5,10,18,22]. However, until now, changes in guanidino compound concentration in both plasma and urine have never been correlated with the degree of renal damage.

The present work was designed to study the metabolism of guanidino compounds in rats subjected to 20–90 per cent nephrectomy and allowed to develop chronic renal failure for 3 weeks. We were interested (1) to establish correlations between either plasma levels or urinary excretion and the degree of renal failure, and (2) to investigate the renal balance of each guanidino compound.

## Materials and methods

### Experimental procedures

Male Sprague-Dawley rats (Iffa Credo, France) weighing 200–220 g were used. Rats were divided at random into three groups before operation.

Rats were anesthetized with intraperitoneal pentobarbital (6 per cent, 0.1 ml/100 g body weight). Control rats (n = 10 ) were sham operated . In a second group rats (21; moderate NX), 30 –50 per cent of the functional renal mass was destroyed by ligating branches of the right renal artery (< 50 per cent NX) or removing the right kidney (50 per cent NX). In a third group rats (14; severe NX), 70–90 per cent of the renal mass was destroyed by removing the right kidney and ligating some branches of the left renal artery.

The rats were kept in individual cages for 3 weeks and had free access to tap water and standard chow (M25 Extralabo, France). Three days before the end of the experiment, 7, 16 and 9 rats, respectively, were kept for an adaptation period in individual metabolic cages for urine collection; they received food and water. Urine was collected for a period of 24 h before the animals were killed.

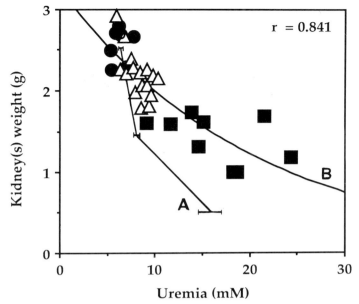

*Fig. 2. Kidney weight as a function of uremia. Line A corresponds to the theoretical kidney weight calculated from the value measured in the control group, and curve B corresponds to the experimental values. Each symbol represents an individual value. corresponding with 0, 42 or 82 per cent nephrectomy, respectively.*

Before being killed, the rats were anesthetized with 0.1 ml pentobarbital per 100 g body weight. Samples of arterial blood were taken from the abdominal aorta and the remaining kidney(s) were removed, decapsuled and weighed. Blood samples were centrifuged at 4000g for 20 min at 4 °C. A fraction of the plasma was frozen at ⁻50 °C until determination of guanidino compound concentration. Uremia and creatininemia were measured on the remaining plasma with an Astra 8 Beckman analyzer.

Urines were centrifuged at 4000g for 20 min at 4 °C, then urine volume was measured. Samples of urine were frozen at –50 °C until determination of the guanidino compounds. Urea was measured on a Technicon SMA 6 and creatinine was determined by a modified Jaffé method.

### Determination of guanidino compounds

Guanidino compounds were determined on deproteinized plasma and urine. Equal volumes of plasma and urine were mixed with a 200 g/l trichloroacetic acid solution. The samples were centrifuged (8000 g) for 5 min. The clear supernatant was diluted with 0.02 N HCl and used for guanidino compound analysis in urine; 200 μl supernatant was used for the determinations in plasma. Guanidino compound concentration was determined with a Biotronic LC 5001 (Biotronik, Maintal, Germany) amino acid analyzer adapted for GC determination of guanidino compounds[11].

### Calculations

The degree of necrosis was deduced *de visu* during the surgical procedure by estimating the damage following ligature of the branches of the renal artery[7,10]. Removal of one kidney corresponds to a value of 50 per cent NX[7,10].

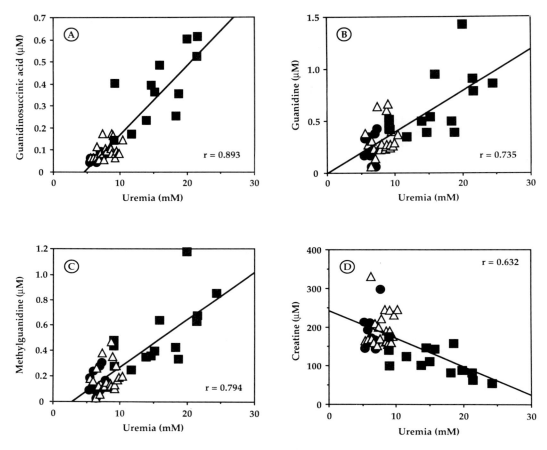

*Fig. 3. Concentration of plasma guanidino compounds as a function of uremia. n = 45 rats were subdivided into 10 controls (closed circles), 21 with 42 per cent NX (open triangles) and 14 with 80 per cent NX (closed squares). Heavy lines are lines of regression and r is the coefficient of correlation; in each case, P < 0.01.*

The renal balance, $\underline{B}$, corresponds to the fraction of the guanidino compound (GC) reabsorbed: the difference between the renal entry (amounts filtered) and the renal exit (amounts excreted) compared with the amount filtered. $\underline{B}$ was determined for each guanidino compound by the following equation:

$$B = \frac{[(GFR \times [GC]_{plasma}) - (V_{urine} \times [GC]_{urine})]}{(GFR \times [GC]_{plasma})}$$

where GFR is the glomerular filtration rate (ml/day) determined by creatinine clearance, [GC] the concentration of one guanidino compound in the plasma or urine, and $V$ the volume of urine collected during 24 h. The $\underline{B}$ value indicates the renal movement of each guanidino compound (either a reabsorption, when $B$ is positive, or an excretion, when $B$ is negative).

**Statistical analysis**

The data were compared by the analysis of variance (ANOVA, statview II SE), and the Fisher test was used to determine the $P$ value at the 95 and 99 per cent levels of significance. For the correlation

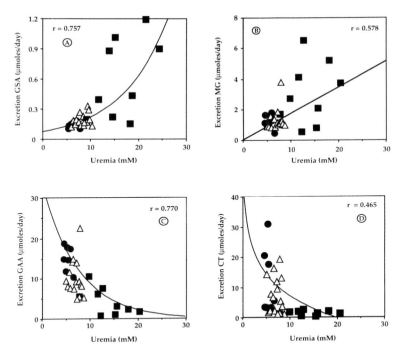

*Fig. 4. Urinary excretion of guanidino compounds as a function of uremia. For each rat, urine was collected for 24 h and results are expressed in μmol excreted per day. The 32 rats were subdivided into seven controls (closed circles), 16 with 42 per cent NX (open triangles) and nine with 80 per cent NX (closed squares). Heavy lines are lines of regression and r is the coefficient of correlation; in each case, P < 0.01.*

analysis, the coefficient r was calculated and the *P* value was determined from the table of correlation at the level 95 and 99 per cent significance.

## Results

In the group of rats with 'moderate' and 'severe' NX, the average percentage (mean ± SE) of renal tissue destroyed was $41.8 \pm 1.7$ and $80.3 \pm 1.4$, respectively.

In rats with 42 per cent nephrectomy, uremia was not statistically different from that of control rats. Figure 1 depicts the exponential increase of uremia in rats with 80 per cent NX ($r = 0.835, P < 0.01$). There is a correlation between changes in GC concentration and uremia.

Figure 2 shows the decrease of the absolute kidney weight correlated with uremia ($r = 0.841, P < 0.01$). The difference between the two curves (A theoretical and B experimental) corresponds to renal compensatory hypertrophy. Indeed, the mean kidney weight of rats with 80 per cent NX was only 42 per cent lower than that of the controls[10].

Urine volume was increased by 38 per cent and 120 per cent in rats with 42 per cent and 80 per cent NX, respectively, and urine osmolality was decreased in rats with moderate (−13 per cent) and severe (−48 per cent) renal damage. These results indicate a net reduction in capacity to concentrate urine[10].

Fig. 5. The guanidine cycle. In uremic rats, the guanidine cycle leads to significant production of GSA and G; however, presumably no GAA, CTN and CT is produced.

**Guanidino compounds in plasma**

Values for the plasma concentration of 12 guanidino compounds in normal and uremic rats have been published in detail recently[10]. Removal of 42 per cent of the renal tissue did not significantly change the total concentration of guanidino compounds (controls $429 \pm 20$ μM vs moderate NX $450 \pm 10$ μM). In contrast, in rats with 80 per cent NX, this concentration decreased to $406 \pm 11$ μM ($P < 0.05$ vs moderate NX) because of the sharp decrease in creatinemia. When the creatine value was excluded from our calculation, the plasma concentration of GC was increased in uremic rats ($240 \pm 10$ μM, $253 \pm 7$ μM, $293 \pm 9$ μM, in controls, moderate NX and severe NX, respectively, $P < 0.01$ in the group of rats with severe NX).

In rats with 42 per cent NX, β-guanidinopropionic acid (β-GPA), which was undetectable in control rats, was significantly increased and remained at the same final level in rats with severe nephrectomy ($0.061 \pm 0.002$ μM, $P < 0.01$).

In rats with severe renal damage (80 per cent NX), we focused our attention on GSA, G, MG and CT metabolism; for the other compounds see Levillain et al.[10]. GSA and urea concentrations increased in parallel as illustrated by the linear correlation as depicted in Fig. 3A ($r = 0.893$, $P < 0.01$). This suggests that the metabolism of urea and GSA might be linked. The same observations and conclusions could be proposed for guanidine (Fig. 3B, $r = 0.735$, $P < 0.01$) and methylguanidine (Fig. 3C, $r = 0.794$, $P < 0.01$), which are also positively correlated with uremia. In contrast, Fig. 3D shows a correlation between the parallel decrease of creatinemia and of uremia ($r = 0.632$, $P < 0.01$).

**Guanidino compounds in urine**

In rats with 42 per cent NX, urinary excretion of GC did not differ from that of controls except for GAA, which was lower ($12.53 \pm 1.33$ vs $18.40 \pm 1.48$ μmol/day, respectively, $P < 0.01$), and both ArgA and HArg, which were significantly increased ($0.092 \pm 0.005$ vs $0.071 \pm 0.005$, $P < 0.05$ and $0.086 \pm 0.009$ vs $0.041 \pm 0.007$, $P < 0.01$), respectively.

In rats with 80 per cent NX, excretion of several guanidino compounds such as GAA and CT, was decreased whereas the amounts of GSA, HArg, G and MG excreted per day were increased but not in the same manner (for more details see Levillain *et al.*[10]). Figure 4A shows that GSA excretion increased exponentially in rats with severe renal damage ($r = 0.757, P < 0.01$), suggesting an important production of GSA. Methylguanidine concentrations linearly correlated with urea levels (Fig. 4B, $r = 0.578, P < 0.01$). Figures 4C and 4D depict the exponential decrease of the daily urinary excretion of GAA and CT as a function of uremia. In both cases, the correlation was statistically significant (GAA: $r = 0.770, P < 0.01$ and CT: $r = 0.465, P < 0.01$).

### Renal balance of guanidino compounds

In rats with severe renal damage, the daily amounts of guanidino compounds filtered by the kidney were approximately half of those in both controls and rats with 42 per cent NX. In rats with 80 per cent NX, GAA, CT and HArg were less filtered whereas GSA and MG were 2.6-fold more filtered (for more details see Levillain *et al.*[10]).

We calculated that, in control rats, GSA, CT and HArg were reabsorbed and CTN was excreted. About 16 per cent of the filtered GSA was reabsorbed in control rats. In uremic rats, GSA and CT reabsorption remained unchanged. In contrast, reabsorption of HArg was proportionally decreased to the degree of renal failure[10].

We also found that the amount of some compounds (i.e. GAA, G and MG) excreted in urine of control rats was higher than the amount filtered, suggesting a net addition in the tubular lumen. After 80 per cent nephrectomy, excretion of GAA, α-keto- δ-guanidinovaleric acid and γ-guanidinobutyric acid was decreased, suggesting either a net reabsorption or a decrease in synthesis. Excretion of G and MG remained unchanged in uremic animals.

### Discussion

Only some guanidino compounds had significantly modified levels in plasma and urine of rats with moderate NX, indicating that only some metabolic pathways were disturbed. We propose that the renal hypertrophy partially compensates for changes in GC metabolism and thus masks the real effects of moderate nephrectomy. This might explain why the modifications in plasma and urine were not statistically significant in this group. In contrast, the composition of plasma and urine from rats subjected to severe renal damage (80 per cent NX) exhibited important changes, suggesting that the metabolism of nearly all guanidino compounds was strongly affected. We conclude that, under our experimental conditions, the increase of guanidino compounds in plasma was not proportional to the degree of nephrectomy.

The kidney is involved in arginine synthesis, the precursor of several guanidino compounds, thus a reduction of the renal mass (42 per cent) might alter both Arg and GC production. Thus, guanidino compounds could be potent markers of the onset of renal failure. Indeed, the plasma concentration of β-GPA was significantly increased in uremic rats[10] and patients[21]; unfortunately, in rats, β-GPA is not specific for uremic states because the same increase is observed in fasted rats [12]. Recently, it has been proposed that β-GPA could be an analog of creatine[4]. Fitch *et al.*[4] reported that the phosphorylated form of β-GPA (β-GPAP, an analog of creatine phosphate) could be used by the creatine kinase to produce energy for the muscles.

The plasma concentration of GSA and its excretion were increased in rats with severe renal failure, indicating that GSA biosynthesis was enhanced. Our data are consistent with those of Cohen et *al.*[1] and Perez *et al.*[18] who discovered that the liver of uremic rats produces more GSA than that of controls. It has been proposed that, in the liver, the urea cycle could be linked to the guanidine cycle (Fig. 5)[16,17]. Urea produced by hepatic arginase could be recycled under special physiologic and pathophysiologic conditions to produce GSA, G, GAA (the precursor of CTN) and CT to produce creatine phosphate, an essential energetic compound for muscles. Our results seem to be consistent with this (new) pathway, as GSA and G synthesis were increased in uremic rats. From our calculations (Fig. 3A and

3B), we observe that the coefficient of correlation between GSA and urea ($r = 0.893$) is higher than that found for G and urea ($r = 0.735$), suggesting that the preferential pathway of the guanidine cycle is to produce GSA (Fig 5.). No significant correlation could be obtained between GAA and urea ($r = 0.049$) suggesting that, in uremic rats, the guanidine cycle does not produce GAA to compensate for the very low renal production of GAA (Fig. 4C). Indeed, our data indicate that, in spite of the renal compensatory hypertrophy, GAA synthesis was very low in rats with either moderate or severe renal failure. Consequently, in the proximal convoluted tubule, Arg and GAA synthesis are strongly disturbed; the total activity of the renal glycine amidino-transferase is lower in uremic rats than in control rats[6]. The negative renal balance of GAA in control rats indicates that GAA is excreted, but in uremic rats the balance becomes positive indicating a renal reabsorption and/or a decrease in GAA synthesis[10]. Because GAA production is very low, the kidney reabsorbs GAA, probably for creatine synthesis. Indeed, the low production of GAA is confirmed by the sharp decrease of CT in both the plasma and the urine of rats.

Methylguanidine is proposed to be a toxic compound that accumulates in body fluids mainly in the intracellular compartment, where it exerts deleterious effects. The higher production of MG in uremic rats or patients is compatible with the fact that creatinine is the precursor of MG[13]. In the presence of superoxide radicals, hydrogen peroxide and hydroxyl radicals, MG synthesis is enhanced, whereas scavengers such as glutathione reduce this process[14].

In conclusion, we demonstrate that there was a good correlation between uremia and guanidino compounds (GSA and G) belonging to the guanidine cycle.

### Acknowledgements

Thanks are due to the UIA, NFWO (grants 3.0044.92 and 3.0064.93), the Born-Bunge Foundation, the United Fund of Belgium and the OCMW Medical Research Foundation for support. Dr O. Levillain received a scholarship from the Fondation de la Recherche Médicale.

## References

1.    Cohen, B.D., Stein, I.M. & Bonas, J.E. (1968): Guanidinosuccinic aciduria in uremia: a possible alternate pathway for urea synthesis. *Am. J. Med.* **45,** 63–68.

2.    Dhanakoti, S.N., Brosnan, J.T., Herzberg, G.R. & Brosnan, M.E. (1990): Renal arginine synthesis: studies *in vitro* and *in vivo. Am. J. Physiol.* **259** (*Endocrinol. Metab.* **22**), E437–E442.

3.    Featherston, W.R., Rogers, Q.R. & Freedland, R.A. (1973): Relative importance of kidney and liver in synthesis of arginine by the rat. *Am. J. Physiol.* **224,** 127–129.

4.    Fitch, C.D., Jellinek, M., Fitts, R.H., Baldwin, K.M. & Holloskzy, J.O. (1975): Phosphorylated β-guanidinopropionate as a substitute for phosphocreatine in rat muscle. *Am. J. Physiol.* **228,** 1123–1125.

5.    Giovannetti, S., Balestri, P.L. & Barsotti, G. (1973): Methylguanidine in uremia. *Arch. Intern. Med.* **131,** 709–713.

6.    Inouchi, M., Fujino, T., Sato, T., Yasuda, T., Tomita, H., Kanazawa, T., Shiba, C., Ozawa, S., Ohwada, S. & Ishida, M. (1989): Disturbance of creatine metabolism in rats with chronic renal failure. In: *Guanidines 2. Further explorations of the Biological and Clinical Significance of Guanidino Compounds*, Eds. A. Mori, B.D. Cohen & H. Koide, pp. 277–288. New York: Plenum Press.

7.    Kaufman, J.M., Dimeola, H.J., Siegel, N.J., Lytton, B., Kashgarian, M. & Hayslett, J.P. (1974): Compensatory adaptation of structure and function following progressive renal ablation. *Kidney Int.* **6,** 10–17.

8.    Levillain, O., Hus-Citharel, A., Morel, F. & Bankir, L. (1990): Localization of arginine synthesis along rat nephron. *Am. J. Physiol.* **259** (*Renal Fluid Electrolytes Physiol.* **28**), F916–F923.

9.    Levillain, O., Hus-Citharel, A., Morel, F. & Bankir, L. (1993): Arginine synthesis in mouse and rabbit nephron: localization and functional significance. *Am. J. Physiol.* **264** (*Renal Fluid Electrolytes Physiol.* **32**), F1038–F1045.

10.   Levillain, O., Marescau, B. & De Deyn, P.P. (1995): Guanidino compound metabolism in rats with induced 20 per cent to 90 per cent nephrectomy. *Kidney Int.* **47** (464-472).

11. Marescau, B., Deshmukh, D.R., Kockx, M., Possemiers, I., Qureshi, I.A., Wiechert, P. & De Deyn, P.P. (1992): Guanidino compounds in serum, urine, liver, kidney, and brain of man and some ureotelic animals. *Metabolism* **41,** 526–532.

12. McGuire, D.M., Gross, M.D., Elde, R.P. & Van Pilsum, J.F. (1986): Localization of l-arginine-glycine amidinotransferase protein in rat tissues by immunofluorescence microscopy. *Histochem. Cytochem.* **34,** 429–435.

13. Nagase, S., Aoyagi, K., Narita, M. & Tojo, S. (1985): Biosynthesis of methylguanidine in isolated rat hepatocytes and *in vivo*. *Nephron* **40,** 470–475.

14. Nagase, S., Aoyagi, K., Narita, M. & Tojo, S. (1986): Active oxygen in methylguanidine synthesis. *Nephron* **44,** 299–303.

15. Nakayama, S., Junen, M., Kiyatake, I. & Koide, H. (1989): Urinary guanidinoacetic acid excretion as an indicator of gentamicin nephrotoxicity in rats. In: *Guanidines 2. Further Explorations of the Biological and Clinical Significance of Guanidino Compounds*. Eds. A. Mori, B.D. Cohen & H. Koide, pp. 303–311. New York: Plenum Press.

16. Natelson, S. (1984): Metabolic relationship between urea and guanidino compounds as studied by automated fluorimetry of guanidino compounds in urine. *Clin. Chem.* **30,** 252–258.

17. Natelson, S. & Sherwin, J.E. (1975): Proposed mechanism for urea nitrogen re-utilization: relationship between urea and proposed guanidine cycles. *Clin. Chem.* **25,** 1343–1344.

18. Perez, G.O., Rietberg, B., Owens, B. & Schiff, E.R. (1977): Effect of acute uremia on arginine metabolism and urea and guanidino acid production by perfused rat liver. *Pflügers Arch.* **372,** 275–278.

19. Ratner, S. & Petrack, B. (1953): The mechanism of arginine synthesis from citrulline in kidney. *J. Biol. Chem.* **200,** 175–185.

20. Robin, Y. & Marescau, B. (1985): Natural guanidino compounds. In: *Guanidines*. Eds. A. Mori, B.D. Cohen & A. Lowenthal, pp. 383–438. New York: Plenum Press.

21. Shainkin, R., Giatt, Y. & Berlyne, G.M. (1975): The presence and toxicity of guanidinopropionic acid. *Kidney Int.* **7,** S302–S305.

22. Stein, I.M., Perez, G., Jonhson, R. & Cummings, N.B. (1971): Serum levels and urinary excretion of methylguanidine in chronic renal failure. *J. Lab. Clin. Med.* **77,** 1020–1024.

23. Takeda, M., Kiyatake, I., Koide, H., Jung, K.J. & Endou, H. (1992): Biosynthesis of guanidinoacetic acid in isolated renal tubules. *Eur. J. Clin. Chem. Clin. Biochem.* **30,** 325–331.

24. Takeda, M., Koide, H., Jung, K.J. & Endou, H. (1992): Intranephron distribution of glycine-amidinotransferase activity in rats. *Renal Physiol. Biochem.* **15,** 113–118.

*Guanidino Compounds : 2*, eds. by P.P. De Deyn, B. Marescau, I.A. Qureshi and A. Mori.
©1997 John Libbey & Company Ltd., pp. 299–302.

# Chapter 30

# ¹H–NMR study of cerebrum in chronic hemodialysis patients

G. OGIMOTO, S. OZAWA, T. SATOH, T. MAEBA, S. OWADA, M. ISHIDA,
K. IMAMURA,[*] and T. ISHIKAWA.[*]

*The first department of Internal Medicine , St. Mariannna University School of Medicine; [*]Department of
Radiology, Kawasaki, Japan*

## Summary

In patients with chronic renal failure, nervous system dysfunction is one of the well-known complications. Its pathogenisis, however, has not been clarified yet. Magnetic Resonace Spectroscopy (MRS) is an excellent technique for determination of chemical compounds *in vivo*. Proton MRS (¹H–MRS) shows prominent resonances from N-acetyl aspartate (NAA), chlorine-containing compouns (Chol) and phosphocreatine + creatine (Cr). The observed decrease of NAA levels in the brain may reflect the reduction in the number of normally functioning neurons. We therefore tried to detect the chemical compounds of the brain by ¹H–MRS in chronic hemodialysis (HD) patients. Six HD patients (2 males, 4 females) and 4 normal controls (2 males, 2 females) were examined using NMR apparatus for the ¹H-MRS. Occipital lobe of the brain was chosen for the examination that use 1.5–T whole body MR imager (Philips, Gyroscan S–15), operating at 63.9 MHz for ¹H–MRS. The ratio of NAA/Cr, NAA/Chol, Chol/Cr in the brain were calculated from spectra. The results were as follows: (1) The ratios of NAA/Chol in HD patients was significantly lower than in controls ($1.67 \pm 0.14$ *vs* $2.43 \pm 0.01$; $P < 0.05$). (2) The ratio of NAA/Cr and Chol/Cr did not show differences between HD patients and controls.

In conclusion, we obtained the reduction of NAA/Chol ratio in the brain of HD patients and this alteration of chemical compounds in the brain may be related to the nervous system dysfunction in dialysis patients.

## Introduction

In uremic patients, nervous system dysfunction is one of the well-known complications. Uremic encephalopathy may occur in patient with either acute or chronic renal failure when the glomerular filtration decreases below approximately 10 per cent of normal[4]. Available evidences suggest the pathogenesis of uremic encephalopathy as follows; (1) decrease in brain metabolic rate and cerebral oxygen consumption, (2) metabolic acidosis, (3) uremic toxins, (4) excess of parathyroid hormone, (5) disturbances of intra-cerebral amino acid metabolism. In spite of these many contributions, the pathogenesis of uremic encephalopathy is not clarified yet.

Recently, we can utilize the nuclear magnetic resonance spectroscopy (NMR) for the biochemical analysis in vivo. Proton NMR (¹H–NMR) is found to detect N-acetyl aspartate (NAA), cholinecontaining compounds (Chol), and creatine + phosphocreatine (Cr). NAA is the substance that may represent the neuronal activity[7]. The present study, therefore, was designed to determine the neuronal activity in uremia by measuring NAA, Chol and Cr from cerebrum using ¹H–NMR.

## Patients and methods

Six patients, two males and four female, undergoing hemodialysis were studied. All patients had received regular hemodialysis with bicarbonate dialysate three times a week, for a period of 4 h each time. The mean age was 47 years and the mean hemodialysis duration was 79 months. The cause of renal failure at the end stage was chronic glomerulonephritis in all six patients. Control subjects were four non-uremic healthy adults, two male and two female, mean age 43 years.

For $^1$H–NMR determination, a Philips Gyroscan S15 NMR apparatus was used with a superconducting magnet at a field of 1.5 T, giving an operating frequency of 63.9 MHz. A volume of interest (VOI) of $30 \times 40 \times 30$ mm was set up in the occipital lobe as shown in Fig. 1. A double spin-echo (PRESS) sequence was used for VOI localization. For water suppression, a zero crossing inversion-recovery technique comprising adiabatic inversion pulse was used. The repetition time was 2000 msec, echo time 136 msec, sample frequency 1000 Hz, number of sample points 512, and number of measurements 128 or 256. The obtained measurements were averaged and processed by Fourier transformation. A straight baseline was drawn, the peak areas of NAA, Chol and Cr were computed, and the ratios of NAA to Chol, NAA to Cr and Chol to Cr were calculated from the peak areas.

In addition, the serum creatinine, urea nitrogen (BUN), inorganic phosphate (Pi), calcium (Ca), β-2 microglobulin (BMG), intact PTH (i-PTH) aluminum (Al) were measured as biochemical parameters by conventional laboratory methods.

The statistical significance of differences in the data evaluated by the non-paired Wilcoxon's rank was sum test, and significance of correlation by Spearman's test.

## Results

The biochemical data are shown in Table 1. Serum creatine, BUN, Pi, BMG and i-PTH were higher than normal in all patients. Aliminium was higher than normal in all but one patient.

A typical proton spectrum is shown in Fig. 2. The spectrum consists of three major peaks: from the left, Chol, Cr and NAA.

### Table 1. Serum results in uremic patients

|  | Creatinine | BUN | Pi | Ca | BMG | i–PTH | Al |
|---|---|---|---|---|---|---|---|
|  |  |  | (mg/dl) |  | (µg/ml) | (pg/ml) | (µg/ml) |
| Mean | 12.9 | 69.2 | 4.8 | 9.8 | 32.5 | 405.2 | 3,6 |
| SD | 1.2 | 10.8 | 0.7 | 0.3 | 7.5 | 561.0 | 2.0 |
| Normal value | 0.7–1.3 | 7–20 | 2.7–4.3 | 8.9–10.1 | 0.8–19 | 230–560 | below 0.9 |

Table 2 shows the ratios of compounds in the brain of uremic patients and control subjects determined by NMR. The mean ratio of NAA to Chol was significantly lower in uremic patients than in control subjects ($1.67 \pm 0.14$ vs $2.43 \pm 0.01$; $P < 0.05$). In contrast, no significant difference was observed in another parameters, NAA to Cr or Chol to Cr.

### Table 2. Cerebral results in uremic patients and control subjects

|  | n | NAA/Chol | NAA/ Cr | Chol/Cr |
|---|---|---|---|---|
| Patients | 6 | $1.67 \pm 0.14$ | $1.81 \pm 0.40$ | $0.83 \pm 0.04$ |
| Controls | 4 | $2.43 \pm 0.01$ | $1.85 \pm 0.29$ | $0.69 \pm 0.03$ |

In addition, a significant positive correlation was found between the ratio of NAA to Cr and serum BMG ($P < 0.05$).

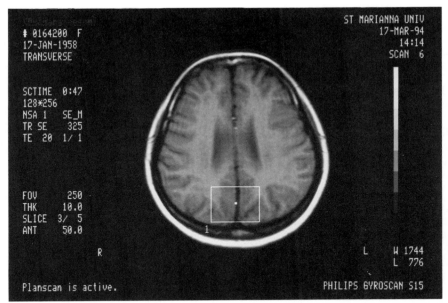

*Fig. 1. Volume of interest in the cerebrum.*

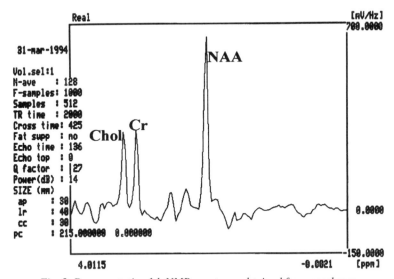

*Fig. 2. Representative 1 h NMR spectrum obtained from cerebrum.*

**Discussion**

*In vivo* NMR spectroscopy is a non-invasive and excellent technique for the determination of metabolic condition in organs. ³¹P-NMR and ¹H-NMR have been used in recent clinical studies. The peak of N-acetylaspartate (NAA) that is recognized on ¹H–NMR spectra is known to be exclusively located in neurons and their branches[1]. Thus, NAA may be a useful neuronal marker. The ratio of

NAA to Cr is also an indicator of neuronal activity[2,3]. The reduction of NAA in Alzheimer's disease may relate to the neuronal loss[2].

Similarly, a reduction in the ratio of NAA to Chol and of NAA to Cr have been reported in patients with HIV infection, suggesting neuronal loss in this disease[3].

Uremic encephalopathy is one of the well known nervous system complications in chronic renal failure. Its pathogenesis, however, is not clear. Clinically, uremic encephalopathy may be improved by dialysis. Abnormal findings on electroencephalography and dysfunction of the autonomic nervous system, also associated with end-stage renal failure, are improved only by renal transplantation (unpublished observation and ref. [5]). These facts may reveal the presence of neuronal function disorders in chronic renal failure. However, neuronal functional disorders cannot be detected by x-ray computed tomography or magnetic resonance imaging – the conventional radiographic approaches. Therefore, for the purpose of functional examination of the cerebrum in patients with chronic renal failure, we used magnetic resonance spectroscopy.

No statistical differences in the ratio of NAA to Cr were found between patients and controls, but a significant decrease was found in the ratio of NAA to Chol in patients in this study. Although statistically not significant, the ratio of Chol to Cr was increased in patients, being in general accord with the recent observation by Menon *et al.*[6]. Since the Chol peak contains several compounds, the mechanism of increase of the ratio of Chol to Cr is hard to elucidated. These biochemical changes in the cerebrum of patients with chronic renal failure may relate to cerebral cellular dysfunction.

Moreover, we found a significant positive correlation between the ratio of NAA to Cr and serum BMG in our patients. This finding suggests that the long-term hemodialysis may itself affect the ratio of NAA to Cr. Thus, we may have to monitor cerebral function in patients on dialysis.

## References

1. Knaap, M.S., Ground, J., Luyten, P.R., Hollander, J.A., Nauta, J.J.P. & Valk, J. (1992): [1]H and [31]P magnetic resonance spectroscopy of the brain in degenerative cerebral disorders. *Ann. Neurol.* **31,** 202–211.

2. Klunk, W.E., Panchaligam, K., Moossy, J., McClure, R.J. & Pettegrew, J.W. (1992): N-acetyl-l-aspartate and other amino acid metabolites in Alzheimer's disease brain. *Neurology* **42,** 1578–1585.

3. Meyerhoff, D.J., MacKay, S., Bachan, L., Poole, N., Dillon, W.P., Weiner, M.W. & Fein, G. (1993): Reduced brain N-acetylaspartate suggests neuronal loss in cognitively impaired human immunodeficiency virus-seropositive individuals. *Neurology* **43,** 509–515.

4. Fraser, C.L. & Arieff, A.I. (1988): Nervous system complications in uremia. *Ann. Intern. Med.* **109,** 143–153.

5. De Deyn, P.P., Saxena, V.K., Abts, H., Borggreve, F., D'Hooge, R., Marescau, B. & Crols, R. (1992): Clinical and pathological aspects of neurological complications in renal failure. *Acta Neurol. Belg.* **92,** 191–206.

6. Menon, D.K., Sanford, R.N., Cassidy, M.J.D., Bell, J.D., Baudouin, C.J., Sargentoni, J. (1991): Proton magnetic resonance spectroscopy in chronic renal failure. *Lancet* **337,** 244–245.

7. Miller, B.L. (1991): A review of chemical issues in [1]H NMR spectroscopy: N-actyl-L-aspartate, creatine and choline. *NMR Biomed.* **4,** 47–52.

*Guanidino Compounds : 2*, eds. by P.P. De Deyn, B. Marescau, I.A. Qureshi and A. Mori.
©1997 John Libbey & Company Ltd., pp. 303–310.

# Chapter 31

# Biosynthesis of methylguanidine in the hepatic peroxisomes

Katsumi TAKEMURA, Kazumasa AOYAGI*, Sohji NAGASE*, Michihiro GOTOH*,
Chie TOMIDA*, Aki HIRAYAMA*, Makiko SATOH* and Akio KOYAMA*

*Department of Internal Medicine, Kamitsuga General Hospital, Tochigi 322, Japan, and * Department of Internal Medicine, Institute of Clinical Medicine, University of Tsukuba, Ibaraki 305, Japan*

## Summary

We have reported that reactive oxygen species (ROS), especially the hydroxyl radical, play an important role in biosynthesis of methylguanidine (MG). In this study, we investigated the synthesis of MG by ROS generated in peroxisomes from two groups of male Wistar rats: one fed with chow containing 0.5 per cent clofibrate to induce hepatic peroxisomes, and one fed with ordinary laboratory chow. Peroxisomal fractions were obtained from liver homogenates by centrifugation, and incubated with creatinine in 0.1 M potassium phosphate buffer (pH 7.4) at 37 °C.

MG synthesis as well as hydrogen peroxide occurs in the peroxisomal fraction. MG synthesis was inhibited by the addition of dimethylsulfoxide, glutathione, catalase or superoxide dismutase. The rate of MG synthesis in clofibrate-treated rats is less than that in untreated rats.

These results suggest that MG is partially synthesized from creatinine by peroxisomes, and that ROS generated through the enzymtic reaction in peroxisomes play a role in MG synthesis. Moreover, the induction of the scavenger system by clofibrate is higher than ROS generation in peroxisomes.

## Introduction

Elevated plasma levels of methylguanidine (⁻MG) are reported in patients with uremia, and are felt to contribute to toxicity in those patients[2,5,7,8,9]. We have reported that MG is synthesized from creatinine in various tissues of rats[11] and that reactive oxygen species (ROS) play an important role in MG synthesis both *in vitro*[13] and in isolated rat hepatocytes[1]. Moreover, we have reported that peroxidative state affects MG synthesis in patients undergoing hyperbaric oxygen therapy[16].

We have also reported that rat liver homogenates had no activity on MG synthesis from creatinine; however, activity appeared in the homogenate after incubation at 60 °C for 15 min[12]. This activity disappeared after heating at 100 °C or the addition of glutathione[13]. These results suggest that some enzyme that participates in the production of ROS may play an important role in MG synthesis.

In this study, we attempted to demonstrate MG synthesis in peroxisomal fractions, which have β-oxidation systems of fatty acids consisting of oxidases and catalase. In addition, we investigated the effect of clofibrate, which induces peroxisomes, on MG synthesis.

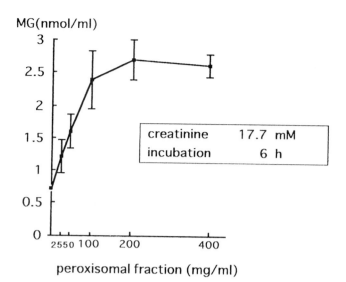

*Fig. 1. Effect of peroxisomal concentration on methylguanidine (MG) synthesis from creatinine. The final concentration of creatinine is 17.7 mM and that of the peroxisomes is indicated in the figure. The incubation period is 6 h. Each point represents the mean of duplicate incubations.*

## Materials and methods

### Preparation of peroxisomal fraction

Two groups of male Wistar rats weighing 250–300 g were used. One group was fed ordinary laboratory chow, and the other was fed with chow containing 0.5 per cent clofibrate for 2 weeks to induce peroxisomes. Rats were anesthetized with ether and the livers were homogenized in six volumes of a 0.25 M sucrose solution using a Potter-Elvehjem Teflon homogenizer after perfusion with heparinized saline. The nuclear fraction was obtained by centrifugation at 600$g$ for 10 min, the mitochondrial fraction at 3300$g$ for 10 min from the supernatant, the peroxisomal fraction obtained at 12 500$g$ for 20 min and the microsomal fraction at 105 000$g$ for 90 min. These procedures are essentially based on the method of C. de Duve *et al.*[6]. Synthesis of hydrogen peroxide in the peroxisomal fraction was determined to confirm the activity of the various oxidases of peroxisomes.

### Reaction mixture

The reaction mixture consisted of 17.7 mM creatinine and the peroxisomal fraction, with or without 0.1 mM palmitoyl CoA. All were brought to a final volume of 1.0 ml with 0.1 M potassium phosphate buffer (pH 7.4 at 37 °C). This mixture was incubated in a 10 ml glass test tube with shaking at 60 cycles/min, and the temperature was kept at 37 °C in the room air. The reaction was terminated and MG was extracted by the addition of 0.1 ml of 100 per cent (w/v) trichloroacetic acid (TCA) followed by sonication.

## Assay of MG

MG in the supernatant following centrifugation at 1700$g$ for 15 min was determined by high-pressure liquid chromatographic analysis on a cation-exchange column using 9,10-phenanthrenequinone as post-column derivatization reagent[17].

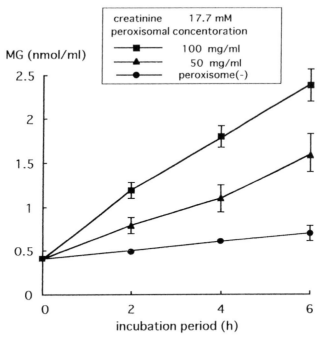

Fig. 2. Time course of methylguanidine (MG) synthesis from creatinine in the
peroxisomal fraction. The final concentration of creatinine is 17.7 mM and of
peroxisomes 50 mg/ml, 100 mg/ml or 0. The incubation period is indicated in
the figure. Each point represents the mean of duplicate incubations.

### Assay of hydrogen peroxide

According to the method using scopoletin[4], the peroxisomal fraction (0.9 mg of peroxisomes) was
added to the reaction mixture containing 0.1 ml horseradish peroxidase (400 mg/ml) in a total volume
of 3 ml with 0.1 M potassium phosphate buffer (pH 7.4 at 37 °C). 3 mM scopoletin was added after
the addition of peroxisomes, and the decay of fluorescence (excitation at 385 nm and emission at 460
nm) was recorded by a spectrofluorometer. The amount of hydrogen peroxide produced was
calculated by using a standard curve obtained in each experiment.

### Statistical analysis

Unpaired data were analyzed by Students two-tailed $t$-test, and $P$ values $< 0.05$ were considered to
be statistically significant.

## Results

### MG synthesis by various organelles

The incubation mixture containing 100 mg/ml of mitochondrial or microsomal fraction and 17.7 mM
creatinine showed no MG production even after 6 h incubation.

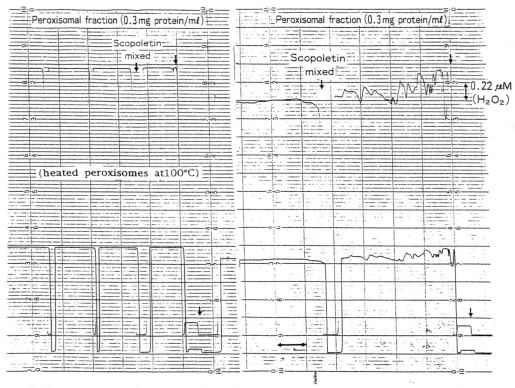

*Fig. 3. The spectrum of hydrogen peroxide production in the peroxisomal fraction. The reaction mixture contains peroxisomes (0.3 mg/ml), horseradish peroxidase (1.33 mg/ml) and 3 mM scopoletin with 0.1M potassium phosphate buffer (pH 7.4 at 37 °C). Left side: heated peroxisomes at 100 °C. Right side: control.*

### MG synthesis in the peroxisomal fraction

MG was synthesized in the reaction mixture containing peroxisomes and 17.7 mM creatinine. The rate of MG synthesis increased as the concentration of the peroxisomal fraction increased to 100 mg/ml (Fig. 1). This reaction proceeded linearly to incubation periods up to 360 min (Fig. 2).

### Production of hydrogen peroxide in the peroxisomal fraction

The production of hydrogen peroxide was confirmed from the peroxisomal fraction (Fig. 3). The production rate of hydrogen peroxide in the peroxisomal fraction was not significantly different in the clofibrate treated and the untreated groups. ($H_2O_2$ production rate: $0.05 \pm 0.01$ µM/min.)

### Effect of palmitoyl CoA on MG synthesis in the peroxisomal fraction

Palmitoyl CoA, which served as a substrate of acyl-CoA oxidase, did not affect MG synthesis (Fig. 4).

### Effect of clofibrate treatment on MG synthesis in the peroxisomal fraction

The rate of MG synthesis in the peroxisomal fraction was significantly less in the clofibrate-treated group than in the untreated group (Fig. 5).

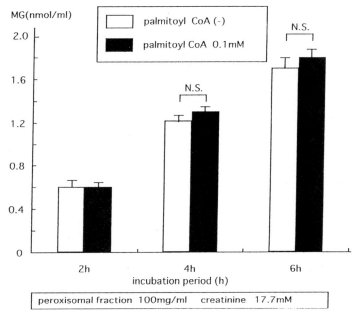

*Fig. 4. Effect of palmitoyl CoA on methylguanidine (MG) synthesis in the peroxisomal fraction. The final concentration of creatinine is 17.7 mM, of palmitoyl CoA 0.1 mM and of the peroxisomes 100 mg/ml. The incubation period is indicated in the figure. Each column represents the mean of duplicate incubations.*

### Effect of various reagents on MG synthesis in the peroxisomal fraction

MG synthesis in the peroxisomal fraction is partially inhibited by the addition of catalase, dimethyl-sulfoxide (DMSO) or gultathione, whereas sodium azide, an inhibitor of catalase, increased MG synthesis slightly. Superoxide dismutase (SOD), a scavenger of superoxide radicals, displayed a small effect at high doses (Fig. 6).

### Discussion

MG synthesis from creatinine disappeared after homogenization of rat liver, though it clearly occurred in isolated hepatocytes in our previous studies[1,12]. In this study, we investigated the intracellular synthesis site of MG with special reference to the peroxisomes and the effect of clofibrate. The results show that MG synthesis occurs only in the peroxisomal fraction. This peroxisomal fraction contains peroxisomes and light mitochondria. However, the activity of MG synthesis in this fraction is mainly derived from peroxisomes because MG synthesis did not occur in the mitochondrial fraction.

The concentration of creatinine in this study is artificially elevated. Our previous studies[1,11,13] revealed that the conversion rate of MG from creatinine is remarkably low, though creatinine is the only possible precursor of MG synthesis. Therefore, a high concentration was required to obtain detectable amounts of MG.

Hepatic peroxisomes have β-oxidation systems of fatty acids independent of carnitine, and generate hydrogen peroxide through an enzymtic reaction[3,10]. The production rate of hydrogen peroxide in the peroxisomal fraction was $0.17 \pm 0.03$ nmol/min per mg of peroxisomes, and hydrogen peroxide was not generated in the heated fraction at 100 °C.

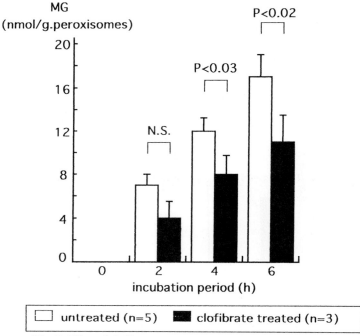

Fig. 5. Effect of clofibrate treatment on methylguanidine (MG) synthesis in the peroxisomal fraction. The final concentration of creatinine is 17.7 mM and of the peroxisomes 100 mg/ml. The incubation period is indicated in the figure. The clofibrate-treated group consisted of three rats and the untreated group of five. Results are expressed as mean ± standard deviation in separate experiments. P values are determined by the unpaired Student's t-test

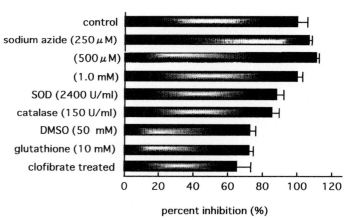

Fig. 6. Effect of various reagents on methylguanidine synthesis from creatinine in the peroxisomal fraction. The final concentration of creatinine is 17.7 mM and that of the peroxisomes 100 mg/ml. The incubation period is 6 hours . Each column represents the mean of duplicate incubations expressed as a percentage of the control value. Bars indicate the range of each determination.

As shown in Figs. 1 and 2, MG synthesis occurred in the peroxisomal fraction without the addition of any substrates. The rate of MG synthesis increased depending on the concentration of the peroxisomal fraction and the incubation period up to 100 mg/ml and 6 h, respectively. On the other hand, our previous study[14] revealed that MG synthesis could occur in microsomes by the addition of NADPH, which is cofactor of microsomal mixed function oxidases. These results suggest that the MG synthesis occurs in the peroxisomal fraction through enzymtic reactions with some endogenous substrates. As the ability of acyl-CoA oxidase to generate hydrogen peroxide is well known, we added palmitoyl CoA (which is a substrates of acyl-CoA oxidase) to the peroxisomal reaction mixture. Our results (Fig. 4) showed that palmitoyl CoA was not a limiting factor for MG synthesis. This suggests that endogenous substrate such as fatty acids rich in the crude peroxisomal fraction.

**Table 1 . Total amount of MG production in whole peroxisomes.**

| | Weight of liver (g) | Peroxisomal pellets (g) | MG synthesis in whole peroxisomes ($\mu$mol/2 h) |
|---|---|---|---|
| Clofibrate treated | $15.2 \pm 2.6$ | $3.7 \pm 0.6$ | $14.8 \pm 2.4$ |
| Untreated (control) | $13.2 \pm 2.0$ | $2.7 \pm 0.4$ | $16.2 \pm 2.4$ |

The clofibrate-treated group consists of three rats and the untreated group consists of five. Results are expressed as the mean ± the standard deviation in separate experiments.

The production rate of MG in the peroxisomal fraction is 0.04 nmol/g liver per h per mM creatinine. This value accounts for about 5–10 per cent of MG synthesis in isolated rat hepatocytes[1]. The remaining MG synthesis in hepatocytes may occur in other intracellular fractions, but the integrity of the cell should be necessary for detection of MG synthesis. Peroxisomes are induced by various substances and clofibrate is well known to be one of the typical inducers[10]. Contrary to our expectations, MG synthesis in clofibrate-treated peroxisomes is less than that in control peroxisomes (Fig. 5). In addition, it has been reported that clofibrate induces catalase in peroxisomes[10] and cytosolic glutathione peroxidase activity[15]. As shown in Fig. 6, the promotive effect of sodium azide and the inhibitory effect of catalase on MG synthesis suggest that hydrogen peroxide participates in this activity. Moreover, the inhibitory effect of DMSO, glutathione and SOD suggests that hydroxyl radicals or perferryl iron should be generated from endogenous iron and play some role in MG synthesis. Based on these results, the induction of scavenger systems is more important than ROS generation in clofibrate-treated peroxisomes. However, the calculated total amount of MG production in the whole peroxisomal fraction showed no significant difference between the clofibrate-treated and the untreated groups. Indeed, the total amount of peroxisomal fraction obtained from the clofibrate-treated group increased only 1.3–1.4-fold compared with the untreated group (Table 1). Further studies are required to clarify the beneficial effect of clofibrate, a well known hypolipidemic agent commonly used in clinical practice, *in vivo* from the point of view of peroxidative state.

**Acknowledgements**

This study was supported in part by reseach grants from the University of Tsukuba Project Reseach Fund and the Intractable Disease Division, Public Health Bureau, Ministry of Health and Welfare, Japan. We are indebted to Mrs Satomi Kawamura for her valuable assistance.

## References

1.  Aoyagi, K., Nagase, S., Narita, M. & Tojo, S. (1987): Role of active oxygen on methylguanidine synthesis in isolated rat hepatocytes. *Kidney Int.* **22**, S229–S233.

2.  Barsotti, G., Bevilacqua, G., Morelli, E., Cappelli, P., Balestri, P.L. & Giovannetti, S. (1975): Toxicity arising from guanidine compounds: role of methylguanidine as a uremic toxin. *Kidney Int.* **7**: S229–S301.

3.  Boveris, A., Oshino, N. & Chance, B. (1972): The cellular production of hydrogen peroxide. *Biochem. J.* **128**, 617–630.

4.    Boveris, J., Martino, E. & Stoppani, O.M. (1973): Evaluation of the horseradish peroxidase-scopoletin method for measurement of hydrogen peroxide formation in biological systems. *Ann. Biochem.* **80**, 145–158.

5.    De Deyn, P.P. & Macdonald, R.L. (1990): Guanidino compounds that are increased in uremia inhibit GABA- and glycine-responses on mouse neurons in cell culture. *Ann. Neurol.* **28**, 627–633.

6.    De Duve, C., Pressman, B.C., Gianetto, R., Watiaux, R. & Applemans, F. (1955): *Biochem. J.* **60**, 604–617.

7.    D'Hooge, R., Pei, Y.-Q., Marescau, B. & De Deyn, P.P. (1992): Convulsive action and toxicity of uremic guanidino compounds: behavioral assessment and relation to brain concentration in adult mice. *J. Neurol. Sci.* **112**, 96–105.

8.    Giovannetti, S., Biagni, M., Barestri, P.L., Navalesi, R., Giagnoni, P., deMatteis, A., Ferro-Milone, P. & Perfetti, C. (1969): Uremia-like syndrome in dogs chronically intoxicated with methylguanidine and creatinine. *Clin. Sci.* **36**, 445–452.

9.    Giovannetti, S., Balestri, P.L. & Barsotti, G. (1973): Methylguanidine in uremia. *Arch. Intern. Med.* **131**, 709–713.

10.   Lazarow, P.B. & De Duve, C. (1976): A fatty acyl CoA oxidizing system in rat liver peroxisomes: enhancement by clofibrate, a hypolipidemic drug. *Proc. Natl Acad. Sci. USA* **73**, 2043–2046.

11.   Nagese, S., Aoyagi, K., Narita, M. & Tojo, S. (1985): Biosynthesis of methylguanidine in isolated rat hepatocytes and *in vivo*. *Nephron* **40**, 470–475.

12.   Nagese, S., Aoyagi, K., Narita, M. & Tojo, S. (1985): Stimulatory and inhibitory factors of methylguanidine synthesis in rat organs. *Jap. J. Nephrol.* **27**, 1141–1147.

13.   Nagese, S., Aoyagi, K., Narita, M. & Tojo, S. (1986): Active oxygen in methylguanidine synthesis. *Nephron* **44**, 299–303.

14.   Nagase, S., Aoyagi, K., Sakamoto, M., Takemura, K., Ishikawa T. & Narita, M. (1992): Biosynthesis of methylguanidine in the hepatic microsomal fraction. *Nephron* **62**, 182–186.

15.   Nakagawa, M., Ishihara, N., Smokawa, T. & Kojima, S. (1987): Effect of clofibrate on lipid peroxidation in rat treated with aspirin and 4-pentenoic acid. *J. Biochem.* **101**, 81–88.

16.   Takemura, K., Aoyagi, K., Nagase, S., Sakamoto, M., Ishikawa, T. & Narita, M. (1992): Effect of hyperbaric oxygen therapy on urinary methylguanidine excretion in patients with or without renal failure. In: *Guanidino Compounds in Biology and Medicine*. Eds. P.P. De Deyn, B. Marescau, V. Stalon & I.A. Qureshi, pp. 301–307. London: John Libbey.

17.   Yamamoto, Y., Manji, T., Saito, A., Maeda, K. & Ohta, K. (1979): Ion-exchange chromatographic separation and fluorometric detection of guanidino compounds in physiologic fluids. *J. Chromatogr.* **162**, 327–340.

*Guanidino Compounds : 2*, eds. by P.P. De Deyn, B. Marescau, I.A. Qureshi and A. Mori.
©1997 John Libbey & Company Ltd., pp. 311–316.

# Chapter 32

# Inhibition of methylguanidine synthesis in isolated rat hepatocytes by protein kinase C inhibitor and calmodulin antagonist

K. AOYAGI, K. TAKEMURA, S. NAGASE, K. AKIYAMA, C. TOMIDA,
M. GOTOH, A. KOYAMA and M. NARITA[*]

*Institute of Clinical Medicine, University of Tsukuba, Tsukuba City, Ibaraki 305, and *Mito Central Hospital, Japan*

## Summary

In this study we investigated the effect of protein kinase C (PKC) inhibitor and calmodulin antagonist on methylguanidine (MG) synthesis.

Isolated hepatocytes were incubated in Krebs–Henseleit bicarbonate buffer with 3 per cent bovine serum albumin and 200 mg/dl creatinine and reagents. MG was determined by high-performance liquid chromatography. 1-(5-Isoquinolinesulfonyl)-2-methylpiperazine dihydrochloride (H–7) was used as protein kinase C inhibitor. N-(6-aminohexyl)-5-chloro-1-naphthalenesulfonamide (W–7) was used as calmodulin antagonist.

MG synthesis in isolated rat hepatocytes (78 nmol/g per 4 h) was increased by 1.9 mM puromycin aminonucleoside (PA) by 160 per cent. 25 µM H–7 inhibited MG synthesis increased by PA by 45 per cent. 100 µM H–7 completely inhibited MG synthesis. The inhibition was observed during the tested time (from 2 to 6 h). W–7 (100 µM) also inhibited increase of MG synthesis by PA, completely. PKC may play an important role in ROS generation of reactive oxygen species induced by PA.

## Introduction

We have reported that methylguanidine (MG) is formed from creatinine by reactive oxygen species in (ROS) in *in vitro* experiments[11], in human leukocytes[15] and in isolated hepatocytes[2]. Underhyperbaric oxygen therapy, MG excretion in urine increases in humans[16,17]. Puromycin aminonucleoside (PA), which induces heavy proteinuria, increased MG synthesis in isolated rat hepatocytes[2,5] and in PA administered rats[7].

We have also reported that adenosine and its potentiators (dipyridamole) inhibited MG synthesis in isolated rat hepatocytes[4] and in rats *in vivo*[7]. Adenosine and its potentiators have been known to increase cAMP levels[1,6] and inhibit protein kinase C (PKC)[14]. In this study we investigated the effects of PKC inhibitor and calmodulin antagonist on MG synthesis in isolated rat hepatocytes.

## Methods

### Preparation of isolated rat hepatocytes

Male Wistar rats weighing 300–350 g were used in all experiments. The rats were allowed free access to water and laboratory chow containing 25 per cent protein. Isolated hepatocytes were prepared essentially according to the method of Berry & Friend[9] as described previously[1]. We calculated that $9.8 \times 10^7$ cells correspond to 1 g liver (wet weight)[19].

### Incubation of cells

Cells were incubated in 6 ml of Krebs–Henseleit bicarbonate buffer containing 3 per cent bovine serum albumin, 10 mM sodium lactate and 200 mg/dl of creatinine and indicated substances. The incubation mixture was shaken at 60 cycles/min in a 30 ml conical flask with a rubber cap under 95 per cent oxygen and 5 per cent carbon dioxide at 37 °C for 4 h (except for the time dependence experiment). Equilibration of the buffer was repeated every hour. To measure the rate of non-biological conversion of creatinine to MG, incubations were carried out without cells. Incubation was arrested by the addition of 0.6 ml of 100 per cent (w/v) trichloroacetic acid. After sonication, the supernatant of cells and medium was obtained by centrifugation at 1700g for 15 min at 0 °C, and 0.2 ml of the extract was used for MG measurements. MG was determined by high-performance liquid chromatographic analysis using 9,10-phenanthrenequinone for the post-labeling method as described previously[1]. Dimethylformamide for fluorometrical use was purchased from Wako, Japan.

PA was purchased from Sigma Chemical, St Louis. 1-(5-lsoquinolinesulfonyl)-2-methylpiperazine dihydrochloride (H–7), a protein kinase inhibitor, and N-(6-aminohexyl)-5-chloro-1-naphthalenesulfonamide (W–7), a calmodulin antagonist, were purchased from Seikagaku Kogyo Co, Japan.

## Results

### Inhibition of MG biosynthesis by H–7 in isolated rat hepatocytes

MG synthesis was stimulated by PA from 156 nmol/g per 4 h to 206 nmol/g per 4 h in these experiments. H–7, a PKC inhibitor[13], significantly inhibited MG synthesis by isolated rat hepato-

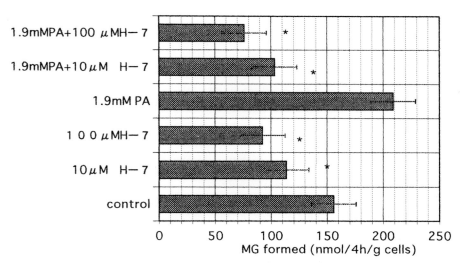

*Fig. 1. Inhibition of MG biosynthesis by H–7 in isolated rat hepatocytes. Each column represents the mean of five incubations. Bars indicate the standard error. * P < 0.05 vs control value.*

cytes at 10 μM by 27 per cent in the absence of PA and by 50 per cent in the presence of PA. 100 μM H–7 also inhibited MG synthesis; however, the inhibition rate was not so greatly increased, as shown in Fig. 1.

### Effect of concentration of H–7 on MG synthesis in isolated rat hepatocytes

H–7 at a concentration of 1–100 μM was investigated in the hepatocyte system. As shown in Fig. 2, up to a concentration of 25 μM, inhibition of H–7 was stronger than that at high concentration. The rate of inhibition in the presence of PA is greater than that in the absence of PA, as shown in Fig. 1.

### Time course of inhibition of MG synthesis by H–7

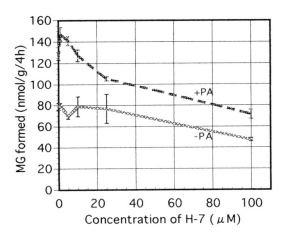

*Fig. 2. Effect of concentration of H–7 on MG synthesis in isolated rat hepatocytes. Each point represents the mean of duplicate incubations. Bars indicate the range of each determination.*

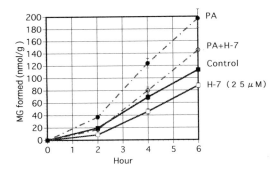

*Fig. 3. Time course of inhibition of MG synthesis by H–7 (25 μM). Each point represents the mean of duplicate incubations. Bars indicate the range of each determination.*

The effect of the inhibition of MG synthesis by H–7 was investigated for up to 6 h at 25 (Fig. 3) or 100 μM (Fig.4) H–7. Inhibition of MG synthesis was observed throughout the tested hours at the both concentrations. Marked inhibition was observed at 100 μM, as shown in Fig. 4.

### Effect of the $Ca^{2+}$ ions in the incubation medium

Activation of PKC depends on the $Ca^{2+}$ concentration. When, we varied the $Ca^{2+}$ concentration added to the incubation medium, no change of MG synthesis was observed. However, addition of ethylene glycol bis (β - aminoethyl ether – N,N,N'N'– tetraacetic acid (EGTA) (which binds $Ca^{2+}$) inhibited MG synthesis from 1 mM, as shown in Fig. 5. Intracellular $Ca^{2+}$ may be involved in MG synthesis.

### Effect of H–7 on MG synthesis from creatol

Recently, Ienaga et al reported that hydroxyl radicals act on creatinine and form creatol. Creatol changes into creaton A and B and then forms MG[12,18].

We investigated the effect of H–7 on the synthesis of MG from creatol. MG synthesis from 4 μM creatol by isolated rat hepatocytes was not affected. About 30 per cent of the creatol was converted into MG by 0.2 g of liver cells after 6 h.

### Effect of W–7 on MG synthesis

To investigat the role of calmodulin on PA-stimulated MG synthesis we tested W–7, a calmodulin antagonist. W–7 at 100 μM completely inhibited MG synthesis stimulated by PA. These data suggested that calmodulin may be involved in PA-stimulated MG synthesis.

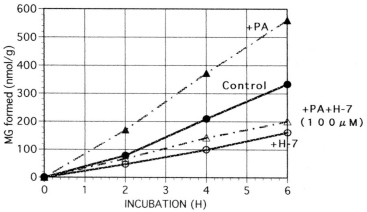

*Fig. 4. Time course of inhibition of MG synthesis by H–7 (100 μM). Each point represents the mean of duplicate incubations. Bars indicate the range of each determination.*

## Discussion

Recently, Nishizuka *et al.* reported that PKC plays an important role in cell function[14]. In leukocytes, activation of PKC resulted in the generation of reactive oxygen[10]. We have proposed that MG is a good marker of reactive oxygen generation in the cell[2] and have tested this hypothesis in tissue cells[2,6] as well as in leukocytes[15]. An increase in cAMP also decreased MG synthesis in isolated rat hepatocytes[4,5], and an interaction of cAMP and PKC activation has been reported (Fig. 6)[14].

In this paper, we have shown that PKC inhibitor inhibits MG synthesis at low concentration. At 10 μM, HA1004, (an inactive reagent[8]) does not affect MG synthesis. However, MG synthesis in isolated hepatocytes from creatinine is not so simple. The role of creatol, an intermediate for MG, should be clarified to understand what is occurring in the cells.

In this paper, we have shown that reactive oxygen generation by PA, which causes heavy proteinuria in rats is involved in the activation of PKC.

### Effect of EGTA on MG synthesis by isolated rat hepatocytes

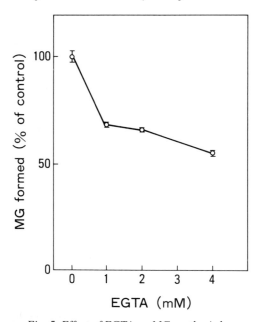

*Fig. 5. Effect of EGTA on MG synthesis by isolated rat hepatocytes. Each point represents the mean of duplicate incubations. Bars indicate the range of each determination.*

### Acknowledgements

This study was supported in part by the Scientific Research Funds of the Ministry of Education, Science and Culture of Japan (C–05670942, C–06671126), a Grant for Scientific Research Expenses for Health and Welfare Programs, Ministry of Health and Welfare, Japan and a Grant for Research Projects, University of Tsukuba, Japan.

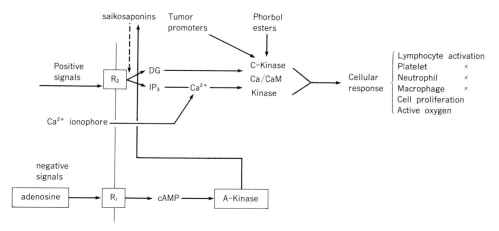

*Fig. 6. Interaction of PKC and cAMP dependent protein kinase (A-kinase).*

## References

1.  Aoyagi, K., Ohba, S., Narita, M. & Tojo, S. (1983): Regulation of biosynthesis of guanidinosuccinic acid in isolated rat hepatocytes and *in vivo*. *Kidney Int.* **24**, S224–S228.

2.  Aoyagi, K., Nagase, S., Narita, M. & Tojo, S. (1987): Role of active oxygen on methylguanidine synthesis in isolated rat hepatocytes. *Kidney Int.* **32**, S229–S233.

3.  Aoyagi, K., Nagase, S., Sakamoto, M., Narita, M. & Tojo, S. (1989): Puromycin aminonucleoside stimulates the synthesis of methylguanidine: a possible marker of active oxygen generation in isolated rat hepatocytes. In: *Guanidines 2.* Eds. A. Mori, B.D. Cohen & H. Koide, pp. 71–77. New York: Plenum Press.

4.  Aoyagi, K., Nagase, S., Sakamoto, M., Narita, M. & Tojo, S. (1989): Adenosine, adenosine analogues and their potentiators inhibit methylguanidine synthesis, a possible marker of active oxygen in isolated rat hepatocyte. *Guanidines 2.* Eds. A. Mori, B.D. Cohen & H. Koide, pp. 123–128. New York: Plenum Press.

5.  Aoyagi, K. & Narita, M. (1990): Active oxygen toxicity in renal diseases. *Jpn. J. Med.* **29**, 681–682.

6.  Aoyagi, K. & Narita, M. (1991): Mechanism of abnormal active oxygen generation in tissue cells by puromycin aminonucleoside. *Acta. Med. Biol.* **39** (Suppl), 53–62.

7.  Aoyagi, K., Nagase, S., Takemura, K., Ohba, S. & Narita, M. (1992): Dipyridamole decreased urinary excretion of methylguanidine increased by puromycin aminonucleoside *in vivo*. In: *Guanidino Compounds in Biology and Medicine.* Eds. P.P. De Deyn, B. Marescau, V. Stalon & I.A. Qureshi, pp. 309–313. London: John Libbey.

8.  Asano T., Hidaka, H. (1984): Vasodilatory action of HA1004 [N-(2-guanidinoethyl)-5-isoquinolinesulfonamide], a novel calcium antagonist with no effect on cardiac function. *J. Pharmacol. Exp. Ther.* **231**, 141–145.

9.  Berry, M.N. & Friend, D.S. (1969): High-yield preparation of isolated liver cells. *J. Cell. Biol.* **43**, 506–520.

10. Cox, J.A., Jeng, A.Y., Sharkey, N.A., Blumberg, P.M. & Tauber, A.I. (1985): Activation of neutrophil nicotinamide adenine dinucleotide phosphate (NADPH)-oxidase by protein kinase. *J. Clin. Invest.* **76**, 1932–1938.

11. Nagase, S., Aoyagi, K., Narita, M. & Tojo, S. (1986): Active oxygen in methylguanidine synthesis. *Nephron* **44**, 299–303.

12. Nakamura, K., Ienaga, K., Yokozawa, T., Fujitsuka, N. & Oura, H. (1991): Production of methylguanidine from creatinine via creatol by active oxygen species: analyses of the catabolism *in vitro*. *Nephron* **58**, 42–46.

13. Nishikawa, M., Uemura, Y., Hidaka, H., Shirakawa, S. (1986): 1-(5-Isoquinolinesulfonyl)-2-methylpiperazine dihydrochloride (H–7), a potent inhibitor of protein kinase, inhibits the differentiation of HL–60 cells induced by phorbol diester. *Life Sci.* **39**, 1101–1107.

14. Nishizuka, Y. (1984): The role of protein kinase C in cellular signal transduction and tumor promotion. *Nature* **308**, 693–698.

315

15.   Sakamoto, M., Aoyagi, K., Nagase, S., Ishikawa, T., Takemura, K., & Narita, M. (1989): [Methylguanidine synthesis by active oxygen generated by stimulated human neutrophils] (in Japanese). *Nippon Jinzo Gakkai Shi* **31,** 851–858.

16.   Takemura, K., Aoyagi, K., Nagase, S., Sakamoto, M., Ishikawa, T., & Narita, M. (1990): [Urinary excretion rate of methylguanidine as a new marker of active oxygen *in vivo*: demonstration in hyperbaric oxygen therapy] (in Japanese). *Nippon Jinzo Gakkai Shi* **32,** 1195-1201.

17.   Takemura, K., Aoyagi, K., Nagase, S., Sakamoto, M., Ishikawa, T., & Narita, M. (1992): Effect of hyperbaric therapy on urinary methylguanidine excretion in normal humans and patients with renal failure. In: *Guanidino Compounds in Biology and Medicine.* Eds. P. P. De Deyn, B. Marescau, V. Stalon & I.A. Qureshi, pp. 301–307. London: John Libbey.

18.   Yokozawa, T., Fujitsuka, N., Oura, H., Ienaga, K. & Nakamura, K. (1991): Comparison of methylguanidine production from creatinine and creatol *in vivo*. *Nephron* **58,** 125–126.

19.   Zahlten, R.N., Stratman, F.W. & Lardy, H.A. (1973): Regulation of glucose synthesis in hormone-sensitive isolated rat hepatocytes. *Proc. Natl Acad. Sci.* USA **70,** 3213–3218.

*Guanidino Compounds : 2*, eds. by P.P. De Deyn, B. Marescau, I.A. Qureshi and A. Mori.
©1997 John Libbey & Company Ltd., pp. 317–324.

# Chapter 33

# Serum effects on methylguanidine synthesis by the hydroxyl radicals, and decreased serum antioxidant activity in hemodialysis patients

Sohji NAGASE, Kazumasa AOYAGI, Michihiro GOTOH, Katsumi TAKEMURA, Aki HIRAYAMA, Yasushi NAGAI*, Chie TOMIDA, Makiko SATOH, Toshiko HIBINO, Hiroshi KIKUCHI, Akio KOYAMA

*Department of Internal Medicine, Institute of Clinical Medicine, University of Tsukuba, 1–1–1 Ten-nodai, Tsukuba, Ibaraki 305, Japan, and *Tsukuba Research Center, Eisai Co. Ltd., 5–1–3 Tokodai, Tsukuba, Ibaraki 300–26, Japan*

## Summary

Various evidence implies that uremic patients are in a highly peroxidative state. Methylguanidine is a toxin present in increased amount in uremia, and we have shown that methylguanidine is a peroxidative product of creatinine by the action of active oxygen. Our findings suggest that methylguanidine may be a useful indicator of peroxidative state in uremia because serum methylguanidine can be detected only in patients with renal failure. In this study we investigated the serum antioxidant activity, which may regulate or represent the redox state *in vivo,* by the addition of sera to a chemical system synthesizing methylguanidine from creatinine. In addition, we compared the serum antioxidant activity indicated in this system by measuring the signal intensity of electron spin resonance during the reaction of sera with the hydroxyl radical generating system. The decreased serum antioxidant activity of hemodialysis patients was demonstrated by both systems. Furthermore, there was a statistically significant positive correlation between the amount of methylguanidine synthesized and the signal intensity in the reaction mixtures containing healthy sera.

## Introduction

**H**ealthy organisms protect themselves against oxygen toxicity with a variety of defenses. In pathologic states these protective mechanisms appear to be attenuated, leading to a number of important sequelae. These include inflammation, fibrosis, hemolysis, arteriosclerosis, cancer, and the concomitants of aging, all of which may be a consequence of this peroxidative tendency[3,4,14,15,19]. In a previous study, we reported that methylguanidine (MG), a potent uremic toxin, is a peroxidative product of creatinine[1,11,12], a substance once thought to be biologically inert. Several additional reports suggest that uremia is a state of oxidative stress. These include findings such as the increased level of lipid peroxides in red blood cell membranes[5] and the decrease in serum antioxidant activity (AOA) as shown by an increased rate of ox brain autoxidation in the presence of uremic serum[8]. In this study, we investigated serum AOA by measuring the inhibition rate of a chemical system synthesizing MG from creatinine[11].

Recent progress in electronics has made it possible to determine and quantify the species of active oxygen involved in a reaction by electron spin resonance (ESR)[6,16], and this is thought to be a reliable

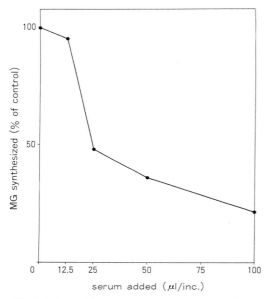

Fig. 1. Effect of the serum concentration on MG synthesis. The reaction mixture consisted of 1 ml of 30 mM phosphate buffer (pH 7.4 at 37 °C) containing 10 mM creatinine, 0.5 mM FeCl2, 5 mM hydrogen peroxide and indicated amounts of serum, from healthy controls.

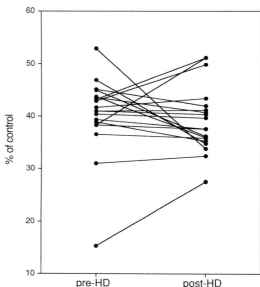

Fig. 2. Comparison of MG synthesis between the reaction mixtures containing sera before and after HD. The reaction mixture consisted of 1 ml of 30 mM phosphate buffer (pH 7.4 at 37 °C) containing 10 mM creatinine, 0.5 mM FeCl2, 5 mM hydrogen peroxide and 100 μl of serum. The chromatogram obtained from the reaction mixture without serum was used as a control, and the rate of MG synthesis was expressed as a percentage of the control value.

indication of active oxygen presence. We compared the serum AOA indicated by the inhibition rate of MG synthesis with that of ESR signal intensities to evaluate the importance of MG as an indicator of active oxygen generation. This was accomplished by adding serum to the system containing creatinine or spin trapping reagent and the hydroxyl radical generator. We also compared the serum AOA in patients before and after hemodialysis (HD) and healthy controls in both systems to evaluate the decreased AOA of HD patients after and the effect of HD on peroxidative state.

## Materials and methods

### Subjects

This study was carried out on 23 healthy controls (20 male, three female) aged 44.65 ± 8.50 years (mean ± standard deviation) and 19 patients (14 male, five female) aged 44.84 ± 14.80 years undergoing chronic maintenance HD 545.63 ± 422.81 times. The etiologic diagnoses were chronic glomerulonephritis (m=11), diabetic nephropathy (n = 3), nephrosclerosis (n = 2), polycystic kidney disease (n = 2) and multiple myeloma kidney (n = 1).

Blood samples were drawn from the antecubital veins of healthy subjects and from the arterial side of the arteriovenous fistula of patients before and after HD. Sera were separated by centrifugation at 1,100g for 15 min at 4 °C and kept refrigerated at –20 °C until analysis.

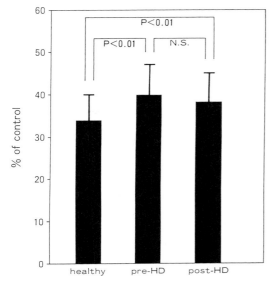

Fig. 3. Comparison of MG synthesis in reaction mixtures containing healthy, pre-HD and post-HD sera. The reaction mixture consisted of 1 ml of 30 mM phosphate buffer (pH 7.4 at 37 °C) containing 10 mM creatinine, 0.5 mM FeCl₂, 5 mM hydrogen peroxide and 100 μl of serum. The chromatogram obtained from the reaction mixture without serum was used as a control, and the rate of MG synthesis was expressed as a percentage of the control value. The columns and bars indicate means and standard deviations, respectively. N.S. = not significant.

### Clinical parameters

Serum total protein, albumin, urea nitrogen, creatinine, uric acid, total and direct bilirubin, glutamic-oxaloacetic transaminase (GOT), glutamic-pyruvic transaminase (GPT), lactate dehydrogenase (LDH), alkaline phosphatase, total and high-density lipoprotein cholesterol, triglyceride, glucose, total calcium, inorganic phosphorus, sodium, potassium and chlorine were determined by an autoanalyzer (Hitachi 736-15, Japan), and tocopherols were measured by high-pressure liquid chromatography (HPLC). Serum iron was measured colorimetrically and serum ferritin by a commercial radioimmunoassay kit (Baxter-Travenol, USA). Red and white blood cells, hemoglobin, hematocrit and platelets were determined by an autoanalyzer (Coulter STKS, USA).

### Measurement of the inhibitory activity of serum on MG synthesis by hydroxyl radicals

The inhibitory action of serum on chemical MG synthesis by the hydroxyl radical derived from Fenton's reaction[7] was measured. The reaction mixture consisted of 1 ml 30 mM phosphate buffer (pH 7.4 at 37 °C) containing indicated concentrations of serum, 10 mM creatinine and as a generator of the hydroxyl radical. Hydrogen peroxide was added to achieve a final concentration of 0.5 mM or 5 mM. The reaction was initiated by the addition of hydrogen peroxide and incubated in a 10 ml glass test tube with shaking at 60 cycles/min for 15 min at 37 °C in room air. Termination of the reaction occurred with the addition of trichloroacetic acid to a final concentration of 10 per cent. The MG concentration in the supernatant following centrifugation at 1,700g for 15 min was determined by HPLC using 9,10-phenanthrenequinone as described previously[13]. The MG in each reaction mixture without incubation was measured in the same manner and subtracted from the incubated one to account for the amount of MG in the serum and creatinine. The chromatogram obtained from the reaction mixture without serum was used as a control, and the rate of MG synthesis was expressed as a percentage of the control value.

### Measurement of the ESR signal

The serum scavenging activity for the hydroxyl radical was measured directly in terms of the inhibition rate of ESR signals by the addition of sera to a system generating the hydroxyl radical. This is a very reactive and short-lived free radical, which reacts with spin trapping reagent and produce a relatively long-lived product (spin adduct) that can be detected by ESR. The ESR spectrum allows the identification and quantification of the original reactive radical. The reaction mixture consisted of 1 ml of distilled water containing 10 per cent serum, 0.5 mM FeCl₂, 5 mM hydrogen peroxide and 100 mM 5,5-dimethyl-1-pyrroline-N-oxide (DMPO) as a spin trapping reagent. The ESR spectra were recorded 1 min after the initiation of the reaction by the addition of hydrogen peroxide using a JES-FE

319

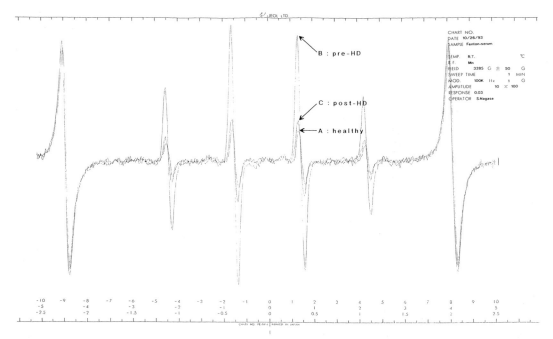

*Fig. 4. Typical spectra of ESR signals. The reaction mixture consisted of 1 ml of distilled water containing 10 per cent serum, 0.5 mM FeCl₂, 5 mM hydrogen peroxide and 100 mM 5,5-dimethyl-1-pyrroline-N-oxide (DMPO).*

2XG ESR apparatus (Nippon Denshi, Tokyo, Japan) at room temperature (25 °C). The spectrum of the reaction without serum was used as a control, and the signal intensities were expressed as percentages of the control values.

### Measurement of serum catalase activity

The rate of serum degradation of hydrogen peroxide was measured spectrophotometrically by following the decrease in ultraviolet absorption at 250 nm. [2] The reaction mixture consisted of 1 ml 50 mM phosphate buffer (pH 7.0 at 25 °C) containing 10 mM hydrogen peroxide and 25 μl of serum. The breakdown of hydrogen peroxide was calculated from the absorption before and after 10 min incubation at 25 °C. One unit of catalase activity was defined as the amount of enzyme that consumed 1 μmol of hydrogen peroxide per min.

## Results

### Inhibitory effect of serum on MG synthesis

The rate of MG synthesis decreased as the amount of serum from healthy controls added to the incubation mixture rose (Fig. 1). Based on these results, 100 μl of serum was added in 1 ml of the reaction mixture to determine its inhibitory effect on MG synthesis in the subsequent experiments. Figure 2 shows the individual results of 19 pairs of pre-HD and post-HD sera. In addition to these results, we examined the effects of serum from 23 healthy controls and it was clear that the inhibitory effects on MG synthesis are significantly weaker in both pre-HD and post-HD sera than in healthy sera ($P < 0.01$), as shown in Fig. 3.

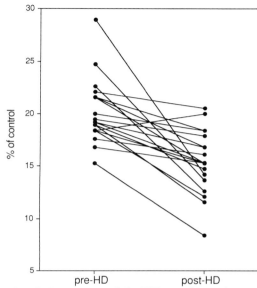

*Fig. 5. Comparison of the ESR signal intensity of reaction mixtures containing pre-HD and post-HD sera. The reaction mixture consisted of 1ml distilled water containing 10 per cent serum, 0.5 mM FeCl₂, 5 mM hydrogen peroxide and 100 mM 5,5-dimethyl-1-pyrroline-N-oxide (DMPO). The spectrum of the reaction without serum was used as a control and the signal intensities were expressed as a percentage of the control value.*

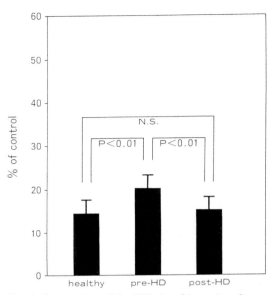

*Fig. 6. Comparison of the ESR signal intensity of reaction mixtures containing healthy, pre-HD and post-HD sera. The reaction mixture consisted of 1 ml distilled water containing 10 per cent serum, 0.5 mM FeCl₂, 5 mM hydrogen peroxide and 100 mM 5,5-dimethyl-1-pyrroline-N-oxide (DMPO). The spectrum of the reaction without serum was used as a control, and the signal intensities were expressed as a percentage of the control value. The columns and bars indicate means and standard deviations, respectively. N.S. = not significant.*

### Direct proof of serum reducing effect on the hydroxyl radical by ESR

The reduced serum AOA is suggested by the aforementioned results in MG synthesis. We investigated the scavenging activity against the hydroxyl radical directly by using ESR in the same three sets of sera. The typical ESR spectra are shown in Fig. 4. The signal intensity of the reaction mixture containing pre- HD serum (B) is stronger than that obtained with serum from a healthy control individual (A) or that recovered to a healthy value after HD (C). Figure 5 shows the individual results of 19 pairs of pre-HD and post-HD sera and the antioxidant activity of pre-HD sera increased after HD in almost every patient. Figure 6 clearly demonstrates that the ESR signals of the reaction mixture containing pre-HD sera are significantly stronger than those obtained with healthy sera ($P < 0.01$). After HD treatment, the signals recover to values found in healthy individuals. These results directly indicate the decreased AOA of pre-HD sera and that HD improves it to healthy levels.

### Correlation among MG synthesis, ESR signal intensity and various parameters

Blood urea nitrogen, creatinine and uric acid changed in the same manner as MG synthesis or ESR signal; however, the other clinical parameters (listed under Materials and methods), tocopherol and catalase activity did not change in the same manner. There is no significant difference among pre-HD, post-HD and healthy sera in catalase activity. However, there is a significant negative correlation between MG synthesis in the reaction mixture containing post-HD sera and post-HD serum catalase activity ($r = -0.522$), though no parameters correlate with healthy or pre-HD serum catalase activity.

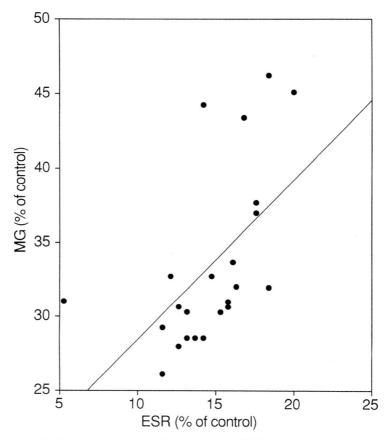

*Fig. 7. Correlation between MG synthesis and ESR signal intensity in the healthy group.*

Interestingly, there is a significant positive correlation between the amount of MG synthesized and the ESR signal intensity in the reaction mixture containing healthy sera ($r = 0.566$; $n = 19$), as indicated in Fig.7.

## Discussion

The results of this study indicate that the inhibition of MG synthesis by healthy serum positively correlates with ESR signal intensity. This suggests that MG is a useful parameter for measuring the generation of active oxygen, especially the hydroxyl radical. In other words, the conversion of creatinine to MG can be used as a marker to predict the generation of active oxygen or the peroxidative state *in vivo*. However, additional study is required to clarify why a positive correlation cannot be found in the reaction with pre-HD or post-HD sera.

The results of the experiments using ESR show that the serum AOA of patients on HD is significantly lower than that of healthy controls, and this reduced AOA recovers to a healthy level after HD. However, the effect of HD could not be recognized in the experiments measuring the inhibition of MG synthesis. There may be some differences in the specificity in the detection of active oxygen between the conversion of creatinine to MG and ESR determination. There have been several reports concerning the peroxidative state in patients with chronic renal failure or undergoing HD [5,8,10,17,18,21]. Conversely, it is reported that serum superoxide activity measured by ESR in patients with chronic glomerulonephritis or chronic renal failure is significantly higher than that in healthy controls[20]. The conclusion of our study does not conflict with this report because the species of active oxygen is different.

The reduced serum AOA shown in this experiment was possibly derived from two factors. One is increased production of the hydroxyl radical via an increased concentration of the substrates of Fenton's reaction, i.e. decreased hydrogen peroxide degradation and/or the chelating of free iron. However, neither is likely because the serum iron binding capacity is around 0.05 mM and the serum concentration in this experiment is 10 per cent, so the iron binding capacity in the reaction mixture is 0.005 mM. This is only 1 per cent of the iron present in the reaction mixture. In addition, we measured the serum capacity to degrade hydrogen peroxide, and apparently there are no significant differences among healthy controls and patients before and after HD. The second possibility is reduced serum scavenging action against the hydroxyl radical. In this reaction mixture, the hydroxyl radical is detected by MG synthesis from the creatinine or by ESR signal after binding with DMPO. Therefore, a scavenger in the serum must compete with creatinine or DMPO for binding of the hydroxyl radicals to reduce the MG production or ESR signals. The concentration of creatinine or DMPO in the reaction mixture is relatively high and there are no substances in the serum of comparable concentration.   However, the MG synthesis in the reaction mixture with post-HD sera decreased as the serum catalase activity became higher. This implies that serum AOA is possibly derived from a collaboration of several scavengers of active oxygen as serum catalase activity. In patients with chronic renal failure, these activities may be attenuated by the effects of toxic substances that accumulate in the disease condition not clarified in this study.

## Acknowledgements

We are grateful to Ms Y. Shimozawa and Ms C. Horiuchi for their excellent technical assistance, Dr M. Igarashi for his assistance in statistical analysis using Statview 4.0, and Dr and Ms Fons for their valuable criticism in preparing the manuscript. This study was supported in part by the Scientific Research Funds of the Ministry of Education, Science and Culture of Japan (C–05670942, C–06671126), a Grant for Scientific Research Expenses for Health and Welfare Programs, Ministry of Health and Welfare, Japan and a Grant for Research Projects, University of Tsukuba, Japan.

## References

1.  Aoyagi, K., Nagase, S., Narita, M. & Tojo, S. (1987): Role of active oxygen on methylguanidine synthesis in isolated hepatocytes. *Kidney Int.* **22**, S229–S232.

2.  Beers, R.F. & Silver, I.W. (1952): A spectrophotometric method for measuring the breakdown of hydrogen peroxide by catalase. *J. Biol. Chem.* **195**, 133–140.

3.  Cohen, M.V. (1989): Free radicals in ischemic reperfusion myocardial injury: is this time for clinical trials? *Ann. Intern. Med.* **111**, 918–931.

4.  Freeman, B.A., & Crapo, J.D. (1982): Free radical and tissue injury. *Lab. Invest.* **47**, 412–426.

5.  Giardini, O., Taccone-Gallucci, M., Lubrano, R., Riccardi-Tenore, G., Bandino, D. & Silvi, I. (1984): Evidence of red blood cell membrane lipid peroxidation in haemodialysis patients. *Nephron* **36**, 235–237.

6.  Green, M.R., Hill, H.A.O., Okolow-Zubkowska, M.J. & Segal, A.W. (1979): The production of hydroxyl and superoxide radicals by stimulated human neutrophils: measurement by EPR spectroscopy. *FEBS Lett.* **100**, 23–26.

7.  Gutteridge, J.M., Maidt, L. & Poyer, L. (1990): Superoxide dismutase and Fenton chemistry. *Biochem. J.* **269**, 169–174.

8.  Kuroda, M., Asaka, S., Tofuku, Y. & Takeda, R. (1985): Serum antioxidant activity in uremic patients. *Nephron* **41**, 293–298.

9.  Lucchi, L., Cappelli, G., Angela Acerbi, M., Spattini, A. & Lusvarghi, E. (1989) Oxidative metabolism of polymorphonuclear leukocytes and serum opsonic activity in chronic renal failure. *Nephron* **51**, 44–50.

10. Lucchi, L., Banni, S., Botti, B., Cappelli, G., Medici, G., Paolo Melis, M., Tomasi, A., Vannini, V. & Lusvarghi, E. (1993): Conjugated diene fatty acids in patients with chronic renal failure: evidence of increased lipid peroxidation. *Nephron* **65**, 401–409.

11. Nagase, S., Aoyagi, K., Narita, M. & Tojo, S. (1986): Active oxygen in methylguanidine synthesis. *Nephron* **44**, 299–303.

12. Nagase, S., Aoyagi, K., Sakamoto, M., Takemura, K., Ishikawa, T. & Narita, M. (1992): Biosynthesis of methylguanidine in the hepatic microsomal fraction. *Nephron* **62**, 182–186.

13. Nagase, S., Aoyagi, K., Narita, M. & Tojo, S. (1985): Biosynthesis of methylguanidine in isolated rat hepatocytes and *in vivo*. *Nephron* **40**, 470–475.

14. Paller, M.S., Hoidal, J. R. & Ferris, T.F. (1984): Oxygen free radicals in ischemic acute renal failure in the rat. *J. Clin. Invest.* **74**, 1156–1164.

15. Rehan, A., Johnson, K.J., Wiggins, R.C., Kunkel, R.G. & Ward, P.A. (1984): Evidence for the role of oxygen radicals in acute nephrotic nephritis. *Lab. Invest.* **51**, 396–403.

16. Rosen, H. & Klebanoff, J. (1979): Hydroxyl radical generation by polymorphonuclear leukocytes measured by electron spin resonance spectroscopy. *J. Clin. Invest.* **64**, 1725–1729.

17. Shainkin-Kestenbaum, R., Caruso, C. & Berlyne, G.M. (1990): Reduced superoxide dismutase activity in erythrocytes of dialysis patients: a possible factor in the etiology of uremic anemia. *Nephron* **55**, 251–253.

18. Toborek, M., Wasik, T., Drozzdz, M., Klin, M., Magner-Wrobel, K. & Kopieczna-Grzebieniak, E. (1992) Effect of hemodialysis on lipid peroxidation and antioxidant system in patients with chronic renal failure. *Metabolism* **41**, 1229–1232.

19. Weitzman, S.A. & Stossel, T.P. (1982): Effect of oxygen radical scavengers and antioxidants on phagocyte-induced mutagenesis. *J. Immunol.* **128**, 2770–2772.

20. Yokoyama, K., Tomino, Y., Yaguchi, Y., Koide, H., Ohmori, D. & Yamakura, H. (1993): Serum superoxide dismutase (SOD) activity in patients with renal disease by a spintrap method using electron spin resonance (ESR). *Jpn. J. Nephrol.* **35**, 809–814 (in Japanese).

21. Zachee, P., Ferrant, A., Daelemans, R., Coolen, L., Goosens, W., Lins, R.L., Couttenye, M., De Broe, M.E. & Boogaerts, M.A. (1993): Oxidative injury to erythrocytes, cell rigidity and splenic hemolysis in hemodialyzed patients before and during erythropoietin treatment. *Nephron* **65**, 288–293.

# Section VIII
## Guanidino compounds in liver failure, hyperammonemia and diabetes

*Guanidino Compounds : 2*, eds. by P.P. De Deyn, B. Marescau, I.A. Qureshi and A. Mori.
©1997 John Libbey & Company Ltd., pp. 327–333.

# Chapter 34

---

# Serum and urinary guanidino compound levels in patients with cirrhosis

---

Bart MARESCAU, Peter P. DE DEYN, Jan HOLVOET, Ilse POSSEMIERS,
Guy NAGELS, Vici SAXENA and Charles MAHLER

*Department of Medicine-UIA, Laboratory of Neurochemistry and Behavior, Born-Bunge Foundation, University of Antwerp, 2610 Wilrijk, Belgium, and Department of Neurology and Internal Medicine, General Hospital Middelheim, 2020 Antwerp, Belgium*

## Summary

To investigate the metabolic relationship between urea and guanidinosuccinic acid we determined the levels of the guanidino compounds, including guanidinosuccinic acid, and urea in serum and urine of cirrhotic patients. Linear correlation studies between serum urea and guanidinosuccinic acid levels were performed. Positive linear correlation coefficients were found in the Child C subgroup ($r = 0.847$; $P < 0.001$) and in the total subgroup including Child B and C patients ($r = 0.848$; $P < 0.0001$).

Serum guanidinoacetic acid levels are significantly increased in the Child C subgroup ($P < 0.0001$ for men and $P < 0.001$ for women). However, the guanidinosuccinic acid levels are significantly ($P < 0.0001$) decreased in the three studied subgroups. Similar results were found for urinary guanidinosuccinic acid excretion levels.

Within each Child subgroup, serum and urinary guanidinosuccinic acid levels were significantly lower in patients with alcohol-induced cirrhosis than in those with non-alcoholic cirrhosis. Similar results were obtained for urea.

The presented findings in cirrhotic patients clearly demonstrate a metabolic relationship between urea and guanidinosuccinic acid. They also show that the urea and guanidinosuccinic acid biosynthesis is significantly lower in patients with cirrhosis of an alcoholic origin than in patients with cirrhosis of a non-alcoholic origin and controls.

## Introduction

One of the main metabolic systems disturbed in cirrhotic patients is certainly ammonia nitrogen metabolism. Indeed, in patients with severe liver disease, high ammonia levels are frequently found and not easy to lower because the cirrhotic liver is failing to metabolize ammonia by the urea cycle[7]. So the capacity for urea synthesis is reduced and the daily urea nitrogen synthesis rate is significantly decreased in cirrhotic patients[8,11].

The pathobiochemistry of the guanidino compounds in uremia and hyperargininæmia has shown a metabolic relationship between urea and guanidinosuccinic acid[9]. Two hypotheses for the biosynthesis of guanidinosuccinic acid have been proposed. In 1970, Cohen suggested that guanidinosuccinic acid could be formed by transamidination of arginine to aspartic acid[5]. According to this hypothesis guanidinosuccinic acid would be a direct catabolite of arginine, formed through one enzymatic reaction. In 1979, Natelson & Sherwin proposed an alternative hypothesis: guanidinosuccinic acid would be formed through different enzymatic steps from urea[12]. The pathobiochemistry of guanidinosuccinic acid in uremia and hyperargininemia could be in favor of the last hypothesis: uremic patients, characterized by increased serum urea levels, are also characterized by increased serum guanidinosuccinic acid levels. Patients with hyperargininemia characterized by a disturbed urea cycle and decreased urea biosynthesis, certainly while under treatment with protein restriction together with supplementation of essential amino acids with or without sodium benzoate, also display decreased biosynthesis of guanidinosuccinic acid[9].

We performed this study to investigate whether the metabolic relationship between urea and guanidinosuccinic acid is also seen in cirrhotic patients. Therefore, we determined the levels of the guanidino compounds, including guanidinosuccinic acid, and urea in serum and urine of a cirrhotic population. Correlation studies were performed.

## Materials and methods

### Patients

This study considers 64 cirrhotic patients (37 men, 27 women) with varying degrees of liver failure. Their ages ranged from 30 to 79 years. The etiologies were alcoholic (n = 49), chronic active hepatitis (n = 9), Budd–Chiari syndrome (n = 1) and iatrogenic (n = 1). In addition, there were one case of primary biliary cirrhosis and three cases of unknown origin. Cirrhotic patients with renal insufficiency were not considered for this study. Diagnosis was based on history, clinical findings, biochemical data, radiologic examinations and in some cases histologic data.

All cirrhotic patients were classified according to Child (Child and Turcotte)[4]. Following items were scored: serum bilirubin mg per cent (< 2.0 = 1; 2.0–3.0 = 2; > 3 = 3), serum albumine g per cent (> 3.5 = 1; 3.0–3.5 = 2; <3 = 3), ascites (none = 1; easily controlled = 2; poorly controlled = 3), encephalopathy (none = 1; minimal = 2; advanced = 3), nutrition (excellent = 1; good = 2; wasting = 3), prothrombin time (1 s prolonged = 1; > 1 s < 4 s prolonged = 2; 4 s prolonged = 3). Total scores of 6–9 corresponded to Child A; scores between 10–14 were rated as Child B and Child C was given to patients with scores between 15 and 18 or > 12 if coma, tense ascites, or substantial encephalopathy was present.

### Collection and preparation of samples

Fasting sampling was done in the morning. After clotting, the blood was centrifuged at $2200 \times g$ at 6 °C for 10 min. A portion of the serum was reserved for urea determination. The remaining serum was stored at –75 °C until analysed. Fasting morning urine was also collected. For the determination of the guanidino compounds, serum and urine samples were deproteinized by mixing equal volumes of a 200 g/l trichloroacetic acid solution with serum or urine. The proteins were centrifuged in a Beckman microfuge (Beckman Instruments International, CH-1207 Geneva, Switzerland). Two hundred microliters supernatant was used for analysis.

### Guanidino compounds and other chemicals

Standard guanidino compounds were acquired from Sigma Chemical Company (St. Louis, MO), creatine and creatinine from Merck (Darmstadt, Germany). α-Keto-δ-guanidinovaleric acid was synthesized enzymatically as described earlier[10]. All other reagents employed were obtained from Merck and of analytical grade.

### Laboratory Methods

The concentration of the guanidino compounds was determined using a Biotronik LC 5001 (Biotronik, Maintal, Germany) amino acid analyser adapted for guanidino compound determination. The guanidino compounds were separated over a cation exchange column using sodium citrate buffers and were detected with the fluorescence ninhydrin method as has been reported in detail earlier[10]. Serum urea nitrogen was determined with diacetylmonoxime as described by Ceriotti[3].

The guanidino compound levels were given as mean ± standard deviation for guanidino compounds present at detectable levels. If, in addition to detectable levels of a particular guanidino compound, no levels could be detected in some samples, the results were given as a range from lower than the detection limit to the highest level obtained in this group. In urine the levels of creatine have a large dispersion, and therefore these results were also expressed as a range of the lowest to the highest value. Results were compared using analysis of variance with LSD *post hoc* comparison (ANOVA, SPSS). The interrelationship of individual serum guanidinosuccinic acid levels with their corresponding serum urea nitrogen levels was assessed by linear correlation studies.

## Results

Levels of serum arginine and homoarginine in cirrhotic patients were not significantly different from control levels (Table 1). Serum levels of guanidinoacetic acid, the first component of the creatine–creatinine biosynthesis pathway, were significantly ($P < 0.0001$ for men and $P < 0.001$ for women) increased in Child C patients. The serum guanidinosuccinic acid levels were very significantly ($P < 0.0001$) decreased in all studied cirrhotic patients independent of their Child score. Urea levels, although to a lesser extent, are also decreased in all cirrhotic patients. The decrease is significant in Child B and C.

**Table 1. Serum guanidino compound (μmol/l) and urea (mmol/e) levels in controls and cirrhotic patients classified according to Child**

| Compound | Controls (n = 66) | Child A (n = 29) | Child B (n = 22) | Child C (n = 13) |
|---|---|---|---|---|
| α-Keto-δ-guanidinovaleric acid | < 0.035 – 0.200 | < 0.035 – 0.220 | < 0.035 | < 0.035 – 0.150 |
| Guanidinosuccinic acid | 0.259 ± 0.096 | 0.139 ±0.149**** | 0.118±0.106**** | 0.097 ± 0.133**** |
| Creatine in men | 30.1 ± 12.3 | 29.8 ± 22.2 | 30.8 ± 27.4 | 16.5 ± 11.5 |
| Creatine in women | 54.8 ± 21.0 | 28.3 ± 16.2** | 35.9 ± 30.7* | 56.6 ± 50.2 |
| Guanidinoacetic acid in men | 2.61 ± 0.517 | 3.10 ± 1.70 | 3.46 ± 1.44 | 5.20 ± 2.95**** |
| Guanidinoacetic acid in women | 2.01 ± 0.572 | 2.67 ± 1.71 | 2.68 ± 1.24 | 4.36 ± 2.64*** |
| α-N-Acetylarginine | < 0.015 – 0.620 | < 0.015 – 0.420 | < 0.015 – 0.500 | < 0.015 – 0.600 |
| Argininic acid | < 0.015 – 0.440 | < 0.015 – 0.410 | < 0.015 – 0.500 | < 0.015 – 0.130 |
| Creatinine in men | 80.8 ± 17.7 | 56.5 ± 13.7**** | 56.0 ± 13.6*** | 65.5 ± 18.3* |
| Creatinine in women | 65.3 ± 19.7 | 56.9 ± 15.0 | 64.5 ± 28.3 | 54.2 ± 19.8 |
| γ-Guanidinobutyric acid | < 0.013 - 0.055 | < 0.013 – 0.09 | < 0.013 – 0.08 | < 0.013 |
| Arginine | 110 ± 23.9 | 105 ± 37.2 | 97.0 ± 29.1 | 98.7 ± 41.7 |
| Homoarginine in men | 1.98 ± 0.634 | 1.87 ± 0.788 | 1.65 ± 0.785 | 1.79 ± 0.670 |
| Homoarginine in women | 1.51 ± 0.609 | 1.52 ± 0.761 | 1.24 ± 0.382 | 1.62 ± 1.03 |
| Guanidine | < 0.06 – 0.210 | < 0.06 – 0.300 | < 0.06 – 0.350 | < 0.06 – 0.400 |
| Urea | 4.42 ± 1.10 | 4.04 ± 1.93 | 3.51 ± 1.53* | 3.18 ± 2.06** |

The levels of β-guanidinopropionic acid and methylguanidine were lower than the detection limit, i.e. < 0.013 and < 0.02 μmol/l, respectively.
The control group considered 33 men and 33 women, the Child A group 19 and 10.
The Child B group 9 and 13, and the Child C group 8 and 5. The data were compared using analysis of variance with LSD *post hoc* comparison (ANOVA, SPSS).
\* Significantly different from control levels ($P < 0.05$).
\*\* Significantly different from control levels ($P < 0.01$).
\*\*\* Significantly different from control levels ($P < 0.001$).
\*\*\*\* Significantly different from control levels ($P < 0.0001$).

As in serum, the urinary guanidinosuccinic acid excretion levels are significantly decreased ($P < 0.01$ for the Child A and $P < 0.0001$ for the Child B and C subgroups) in all the studied cirrhotic patients (Table 2). Urinary excretion levels of γ-guanidinobutyric acid are also significantly decreased.

Within each Child subgroup, serum and urinary guanidino compound levels of the patients with cirrhosis of an alcoholic origin were compared with those of patients with cirrhosis of non-alcoholic origin. This comparative study demonstrated that levels of guanidinosuccinic acid, in both serum and urine, were significantly lower in patients with alcoholic cirrhosis than in others (Table 3). Also, the serum urea levels in alcoholic cirrhosis patients are significantly lower than those seen in non-alcoholic cirrhosis patients. The urinary urea excretion levels found in patients with alcohol-induced cirrhosis were also marginally significantly lower than those observed in non-alcoholic cirrhosis patients except for the Child B subgroup.

**Table 2. Urinary guanidino compound ($\mu$mol/g CTN) and urea ($\mu$mol/g CTN) levels in controls and cirrhotic patients classified according to Child**

| Compound | Controls (n = 34) | Child A (n = 29) | Child B (n = 23) | Child C (n = 13) |
|---|---|---|---|---|
| $\alpha$-Keto-$\delta$-guanidinovaleric acid | < DL – 35 | < DL – 35 | < DL – 35 | < DL – 35 |
| Guanidinosuccinic acid | 25.0 ± 9.03 | 16.3 ± 13.9** | 11.7 ± 8.52*** | 9.87 ± 14.5*** |
| Creatine in men | 30 – 1700 | 43 – 1240 | 14 – 4720 | 25 – 170 |
| Creatine in women | 30 – 3200 | 37 – 3440 | 33 – 1610 | 30 – 4680 |
| Guanidinoacetic acid | 223 ± 128 | 336 ± 326 | 206 ± 201 | 205 ± 152 |
| $\alpha$-N-Acetylarginine | 22.2 ± 9.63 | 28.4 ± 13.7 | 23.9 ± 10.5 | 21.7 ± 9.32 |
| Argininic acid | 6.73 ± 2.80 | 6.44 ± 2.73 | 6.66 ± 5.22 | 5.33 ± 2.76 |
| $\beta$-Guanidinopropionic acid | < DL – 1.00 | < DL – 1.00 | < DL – 1.00 | < DL – 0.620 |
| $\gamma$-Guanidinobutyric acid | 11.8 ± 9.29 | 9.69 ± 7.04 | 7.51 ± 4.60* | 4.14 ± 2.43** |
| Arginine | 20.9 ± 15.9 | 14.9 ± 8.29 | 15.1 ± 7.18 | 17.4 ± 35.3 |
| Homoarginine | < DL – 6 | < DL –1.4 | < DL – 2.4 | < DL – 15 |
| Guanidine | 9.75 ± 3.38 | 12.5 ± 7.73 | 12.0 ± 5.94 | 10.8 ± 4.23 |
| Methylguanidine | 2.99 ± 1.24 | < DL – 10 | < DL – 10 | < DL – 10 |
| Urea | 185 ± 48.6 | 189 ± 68.6 | 192 ± 93.6 | 145 ± 71.8 |

The control group consisted of 15 men and 19 women, the Child A group 19 and 10, the Child B group 10 and 13, and the Child C group 8 and 5. The data were compared using analysis of variance with LSD *post hoc* comparison (ANOVA, SPSS). DL, detection limit.
* Significantly different from control levels ($P < 0.05$).
** Significantly different from control levels ($P < 0.01$).
*** Significantly different from control levels ($P < 0.0001$).

Figure 1 shows the relation between the serum urea levels of Child B and C patients and their corresponding levels of serum guanidinosuccinic acid. A significant positive linear correlation ($r = 0.848$, $P < 0.0001$) shows that patients with lower serum urea levels also had lower serum guanidinosuccinic acid levels.

## Discussion

The urea cycle incorporates nitrogen not required for net biosynthetic purposes into urea, which serves as a waste nitrogen product in ureotelic animals. The urea cycle also forms part of those biochemical reactions required for the *de novo* biosynthesis and degradation of arginine. In ureotelic animals a large proportion of arginine biosynthesis takes place in the liver. Arginine and homoarginine, both hydrolyzed to urea by arginase, had normal serum and urinary excretion levels in our studied cirrhotic patients, thus suggesting that their arginine and homoarginine hepatic biosynthesis is not disturbed. However, another explanation for these normal levels could be that a normally functioning kidney in cirrhotic patients is itself capable of synthesizing enough arginine and homoarginine. Indeed, renal biosynthesis of arginine has been long recognized[1,13].

The biosynthesis of guanidinoacetic acid occurs primarily in the kidney, whereas its methylation to creatine takes place mainly in the liver[15]. Our results clearly show that serum guanidinoacetic acid levels are increased with increased liver damage. This suggests that the biosynthesis of creatine is disrupted in the damaged liver. The increased retention of guanidinoacetic acid in blood could be a consequence of liver failure. In addition, it is noteworthy that in renal failure decreased levels of guanidinoacetic acid are found in serum[14].

**Table 3. Serum guanidinosuccinic acid (mmol/l) and urea (mmol/l) and urinary guanidinosuccinic acid (mmol/g CTN) and urea (mmol/g CTN)levels in alcoholic and non-alcoholic cirrhotic patients classified according to Child and controls**

| Serum | Controls n = 66 | Child A ethylic n = 22 | Child A non-ethylic n = 7 | Child B ethylic n = 16 | Child B non-ethylic n = 6 | Child C ethylic n = 11 | Child C non-ethylic n = 2 |
|---|---|---|---|---|---|---|---|
| GSA | 0.259 ± 0.096 | 0.085 ± 0.058[h] | 0.307 ± 0.222[d] | 0.086 ± 0.093[h] | 0.203 ± 0.094[a] | 0.043 ± 0.030[h] | 0.390 ± 0.028[d] |
| Urea | 4.42 ± 1.10 | 3.53 ± 1.58[f] | 5.62 ± 2.21[c] | 3.12 ± 1.49[g] | 4.52 ± 1.24[a] | 2.50 ± 1.31[h] | 6.92 ± 0.877[d] |
| Urine | Controls n = 34 | Child A ethylic n = 21 | Child A non-ethylic n = 7 | Child B ethylic n = 17 | Child B non-ethylic n = 6 | Child C ethylic n = 11 | Child C non-ethylic n = 2 |
| GSA | 25.0 ± 9.03 | 11.8 ± 8.30[h] | 29.8 ± 19.0[d] | 8.19 ± 5.41[h] | 21.8 ± 7.82[b] | 4.46 ± 1.72[h] | 39.7 ± 19.6[d] |
| Urea | 185 ± 48.6 | 175 ± 58.4 | 231 ± 84.2 | 188 ± 102 | 216 ± 65.5 | 130 ± 61.6[e] | 231 ± 81.3 |

The data were compared using analysis of variance with LSD *post hoc* comparison (ANOVA, SPSS).
[a] Significantly higher than levels found in alcoholic cirrhotic patients of the same subgroup ($P < 0.05$).
[b] Significantly higher than levels found in alcoholic cirrhotic patients of the same subgroup ($P < 0.01$).
[c] Significantly higher than levels found in alcoholic cirrhotic patients of the same subgroup ($P < 0.001$).
[d] Significantly higher than levels found in alcoholic cirrhotic patients of the same subgroup ($P < 0.0001$).
[e] Significantly different from control levels ($P < 0.05$).
[f] Significantly different from control levels ($P < 0.01$);
[g] Significantly different from control levels ($P < 0.001$);
[h] Significantly different from control levels ($P < 0.0001$)

It has long been accepted that urea biosynthesis in cirrhotic patients is decreased. Serum and urinary excretion levels of guanidinosuccinic acid in our cirrhotic patients are clearly and significantly decreased compared with the control levels, suggesting decreased biosynthesis. This could denote a metabolic relationship between urea and guanidinosuccinic acid. Moreover, Fig. 1 clearly illustrates a positive linear correlation between the serum urea and guanidinosuccinic acid levels. A significant positive linear correlation is also found in the Child C subgroup ($r = 0.847, P < 0.001$). The correlation coefficient in the total studied cirrhotic patient group (Child A, B and C) is lower ($r = 0.608, P < 0.0001$). This could perhaps be explained by the fact that in the Child A subgroup some patients still have a normal residual urea production capacity, along with those with a subnormal or decreased capacity. Because there is a good correlation between serum urea levels and serum guanidinosuccinic acid levels in our studied cirrhotic patients, we will investigate whether guanidinosuccinic acid can be used as a marker for residual urea production capacity. Could guanidinosuccinic acid levels be used as a parameter or an indicator for liver dysfunction? Brewer *et al.*[2] and Hansen & Poulsen[6] demonstrated that the urea production capacity or the daily urea biosynthesis rate is a marker for residual functional liver mass in rats. Müting *et al.*[11] proposed the application of this test in humans. Because there is a correlation between urea and guanidinosuccinic acid, guanidinosuccinic acid can perhaps be used as a complementary test for residual liver detoxification capacity.

Maier & Gerok[8] and Müting *et al.*[11] showed that the daily urea nitrogen synthesis rate was decreased in cirrhotic patients. Subdividing our studied cirrhotic patients into an alcoholic and a non-alcoholic group, the results presented in Table 3 clearly show lower urea levels in the alcoholic group. Moreover,

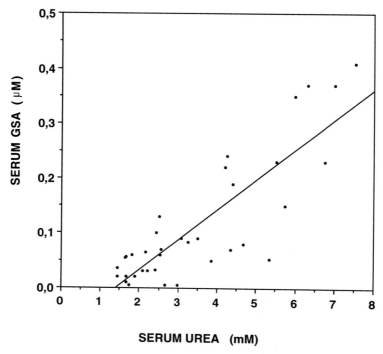

*Fig. 1. Relation between serum urea levels and the corresponding serum guanidinosuccinic acid levels in Child B and Child C cirrhotic patients.*

whereas in the total cirrhotic subgroups only the Child B and C patients showed significantly lower serum urea levels compared with control levels (Table 1), the serum urea levels of alcoholic cirrhotic patients are all significantly lower than the control urea levels (Table 3). The alcoholic cirrhotic patients also have significantly lower guanidinosuccinic acid levels than those found in non-alcoholic cirrhotic patients. The differences seem even more pronounced than those found for urea. These results clearly show that urea biosynthesis, and perhaps ammonia-nitrogen metabolism, is more disturbed and certainly more decreased in patients with cirrhosis of alcoholic origin than in those with cirrhosis of non-alcoholic origin. As a consequence of decreased urea biosynthesis and the metabolic relationship between urea and guanidinosuccinic acid, we also found lower guanidinosuccinic acid levels in alcoholic cirrhotic patients than in non-alcoholic ones.

To conclude, the metabolic relationship between urea and guanidinosuccinic acid demonstrated earlier in the pathobiochemistry of uremia and hyperargininemia is also seen in cirrhotic patients. The results of this study also show that the biosynthesis of urea and guanidinosuccinic acid is significantly lower in patients with cirrhosis of alcoholic origin than in patients with cirrhosis of non-alcoholic origin and controls.

### Acknowledgements

Supported by the Ministerie van Nationale Opvoeding en Nederlandse Cultuur, the University of Antwerp, OCMW Medical Research Foundation, Neurosearch Antwerp, the Born-Bunge Foundation, the United Fund of Belgium and NFWO grants 3.0044.92 and 3.0064.93.

# References

1.  Borsook, H. & Dubnoff, J.W. (1941): The conversion of citrulline to arginine in kidney. *J. Biol. Chem.* **140,** 717–738.

2.  Brewer, T.G., Berry, W.R., Harmon, J.W., Walker, S.H. & Dunn, M.A. (1984): Urea synthesis after protein feeding reflects hepatic mass in rats. *Hepatology* **4,** 905–911.

3.  Ceriotti, G. (1971): Ultramicro determination of plasma urea by reaction with diacetylmonoxime antipyrine without deproteinization. *Clin. Chem.* **17,** 400–402.

4.  Child, C.G. & Turcotte, J.G. (1964): Surgery and portal hypertension. In: *The Liver and Portal Hypertension.* Ed. J.E. Dunphy, pp. 49–51. Philadelphia: W.B. Saunders.

5.  Cohen, B. (1970): Guanidinosuccinic acid in uremia. *Arch. Intern. Med.* **126,** 846–850.

6.  Hansen, B.A. & Poulsen, H.E. (1986): The capacity of urea-N synthesis as a quanititative measure of the liver mass in rats. *J. Hepatol.* **2,** 468–474.

7.  Lockwood, A.H., McDonald, J.M. & Reiman, R.E. (1979): The dynamics of ammonia metabolism in man: effects of liver disease and hyperammonemia. *J. Clin. Invest.* **63,** 449–460.

8.  Maier, K.P. & Gerok, W. (1980): Maximal rates of urea-synthesis in liver biopsies of normal and cirrhotic subjects (abstract). *Gastroenterology* **79,** 1115.

9.  Marescau, B., De Deyn, P.P., Qureshi, I.A., De Broe, M.E., Antonozzi, I., Cederbaum, S.D., Cerone, R., Chamoles, N., Gatti, R., Kang, S.-S., Lambert, M., Possemiers, I., Snyderman, S.E. & Yoshino, M. (1992): The pathobiochemistry of uremia and hyperargininemia further demonstrates a metabolic relationship between urea and guanidinosuccinic acid. *Metabolism* **41,** 1021–1024.

10. Marescau, B., Deshmukh, D.R., Kockx, M., Possemiers, I., Qureshi, I.A., Wiechert, P.P. & De Deyn, P.P. (1992): Guanidino compounds in serum, urine, liver, kidney and brain of man and some ureotelic animals. *Metabolism* **41,** 526–532.

11. Müting, D., Paquet, K.J. & Koussouris, P. (1988): Plasma ammonia and urea-N synthesis rate in human liver cirrhosis. In: *Advances in Ammonia Metabolism and Hepatic Encephalopathy.* Eds. P.A. Soeters, J.H.P. Wilson & A.J. Meijer, pp. 91–98. Amsterdam: Excerpta Medica.

12. Natelson, S. & Sherwin, J.E. (1979): Proposed mechanism for urea nitrogen reutilization: relationship between urea and proposed guanidine cycles. *Clin. Chem.* **25,** 1343–1344.

13. Ratner, S. & Petrack, B. (1953): The mechanism of arginine synthesis from citrulline in kidney. *J. Biol. Chem.* **200,** 175–185.

14. Tofuku, Y., Muramoto, H., Kuroda, M. & Takeda, R. (1985): Impaired metabolism of guanidinoacetic acid in uremia. *Nephron* **41,** 174–178.

15. Van Pilsum, J.F., Stephens, G.C. & Taylor, D. (1972): Distribution of creatine, guanidinoacetic acid and the enzymes for their biosynthesis in the animal kingdom. *Biochem. J.* **126,** 325–345.

*Guanidino Compounds : 2*, eds. by P.P. De Deyn, B. Marescau, I.A. Qureshi and A. Mori.
©1997 John Libbey & Company Ltd., pp. 335–340.

# Chapter 35

# Guanidino compounds in the brain and peripheral organs of ornithine transcarbamylase deficient *spf* mutant mice

I.A. QURESHI[1],[*] L. RATNAKUMARI[1], B. MARESCAU[2] and P.P. DE DEYN[2]

[1]*Division of Medical Genetics, Sainte-Justine Hospital, Montreal, QC, Canada H3T 1C5, and* [2]*Laboratory of Neurochemistry and Behavior, University of Antwerp, Boon-Bunge Foundation, UIA, 2610 Antwerp, Belgium*

## Summary

The spf mouse with an X-linked hepatic ornithine transcarbamoylase (OTC: E.C. 2.1.3.3) deficiency exhibits significantly lower levels of arginine and citrulline in the serum than normal controls. In the present study, the effect of sustained lower arginine levels on guanidino compound metabolism was studied by measuring the levels of several arginine-related guanidino compounds in spf mouse brain along with liver and kidney. The cerebral levels of $\gamma$-guanidinobutyric acid ($P < 0.01$), $\alpha$-N-acetylarginine ($P < 0.001$), argininic acid ($P < 0.001$), guanidinoacetic acid ($P < 0.001$), creatine ($P < 0.01$), and guanidinosuccinic acid ($P < 0.01$) were lower in *spf* mice than in normal controls. In the liver, $\alpha$-keto-$\delta$-guanidinovaleric acid and argininic acid were significantly lower than in control mice. The renal levels of $\gamma$-guanidinobutyric acid and guanidinosuccinic acid were significantly higher whereas guanidinoacetic acid were lower in mutant *spf* mice than in normal controls. Because arginine and guanidino compounds play an important role in nitric oxide synthesis, the observed decrease in cerebral levels of these compounds may contribute to the decreased production of nitric oxide in congenital OTC deficiency.

## Introduction

Guanidino derivatives accumulate significantly in certain pathologic conditions, such as uremia[1,2], hyperargininemia[8] and epilepsy[4], and contribute to the pathophysiologic manifestations of these disorders. Arginine, the conditionally essential amino acid, plays an important role in the urea cycle and some other metabolic pathways by providing the guanidino group. $\alpha$-Keto-$\delta$-guanidinovaleric acid (GVA), $\alpha$-N-acetylarginine (NAA), argininic acid (ArgA) and $\gamma$-guanidinobutyric acid (GBA) are the catabolic products of arginine resulting from the action of various enzymes, namely a transaminase, an acetylase, a dehydrogenase and a transamidinase. The enzyme arginine-glycine amidinotransferase catalyzes the transfer of amidino groups from arginine to glycine to form guanidinoacetic acid (GAA)[17]. Guanidinoacetate is a precursor for creatine, which serves as an important energy metabolite in muscle and nervous tissue.

The spf mutant mouse with an X-chromosomal defect of hepatic ornithine transcarbamylase (OTC) serves as a useful animal model to study the neurochemical consequences of congenital urea cycle disorders[12,13]. With less than 10 per cent of normal liver OTC activity and a significantly increased urinary orotate excretion, the sparse-fur mutant mouse very closely resembles the congenital hyperammonemia type II syndrome with metabolic arginine deficiency seen in children[6]. The serum, urinary and cerebral arginine levels are significantly lower in spf mutant mice than in control mice[5,14,15]. In the present study, we investigated the effect of sustained lower arginine levels on the concentration of arginine-related guanidino compounds in OTC deficiency. The cerebral levels of several of these guanidino compounds were significantly decreased in spf mutant mice as compared with normal controls.

## Materials and methods

### Animals

The parent stock for the colony of sparse-fur mice was originally obtained from Dr L.B. Russel of the Oak Ridge National Laboratories, Oak Ridge, TN[3]. Male spf mice were the progeny of matings of homozygous affected spf/spf females with spf males[12]. All male progeny of these matings, being spf/Y, were separated by simple sexing. However, at the end of each experimental study, liver ornithine transcarbamylase activity was determined to check the mutant status of the animals. CD–1 strain mice obtained from Canadian Breeding Farms, St Constant, Quebec, were used as normal controls. All animals were kept in a controlled environment (12 h dark/12 h light) with free access to water and food (Purina mouse chow, Ralston Purina, St Louis, MO) at the animal house of the Sainte-Justine Hospital. The animals were kept and experimented on according to the guidelines of the Canadian Council on Animal Care (Guide to the care and use of experimental animals, Vol. 2, 1984).

### Hepatic OTC activity

Liver OTC activity was measured as described previously[12] to verify the mutant status of the spf animals.

### Measurement of guanidino compounds

Brain tissue ($\pm 300$ mg) was homogenized in 1 ml of distilled water at $0°C$ with a tissue tearor (Biospec Products, Bartlesville, USA), model 985. The probe was washed with 1 ml of 30 per cent trichloroacetic acid and added to the homogenate. After deproteinization the samples were centrifuged (10 0 000 x $g$ for 30 min at 6 °C) and the supernatant was used for guanidino compound analysis using a Biotronic LC 5001 (Biotronik, Maintal, Germany) amino acid analyzer adapted for guanidino compound determination. The guanidino compounds were separated over a cation exchange column using sodium citrate buffers and were detected with the fluorescence ninhydrin method as has been reported in detail elsewhere[10].

### Measurement of amino acids

Another fraction of the supernatant obtained for guanidino compound determination (see above) was used for the determination of citrulline and ornithine using a Biotronic LC 6001 (Biotronik, Maintal, Germany) amino acid analyzer. They were separated over a cation exchange column using lithium citrate buffers and were detected with the colorimetric ninhydrin method. Chromatographic conditions and characteristics have been described in detail by Pei.[11]

Protein concentrations were determined by the method of Lowry et al.[7] Statistical analysis was done by Student's $t$-test. All the data are expressed as mean $\pm$ SE.

## Results and discussion

The hepatic OTC activity levels of spf mutant mice were approximately one tenth of the activity levels of control mice (6.23 $\pm$ 0.8 µmol citrulline/min per mg protein in spf mice $vs$ 64.2$\pm$ 4.0 in control mice).

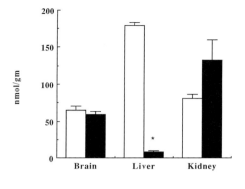

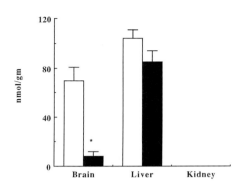

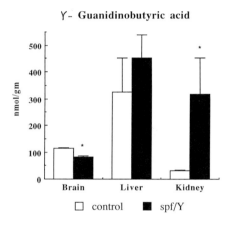

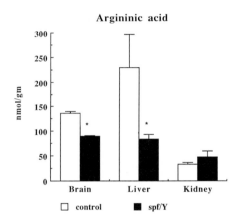

Fig. 1. Effect of congenital OTC deficiency on α-keto-δ-guanidinovaleric acid and γ-guanidinobutyric acid. Each value represents mean ± SE of five mice. * P<0.01 vs control (unpaired t-test)

Fig. 2. Effect of congenital OTC deficiency on α-N-acetylarginine and argininic acid. Each value represents mean ± SE of five mice. *P < 0.001 vs control (unpaired Student's t-test).

Because the metabolism of guanidines in different tissues could be interrelated, we measured the levels of various guanidino compounds in brain, liver and kidney of control and OTC deficient spf mutant mice, to evaluate the effect of sustained lower arginine levels on guanidino compound metabolism. The cerebral concentrations of GBA, NAA, ArgA, GAA, creatine and guanidinosuccinic acid were significantly lower in OTC deficient spf mutant mice than in normal controls (Fig. 1–4). Cerebral levels of homoarginine and β-guanidinopropionic acid were below the level of detection in both control and spf mutant mice (results not shown). The decrease observed in cerebral arginine, citrulline and ornithine levels of spf mutant mice compared with normal controls (Table 1) may be the consequence of decreased levels of these amino acids in the serum[14]. Several catabolic products of arginine, like GBA, ArgA and NAA, were significantly decreased in the spf mutant mouse brain, which may be caused by the lower levels of cerebral arginine. The GAA which is formed by the arginine-glycine transamidinase was also decreased, with the creatine. Though the liver creatine levels are in the normal range, both serum[14] and brain creatine (Fig.4) levels were significantly decreased in spf mutant mice. Because creatine also serves as an energy source in the nervous tissue, this decrease in creatine may contribute to the pathophysiology of the OTC deficiency.

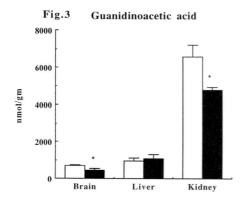

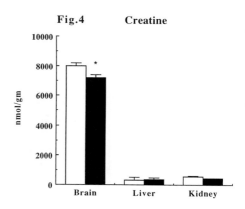

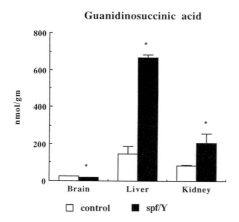

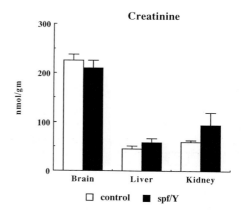

*Fig. 3. Effect of congenital OTC deficiency on guanidinoacetic acid and guanidinosuccinic acid. Each value represents mean ± SE of five mice. *P < 0.001 vs control (unpaired Student's t-test).*

*Fig. 4. Effect of congenital OTC deficiency on creatine and creatinine. Each value represents mean ± SE of five mice. *P < 0.01 vs control (unpaired Student's t-test).*

In liver, the direct catabolic products of arginine, like GVA and argininic acid, were significantly decreased (Fig. 1–2) in OTC deficient spf mutant mice as compared with normal controls. However, the significant increase in spf mouse liver guanidinosuccinic acid levels was intriguing as, in many human OTC deficient patients, there has been an absence of urinary guanidinosuccinic acid[16]. The levels of other guanidino compounds in livers of spf mice were not statistically different from those of control mice .

In the kidney, GBA and guanidinosuccinic acid levels were significantly higher (Fig 1,3) in spf mutant mice than in normal controls. The GAA levels of kidney were significantly lower in spf mice than in normal controls (Fig. 3), indicating the lower transamidinase activity in spf mice. This agrees well with the reported decreases in serum and urinary GAA levels in spf mice[14]. The concentration of α-N-acetylarginine was below the level of detection in kidney samples of both control and spf mutant mice.

The exact regulatory mechanism for guanidino compound synthesis in brain is not yet clear. Earlier studies of Watanabe *et al.*[18] reported that GAA levels in both liver and kidney reach maximum levels 3 weeks after birth whereas in the brain the highest levels are observed 1 week following birth, suggesting a different regulatory mechanism for GAA in the brain. It is possible that some of the changes seen in the cerebral levels of guanidino compounds in this study are the result of altered arginine, ornithine and citrulline metabolism in other organs, such as liver, intestine and kidney. Earlier studies from our laboratory have demonstrated a significant reduction in small intestinal OTC activity in spf mutant mice[13].

**Table 1. Amino acid levels in brain, liver and kidney of spf mutant mouse**

| Amino acid | Controls (nmol/g) | *spf* mutant mice (nmol/g) |
|---|---|---|
| *Brain* | | |
| Arginine | 128 ± 3.1 | 78 ± 7.5** |
| Citrulline | 12 ± 0.29 | 8.0 ± 0.1** |
| Ornithine | 11.7 ± 3.1 | 6.9 ± 0.7* |
| *Liver* | | |
| Arginine | 15 ± 0.93 | 24.2 ± 4.8 |
| Citrulline | 79 ± 25 | 17 ± 0.8** |
| Ornithine | 727 ± 115 | 320 ± 19** |
| *Kidney* | | |
| Arginine | 189 ± 15 | 183 ± 12 |
| Citrulline | 20 ± 1.1 | 14.7 ± 1.9* |
| Ornithine | 17 ± 0.9 | 35 ± 4** |

Each value is mean ± SE of five animals. *$P < 0.05$; **$P < 0.001$ *vs* control (unpaired Student's *t*-test).

In conclusion, the results of the present study would indicate that, during congenital OTC deficiency, the sustained low levels of cerebral arginine result in decreased production of several guanidino compounds especially in the brain. Because arginine and guanidino compounds, play an important role in nitric oxide synthesis, the observed decrease in cerebral levels of these compounds may contribute to the decreased cerebral nitric oxide production in OTC deficiency.

### Acknowledgements

The studies described here were funded by The Medical Research Council of Canada (MT–9124), UIA, NFWO (grants 3.0044.92 and 3.0064.93), Born–Bunge Foundation, the United Fund of Belgium and the Antwerp OCMW Medical Research Foundation. We would like to thank Mr Michel Leblanc for his help in breeding the spf mutant mice.

### References

1. Cohen, B.D., Stein, I.M. & Bonas, J.E. (1968): Guanidinosuccinic aciduria in uremia: a possible alternate pathway for urea synthesis. *Am. J. Med.* **45**, 63–68.

2. De Deyn, P.P., Marescau, B., Lornoy, W., Becaus, I. & Lowenthal, A. (1986): Guanidino compounds in uraemic dialysed patients. *Clin. Chim. Acta* **157**, 143–150.

3. DeMars, R., LeVan, S.L., Trend, B.L. & Russel, L.B. (1976): Abnormal ornithine carbamoyltransferase in mice having the sparse-fur mutation. *Proc. Natl Acad. Sci. USA.* **73**, 1693–1697.

4. Hirayasu, Y., Morimoto, K., Otsuki, S. & Mori, A. (1994): Increase of guanidinoacetic acid and methylguanidine in the brain following amygdala kindling in rats. In: *Guanidino Compounds in Biology and Medicine*. Eds. P.P. De Deyn., B. Marescau., V. Stalon. & I.A. Qureshi, pp. 419–424. London: John Libbey.

5. Inoue, I., Gushiken, T., Kobayashi, K. & Saheki, T. (1987): Accumulation of large neutral amino acids in the brain of sparse-fur mice at hyperammonemic state. *Biochem. Med. Metab. Biol.* **38**, 378–386.

6. Levin, B., Oberholzer, V.G. & Sinclair, R.L. (1969): Biochemical investigation of hyperammonemia. *Lancet* **ii**, 170–174.

7. Lowry, O.H., Rosebrough, N.J., Farr, A.L. & Randall, R.J. (1951): Protein measurement with the folin-phenol reagent. *J. Biol. Chem.* **193**, 265–275.

8.    Marescau, B., Qureshi, I.A., De Deyn, P., Letarte, J., Ryba, R. & Lowenthal, A. (1985): Guanidino compounds in plasma, urine and cerebrospinal fluid of hyperargininemic patients during therapy. *Clin. Chim. Acta* **146,** 21–27.

9.    Marescau, B., De Deyn, P.P., Wiechert, P., Van Gorp, L. & Lowenthal, A. (1986): Comparative study of guanidino compounds in serum and brain of mouse, rat, rabbit and man. *J. Neurochem.* **46,** 717–720.

10.   Marescau, B., Deshmukh, D.R., Kockx, M., Possemiers, I., Qureshi, I.A., Wiechert, P. & De Deyn, P.P. (1992): Guanidino compounds in serum, urine, liver, kidney and brain of man and some urotelic animals. *Metabolism.* **41,** 526–532.

11.   Pei, H. (1994): The use of cation-exchange resin in analytical research of amino acids. Dissertation to obtain the degree of Master in Biomedical Sciences, University of Antwerp (UIA), Belgium.

12.   Qureshi, I. A., Letarte, J. & Ouellet, R. (1979): Ornithine transcarbamylase deficiency in mutant mice. 1. Studies on the characterization of enzyme defect and suitability as an animal model of human disease. *Pediatr. Res.* **13,** 807–811.

13.   Qureshi, I. A., Letarte, J. & Ouellet, R. (1985): Expression of ornithine transcarbamylase deficiency in the small intestine and colon of sparse-fur mutant mice. *J. Pediatr. Gastroenterol. Nutr.* **4,** 118–124.

14.   Qureshi, I.A., Marescau, B., Levy, M., De Deyn, P.P., Letarte, J. & Lowenthal, A. (1989): Serum and urinary guanidino compounds in sparse-fur mutant mice with ornithine transcarbamylase deficiency. In: *Guanidines 2.* Eds. A. Mori, B.D. Cohen & H. Koide, pp. 45–51. New York: Plenum Press.

15.   Seiler, N., Grauffel, C., Daune-Anglard, G., Sarhan, S. & Knodgen, B. (1994): Decreased hyperammonemia and orotic aciduria due to inactivation of ornithine aminotransferase in mice with a hereditary abnormal ornithine carbamoyltransferase. *J. Inherit. Metab. Dis.* **17,** 691–703.

16.   Stein, J.M., Cohen, B.D. & Kornhauser, R.S. (1969): Guanidinosuccinic acid in renal failure, experimental azotemia and inborn errors of the urea cycle. *N. Engl. J. Med.* **280,** 926–930.

17.   Van Pilsum, J.F., Stephens, G.C. & Taylor, D. (1972): Distribution of creatine, guanidinoacetate and the enzymes for their biosynthesis in the animal kingdom: implication for phylogeny. *Biochem. J.* **126,** 325–345.

18.   Watanabe, Y., Shindo, S. & Mori, A. (1985): Developmental changes in guanidino compound levels in mouse organs. In: *Guanidines.* Eds. A. Mori, B.D. Cohen & A. Lowenthal, pp. 49–58. New York: Plenum Press.

*Guanidino Compounds : 2*, eds. by P.P. De Deyn, B. Marescau, I.A. Qureshi and A. Mori.
©1997 John Libbey & Company Ltd., pp. 341–346.

# Chapter 36

# Urinary guanidinoacetic acid excretion in streptozotocin-induced diabetic rats

## H. KOIDE and I. KIYATAKE

*Division of Nephrology, Department of Medicine, Juntendo University School of Medicine, Hongo, Bunkyo-ku, Tokyo 113, Japan*

## Summary

Guanidinoacetic acid is considered to be synthesized by glycine amidinotransferase in the proximal tubular epithelial cells. The aim of the present study was to determine serum, urinary and renal cortical guanidinoacetic acid levels in streptozotocin-induced diabetic rats with or without insulin treatment to assess whether guanidinoacetic acid can be a clinical marker for renal tubular dysfunction in diabetes.

Serum, urine and renal cortex were obtained 1, 2 and 3 weeks after injection of 65 mg/kg body weight of streptozotocin. Guanidinoacetic acid and 1,5-anhydroglucitol levels were measured by high-performance liquid chromatography. N-Acetyl-β-D-glucosaminidase (NAG) activity was determined by the enzymatic method.

Guanidinoacetic acid levels in serum, urine and renal cortex from diabetic rats were significantly decreased compared with levels in control rats. In contrast, urinary 1,5-anhydroglucitol levels and NAG activity increased significantly in diabetic rats. A decrease in serum, urinary and renal cortical guanidinoacetic acid levels was attenuated by insulin treatment.

These results indicate that a high serum glucose level may affect guanidinoacetic acid synthesis in renal cortex, and that urinary guanidinoacetic acid is a useful indicator for renal tubular injury in the early stage of diabetic nephropathy.

## Introduction

Guanidinoacetic acid is synthesized from arginine and glycine by glycine amidinotransferase mainly in the kidney[3]. A part of guanidinoacetic acid synthesized in the kidney is transported to the liver, where it is methylated to creatine by guanidinoacetate methyltransferase[5], and the remainder is excreted in the urine[11]. Using isolated renal tubules from the rat kidney, we have demonstrated previously that guanidinoacetic acid is synthesized from arginine or canavanine and glycine, but not from canaline, hydroxyurea, citrulline and argininosuccinic acid[14], suggesting that the guanidine cycle proposed by Natelson & Sherwin[10] may not function fully in the renal tubules. Which nephron segment is the site of guanidinoacetic acid synthesis in the kidney? We have reported previously that, in microdissected individual nephron segments from the rat kidney, guanidinoacetic acid synthesis was limited to the first portion (S1) and the second portion (S2) of the proximal tubule using arginine and glycine as substrate, and that the S1 portion showed significantly higher activity for guanidinoacetic acid synthesis than did the S2 portion[15]. These results suggest that urinary guanidinoacetic acid levels may reflect the metabolic function of the proximal tubule in the kidney. We have reported previously that urinary excretion of guanidinoacetic acid decreased in gentamicin

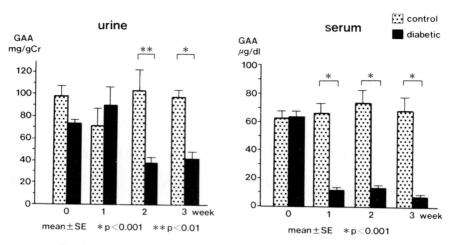

Fig. 1. *Urinary and serum guanidinoacetic acid levels in diabetic rats.*

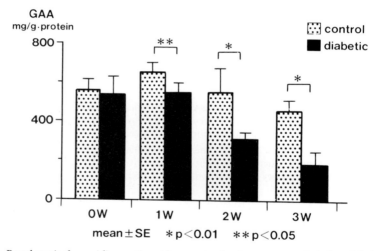

Fig. 2. *Renal cortical guanidinoacetic acid concentration in streptozotocin-induced diabetic rats.*

nephrotoxicity[9]. A small dose of 5 mg/kg body weight of gentamicin was enough to reduce urinary guanidinoacetic acid levels in rats. Thus, it is highly likely that the measurement of urinary guanidinoacetic acid excretion is a sensitive tool for the early diagnosis of the proximal tubular injury such as drug nephrotoxicity, hypertensive renal diseases[13] or rejection crises in kidney transplantation[7].

The aim of the present study was to determine guanidinoacetic acid levels in serum, urine and renal cortex from streptozotocin-induced diabetic rats with or without insulin treatment to assess whether urinary guanidinoacetic acid can be a sensitive marker for renal tubular dysfunction in the early stage of diabetic nephropathy.

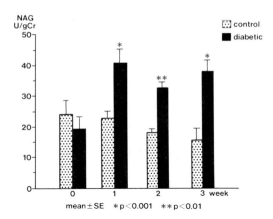

*Fig. 3. Urinary N-acetyl-β-D-glucosaminidase in rats.*

## Materials and methods

Eighty male Sprague-Dawley rats (Charles River Japan, Tokyo, Japan), weighing 150 g, were used in this study. Diabetes was induced in 60 rats by a single tail-vein injection of streptozotocin(65 mg/kg body weight). Streptozotocin was obtained from Sigma Chemical (St. Louis, MO, USA). After 24 h, diabetes was confirmed in streptozotocin-treated rats by measurement of the tail-vein plasma glucose level by the o-toluidine method[6]. Rats were excluded if their plasma glucose levels were less than 250 mg/dl. Age-matched control rats were injected with the same volume of saline solution. The diabetic and control rats were killed 0, 1, 2 and 3 weeks after streptozotocin injection. Six rats from each group were used for determination of guanidinoacetic acid in serum, urine and renal cortex at each time. Seven days after the injection of streptozotocin, 20 rats received NPH insulin (Nordisk, Copenhagen, Denmark). The appropriate dose of insulin was adjusted between 8 and 14 units daily to maintain the plasma glucose level at approximately 200 mg/dl. Seven insulin-treated diabetic rats were studied 1 week after insulin injection. A 24 h urine specimen and blood samples were collected for the determination of glucose, creatinine and guanidinoacetic acid. The kidney was removed and the renal cortex was isolated for guanidinoacetic acid assay.

Guanidinoacetic acid was determined by high-performance liquid chromatography using a Guanidinopack II (Japan Spectroscopic, Tokyo, Japan) and a fluorescence detector (Model FP 110-C, Japan Spectroscopic, Tokyo, Japan)[19]. N-Acetyl-β-D-glucosaminidase (NAG) activity and 1,5-anhydroglucitol concentration were analyzed in the remaining urine specimens. NAG activity was measured using m-cresolsulfophthalyl-N-acetyl-β-D-glucosaminide as substrate (NAG test Shionogi, Shionogi, Tokyo, Japan). Serum and urinary 1,5-anhydroglucitol levels were measured by a enzymatic method after the isolation of polyols by liquid chromatography[1].

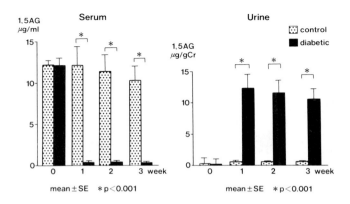

*Fig. 4. 1,5-Anhydroglucitol levels in serum and urine in diabetic rats.*

## Results and discussion

Table 1 shows the serum glucose and creatinine levels in diabetic rats. Although the serum glucose levels increased markedly in diabetic rats, there was no elevation in serum creatinine levels in diabetic rats compared with those in control rats. Urinary guanidinoacetic acid levels were decreased significantly 2 and 3 weeks after streptozotocin injection (Fig. 1). Serum guanidinoacetic acid concentration was also decreased significantly 1 week after streptozotocin injection, and thereafter decreased with the progression of diabetic nephropathy (Fig. 1). There was a time-lag in the reduction of guanidinoacetic acid between serum and urine. The decreased excretion of guanidinoacetic acid in diabetic rats may be due to proximal tubule impairment and due to a decrease in serum guanidinoacetic acid levels. It is likely that a decrease in serum guanidinoacetic acid levels is due to proximal tubular injury in the early stage of diabetic nephropathy but not due to chronic renal failure, because serum creatinine levels were not elevated. Another possible cause of the reduction in serum guanidinoacetic acid levels is dysfunction of the pancreas in streptozotocin-induced diabetic rats. The pancreas is generally considered to be one of the organs with guanidinoacetic acid synthesis[16].

**Table 1. Serum glucose and creatine levels in diabetic rats**

| Week | Glucose (mg/dl) | | Creatinine (mg/dl) | |
| | Control | Diabetes | Control | Diabetes |
|---|---|---|---|---|
| 0 | 122.3 ± 5.6 | 110.5 ± 5.7 | 0.57 ± 0.07 | 0.55 ± 0.05 |
| 1 | 112.2 ± 4.6 | 265.1 ± 23.5* | 0.62 ± 0.02 | 0.51 ± 0.03 |
| 2 | 133.4 ± 4.6 | 333.0 ± 15.7* | 0.64 ± 0.02 | 0.64 ± 0.02 |
| 3 | 101.8 ± 2.3 | 400.7 ± 23.4* | 0.64 ± 0.02 | 0.58 ± 0.05 |

Values are means ± SE; *$P < 0.001$.

Renal cortical guanidinoacetic acid levels were significantly decreased at 1 week in diabetic rats and exhibited a progressive decrease until 3 weeks (Fig. 2). As guanidinoacetic acid synthesis occurs only at the proximal tubule of the kidney, this decrease may reflect metabolic disturbance of the proximal tubules in the early stage of diabetic nephropathy.

Increased urinary NAG activity has been reported in patients with renal parenchymal diseases[4,17]. In the present study, we demonstrated an increase in urinary NAG activity 1 week after diabetes was

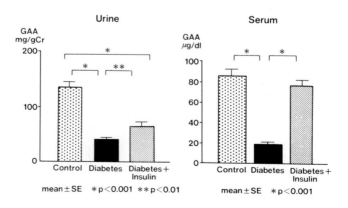

*Fig. 5. Effect of insulin on urinary and serum guanidinoacetic acid levels in diabetic rats.*

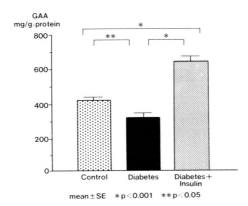

GAA
mg/g·protein

*Fig. 6. Effect of insulin on guanidinoacetic acid levels in renal
cortex from diabetic rats.*

induced (Fig. 3). A high level of urinary NAG was maintained until 3 weeks. According to recent knowledge of lysosomal enzymes, the increased release of NAG from the cells to the extracellular fluid can result not only from cell lysis but also from extracellular secretion or exocytosis. In contrast, urinary guanidinoacetic acid levels may reflect net synthesis of guanidinoacetic acid in the cells. Therefore, there may be a difference in clinical significance between urinary guanidinoacetic acid level and NAG activity.

A decrease in plasma 1,5-anhydroglucitol has been reported in experimental diabetic animals and diabetic humans[1,12,20]. This reduction has been sensitively and specifically demonstrated in diabetes mellitus. It was suggested that the mechanism of 1,5-anhydroglucitol uptake into the renal tubular brush border is common to that of glucose uptake in the kidney[8]. In the present study, serum 1,5-anhydroglucitol levels were decreased markedly at 1 week, and urinary excretion of 1,5-anhydroglucitol was markedly increased at 1 week after streptozotocin injection (Fig. 4). These changes may be due mainly to high concentration of glucose in urine and also, at least partly, to proximal tubular damage where 1,5-anhydroglucitol is reabsorbed competitively with glucose.

Insulin treatment blunted a decrease in urinary and serum guanidinoacetic acid levels in diabetic rats (Fig. 5), suggesting that serum glucose concentration may affect the synthesis of guanidinoacetic acid in the kidney. Figure 6 shows the effects of insulin on guanidinoacetic acid levels in the renal cortex of diabetic rats: insulin attenuated a decrease in guanidinoacetic acid levels in the renal cortex. In insulin-treated rats, renal cortical guanidinoacetic acid concentration increased to higher levels than in control rats. Guanidinoacetic acid has been reported to stimulate insulin secretion in the pancreas[2]. However, the reason why guanidinoacetic acid levels were increased by insulin treatment remains unknown.

In summary, guanidinoacetic acid levels in serum, urine and renal cortex from diabetic rats were significantly decreased compared with those in control rats. Urinary NAG activity and 1,5-anhydroglucitol excretion increased significantly in diabetic rats. A decrease in serum, urinary and renal cortical guanidinoacetic acid levels was partially antagonized by insulin treatment.

We conclude from this study that serum and urinary guanidinoacetic acid is a useful indicator for renal tubular dysfunction in the early stage of diabetic nephropathy.

# References

1. Akanuma, H., Ogawa, K., Lee, Y. & Akanuma,Y. (1981): Reduced levels of plasma 1,5-anhydroglucitol in diabetic patients. *J. Biochem.* **90,** 157–162.

2. Alsever, R.N., Georg, R.H. & Sussman, K.E. (1970): Stimulation of insulin secretion by guanidinoacetic acid and other guanidine derivatives. *Endocrinology* **86,** 332–336.

3. Borsook, H. & Dubnoff, J.W. (1941): The formation of glycocyamine in animal tissues. *J. Biol. Chem.* **138,** 389–403.

4. Dance, N. & Price, R.G. (1970): The excretion of N-acetyl-β-glucosaminidase and β-galactosidase by patients with renal disease. *Clin. Chim. Acta* **27,** 87–92.

5. Gerber, G.B., Gerber, G., Koszalka, T.R. & Miller, L.L. (1962): The rate of creatine synthesis in the isolated, perfused rat liver. *J. Biol. Chem.* **237,** 2246–2250.

6. Hyvarinen, A. & Nikkila, E.A. (1962): Specific determination of blood glucose with o-toluidine. *Clin. Chem.* **7,** 140–143.

7. Ishizaki, M., Kitamura, H., Takahashi, H., Asano, H., Mimura, K. & Okazaki, H. (1985): Evaluation of the efficacy of antirejection therapy using the quantitative analysis of guanidinoacetic acid (GAA) urinary excretion as a guide. In: *Guanidines.* Eds. A. Mori, B.D. Cohen & A. Lowenthal, pp. 353-363. New York: Plenum Press.

8. Kametani, S., Hashimoto,Y., Yamanouchi, T., Akanuma, Y. & Akanuma, H. (1987): Reduced renal reabsorption of 1,5-anhydro-D-glucitol in diabetic rats and mice. *J. Biochem.* **102,** 1599–1607.

9. Nakayama, S., Junen, M., Kiyatake, I. & Koide, H.( 1989): Urinary guanidinoacetic acid excretion as an indicator of gentamicin nephrotoxicity in rats. In: *Guanidines* 2. Eds. A. Mori, B.D. Cohen & H. Koide, pp. 303–311. New York: Plenum Press.

10. Natelson, S. & Sherwin, J.E. (1979): Proposed mechanism for urea nitrogen re-utilization: relationship between urea and proposed guanidine cycles. *Clin. Chem.* **25,** 1343–1344.

11. Sasaki, M., Takahara, K. & Natelson, S. (1973): Urinary guanidinoacetate/guanidinosuccinate ratio: an indicator of kidney dysfunction. *Clin. Chem.* **19,** 315–321.

12. Servo, C. & Pitkänen, E. (1975):Variation in polyol levels in cerebrospinal fluid and serum in diabetic patients. *Diabetologia* **11,** 575–580.

13. Takano,Y., Aoike, I., Gejo, F. & Arakawa, M. (1989): Urinary excretion rate of guanidinoacetic acid as a new marker in hypertensive renal damage. *Nephron* **52,** 273–277.

14. Takeda, M., Kiyatake, I., Koide, H., Jung, K.Y. & Endou, H. (1992): Biosynthesis of guanidinoacetic acid in isolated renal tubules. *Eur. J. Clin. Chem. Clin. Biochem.* **30,** 325–331.

15. Takeda, M., Koide, H., Jung, K.Y. & Endou, H. (1992): Intranephron distribution of glycine-amidinotransferase activity in rats. *Renal Physiol. Biochem.* **15,** 113–118.

16. Walker, J.B. (1958): Role for pancreas in biosynthesis of creatine. *Proc. Soc. Exp. Biol. Med.* **98,** 7–9.

17. Wellwood, J.M., Ellis, B.G., Price, R.G., Hammond, K., Thompson, A.E. & Jones, N.F. (1975): Urinary N-acetyl-β-D-glucosaminidase activities in patients with renal disease. *BMJ* **3,** 408–411.

18. Yabuuchi, M., Masuda, M., Katoh, K., Nakamura, T. & Akanuma, H. (1989): Simple enzymic method for determining 1,5-anhydro-D-glucitol in plasma for diagnosis of diabetes mellitus. *Clin. Chem.* **35,** 2039–2043.

19. Yamamoto, Y., Manji, T., Sato, A., Maeda, K. & Ohta, K. (1979): Ion-exchange chromatographic separation and fluorometric detection of guanidino compounds in physiologic fluids. *J. Chromatogr.* **162,** 327–340.

20. Yamanouchi, T., Akanuma, H., Takaku, F. & Akanuma, Y. (1986): Marked depletion of plasma 1,5-anhydroglucitol, a major polyol, in streptozotocin-induced diabetes in rats and the effect of insulin treatment. *Diabetes* **35,** 204–209.

## Section IX

## Electrophysiological and neurochemical studies of guanidino compounds

*Guanidino Compounds : 2*, eds. by P.P. De Deyn, B. Marescau, I.A. Qureshi and A. Mori.
©1997 John Libbey & Company Ltd., pp. 349–357.

# Chapter 37

# Neural excitation by uraemic guanidino compounds

Rudi D'HOOGE[1], Adam RAES[2], Philippe LEBRUN[3], Marc DILTOER[3],
Pierre-Paul VAN BOGAERT[2], Jacqueline MANIL[3], Fernand COLIN[4] and
Peter P. DE DEYN[1]

[1]*Laboratory of Neurochemistry and Behaviour, Born-Bunge Foundation and Department of Neurology, Middelheim
General Hospital, University of Antwerp (UIA),* [2]*Laboratory of Electrobiology, Born-Bunge Foundation, University
of Antwerp (RUCA),* [3] *Laboratory of Physiology and Pathophysiology, Free University of Brussels (VUB),* [4] *and
Laboratory of Electrophysiology, Free University of Brussels (ULB) Belgium*

## Summary

Levels of methylguanidine (MG) and guanidinosuccinate (GSA) are highly increased in uraemic patients. In this chapter, the effects of these uraemic guanidino compounds on the excitatory amino acid system are reported. Firstly, *N*-methyl-D-aspartate (NMDA) receptor blockade antagonized convulsions induced by intracerebroventricular GSA injection in mice, but not those induced by MG. Secondly, application of GSA (between 25 and 10 000 µM) to mouse spinal cord neurons in primary dissociated cell cultures evoked depolarizing inward whole-cell currents dose-dependently. Even 10 mM doses of MG did not produce such currents. GSA-induced whole-cell currents were apparently caused by NMDA receptor activation because NMDA antagonists (2-amino-5-phosphonovalerate and $Mg^{2+}$) blocked GSA-evoked whole-cell currents completely and reversibly, and co-application of a non-NMDA glutamate receptor antagonist (6-cyano-7-nitroquinoxaline-2,3-dione) did not affect these currents. Finally, evoked field potentials in the CA1 region of rat hippocampal slices were completely but reversibly abolished by GSA, and this effect was antagonized by NMDA receptor blockade. Following the application of 1 mM GSA, evoked responses were lastingly potentiated. All data were consistent with selective agonistic action of GSA on the NMDA receptor in a similar manner to aspartate. These results also suggest that NMDA receptors might contribute to the neurological complications of renal failure through GSA-induced inappropriate or excessive activation of NMDA receptors.

## Introduction

The levels of the endogenous protein and amino acid metabolites guanidinosuccinate (GSA) and methylguanidine (MG) have been shown to be highly increased in uraemic patients. MG has been shown to induce convulsions and epileptiform electrocorticographic discharges[18]. Lapin[15] reported that succinate derivatives induce convulsions in mice. D'Hooge *et al.*[9] described GSA-induced convulsions and concomitant epileptiform electrocorticographic discharges. These convulsant properties of both MG and GSA may be due to blockade of chloride channels associated with inhibitory glycine and γ-aminobutyric acid (GABA) receptors[6].

At least one of these endogenous convulsants may have direct excitatory effects as well. A number of analogues of aspartate and glutamate, including GSA, have been found to depolarize cat spinal motoneurons *in vivo*[5]. Reynolds & Rothermund[30] have applied MG and GSA on rat forebrain neurons *in vitro,* and observed an increase in [³H]dizocilpine binding and intracellular $Ca^{2+}$ concentration in the case of GSA but not MG. *In vivo*, GSA selectively potentiated convulsions induced by N-methyl-D-aspartate (NMDA) in mice, and GSA-induced convulsions were blocked by NMDA receptor antagonists[11]. These findings seem to suggest that GSA, but not MG, activates the NMDA receptor, and that NMDA receptors might be involved in GSA-induced convulsions. This hypothesis may have important pathophysiological consequences. Indeed, NMDA receptors are supposed to play a crucial role in many human pathologies, and increased levels of NMDA receptor agonists, either endogenous or exogenous, were shown to produce neurotoxic/excitotoxic lesions[20].

In this chapter, the effects of the uraemic guanidino compounds MG and GSA on the excitatory amino acid system are described. The effect of NMDA receptor antagonist co-application on MG- and GSA-induced convulsions was studied. Different concentrations of MG and GSA were applied to mouse spinal neurons in cell culture, the resulting whole-cell currents were measured, and excitatory amino acid receptor antagonists were co-applied in an effort to block these GSA-induced currents. Finally, the effect of GSA and concomitant NMDA receptor blocking on field potentials in the rat hippocampus was investigated.

## Convulsive dose determination

Male and female Swiss mice (30–40 g) were housed under standard conditions (12 h light, 12 h dark cycle, constant room temperature and humidity) and had free access to food and water. All tested products were administered by intracerebroventricular (i.c.v.) injection according to a technique described elsewhere[10]. They were dissolved in saline and injected in a volume of 5 µl into the left lateral brain ventricle; 50 µl of a 1 per cent lidocaine solution was used as local anaesthetic. Doses inducing convulsions in 50 per cent of the animals ($CD_{50}$) and their 95 per cent confidence limits were determined by Weil's method[35]; significance of differences between $CD_{50}$ values were assessed according to Litchfield and Wilcoxon[16]. The level of significance was set at 5 per cent. Doses of 10 and 20 nmol 2-amino-5-phosphonovalerate were co-injected i.c.v. with the convulsants tested. After administration of their i.c.v. injection, the animals were placed in individual cylindrical plastic cages (see Fig. 1, top) for the assessment of epileptiform activity within a 30 min observation period using criteria described elsewhere[10].

MG as well as GSA induced generalized clonic and tonic convulsions dose-dependently[10]. Minimal seizures appeared as jumping and running fits, myoclonus and grasping forelimb movements (automatisms). Co-application of APV increased the $CD_{50}$ of GSA but not that of MG, dose-dependently and significantly (Fig. 1, bottom). Without APV, $CD_{50}$ (and 95 per cent confidence interval) values for the induction of minimal seizures were 7.7 µg (5.8–10.2 µg; n = 25) for MG and 3.3 µg (2.4–4.7 µg; n = 25) for GSA. With 20 nmol APV co-injection, $CD_{50}$ values were 8.8 µg (4.8–16.1 µg; n = 30) for MG and 10.9 µg (8.8–13.4 µg; n = 40) for GSA (i.e. an increase with GSA of 3.3 times compared to 1.1 times with MG).

## Whole-cell current recording

Whole-cell currents were recorded on mouse spinal cord neurons in primary dissociated cell cultures. Spinal cord cells were obtained from albino Swiss mouse foetuses (12–14 days of gestation). Neurons were grown and maintained *in vitro* as described elsewhere[6]; cultures were 4–6 weeks old when used in patch clamp experiments.

We adapted the technique of whole-cell voltage clamp recording for NMDA-evoked currents[13,14]. GSA, MG, APV and ketamine (all purchased from Sigma Chemicals) and 6-cyano-7-nitroquinoxaline-2,3-dione (CNQX, Tocris) were dissolved in bath solution or saline (pH re-adjusted to 7.4 with NaOH where necessary). All other chemicals were from Merck and of analytical grade. Glass capillaries (Jencons H10/15) were used to make heat-polished patch pipettes (resistance, 3–5 MΩ) with a Flaming Brown Micropipette Puller P-87 (Sutter Instruments). They were filled with pipette solution containing (in mM): KCl, 30; KOH, 110; aspartic acid, 110; N-2-hydroxyethylpiperazine-N'-

## observation of epileptiform behaviour in mice

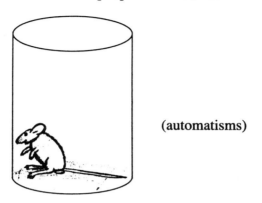

(automatisms)

## effect of APV co-injection on CD$_{50}$ values of GSA and MG

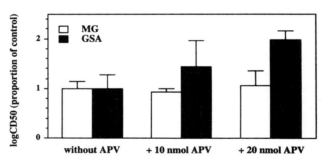

*Fig. 1. Convulsive dose determination. After i.c.v. injection of the compounds, the animals were placed in individual cylindrical plastic cages for the assessment of epileptiform activity within a 30 min observation period using criteria described [10]. There was a dose-dependent increase in the CD$_{50}$ values of GSA, but not in those of MG, under the influence of APV co-injection. White bars represent log CD$_{50}$ values and 95 per cent confidence intervals for minimal convulsions following i.c.v. MG injection; grey bars represent those of GSA. Each value is the result of an experiment using 25–40 animals. The difference between the control GSA value and the 20 nmol APV co-injection value is statistically significant according to the method of Litchfield & Wilcoxon [16]; all other differences between control values and co-injection values are not significant.*

2-ethanesulphonic acid (HEPES), 10; ethylene glycol bis(β-aminoethyl ether)-N, N, N', N'-tetraacetic acid (EGTA), 2; Na-ATP, 5; MgCl$_2$, 2; CaCl$_2$, 1; cyclic AMP, 0.1 (315–320 mOsm with sucrose; pH 7.4 with KOH). Cells were superfused with bath solution containing (in mM): NaCl, 140; KCl, 3.5; CaCl$_2$, 1; HEPES, 10; glucose, 5; glycine, $10^{-3}$; tetrodotoxin, $5 \times 10^{-4}$ (315–320 mOsm with sucrose; pH 7.4 with NaOH). A lab-made fast-perfusion system was used to apply bath and drug solution to

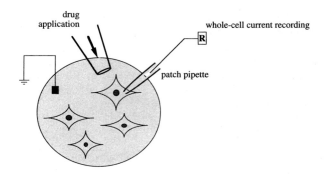

A: current tracing showing dose-depent increase in GSA-evoked whole-cell currents

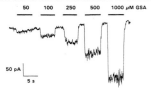

B: dose-response relationship of GSA-evoked whole-cell currents

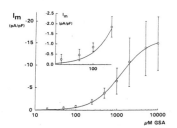

C: current tracing showing the inability of MG to evoke whole-cell currents

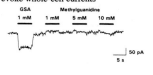

D: current tracing showing APV and Mg blocking GSA-evoked currents

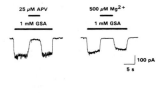

Fig. 2. Whole-cell current recording. Whole-cell currents were recorded on mouse spinal cord neurons in primary dissociated cell cultures. A lab-made fast-perfusion system was used to apply bath and drug solution to the cells. It was shown that application of GSA in concentrations between 25 and 10 000 μM evokes inward whole-cell currents on cultured spinal cord neurons in a dose-dependent fashion. (A) A tracing of the application of increasing concentrations of GSA (between 50 and 1000 μM) on one spinal neuron clamped at −60 mV with scale bars for current magnitude and time. (B) Dose–response relationship of GSA-evoked whole-cell currents. Each data point represents mean current (± standard deviation) in 5–7 cells. Currents were scaled to cell capacity (in pA/pF). As the insert clearly shows, doses between 25 and 250 μM evoked distinctly measurable inward currents. (C) A tracing showing that application of 1 mM GSA induces an inward current whereas doses of MG upto 10 mM do not induce such currents. (D) Effect of competitive and non-competitive N-methyl-D-aspartate receptor antagonists on whole-cell currents evoked by 1 mM GSA. Co-application of both the competitive NMDA receptor antagonist APV and the non-competitive antagonist $Mg^{2+}$ blocked GSA-evoked whole-cell currents completely and reversibly.

the cells, allowing a solution switch within 500 ms (flow rate ~ 700 µl/min). The flow outlet of the perfusion system (diameter ~ 1 mm) was placed within 1 mm of a selected spinal neuron.

Patch pipettes, mounted on Headstage HS-2 (Axon Instruments), were manoeuvred by using a Leitz micromanipulator and a Zeiss inverted microscope. A membrane seal of 3–4 GΩ was made as described by Hamill et al.[13], and clamping currents were measured in whole-cell configuration. All spinal neurons were clamped at the mean resting membrane potential of these cells (–60 mV), except where indicated differently. Voltage clamp was performed by Axoclamp-2A (Axon Instruments) in discontinuous mode (2–3 kHz); currents were recorded via Labmaster TL 125 interface on IBM-compatible computer, and analysed by pClamp 5.6 software package and conventional software. For dose–response relationships, currents were scaled to cell capacity. Cell capacity was determined by applying a 5 mV voltage step and measuring the surface of the resulting capacitive spike.

Application of GSA in concentrations between 25 and 10 000 µM evokes inward whole-cell currents on cultured spinal cord neurons in a dose-dependent fashion (Fig. 2A & B). These currents lead to depolarization of the neurons. Doses between 25 and 250 µM also evoke distinctly measurable inward currents, whereas MG doses of up to 10 mM do not induce whole-cell currents (Fig. 2C). A dose of 1 mM GSA was used to investigate the effect of NMDA receptor antagonists on GSA-induced currents. In mean sized cells clamped at –60 mV, this concentration evokes currents of ~200 pA. All displayed tracings represent typical results of at least five independent recordings. NMDA receptor antagonists APV and $Mg^{2+}$ block GSA-evoked whole-cell currents completely and reversibly (Fig. 2D). The non-NMDA glutamate receptor antagonist CNQX, 4 µM, had no effect on GSA-evoked currents (data not shown in figure).

## Hippocampal field potential recording

Field potentials were recorded from rat hippocampal slices according to a method modified from Alger et al.[1]. Adult male Wistar rats were decapitated under ether anaesthesia and hypothermia. Transverse hippocampal slices (300 µm thick) were cut from dissected brain tissue blocks with a vibratome, and maintained in gassed (95 per cent $O_2$/5 per cent $CO_2$) artificial cerebrospinal fluid (ACSF) solution at 35 °C. The ACSF solution contained (in mM): NaCl, 124; KCl, 2.5; $KH_2PO_4$, 1.25; $MgSO_4$, 2; $CaCl_2$, 2; $NaCO_3$, 18; glucose, 20. Recordings were made on slices at the gas–liquid interface in a continuously perfused Oslo chamber at 35 °C. Products were applied through the bath perfusion system by switching ACSF solution. Schaffer collaterals were stimulated at 2 Hz by bipolar nickel-chrome electrodes. Evoked field potentials were recorded on the CA1 hippocampal region by extracellular pipettes containing 4 mM NaCl. Responses were amplified by a conventional AC-coupled high-impedance amplifier (band pass from 1 Hz to 10 kHz), and displayed and stored on an IBM-compatible computer.

Excitatory post-synaptic field potentials (field EPSPs) were recorded from the pyramidal cell layer in the CA1 region of rat hippocampal slices following stimulation of Schaffer collaterals (Fig. 3). Application of 1 mM GSA to the preparation abolished the excitatory response completely and reversibly (Fig. 3A). In the presence of 10 µM APV, however, the same dose of GSA did not produce such an effect (Fig. 3B). A 6 min application of 1 mM GSA was shown to have lasting effects on the CA1 field EPSPs (Fig. 3C). After an initial small increase in the surface of the field EPSP, GSA application elicits complete abolition of the evoked response. When the compound is withdrawn, the EPSP slowly returns to a surface larger than that before GSA application, and it remains thus potentiated up to at least 1 h after GSA application.

## Discussion

Renal failure results in the accumulation of a number of endogenous compounds in tissues and fluids of patients suffering from this disorder, and some of these are thought to contribute to the uraemia-associated neurological complications[8,28]. Guanidino compounds, some of which are highly increased

extracellular field potentials recorded from CA1 region
in rat hippocampal slices

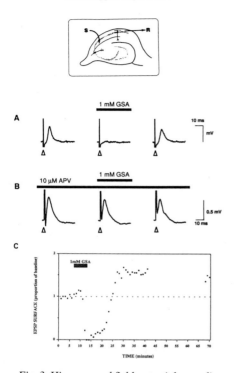

Fig. 3. Hippocampal field potential recording.
*Extracellular field potentials were recorded from the CA1
region in rat hippocampal slices following Schaffer
collateral stimulation. The insert at the top of the figure
depicts a rat hippocampal slice with an enlargement of
the studied synapse. Stimulation (S; also represented as
triangles beneath the tracings in parts A and B of the
figure) was applied to the Schaffer collaterals; the
resulting field potentials were recorded (R) from the
CA1 pyramidal cell layer. (A) Application of 1 mM
GSA completely abolished the potential. (B) In the
presence of the NMDA receptor antagonist APV, field
potentials were not at all affected by GSA application.
(C) The effect of 1 mM GSA application on the surface of the extracellular field potentials in the
CA1 region.
Application of the compound (6 min) is represented
by a grey bar. After an initial small increase in the
surface of the field EPSP, GSA application elicits
complete abolition of the evoked response. When the
chemical is withdrawn, the EPSP slowly returns to a
surface larger than that before the GSA application.
These are the results of a typical experiment.*

in uraemia[7], may play leading roles in these
complications as many of them are known
neurotoxins[22,23].

GSA and MG are probably the two most important uraemic guanidino compounds with
regard to both their levels in patients and their
toxicity[37]. Both compounds were shown to
induce convulsions as well as epileptiform
discharges experimentally[9,18], and these effects may be explained by inhibition of GA-
BAergic neurotransmission[6]. However, there
appears to be more to it, and some authors have
described direct depolarizing or excitatory actions of GSA but never of MG[5,30]. In this
chapter, we have presented a more detailed
investigation of the differential effects of MG
and GSA on excitatory amino acid neurotransmission.

Most central neurons use excitatory amino
acids as neurotransmitters, and at least five
different receptor types are involved in the
process[2,21]. One particular type, the NMDA
receptor, named after its prototypical agonist
N-methyl-D-aspartate, has been the focus of a
large body of scientific literature, and it has
been implemented in physiological and pathophysiological processes such as occur in memory, nervous system development, seizure
disorders, and ischaemic and excitotoxic brain
damage[34].

As we have described previously[11], GSA
potentiates convulsions induced by NMDA
selectively, and NMDA receptor antagonists
like APV, CGP 37849 and ketamine block
GSA-induced generalized convulsions in
mice dose-dependently. Here we show that
co-injection of the competitive NMDA receptor antagonist APV increases the $CD_{50}$ of GSA
following i.c.v. injection but not that of MG.
Although competitive NMDA receptor antagonists have been shown to block clonic convulsions not primarily related to NMDA
receptor activation[4], these findings suggest
that NMDA receptors are somehow involved
in GSA- but not in MG-induced convulsions.

GSA, but again not MG, induced inward
whole-cell currents dose-dependently when
applied to cultured spinal neurons. Both the
competitive NMDA receptor antagonist APV
and the non-competitive antagonist $Mg^{2+}$

blocked the GSA-evoked whole-cell currents completely and reversibly. Apparently, GSA-induced whole-cell currents were caused by NMDA receptor activation, whereas MG did not activate the receptor. The NMDA receptor channel complex possesses different ligand binding sites[17,26,29,36]. APV selectively binds at the agonist site; $Mg^{2+}$, on the other hand, blocks the open NMDA receptor-associated ionic channel rapidly and reversibly. Neuman *et al.*[25] showed that 4 µM CNQX blocked mossy fibre field potentials and currents induced by quisqualate but not those induced by NMDA. The same dose of CNQX did not appear to affect GSA-induced whole-cell current either. These data suggest that GSA selectively opens the NMDA receptor-associated ionic channel in a similar manner to NMDA. Thus, the GSA-evoked increase of intracellular $Ca^{2+}$ observed by Reynolds and Rothermund[30] was most probably due to influx of $Ca^{2+}$ through the activated NMDA receptor-associated ionophore.

GSA abolishes excitatory post-synaptic response in rat hippocampal slices dose-dependently. This effect is antagonized by selective blockade of the NMDA receptor and leads to a long-lasting potentiation of the post-synaptic response. Our findings suggest that GSA activates NMDA receptors in hippocampal CA1 region in a similar manner to aspartate[12]. Collingridge *et al.*[3] proposed that the reduction of field excitatory post-synaptic potentials by excitants like aspartate is due to depolarization and/or increased conductance of CA1 neurons. The long-lasting enhancement of the response after withdrawal of GSA might be due to the activation of intracellular $Ca^{2+}$-dependent processes that underlie long-term potentiation (LTP)[2]. The presence of GSA might thus interfere with physiological LTP in hippocampus. It is widely accepted that the hippocampal neural circuit as well as LTP processes play an important part in cognitive functioning[24,32], and GSA-induced interference might therefore contribute to the cognitive impairment prominent in uraemia[27,31].

In conclusion, the present findings indicate that NMDA receptors are involved in convulsions induced by GSA but not in those induced by MG. The uraemic guanidino compound GSA appears to activate the NMDA receptor ionic channel complex dose-dependently and selectively, which probably also explains the effect of the compound on hippocampal field potentials. Activation of NMDA receptors, in conjunction with previously described antagonism on GABA and glycine receptors[6], might lead to increased activation and neuronal depolarization, and thus to increased $Ca^{2+}$ influx through NMDA receptor-associated ionophores. Increased $Ca^{2+}$ influx not only underlies physiological processes like LTP, but also many pathological processes involved in, for example epilepsy, neurotoxicity and neurodegeneration[2,19,20]. Another endogenous metabolite and selective NMDA agonist, namely quinolinate, did produce neurotoxic/excitotoxic lesions[33]. Although GSA appears to be about twice as potent as quinolinate, it remains to be determined whether or not GSA also has such neurotoxic effects, and whether these effects are produced at pathophysiological concentrations. Elevated GSA levels were indeed observed in patients with renal insufficiency[7], and the concentrations of GSA in cerebrospinal fluid of uraemic patients (up to 30 µM) approach those that activate NMDA receptors.

Excitatory amino acid receptors in general, and those of the NMDA type in particular, might be involved in a variety of neurological disorders, including epilepsy, Huntington's disease, and neuronal damage following hypoxia/ischaemia such as occurs in stroke[19,20]. The results presented here suggest that NMDA receptors might also be involved in the neurological complications of renal failure through GSA-induced inappropriate or excessive activation of NMDA receptors alone or in conjunction with other depolarizing effects[22,23], and GABA/glycine receptor antagonism[6]. Renal failure is attended by various neurological and neuropsychological complications and, modern dialytic therapy notwithstanding, neuronal dysfunction persists in being a major cause of disability in uraemic patients[8,28].

## References

1. Alger, B.E., Dhanjal, S.S., Dingledine, R., Garthwaite, J., Henderson, G., King, G.L., Lipton, P., North, A., Schwartzkroin, P.A., Sears, T.A., Segal, M., Wittingham, T.S. & Williams, J. (1984): Brain slice methods. In: *Brain Slices*. Ed. R. Dingledine, pp. 381–437. New York: Plenum Press.

2.  Collingridge, G.L. & Lester, R.A.J. (1989): Excitatory amino acid receptors in the vertebrate central nervous system. *Pharmacol. Rev.* **40**, 143–210.

3.  Collingridge, G.L., Kehl, S.J. & McLennan, H. (1983): Excitatory amino acids in synaptic transmission in the Schaffer collateral-commissural pathway of the rat hippocampus. *J. Physiol.* **334**, 33–46.

4.  Croucher, M.J., Collins, J.F. & Meldrum, B.S. (1982): Anticonvulsant action of excitatory amino acid antagonists. *Science* **216**, 899–901.

5.  Curtis, D.R., Duggan, A.W., Felix, D., Johnston, A.R., Tebecis, A.K. & Watkins, J.C. (1972): Excitation of mammalian central neurons by acidic amino acids. *Brain Res.* **41**, 283–301.

6.  De Deyn, P.P. & Macdonald, R.L. (1990): Guanidino compounds that are increased in cerebrospinal fluid and brain of uremic patients inhibit GABA and glycine responses on mouse neurons in cell culture. *Ann. Neurol.* **28**, 627–633.

7.  De Deyn, P.P., Marescau, B., Cuykens, J.J., Van Gorp, L., Lowenthal, A. & De Potter, W.P. (1987): Guanidino compounds in serum and cerebrospinal fluid of non-dialysed patients with renal insufficiency. *Clin. Chim. Acta* **167**, 81–88.

8.  De Deyn, P.P., Saxena, V.K., Abts, H., Borggreve, F., D'Hooge, R., Marescau, B. & Crols, R. (1992): Clinical and pathophysiological aspects of neurological complications in renal failure. *Acta Neurol. Belg.* **92**, 191–209.

9.  D'Hooge, R., Pei, Y.Q., Manil, J. & De Deyn, P.P. (1992): The uremic guanidino compound guanidinosuccinic acid induces behavioral convulsions and concomitant epileptiform electrocorticographic discharges in mice. *Brain Res.* **598**, 316–320.

10. D'Hooge, R., Pei, Y.Q., Marescau, B. & De Deyn, P.P. (1992): Convulsive action and toxicity of uremic guanidino compounds: behavioral assessment and relation to brain concentration in adult mice. *J. Neurol. Sci.* **112**, 96–105.

11. D'Hooge, R., Pei, Y.Q. & De Deyn, P.P. (1993): N-methyl-D-aspartate receptors contribute to guanidino succinate-induced convulsions in mice. *Neurosci. Lett.* **157**, 123–126.

12. Fleck, M.W., Henze, D.A., Barrionuevo, G. & Palmer, A.M. (1993): Aspartate and glutamate mediate excitatory synaptic transmission in area CA1 of the hippocampus. *J. Neurosci.* **13**, 3944–3955.

13. Hamill, O.P., Marty, A., Neher, E., Sakmann, B. & Sigworth, F.J. (1981): Improved patch-clamp techniques for high-resolution current recording from cells and cell-free membrane patches. *Pflügers Arch.* **391**, 85–100.

14. Keana, J.F.W., McBurney, R.N., Scherz, M.W., Fischer, J.B., Hamilton, P.N., Smith, S.M., Server, A.C., Finkbeiner, S., Stevens, C.F., Jahr, C. & Weber, E. (1989): Synthesis and characterization of a series of diarylguanidines that are noncompetitive N-methyl-D-aspartate receptor antagonists with neuroprotective properties. *Proc. Natl. Acad. Sci. USA* **86**, 5631–5635.

15. Lapin, I.P. (1982): Convulsant action of intracerebroventricularly administered *l*-kynurenine sulphate, quinolinic acid and other derivatives of succinic acid, and effects of amino acids: structure–activity relationships. *Neuropharmacology* **21**, 1227–1233.

16. Litchfield, J.T & Wilcoxon, F. (1949): A simplified method of evaluating dose–effect experiments. *J. Pharmacol. Exp. Ther.* **96**, 99–113.

17. Lodge, D., Jones, M. & Fletcher, E. (1989): Non-competitive antagonists of N-methyl-D-aspartate. In: *The NMDA Receptor.* Eds. J.C. Watkins & G.L. Collingridge, pp. 37–51. Oxford: Oxford University Press.

18. Matsumoto, M., Kobayashi, K., Kishikawa, H. & Mori, A. (1976): Convulsive activity of methylguanidine in cat and rabbits. *IRCS Med. Sci.* **4**, 65.

19. Meldrum, B.S. (1992): Excitatory amino acids in epilepsy and potential novel therapies. *Epilepsy Res.* **12**, 189–196.

20. Meldrum, B. & Garthwaite, J. (1990): Excitatory amino acid neurotoxicity and neurodegenerative disease. *Trends Pharmacol. Sci.* **11**, 379–387.

21. Monaghan, D.T., Bridges, R.J. & Cotman, C.W. (1989): The excitatory amino acid receptors: their classes, pharmacology, and distinct properties in the function of the central nervous system. *Annu. Rev. Pharmacol. Toxicol.* **29**, 365–402.

22. Mori, A. (1983): Guanidino compounds and neurological disorders. *Neurosciences* **9**, 149–157.

23. Mori, A. (1987): Biochemistry and neurotoxicology of guanidino compounds. History and recent advances. *Pavlov. J. Biol. Sci.* **22**, 85–94.

24. Morris, R.G.M., Kandel, E.R. & Squire, L.R. (1988): The neuroscience of learning and memory: cells, neural circuits and behaviour. *Trends Neurosci.* **11**, 125–127.

25. Neuman, R.S., Ben-Ari, Y., Gho, M. & Cherubini, E. (1988): Blockade of excitatory synaptic transmission by 6-cyano-7-nitroquinoxaline-2,3-dione (CNQX) in the hippocampus *in vitro*. *Neurosci. Lett.* **92**, 64–68.

26. Olverman, H.J. & Watkins, J.C. (1989): NMDA agonists and competitive antagonists. In *The NMDA Receptor.* Eds. J.C. Watkins & G.L. Collingridge, pp. 19–36. Oxford: Oxford University Press.

27. Osberg, J.W., Meares, G.J., McKee, D.C. & Burnett, G.B. (1982): Intellectual functioning in renal failure and chronic dialysis. *J. Chron. Dis.* **35**, 445–457.

28. Raskin, N.H. (1989): Neurological aspects of renal failure. In: *Neurology and General Medicine.* Ed. M.J. Aminoff, pp. 231–246. New York: Churchill Livingstone.

29. Reynolds, I.J. (1990): Modulation of NMDA receptor responsiveness by neurotransmitters, drugs and chemical modification. *Life Sci.* **47**, 1785–1792.

30. Reynolds, I.J. & Rothermund, K. (1992): Multiple modes of NMDA receptor regulation by guanidines. In: *Guanidino Compounds in Biology and Medicine.* Eds. P.P. De Deyn, B. Marescau, V. Stalon & I.A. Qureshi, pp. 441–448. London: John Libbey.

31. Souheaver, G.T., Ryan, J.J. & Dewolfe, A.S. (1982): Neuropsychological patterns in uremia. *J. Clin. Psychol.* **38**, 490–496.

32. Squire, L.R. (1992): Memory and the hippocampus: a synthesis from findings with rats, monkeys, and humans. *Psychol. Rev.* **99**, 195–231.

33. Stone, T.W. & Connick, J.H. (1985): Quinolinic acid and other kynurenines in the central nervous system. *Neuroscience* **15**, 597–617.

34. Watkins, J.C. & Collingridge, G.L. (1989): *The NMDA Receptor.* Oxford: Oxford University Press.

35. Weil, C.S. (1952): Tables for convenient calculation of median-effective dose (LD$_{50}$ and ED$_{50}$) and instructions in their use. *Biometrics* **8**, 249–263.

36. Wong, E.H.F. & Kemp, J.A. (1991): Sites for antagonism on the N-methyl-D-aspartate receptor channel complex. *Annu. Rev. Pharmacol. Toxicol.* **31**, 401–425.

37. Yokozawa, T., Mo, Z.L. & Oura, H. (1989): Comparison of toxic effects of methylguanidine, guanidinosuccinic acid and creatinine in rats with adenine-induced chronic renal failure. *Nephron* **51**, 388–392.

Guanidino Compounds : 2, eds. by P.P. De Deyn, B. Marescau, I.A. Qureshi and A. Mori.
©1997 John Libbey & Company Ltd., pp. 359–371.

# Chapter 38

# Guanidinoethane sulfonate in the investigation of taurine–phospholipid interactions in the developing brain

Ryan J. HUXTABLE, John MURPHY and Pierre-Louis LLEU

*Department of Pharmacology, College of Medicine, University of Arizona, Tucson, Arizona 85724, USA*

## Summary

Guanidinoethane sulfonate has been used to cause taurine depletion in developing rats in order to study the relationship between taurine content and phospholipid methylation. Phospholipid methylation rate in brain synaptosomes is proportional to the taurine concentration. Methylation rate and taurine concentration are both decreased in preparations from guanidinoethane sulfonate-exposed rats such that the proportionality between the two parameters remains unchanged. This relationship, however, does not hold for cardiac membranes. Taurine concentrations in brain synaptosomes undergo a steep fall over the course of development, from 137 µmol/g protein on day 7 of life to 14 µmol/g protein on day 119. It appears that this fall may be connected with the decreasing rate of phospholipid methylation observed in synaptosomes over the same period, and with the neurochemical correlates of altered phospholipid methylation.

## Introduction

Guanidinoethane sulfonate (GES) is the amidino analog of taurine, one of the most abundant free amino acids in mammals (Fig. 1). GES is found in low concentrations in mammalian tissues[22], being formed by transamidination.

Stemming from the observation that GES is a competitive inhibitor of taurine transport[30], GES has become a widely used tool as a taurine-depleting agent, and in the study of the consequences of taurine deficiency[9,11,25,27,31,36–39,43,47,49,53,54,56,62,81]. These consequences include retinal degeneration[3,7,40,63,64,70], cardiac irregularities[41,42], reproductive problems[14] and impaired CNS development and neurochemistry[24,36,39,43,44,54,56,79].

In the adult brain, concentrations of taurine are second only to those of glutamate. Unlike all other neuroactive amino acids, however, taurine levels are highest in developing brain, and decrease during maturation[28]. Synaptosomal taurine concentrations in the rat can vary six-fold, depending on the developmental stage of the brain[29,46].

It has been proposed that the developmental fall in brain taurine concentrations might be related to the changing function and phospholipid composition of membranes[33,34]. In particular, taurine has been considered to be a modifier of the phospholipid methylation pathway that regulates many membrane functions[8]. The methylation route is being increasingly recognized as an important regulator of $Ca^{2+}$ channel opening[50], altered $Ca^{2+}$ binding and $Ca^{2+}$ ATPase activities[59], inhibition of $Na^+$ and $Ca^{2+}$

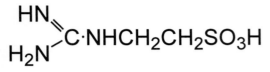

## Guanidinoethane sulfonate

$$H_2NCH_2CH_2SO_3H$$

## Taurine

*Fig. 1. Structures of taurine and its amidino analog, guanidinoethane sulfonate (GES).*

exchange in the heart[13,60], histamine release and various receptor activities[58]. Thus, N-methylation unmasks cryptic α-adrenergic receptors[73,75].

We have considered the possibility, therefore, that membrane phospholipid composition and function might be modified by cellular taurine concentration. We have begun a test of this hypothesis in brain synaptosomes ($P_2B$ fraction) of rats made taurine-deficient by treatment with GES.

## Materials and methods

[$^3$H-*Methyl*]methionine was purchased from New England Nuclear, (Boston, Mass.) GES was synthesized as previously described[30]. Sprague-Dawley rats with timed pregnancies were maintained on Wayne rat chow and were kept in pans on a 12 h photo-period for the duration of the study. Animals were GES-treated during development by maintaining the mothers during pregnancy and lactation on drinking water containing 1 per cent GES. Litter sizes were adjusted at birth to eight male pups.

### Preparation of membrane fractions

Synaptosomal $P_2B$ fractions were isolated from 7, 14, 21, 28, 56 and 119-day-old rats on discontinous sucrose gradients, as previously described[46]. Protein concentrations were determined by the Lowry procedure[48]. $P_2B$ preparations were also prepared using a continuous sucrose gradient from control and GES-treated rats for comparison with the preparations described by Whittaker and co-workers[20,80].

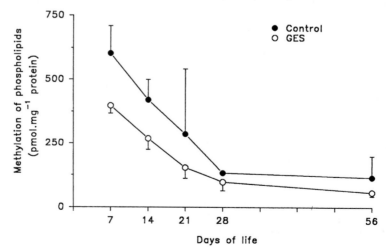

*Fig. 2. Incorporation of [$^3$H-methyl]methionine into synaptosomal phospholipids during brain development in control (closed symbols) and GES-treated rats (open symbols). Incorporations were calculated from the specific activity of the free methionine pool present in synaptosomes. Data are means ± SD for four preparations per point. The treatments yield statistically different values for days 7 and 14 ( P < 0.05).*

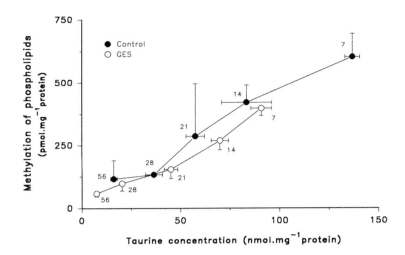

*Fig. 3. Relationship between phospholipid methylation and taurine concentration in synaptosomes from developing brain of control and GES-treated rats. Numbers by each point indicate the day of life. Symbols as for Fig. 2.*

Briefly, gradients were prepared overnight at room temperature by successive stacking of 0.8, 0.9, 1.0, 1.1 and 1.2 M sucrose layers. Synaptosomal $P_2$ fractions were resuspended in 0.32 M sucrose and layered on the top of the gradient and centrifuged at 53 000$g$ for 2 h. Pelleted subcellular fractions were fixed in 2 per cent glutaraldehyde in 0.1 M $PO_{43}$- followed by 2 per cent $OsO_4$ in the same buffer and examined by transmission electron microscopy. Mitochondrial contamination of the synaptosomal fraction was evaluated by determining succinate dehydrogenase activity according to the method of Green *et al.*[21]. Sarcolemmal membranes were prepared from heart as previously described[68].

### Incorporation of [³H-*methyl*] methionine in proteins and phospholipids

At 7, 14, 21, 28 and 56 days of age, male rat pups were randomly selected, one from each litter, weighed and injected intraperitoneally with 300 µCi/kg of [³H-*methyl*]methionine (specific activity 200 mCi/mmol). Nine hours later, rats were decapitated, and their brains and hearts removed and weighed. Brains were pooled such that approximately 2 g were used for subcellular fractionation.

Protein (0.2 ml) was precipitated with 0.2 ml of 7 per cent sulfosalicylic acid. After centrifugation, the precipitated protein was digested with a tissue solubilizer (NEN, Solvable), scintillation cocktail added (ICN, Ecolyte) and the sample counted. Soluble radioactivity was determined on a separate aliquot.

Radioactivity in the protein-free supernatant of tissue homogenates was also determined. This value was used, together with the methionine concentration (see below), to calculate the specific activity of free methionine in heart, liver and brain synaptosomes. The specific activities thus determined were used to calculate incorporation of methionine into protein, and its methyl group into phospholipid.

Lipids were extracted using a modification of the Christiansen method[5]. Membrane fraction (1 ml) was homogenized in a Wheaton glass homogenizer tube with 20 ml of 2:1 $CHCl_3$:$CH_3OH$ containing 0.05 per cent butylated hydroxytoluene. The homogenate was filtered, and the homogenizer tube was rinsed once with an additional 15 ml of the same solvent and once with 15 ml of 1:2 $CHCl_3$:$CH_3OH$ containing 0.05 per cent butylated hydroxytoluene. The rinsings were filtered and added to the original filtrate. The solvent was removed on a rotary evaporator and the residue taken up in 1 ml of 19:1

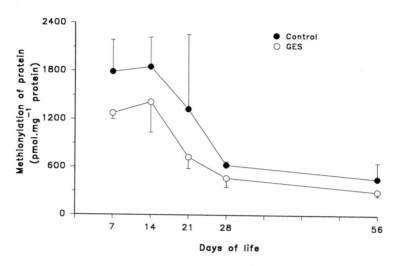

Fig. 4. Incorporation of [³H]methionine into synaptosomal protein during brain development in control (closed symbols) and GES-treated rats (open symbols). Symbols as for Fig. 2. The treatments yield statistically different values for days 7 and 28 (P < 0.05).

$CHCl_3$:$CH_3OH$ and frozen for determination of radioactivity and phosphorus. For the former, aliquots (0.75 ml) were dried in a scintillation vial, scintillation cocktail was added (ICN, cytoscint) and the samples were counted.

### Phospholipid separation and analysis

Phospholipid analysis was performed on extracts of $P_2B$ fractions and heart and liver plasma membranes from 7, 14, 21, 28 and 56-day-old rats. Analysis of $P_2B$ extracts was also carried out on 119-day-old rats. Thin layer chromatography (TLC) separation of the major classes of phospholipids was carried out using two sequential solvent systems on silica gel plates[71]. TLC plates were first developed in a solvent system of chloroform, methanol, acetic acid, formic acid and water (70:40:12:4:2 v/v) at 4 °C for 1 h. After being dried, the TLC plates were further developed in a solvent system of hexane, isopropyl ether and acetic acid (65:35:2 v/v) at room temperature for 40 min[29]. TLC plates were visualized using iodine and the phospholipid bands assigned using standards (Sigma).

The phosphorus analysis was based on the method of Parker & Peterson[61]. Phospholipid-containing areas were scratched from the plate and digested in concentrated sulfuric acid (0.1 ml) for 90 min at 250 °C in tubes placed in heating blocks. Then the tubes were cooled and 50 µl of 30 per cent $H_2O_2$ added. Samples were oxidized for 1 h at 155 °C. Deionized phosphate-free water (2 ml) was added to the cooled tubes. A 1:1 mixture (0.8 ml) of a 10.1 mM ammonium molybdate solution and 280 mM fresh ascorbic acid solution was added to the samples[26], which were then heated in boiling water for 7 min. Tubes were cooled and centrifuged. The optical density of samples was read at 797 nm. Standards tubes contained a known solution of potassium dihydrogen phosphate.

### HPLC analysis of taurine and methionine

Aliquots of $P_2B$ fraction (0.4 ml) were deproteinized with 3.5 per cent sulfosalicylic acid. Taurine and methionine were determined on separate analyses because of their large differences in concentration. Deproteinized samples were filtered and 20 µl aliquots were mixed with 80 µl of o-phthalaldehyde (OPA) solution (prepared by mixing 9 ml of solution of 343 mg of sodium tetraborate decahydrate with a solution of 54 mg OPA in 1 ml HPLC-grade methanol and 50 µl of mercaptoethanol). One minute after the addition of OPA, 20 µl of sample was injected onto an Altex ultrasphere ODS C18 reverse phase column (4.6 mm × 25 cm). A Beckman model 157 fluorescence detector was used. For the separation of the amino acids, a mixture of two solvents was used. Solvent A consisted of 80 per cent 40 mM sodium acetate (adjusted to pH 6.8 with acetic acid) and 20 per cent methanol plus 1 per cent tetrahydrofuran. Solvent B consisted of 20 per cent 10 mM sodium acetate adjusted to pH 6.8 with acetic acid and 80 per cent methanol. A mixture of 80 per cent solvent A and 20 per cent solvent B was run for 10 min, then a linear gradient from 20 to 50 per cent B was run over the next 12 min, followed by a gradient from 50 to 100 per cent B over the next 2 min. Taurine and methionine eluted at 19.4 and 29.2 min, respectively.

## Results and discussion

### Guanidinoethane sulfonate treatment of rats

Pregnant rats were maintained on drinking water containing 1 per cent GES for the final 2 weeks of pregnancy and throughout lactation. The pups were weaned onto the same drinking water. This treatment had no effect on the number of stillbirths or on litter size. No significant effect was observed on body weight gain or on organ weights in the pups (data not shown).

### Comparision of synaptosomes from guanidinoethane sulfonate-treated and control rats

Synaptosomes prepared from control and GES-treated rats showed comparable patterns on sucrose density gradients. No differences in band pattern were seen on either discontinuous or continuous sucrose density gradients.

Electron microscopy showed no difference between the preparations[32]. Regardless of the preparation, the light synaptosomal fraction contained larger synaptosomal fragments than the heavy fraction, with only a few free mitochondria being visible. Most of these were found in pinched-off nerve endings.

The specific activity of succinate dehydrogenase, a marker enzyme for mitochondria, did not differ significantly between control and GES-treated preparations in any of the bands (data not shown).

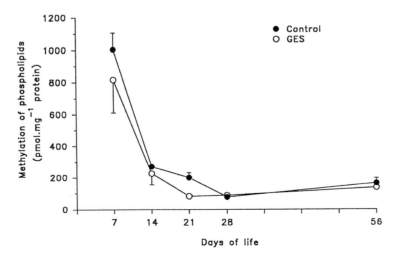

Fig. 5. Incorporation of [³H-methyl]methionine into heart membrane phospholipids during development in control (closed symbols) and GES-treated (open symbols) rats. Symbols as for Fig. 2.

Furthermore, the distribution of enzyme activity along the gradient was the same in the two preparations. Most of the activity (81 per cent) was found in the mitochondrial fraction. The remaining activity was distributed between the two synaptosomal layers, representing mitochondria present in nerve endings.

### Taurine concentrations in brain synaptosomes, heart and liver during development

Taurine concentrations in brain synaptosomes fall uniformly with development (Table 1). At 119 days of life, levels are only around 10 per cent of those found at 7 days. Whole liver shows a less well marked decrease in taurine levels with development over the first 56 days of life. Over the same period, whole heart levels fluctuate but do not change monotonically with time.

These changes are accentuated in GES-treated animals (Table 1). In brain synaptosomes, taurine levels are significantly lower than in control animals at all times points examined, extending through the 119th day of life. At this time, synaptosomes from animals exposed to GES had taurine levels only 46 per cent of those found in control animals. Taurine concentrations in the heart and liver were depleted to an even greater extent than in the brain in rats exposed to GES (Table 1).

**Table 1. Taurine concentrations μmol/g protein in brain synaptosomes, heart and liver of control and guanidinoethane sulfonate-exposed rats during development**

|  | Brain P$_2$B | Heart | Liver |
|---|---|---|---|
| Days of life |  |  |  |
| *Control* |  |  |  |
| 7 | 136.8 ± 4.3 | 79.8 ± 19.7 | 124.5 ± 17.9 |
| 14 | 83.3 ± 14.5 | 60.3 ± 19.2 | 111.7 ± 6.9 |
| 21 | 57.3 ± 5.2 | 15.6 ± 9.2 | 169.4 ± 3.5 |
| 28 | 36.5 ± 4.8 | 6.5 ± 1.0 | 167.5 ± 118 |
| 56 | 16.2 ± 2.4 | 19.2 ± 6.7 | 111.9 ± 4.5 |
| 119 | 13.7 ± 1.1 |  |  |
|  |  |  |  |
| *GES* |  |  |  |
| 7 | 70.7 ± 6.0* | 46.9 ± 7.1* | 73.8 ± 3.5* |
| 14 | 69.8 ± 5.1 | 33.4 ± 4.5* | 81.6 ± 6.7* |
| 21 | 45.1 ± 3.9* | 4.3 ± 1.9 | 78.3 ± 12.2* |
| 28 | 20.4 ± 1.8* | 2.7 ± 0.9* | 71.4 ± 3.9* |
| 56 | 7.52 ± 0.98* | 2.9 ± 0.7* | 27.9 ± 3.8* |
| 119 | 6.78 ± 1.58* |  |  |

*$P < 0.05$ compared with the corresponding control group. Conditions as for Fig. 2.

### Methionine concentrations in brain synaptosomes and heart during development

Compared with taurine, methionine concentrations are much lower (Table 2). However, over the first 56 days of life, methionine levels roughly track those of taurine, falling in brain synaptosomes but not in the heart. In the groups exposed to GES, methionine concentrations were significantly increased in brain synaptosomes on day 14, but significantly decreased on day 28. In the heart, a significant decrease was observed on day 21 (Table 2).

### Neutral phospholipid composition of brain synaptosomes and plasma membranes from heart and liver during development

Phosphatidylethanolamine and phosphatidylcholine compositions are reported in Table 3 for days 7, 14, 21, 28, 56 and 119 of life. In brain synaptosomes, levels of the neutral phospholipids fall steadily with development. In preparations from GES-exposed animals, neutral phospholipid content is unchanged, with an isolated exception (Table 3). In view of the correlation we have found between taurine concentration and neutral phospholipid ratio[29,45,46], the most interesting effect seen in GES-treated synaptosomes is the significant decrease in phosphatidylethanolamine/phosphatidylcholine ratio on days 56 and 119 (Table 3). The lowered synaptosomal taurine concentrations occurring in the developmental period (7-28 days), therefore, are not correlated with any change in neutral phospholipid ratio in this period.

In heart membranes, likewise, there is a developmental decrease relative to protein in the amounts of neutral phospholipids. Preparations from GES-exposed animals showed significant decreases on day 7 and significant increases on day 21 in neutral phospholipid content. No effect was observed, however, on the ratio of the phospholipids.

In liver membranes, the only differences seen in GES-exposed preparations were decreased phosphatidylcholine at day 28 and a decreased neutral phospholipid ratio at day 14.

**Table 2. Methionine concentrations in brain synaptosomes and heart of control and guanidinoethane sulfonate-exposed rats during development**

| | Brain P$_2$B | Heart |
|---|---|---|
| Days of life | (*p*mol / g protein) | (*p*mol / g protein) |
| *Control* | | |
| 7 | 304.1 ± 100.3 | 862.3 ± 126.3 |
| 14 | 146.3 ± 10.9 | 875.5 ± 91.8 |
| 21 | 134.6 ± 10.9 | 684.1 ± 66.8 |
| 28 | 80.3 ± 15.0 | 380.5 ± 61.3 |
| 56 | 59.5 ± 29.3 | 660.7 ± 95.0 |
| | | |
| *GES* | | |
| 7 | 207.9 ± 15.8 | 784.8 ± 13.6 |
| 14 | 228 ± 34.5* | 947.6 ± 206.7 |
| 21 | 114.3 ± 19.2 | 436.3 ± 61.5* |
| 28 | 62.2 ± 11.4* | 425.8 ± 79.2 |
| 56 | 43.9 ± 12.5 | 554.3 ± 77.5 |

*$P < 0.05$ compared with the corresponding control group. Conditions as for Fig. 2a.

## Phospholipid methylation of synaptosomes from guanidinoethane sulfonate-treated and control brain

In a quantitative sense, the major pathway of biosynthesis of phosphatidylcholine is the CDP choline route. However, phosphatidylcholine can also be biosynthesized directly from phosphatidyletha-noamine by the phospholipid methylation route. The substrate and the product of this route have opposing effects on membranes, phosphatidylcholine stabilizing lipid bilayers and phosphatidyl ethanolamine destabilizing[74]. The methyltransferase activities involved in the pathway are high in brain synaptosomes.

Taurine concentration in the brain may be a regulator of phospholipid methylation, as has been reported for the heart[23]. In the brain, however, a positive correlation is observed between taurine concentration and phosphatidylcholine:phosphatidylethanolamine ratio[29].

Phospholipid methylation rates were determined by injection of [$^3$H-*methyl*]methionine *in vivo* followed by isolation of synaptosomal P$_2$B fraction and determination of label incorporation into the phospholipid fraction. Phospholipid methylation rate fell with development (Fig. 2). At all times, however, synaptosomes from GES-treated brains showed lower methylation rates than those from control brains. These methylation rates correlated well with taurine concentration (Fig. 3); i.e. the lower the taurine concentration, the lower the methylation rate. Methylation data from the GES-treated preparations fall on the same curve as those from the control preparations, strongly suggesting that taurine has a direct stimulatory effect on methylation. This is an opposite effect to that reported for taurine in the heart[23].

It appears, therefore, that, although taurine depletion has only an indirect effect on the neutral phospholipid ratio, it may have a direct effect on phospholipid methylation.

## Incorporation of methionine into protein from guanidinoethane sulfonate-treated and control brain synaptosomes

The specificity of the action of GES on phospholipid methylation was checked by determination of the incorporation of methionine into protein (Fig. 4). Protein synthesis as determined by this index decreased with development. Synaptosomes from GES-exposed animals showed lower incorporations at all time points. However, statistical significance was achieved only on days 7 and 28.

**Table 3. Neutral phospholipids in brain synaptosomes and liver and heart plasma membranes of control and guanidinoethane sulfonate-exposed rats during development**

| | Control (nmol phosphorus / mg protein) | | | GES exposed (nmol phosphorus / mg protein) | | |
|---|---|---|---|---|---|---|
| Days of life | PE | PC | Ratio | PE | PC | Ratio |
| *Brain* | | | | | | |
| 7 | 164.1 ± 13.7 | 260.0 ± 30.7 | 0.636 ± 0.052 | 156.1 ± 8.7 | 247.3 ± 15.5 | 0.633 ± 0.038 |
| 14 | 206.2 ± 17.8 | 296.9 ± 31.9 | 0.693 ± 0.041 | 186.4 ± 14.2 | 284.6 ± 16.8 | 0.657 ± 0.058 |
| 21 | 181.2 ± 7.3 | 225.1 ± 11.1 | 0.806 ± 0.028 | 176.1 ± 13.4 | 224.8 ± 9.6 | 0.784 ± 0.020 |
| 28 | 215.7 ± 40.4 | 248.4 ± 46.3 | 0.872 ± 0.023 | 197.2 ± 17.7 | 239.4 ± 26.1 | 0.826 ± 0.033 |
| 56 | 122.0 ± 7.3 | 132.1 ± 6.6 | 0.923 ± 0.010 | 94.2 ± 2.9* | 121.1 ± 6.0 | 0.779 ± 0.020* |
| 119 | 100.2 ± 10.0 | 101.6 ± 9.7 | 0.982 ± 0.018 | 88.0 ± 0.8 | 105.9 ± 2.5 | 0.837 ± 0.014* |
| *Heart* | | | | | | |
| 7 | 124.8 ± 8.7 | 186.8 ± 5.3 | 0.663 ± 0.064 | 89.8 ± 6.2* | 122.0 ± 6.4* | 0.710 ± 0.048 |
| 14 | 67.9 ± 7.5 | 97.2 ± 12.4 | 0.711 ± 0.008 | 57.5 ± 8.3 | 77.7 ± 10.3 | 0.739 ± 0.014 |
| 21 | 80.5 ± 2.5 | 109.8 ± 8.9 | 0.744 ± 0.041 | 105.0 ± 11.1* | 130.8 ± 12.7* | 0.803 ± 0.014 |
| 28 | 77.4 ± 7.7 | 96.0 ± 8.7 | 0.806 ± 0.016 | 106.9 ± 4.6 | 125.3 ± 5.0 | 0.853 ± 0.003 |
| 56 | 74.7 ± 10.9 | 841 ± 9.5 | 0.889 ± 0.049 | 62.7 ± 3.7 | 62.4 ± 5.0 | 1.018 ± 0.057 |
| *Liver* | | | | | | |
| 7 | 90.0 ± 13.7 | 165.7 ± 13.4 | 0.542 ± 0.073 | 108.4 ± 12.0 | 197.5 ± 13.4 | 0.550 ± 0.063 |
| 14 | 75.6 ± 4.0 | 117.8 ± 7.8 | 0.643 ± 0.040 | 56.6 ± 10.1 | 119.5 ± 7.9 | 0.472 ± 0.063* |
| 21 | 68.9 ± 5.5 | 131.8 ± 14.5 | 0.524 ± 0.023 | 67.6 ± 7.0 | 139.9 ± 14.0 | 0.510 ± 0.075 |
| 28 | 55.5 ± 9.7 | 141.0 ± 13.7 | 0.394 ± 0.058 | 46.1 ± 8.3 | 99.5 ± 20.1* | 0.466 ± 0.016 |
| 56 | 40.6 ± 5.2 | 103.0 ± 13.5 | 0.395 ± 0.045 | 43.7 ± 6.2 | 99.2 ± 3.5 | 0.440 ± 0.059 |

PC, phosphatidylcholine; PE, phosphatidylethanolamine; * P <0.05 compared to corresponding control value.

## Phospholipid methylation and protein methionylation in plasma membranes from guanidinoethane sulfonate-treated and control hearts

Schaffer and co-workers[23] have reported that, in the isolated perfused rat heart, taurine reverses the increase in phospholipid methylation induced by the perfusion of methionine (and also reverses the negative inotropy of methionine). These workers also found that 10 mM taurine inhibited the incorporation of [³H]methyl groups into phospholipids on incubation of cardiac sarcolemmal membranes with 300 μM [³H-*methyl*]methionine. Earlier, Cantafora *et al.*[4] had reported that the feeding of a 0.4 per cent taurine-containing diet to guinea pigs led to decreased levels of phosphatidylcholine in liver membranes without change in phosphatidylethanolamine levels.

We find, however, that taurine depletion induced by GES is without effect on phospholipid methylation in the heart *in vivo* (Fig. 5). GES is also without effect on methionine incorporation into protein

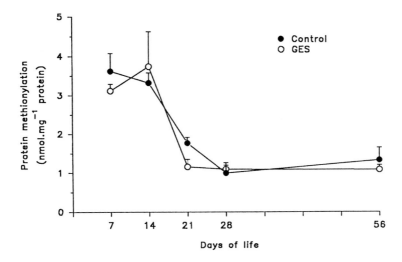

*Fig. 6. Incorporation of [³H]methionine into heart membrane protein during brain development in control (closed symbols) and GES-treated (open symbols) rats. Symbols as for Fig. 2.*

(Fig. 6). The dependency of phospholipid methylation on taurine concentration observed in brain synaptosomes (Fig. 3), therefore, appears to be specific to the brain.

## Conclusion

An enantiostatic function has been proposed for taurine[28]. Enantiostasis implies that taurine helps to maintain constancy of function in a changing biochemical system. In brain synaptosomes, an induced permanent reduction in taurine content has resulted in a reciprocal permanent change in membrane composition and, in the rat, phospholipid methylation rates. We can conceive that the net result of such reciprocal changes under physiological conditions might help to minimize disturbances in synaptosomal function. This, in turn, reflects an enantiostatic function of taurine.

Although GES appears to be the most widely used amidino analog for studying the biochemistry and pharmacology of amino acids, other examples of the use of guanidino analogs are known. Guanidino propionate has been used as a competitive inhibitor of the electrophysiological actions of GABA[12,16,72] and as a depleter and analog of creatine[1,15,17,18,67,69,76–78]. The same compound has been used to investigate mitochondrial energetics and cardiac functioning[6,10,52,66,67,76,77]. Conversely, amino acids have been used as analogs to probe the actions of the corresponding guanidino compound. For example, taurine protects against convulsions induced by GES[35,55]. Other examples are available of guanidino compounds serving as tools or analogs in the investigation of various biological phenomena[2,19,51,57,65,79,82].

Guanidino compounds are not only interesting substances in their own right, exhibiting a wealth of biochemical and pharmacologic actions, but are becoming increasingly useful as analogs in the study of various amino acids and their derivatives.

*Acknowledgement;* Supported by grants from the NIH and Taisho Pharmaceutical Co.

# References

1. Archer, S.L., Nelson, D.P., Zimmer, S., From, A.H.L. & Weir, E.K. (1989): Hypoxic pulmonary vasoconstriction is unaltered by creatine depletion induced by dietary Guanidino propionic acid. *Life Sci.* **45**, 1081–1088.

2. Bommer, M., Nikolarakis, K., Noble, E.P. & Herz, A. (1989): *In vivo* modulation of rat hypothalamic opioid peptide content by intracerebroventricular injection of guanidinoethylmercaptosuccinic acid (GEMSA): possible physiological role of enkephalin convertase. *Brain Res.* **492**, 305–313.

3. Bonhaus, D.W., Pasantes-Morales, H. & Huxtable, R.J. (1985): Actions of guanidinoethane sulfonate on taurine concentration, retinal morphology and seizure threshold in the neonatal rat. *Neurochem. Int.* **7**, 263–270.

4. Cantafora, A., Mantovani, A., Masella, R., Mechelli, L. & Alvaro, D. (1986): Effect of taurine administration on liver lipids in guinea pigs. *Experientia* **42**, 407–408.

5. Christiansen, K. (1975): Lipid extraction procedure for *in vitro* studies of glyceride synthesis with labeled fatty acid. *Anal. Biochem.* **66**, 93–99.

6. Clark, J.F., Khuchua, Z., Kuznetsov, A.V., Vassil'eva, E., Boehm, E., Radda, G.K. & Saks, V. (1994): Actions of the creatine analogue β-guanidinopropionic acid on rat heart mitochondria. *Biochem. J.* **300**, 211–216.

7. Cocker, S. & Lake, N. (1988): Electroretinographic alterations and their reversal in rats treated with guanidinoethyl sulfonate, a taurine depletor. *Exp. Eye Res.* **45**, 977–987.

8. Crews, F. (1985): Phospholipid methylation and membrane function. In: *Phospholipids and Cellular Regulation*, Vol. 1. Ed. K.F. Kuo, p. 131. Boca Raton, FL: CRC Press.

9. De La Rosa, J. & Stipanuk, M.H. (1985): The effect of taurine depletion with guanidinoethanesulfonate on bile acid metabolism in the rat. *Life Sci.* **36**, 1347–1352.

10. De Tata, V., Cavallini, G., Pollera, M., Gori, Z. & Bergamini, E. (1993): The induction of mitochondrial myopathy in the rat by feeding - guanidinopropionic acid and the reversibility of the induced mitochondrial lesions: a biochemical and ultrastructural investigation. *Int. J. Exp. Pathol.* **74**, 501–509.

11. De La Rosa, J. & Stipanuk, M.H. (1984): Effect of guanidinoethane sulfonate administration on taurine levels in rat dams and their pups. *Nutr. Rep. Int.* **30**, 1121.

12. Dudel, J. & Hatt, H. (1976): Four types of GABA receptors in crayfish leg muscles characterized by desensitization and specific antagonist. *Pflügers Archi.* **364**, 217–222.

13. Dyer, J.R. & Greenwood, C.E. (1988): Evidence for altered methionine methyl-group utilization in the diabetic rat's brain. *Neurochem. Res.* **13**, 517–523.

14. Ejiri, K., Akahori, S., Kudo, K., Sekiba, K. & Ubuka, T. (1987): Effect of guanidinoethyl sulfonate on taurine concentrations and fetal growth in pregnant rats. *Biol. Neonate* **51**, 234–240.

15. Ekmehag, B.L. & Hellstrand, P. (1988): Contractile and metabolic characteristics of creatine-depleted vascular smooth muscle of the rat portal vein. *Acta Physiol. Scand.* **133**, 525–533.

16. Feltz, A. (1971): Competitive interaction of β-guanidinopropionic acid and γ-aminobutyric acid on the muscle fibre of the crayfish. *J. Physiol.* **216**, 391–401.

17. Fitch, C.D., Jellinek, M., Fitts, R.H., Baldwin, K.M. & Holloszy, J.O. (1975): Phosphorylated β-guanidino propionate as a substitute for phosphocreatine in rat muscle. *Am. J. Physiol.* **228**, 1123–1125.

18. Fitch, C.D., Jellinek, M. & Mueller, E.J. (1974): Experimental depletion of creatine and phosphocreatine from skeletal muscle. *J. Biol. Chem.* **249**, 1060–1063.

19. Gerzon, K., Humerickhouse, R.A., Besch, H.R. Jr, Bidasee, K.R., Emmick, J.T., Roeske, R.W., Tian, Z., Ruest, L. & Sutko, J.L. (1993): Amino- and guanidinoacylryanodines: basic ryanodine esters with enhanced affinity for the sarcoplasmic reticulum $Ca^{2+}$-release channel. *J. Med. Chem.* **36**, 1319–1323.

20. Gray, E.G. & Whittaker, V.P. (1962): The isolation of nerve endings from brain: an electron microscopic study of cell fragments derived by homogenization and centrifugation. *J. Anat.* **96**, 79–88.

21. Green, D.E., Mii, S. & Kohout, P.M. (1955): Studies on the terminal electron transport system: I. Succinate dehydrogenase. *J. Biol. Chem.* **217**, 551–567.

22. Guidotti, A. & Costagli, P.F. (1970): Occurrence of guanidotaurine in mammals: variation of urinary and tissue concentration after guanidotaurine administration. *Pharmacol. Res. Commun.* **2**, 341-354.

23. Hamaguchi, T., Azuma, J. & Schaffer, S. (1991): Interaction of taurine with methionine: inhibition of myocardial phospholipid methyltransferase. *J. Cardiovasc. Pharmacol.* **18**, 224–230.

24. Herranz, A.S., Solís, J.M., Herreras, O., Menndez, N., Ambrosio, E., Orensanz, L.M. & Martín del Río, R. (1990): The epileptogenic action of the taurine analogue guanidinoethane sulfonate may be caused by a blockade of GABA receptors. *J. Neurosci. Res.* **26**, 98–104.

25. Hiramatsu, M., Niiya-Nishihara, H. & Mori, A. (1982): Effect of taurocyamine on taurine and other amino acids in the brain, liver and muscle of mice. *Neurosciences (Kobe)* **8**, 289–294.

26. Huxtable, R. & Bressler, R. (1973): Determination of orthophosphate. *Anal. Biochem.* **57**, 604–608.

27. Huxtable, R.J. (1982): Guanidinoethane sulfonate and the disposition of dietary taurine in the rat. *J. Nutr.* **12**, 2293–2300.

28. Huxtable, R.J. (1992): Physiological actions of taurine. *Physiol. Rev.* **72**, 101–163.

29. Huxtable, R.J., Crosswell, S. & Parker, D. (1989): Phospholipid composition and taurine content of synaptosomes in developing rat brain. *Neurochem. Int.* **15**, 233–238.

30. Huxtable, R.J., Laird, H.E. & Lippincott, S.E. (1979): The transport of taurine in the heart and the rapid depletion of tissue taurine content by guanidinoethyl sulfonate. *J. Pharmacol. Exp. Ther.* **211**, 465–471.

31. Huxtable, R.J. & Lippincott, S.E. (1981): Comparative metabolism and taurine depleting effects of guanidinoethane sulfonate in cats, mice, and guinea pigs. *Arch. Biochem. Biophys.* **210**, 698–709.

32. Huxtable, R.J., Murphy, J. & Lleu, P.-L. (1994): Developmental effects of taurine depletion on synaptosomal phospholipids in the rat. In: *Taurine in Health and Disease. Eds. R.J. Huxtable & D. Michalk, p. 343. New York: Plenum Press.*

33. Huxtable, R.J. & Sebring, L.A. (1986): Towards a unifying theory for the action of taurine. *Trends Pharmacol. Sci.* **7**, 481–485.

34. Huxtable, R.J. & Sebring, L.A. (1988): Taurine and the heart: the phospholipid connection. In: *Taurine and the Heart. Eds. H. Iwata, J.B. Lombardini & T. Segawa, p. 31. Boston: Kluwer Academic Press.*

35. Izumi, K., Kishita, C., Koja, T., Shimizu, T., Fukuda, T. & Huxtable, R.J. (1985): Effects of guanidinoethane sulfonate and taurine on electroshock seizures in mice. In: *Guanidines: Historical, Biological, Biochemical and Clinical Aspects of the Naturally-Occurring Guanidino Compounds. Eds. A. Mori, B.D. Cohen & A. Lowenthal, p. 227. New York: Plenum Press.*

36. Kontro, P. & Oja, S.S. (1984): Effects of 2-guanidinoethanesulphonic acid, convulsant and anticonvulsant drugs and amino acids on taurine uptake in mouse brain. *Acta Universitatis Tamperensis, Series B* **21**, 30–36.

37. Lake, N. (1981): Depletion of retinal taurine by treatment with guanidinoethyl sulfonate. *Life Sci.* **29**, 445–448.

38. Lake, N. (1982): Depletion of taurine in the adult rat retina. *Neurochem. Res.* **7**, 1385–1390.

39. Lake, N. (1983): Taurine depletion of lactating rats: effects on developing pups. *Neurochem. Res.* **8**, 881–889.

40. Lake, N. (1986): Electroretinographic deficits in rats treated with guanidinoethyl sulfonate, a depletor of taurine. *Exp. Eye Res.* **42**, 87–92.

41. Lake, N., De Roode, M. & Nattel, S. (1987): Effects of taurine depletion on rat cardiac electrophysiology: *in vivo* and *in vitro* studies. *Life Sci.* **40**, 997–1006.

42. Lake, N., Eley, D.W. & ter Keurs, H.E.D.J. (1991): The effects of taurine depletion on excitation–contraction coupling in rat cardiac trabeculae. *Biophys. J.* **59**, 64a.

43. Lehmann, A., Hagberg, H., Huxtable, R.J. & Sandberg, M. (1987): Reduction of brain taurine: effects on neurotoxic and metabolic actions of kainate. *Neurochem. Int.* **10**, 265–274.

44. Lehmann, A., Huxtable, R.J. & Hamberger, A. (1987): Taurine deficiency in the rat and cat: effects on neurotoxic and biochemical actions of kainate. In: *The Biology of Taurine: Methods and Mechanisms. Eds. R.J. Huxtable, F. Franconi & A. Giotti, p. 331. New York: Plenum Press.*

45. Lleu, P.-L., Croswell, S. & Huxtable, R.J. (1992): Phospholipids, phospholipid methylation and taurine content in synaptosomes of developing rat brain. In: *Taurine: Nutritional Value and Mechanisms of Action. Eds. J.B. Lombardini, S.W. Schaffer & J. Azuma, p. 221. New York: Plenum Press.*

46. Lleu, P.-L. & Huxtable, R.J. (1992): Phospholipid methylation and taurine content of synaptosomes from cerebral cortex of developing rat. *Neurochem. Int.* **21**, 109–118.

47. Lombardini, J.B. (1981): Combined effects of guanidinoethanesulfonate, a depletor of tissue taurine levels, and isoproterenol or methoxamine on rat tissues. *Biochem. Pharmacol.* **30**, 1698–1701.

48. Lowry, O.J., Rosebrough, N.J., Farr, A.L. & Randall, R.J. (1951): Protein measurement with folin phenol reagent. *J. Biol. Chem.* **193**, 265–275.

49.    Marnela, K.-M., Kontro, P. & Oja, S.S. (1984): Effects of prolonged guanidinoethanesulphonate administration on taurine and other amino acids in rat tissues. *Med. Biol.* **62,** 239–244.

50.    McGivney, A., Crews, F.T., Hirata, F., Axelrod, J. & Siraganian, R.R. (1981): Rat basophilic leukemia cells defective in phospholipid methyltransferase enzyme, $Ca^{2+}$ influx and histamine release: reconstruction by hybridization. *Proc. Natl Acad. Sci. USA* **78,** 6176–6180.

51.    Meglasson, M.D., Wilson, J.M., Yu, J.H., Robinson, D.D., Wyse, B.M. & De Souza, C.J. (1993): Antihyperglycemic action of guanidinoalkanoic acids: 3-guanidinopropionic acid ameliorates hyperglycemia in diabetic KKA*y* and C57BL6J*ob/ob* mice and increases glucose disappearance in rhesus monkeys. *J. Pharmacol. Exp. Ther.* **266,** 1454–1462.

52.    Mekhfi, H., Hoerter, J., Lauer, C., Wisnewsky, C., Schwartz, K. & Ventura-Clapier, R. (1990): Myocardial adaptation to creatine deficiency in rats fed with β-guanidinopropionic acid, a creatine analog. *Am. J. Physiol. Heart Circ. Physiol.* **258,** H1151–H1157.

53.    Morán, J., Maar, T.E. & Pasantes-Morales, H. (1994): Impaired cell volume regulation in taurine deficient cultured astrocytes. *Neurochem. Res.* **19,** 415–420.

54.    Morán, J. & Pasantes-Morales, H. (1991): Taurine-deficient cultured cerebellar astrocytes and granule neurons obtained by treatment with guanidinoethane sulfonate. *J. Neurosci. Res.* **29,** 533–537.

55.    Mori, A., Katayama, Y., Yokoi, I. & Matsumoto, M. (1981): Inhibition of taurocyamine (guanidinotaurine) induced seizures by taurine. In: *The Effects of Taurine on Excitable Tissues. Eds. S.W. Schaffer, S.I. Baskin & J.J. Kocsis, p. 41. New York: Spectrum Publications.*

56.    Nilsson, M., Lehmann, A. & Hansson, E. (1989): Effects of 2-guanidinoethane sulfonate on glutamate uptake in primary astroglial cultures from the rat cerebral cortex. *Neuropharmacology* **28,** 1415–1418.

57.    Ohira, Y., Ishine, S., Inoue, N. & Yunoki, K. (1991): Reduced growth of Ehrlich ascites tumor cells in creatine depleted mice fed α-guanidinopropionic acid. *Biochim. Biophys. Acta* **1097,** 117–122.

58.    Okumura, K., Panagia, V., Beamish, R.E. & Dhalla, N.S. (1987): Biphasic changes in the sarcolemmal phosphatidylethanolamine N-methylation activity in catecholamine-induced cardiomyopathy. *J. Mol. Cell. Cardiol.* **19,** 356–366.

59.    Panagia, V., Elimban, V., Ganguly, P.K. & Dhalla, N.S. (1987): Decreased $Ca^{2+}$ binding and $Ca^{2+}$ ATPase activities in heart sarcolemma upon phospholipid methylation. *Mol. Cell. Biochem.* **78,** 65–71.

60.    Panagia, V., Makino, N., Ganguly, P.K. & Dhalla, N.S. (1987): Inhibition of $Na^+$–$Ca^{2+}$ exchange in heart sarcolemmal vesicles by phosphatidylethanolamine N-methylation. *Eur. J. Biochem.* **166,** 597–603.

61.    Parker, F. & Peterson, N.F. (1965): Quantitative analysis of phospholipids and phospholipid fatty acids from silica gel thin layer chromatograms. *J. Lipid Res.* **6,** 455–460.

62.    Pasantes-Morales, H., Arzate, M.E., Quesada, O. & Huxtable, R.J. (1987): Higher susceptibility of taurine-deficient rats to seizures induced by 4-aminopyridine. *Neuropharmacology* **26,** 1721–1725.

63.    Pasantes-Morales, H., Quesada, O., Cárabez, A. & Huxtable, R.J. (1983): Effect of the taurine transport antagonists, guanidinoethane sulfonate and beta-alanine, on the morphology of the rat retina. *J. Neurosci. Res.* **9,** 135–143.

64.    Quesada, O., Huxtable, R.J. & Pasantes-Morales, H. (1984): Effect of guanidinoethane sulfonate on taurine uptake by rat retina. *J. Neurosci. Res.* **11,** 179–186.

65.    Rand, M.J. & Li, C.G. (1992): Effects of argininosuccinic acid on nitric oxide-mediated relaxations in rat aorta and anococcygeus muscle. *Clin. Exp. Pharmacol. Physiol.* **19,** 331–334.

66.    Ren, J.-M. & Holloszy, J.O. (1992): Adaptation of rat skeletal muscle to creatine depletion: AMP deaminase and AMP deamination. *J. Appl. Physiol.* **73,** 2713–2716.

67.    Ren, J.-M., Semenkovich, C.F. & Holloszy, J.O. (1993): Adaptation of muscle to creatine depletion: effect on GLUT-4 glucose transporter expression. *Am. J. Physiol. Cell Physiol.* **264,** C146–C150.

68.    Sebring, L.A. & Huxtable, R.J. (1986): Low affinity binding of taurine to phospholiposomes and cardiac sarcolemma. *Biochim. Biophys. Acta* **884,** 559–566.

69.    Shields, R.P. & Whitehair, C.K. (1973): Muscle creatine: *in vivo* depletion by feeding β-guanidinopropionic acid. *Can. J. Biochem.* **51,** 1046–1049.

70.    Shimada, C., Tanaka, S., Hasegawa, M., Kuroda, S., Isaka, K., Sano, M. & Araki, H. (1992): Beneficial effect of intravenous taurine infusion on electroretinographic disorder in taurine deficient rats. *Jpn. J. Pharmacol.* **59,** 43–50.

71. Skipski, V.P., Peterson, R.F. & Barclay, M. (1964): Quantitative analysis of phospholipids by thin-layer chromatography. *Biochem. J.* **90,** 374–378.

72. Snodgrass, S.R. & Lorenzo, A.V. (1973): Transport of GABA from the perfused ventricular system of the cat. *J. Neurochem.* **20,** 761–769.

73. Strittmatter, W.J., Hirata, F. & Axelrod, J. (1979): Phospholipid methylation unmasks cryptic-adrenergic receptors in rat reticulocytes. *Science* **204,** 1207–1209.

74. Sun, G.Y. & Sun, A.Y. (1985): Ethanol and membrane lipids. *Alcohol Clin. Exp. Res.* **9,** 164–180.

75. Tallman, J.F.J., Henneberry, R.C., Hirata, F. & Axelrod, J. (1979): Control of β-adrenergic receptors in Hela cells. In: *Catecholamines: Basic and Clinical Frontiers*, Vol. 1. Eds. E. Usdin, I.J. Kopin & J. Barchas, p. 489. Oxford: Pergamon Press.

76. Unitt, J.F., Radda, G.K. & Seymour, A.-M.L. (1990): Acute replacement of phosphocreatine in the isolated rat heart by perfusion with the creatine analogue β-guanidinopropionic acid. *Biochem. Soc. Trans.* **18,** 606–607.

77. Unitt, J.F., Radda, G.K. & Seymour, A.-M.L. (1993): The acute effects of the creatine analogue, β-guanidinopropionic acid, on cardiac energy metabolism and function. *Biochim. Biophys. Acta* **1143**, 91–96.

78. Van Deursen, J., Jap, P., Heerschap, A., Ter Laak, H., Ruitenbeek, W. & Wieringa, B. (1994): Effects of the creatine analogue β-guanidinopropionic acid on skeletal muscles of mice deficient in muscle creatine kinase. *Biochim. Biophys. Acta* **1185,** 327–335.

79. Watanabe, Y., Watanabe, S., Yokoi, I. & Mori, A. (1991): Effect of guanidinoethanesulfonic acid on brain monoamines in the mouse. *Neurochem. Res.* **16,** 1149–1154.

80. Whittaker, V.P. & Barker, L.A. (1972): The subcellular fractionation of brain tissue with special reference to the preparation of synaptosomes and their component organelles. In: *Methods in Neurochemistry*, Vol. 2. Ed. R. Fried, p. 1. New York: Marcel Dekker.

81. Yan, C.C., Bravo, E. & Cantafora, A. (1990): Rats with taurine-deficiency induced by administration of guanidino ethane sulfonate: an *in vivo* model for studying the physiological role of taurine. *Acta Toxicol. Ther.* **11,** 374–387.

82. Yokoi, I., Itoh, T., Yufu, K., Akiyama, K., Satoh, M., Murakami, S., Kabuto, H. & Mori, A. (1991): Effect of 2-guanidinoethanol on levels of monoamines and their metabolites in the rat brain. *Neurochem. Res.* **16,** 1155–1159.

*Guanidino Compounds : 2*, eds. by P.P. De Deyn, B. Marescau, I.A. Qureshi and A. Mori.
©1997 John Libbey & Company Ltd., pp. 373–377.

# Chapter 39

# The roles of nitric oxide on α-guanidinoglutaric acid induced epileptic discharges

Hitoshi HABU, Isao YOKOI, Hideaki KABUTO, Hiroshi ASAHARA, Junji TOMA, Kazunori INADA and Akitane MORI

*Department of Neuroscience, Institute of Molecular and Cellular Medicine, Okayama University Medical School, 2-5-1 Shikatacho, Okayama 700, Japan*

## Summary

To clarify the role of nitric oxide (NO) in the epileptic seizure mechanism, we examined the effect of L-arginine (L-Arg) on epileptic electrocorticographic discharges induced in rats by α-guanidinoglutaric acid (α-GGA), an NO synthase (NOS) inhibitor. α-GGA (100 mM) alone or was combined with L-Arg (50 or 200 mM) were administered intracerebroventricularly in immobilized and electrocorticographically monitored rats. Recorded electrocorticograms (ECoGs) were classified into four stages according to an index of epileptic discharges. Stages shown in the α-GGA alone group were significantly higher than those in the combined 50 mM or 200 mM L-Arg groups. These results suggest that α-GGA induces epileptic discharges by inhibiting NOS activity in the brain.

## Introduction

Nitric oxide (NO) has been known to play an important role as a neurotransmitter, and is synthesized by NO synthase (NOS) from L-arginine (L-Arg). Recently, we found that α-guanidinoglutaric acid (α-GGA) is a new type of potent NOS inhibitor[19]. α-GGA was first found in cobalt-induced epileptogenic focus tissue in cerebral cortex of cats[13]. Intracerebroventricular α-GGA injection in rats induced convulsions and spikes and polyspikes on ECoGs[16].

We examined effects of L-Arg on the epileptic electrocorticographic discharges induced by α-GGA in order to study the relationship between NO and epileptic seizures.

## Materials and methods

### Experimental animals

Male Sprague-Dawley rats weighing 250–400 g were used. Rats were maintained on a laboratory diet (protein content 24 per cent), MF (Oriental Yeast, Tokyo, Japan) and water ad libitum.

Rats were immobilized with succinylcholine chloride under artificial ventilation with tracheal intubation. Four electrodes were placed epidurally at the sites described elsewhere[18]. ECoGs were recorded with a model EEG-5210 electroencephalograph (Nihon Koden, Japan) from four unipolar leads. For the topical application of examined sample solution, a small trephine hole was made at a point 2.5 mm left and 0.8 mm posterior to the bregma, and the dura mater was removed. A

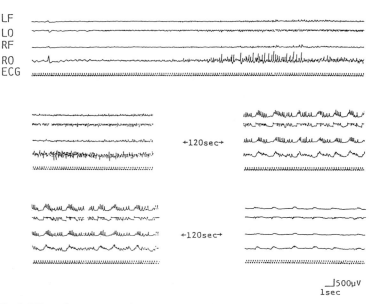

LF
LO
RF
RO
ECG

+120sec→

+120sec→

⌐500µV
1sec

Fig. 1. Effect of intraventricular injection of α-GGA (100 mM) on the rat ECoG.
The ECoG shows desynchronization, recruiting rhythm in stage 3, 30 min
after α-GGA injection and, 120 s after that, clonic phase, i.e. gradual
slowing of discharge frequency and 120 s after that, post-ictal depression.

microsyringe needle was inserted at a depth of 2.5 mm from dura through the hole, and the experimental compounds were injected into the left cerebral ventricle. Rats were allowed to recover for at least 2 h after the preparation in each session, and ECoG was recorded continuously for 3 h. α-GGA and L-Arg were administered as follows:

Group A: α-GGA (100 mM), 10 µl

Group B: α-GGA (100 mM) and L-Arg (50 mM), 10 µl

Group C: α-GGA (100 mM) and L-Arg (200 mM), 10 µl

α-GGA was synthesized in our laboratory according to the method of Zervas & Bergmann[20].

For statistical analysis, the Kruskal-Wallis test and a multiple comparison test was used.

## Criteria of ECoG findings

We regarded sporadic spikes, polyspikes, sharp waves and spike–wave complexes as significant epileptic discharges, but excluded slow wave bursts, small spike–wave complexes, and spiky waves in a wave-like rhythm as insignificant. We classified the severity of ECoG findings into four stages, and defined as follows:

Stage 0: no epileptic discharges

Stage 1: sporadic epileptic discharges, either focal or diffuse in location, associated with normal background activity or with background changes less than 10 sec in duration

Stage 2: continuous epileptic discharges either focal or diffuse in location, associated with background changes greater than 10 sec in duration

Stage 3: electrographic seizures with a sequence of rapid polyspikes, followed by a decrement in frequency and either voltage suppression, or repeated seizures without voltage suppression.

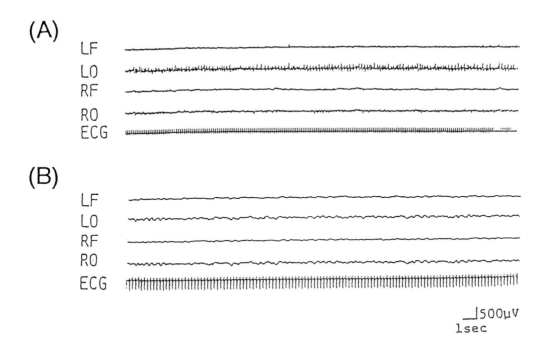

*Fig. 2. Effect of co-injection of α-GGA (100 mM) and L-Arg (50 or 200 mM) on the rat ECoG.*
*(A) Continuous epileptic discharge in stage 2, 30 min after co-injection of α-GGA and L-Arg (50 mM). (B)*
*No epileptic discharge in stage 0, 30 min after co-injection of α-GGA and L-Arg (200mM).*
*Nitric oxide is an important mediator of autoregulation of retinal and coroidal blood flow in newborns.*

## Results

Intracerebroventricular administration of α-GGA alone induced epileptic discharges on ECoG in all rats examined and developed into ictal manifestations in many cases (Fig. 1). Co-administration of L-Arg (50 mM) (Fig. 2A) and L-Arg (200 mM) (Fig. 2B) along with α-GGA (100 mM) prevented the appearance of epileptic discharges in a dose-dependent manner. ECoG scores expressed as stage numbers were higher in group A than in group C (Table 1).

**Table 1. Stage numbers in each group**

|  | 0 | 1 | 2 | 3 |
|---|---|---|---|---|
| α-GGA 100 mM | 0 | 0 | 2 | 4 |
| α-GGA 100 mM + L-Arg 50 mM | 3 | 0 | 1 | 2 |
| α-GGA 100 mM + L-Arg 200 mM | 4 | 1 | 0 | 1 |

Stage scores in group A (α-GGA 100 mM) were higher than those in group C (α-GGA 100 mM+ L-Arg 200 mM).
*$P < 0.05$ by the Kruskal-Wallis test and a multiple comparison test.

## Discussion

Although the electrocorticographic findings were not quantitated, the occurrence of more severe stages in group A, compared with group B or C, suggested that epileptic discharges decreased with an increase of L-Arg dose. Co-injection of L-Arg, however, could not completely prevent the appearance of discharges. Because L-Arg is not an anticonvulsant, its suppression of epileptic discharges induced by α-GGA might be a result of inhibition of an excitatory action. The serotonergic dysfunction[16] and the decrease of catecholamines in brain related to α-GGA-induced seizures[17] are not regarded as the critical factor in convulsive seizures, but one of several important factors. Many papers reported that the mechanism by which guanidino compounds inducing convulsive seizures may be related mainly to the γ-aminobutyric acid receptor system[5,6,11,18].

Recently, glutamate receptors, especially N-methyl-D-aspartate (NMDA) receptor, have been indicated to play a principal role in the evocation of epileptic seizures[1,3,4,9,10,12]. The receptor activation, by an excess influx of $Ca^{2+}$ into the cell body through the receptor channels, may stimulate NO synthesis[7,8]. NO production, as a consequence of such an activation, may be involved in the arrest of seizure activity[2]. Furthermore, serious inhibition of NO synthesis, by suppressing NO-induced NMDA receptor inhibition, could promote neurotoxic effects by activation of these receptors[14, 15]. Suppression of NO production by NOS inhibitors may accelerate the convulsive activities.

From these experiments, it is reasonable to assume that intraventricular administration of α-GGA evoked epileptic discharges in rats by interference with NO production, among others, because co-administration of L-Arg attenuated the appearance of discharges by alleviating the interference.

### Acknowledgements

This work was supported by Grants-in-Aid from the Ministry of Education, Science and Culture of Japan (No. 04454362).

## References

1.  Avioli, M. & Oliver, A. (1987): Bursting in human epileptogenic neocortex is depressed by N-methyl-D-aspartate antagonist. *Neurosci. Lett.* **76**, 249–254.

2.  Buisson, A., Lakhmeche, N., Verrecchia, C., Plotkine, M. & Boulu, R.G. (1993): Nitric oxide: an endogenous anticonvulsant substance. *Neuroreport* **4**, 444–446.

3.  Church, J., Davies, S.N. & Lodge, D. (1986): A correlation between N-methylaspartate antagonist and anticonvulsant properties of PCP/sigma receptor agonists. *Br. J. Pharmacol.* **87**, 270.

4.  Croucher, M.J., Collins, J.F. & Meldrum, B.F. (1982): Anticonvulsant action of excitatory amino acid antagonists. *Science* **216**, 899–901.

5.  De Deyn, P.P. & Macdonald, R.L. (1990): Guanidino compounds that are increased in uremia inhibit GABA- and glycine-responses on mouse neurons in cell culture. *Ann. Neurol.* **28**, 627–633.

6.  De Deyn, P.P., Marescau, B. & Macdonald, R.L. (1990): Epilepsy and the GABA-hypothesis: a brief review and some examples. *Acta Neurol. Belg.* **90**, 65–81.

7.  Garthwaite, J., Charles, S.L. & Chess-Williams, R. (1988): Endothelium-derived relaxing factor release on activation of NMDA receptor suggests role as intercellular messenger in the brain. *Nature* **336**, 385–388.

8.  Garthwaite, J., Garthwaite, G., Palmer, R.M.J. & Moncada, S. (1989): NMDA receptor activation induces nitric oxide synthesis from arginine in rat brain slices. *Eur. J. Pharmacol.* **172**, 413–416.

9.  Hablitz, J.J. & Langmoen, I.A. (1986): N-Methyl-D-aspartate receptor antagonists reduce synaptic excitation in the hippocampus. *J. Neurosci.* **6**, 102–106.

10. Herron, C.E., Williamson, R. & Collingridge, G.L. (1985): A selective N-methyl-D-aspartate antagonist depresses epileptiform activity in rat hippocampal slices. *Neurosci. Lett.* **61**, 255–260.

11. Jinnai, D., Sawai, A. & Mori, A. (1966): γ-Guanidinobutyric acid as a convulsive substance. *Nature* **212**, 617.

12.  Lehmann, J., Schneider, J., McPherson, S., Murphy, D.E., Bernard, P., Tsai, C., Bennett, D.A., Pastor, G., Steel, D.J., Boehm, C., Cheney, D.L., Liebman, J.M., Williams, M. & Wood, P.L. (1987): CPP, a selective N-methyl-D-aspartate (NMDA)- type receptor antagonist: characterization *in vitro* and *vivo. J. Pharmacol. Exp. Ther.* **240,** 737–746.

13.  Mori, A., Akagi, M., Katayama, Y. & Watanabe, Y. (1980): α-Guanidinoglutaric acid in cobalt-induced epileptogenic cerebral cortex of cats. *J. Neurochem.* **35,** 603–605.

14.  Rondouin, G., Lerner-Natoli, M., Manzoni, O., Lafon-Cazal, M. & Bockaert, J. (1992): A nitric oxide (NO) synthase inhibitor accelerates amygdala kindling. *Neuroreport* **3,** 805–808.

15.  Rondouin, G., Bockaert, J. & Lerner-Natoli, M. (1993): L-Nitroarginine, an inhibitor of NO synthase, dramatically worsens limbic epilepsy in rats. *Neuroreport* **4,** 1187–1190.

16.  Shiraga, H., Hiramatsu, M. & Mori, A. (1986): Convulsive activity of α-guanidinoglutaric acid and the possible involvement of 5-hydroxytryptamine in the α-guanidinoglutaric acid-induced seizure mechanism. *J. Neurochem.* **47,** 1832–1836.

17.  Shiraga, H., Hiramatsu, M. & Mori, A. (1989): The involvement of catecholamines in the seizure mechanism induced by α-guanidinoglutaric acid in rats. In: *Guanidines 2.* Eds. A. Mori, B.D. Cohen & H. Koide, pp. 213–222. New York: Plenum Press.

18.  Yokoi, I., Edaki, A., Watanabe, Y., Shimizu, Y., Toda, H. & Mori, A. (1989): Effects of anticonvulsant on convulsive activity induced by 2-guanidinoethanol. In: *Guanidines 2.* Eds. A. Mori, B.D. Cohen & H. Koide, pp. 169–181. New York: Plenum Press.

19.  Yokoi, I., Kabuto, H., Habu, H. & Mori, A. (1994): α-Guanidinoglutaric acid, an endogenous convulsant, as a novel nitric oxide synthase inhibitor. *J. Neurochem.* **63,** 1565–1567.

20.  Zervas, L. & Bergmann, M. (1928):Das Sogen. Arginylarginin von E. Fisher, ein α, δ-Bisguanido -n-valeriansäureanhydred. (25. Mitteilung über Umlagerungen peptidahnlicher Stoffe). *Ber. Dtsch. Chem. Ges.* **61,** 1195–1203.

*Guanidino Compounds : 2*, eds. by P.P. De Deyn, B. Marescau, I.A. Qureshi and A. Mori.
©1997 John Libbey & Company Ltd., pp. 379–386.

# Chapter 40

# Seizure activity and neuronal damage induced by intrahippocampal guanidinosuccinic acid injection in rats

J.C. PAN[1], Y.Q. PEI[1], L. AN[2], L. LAI[2], R. D'HOOGE[3] and P.P. DE DEYN[3]

[1]*Departments of Pharmacology and* [2]*Cell Biology, Beijing Medical University, School of Basic Medical Sciences, Xue Yuan Road, Beijing (100083), China, and* [3]*Laboratory of Neurochemistry and Behaviour*[3]*, Born-Bunge Foundation, University of Antwerp (UIA), Universiteitsplein 1, 2610 Wilrijk-Antwerp, Belgium*

## Summary

Guanidinosuccinic acid (GSA) is a guanidino compound found in mammalian central nervous system and physiological fluids. Its level has been found to be greatly increased in serum and cerebrospinal fluid of patients with renal failure, and the agent is suggested to play a role in uraemic encephalopathy. In this report we examined the behavioural, electrographic and morphological effects of intrahippocampal GSA injection in unanaesthetized rats. Intrahippocampal administration of 2 µl GSA solution (3.5 nM) was followed by behaviour observation, and electrohippocampogram and electrocorticogram activity recording. No behavioural or electrographic canges of any significance were observed in sham-injected animals. GSA-injected animals, on the other hand, showed partial clonic seizures leading to generalized clonic seizures, and eventually full status epilepticus, i.e. continuous clonic convulsions. The behavioural manifestations were accompanied by epileptiform electrographic discharges. During generalized clonic seizures, the electrohippocampogram showed bilateral arrythmic bursting spikes. Epileptiform electric activity persisted even after the generalized clonic convulsions had stopped, and lasted until the animals were killed 5 days following injection. Microscopic examination of slices of these rats' brains revealed severe neuronal damage in the CA1 area of the hippocampus. Treatment of rats with anticonvulsant doses of phenobarbital and phenytoin did not protect against all behavioural effects and epileptiform electrographic discharges produced by intrahippocampal injection of GSA. However, the non-competitive N-methyl-D-aspartate receptor antagonist ketamine not only prevented behavioural signs of both partial and generalized clonic seizures but also prevented epileptiform discharges. Ketamine treatment also prevented GSA-induced hippocampal damage.

## Introduction

Guanidinosuccinic acid (GSA) is a guanidino compound found in mammalian central nervous system (CNS) and physiological fluids. Its level has been found to be greatly increased in serum and cerebrospinal fluid of patients with renal failure, and it is suggested to play a role in uraemic encephalopathy[5,9]. GSA was reported to have a central stimulating action when infused into the lateral ventricles of cannulated urethane-anaesthesized rats[7], and to inhibit γ-aminobutyric acid and glycine responses in mouse spinal cord neurons in cell culture[4]. Earlier, we have reported on the characterization of the GSA-induced seizures in conscious mice[2,11]. The chemical structure of GSA is very similar to the structure of L-glutamate, L-aspartate and N-methyl-D-aspartate (NMDA). Glutamate and aspartate both function as endogenous transmitters in area CA1 of the rat hippocampus[6]. We have also reported that NMDA receptors contribute to GSA-induced convulsions[1]. This study further deals with characterization of seizures, electrographic changes and local neuronal degeneration induced by intrahippocampal injection of GSA into conscious rats.

## Materials and methods

Male and female Wistar rats weighing 200–280 g were used. Animals were housed under controlled environmental conditions, 12 h light 12 h dark cycle, and free access to food and water. GSA and ketamine were purchased from Sigma Chemical Company. Phenobarbital and phenytoin were purchased from Beijing Pharmaceutical Company.

GSA was dissolved in phosphate buffer (pH 7.4) and injected in a volume of 2 µl/rat with a Hamilton microsyringe into the left hippocampus. The injection velocity was 1 µl/min. Phenobarbital, DPH and ketamine were dissolved in saline and injected intraperitoneally (i.p.). The concentrations were calculated in such a manner that each animal received 10 ml solution per kg body weight.

### Stereotaxic procedures (cannulation and electrode placement)

Animals were anaesthetized with sodium pentobarbital (40 mg/kg, i.p.), and 4 per cent xylocaine was used as local anaesthetic. Thereafter, animals were placed in a stereotaxic frame under sterile conditions and a midline scalp incision of approximately 10–12 mm was made. Then the fascia overlying the cranium was cleared away, and the skull surface was cleaned with alcohol-soaked cotton. Small burr holes for placement of electrodes and anchoring screws were made by means of a dental drill. For intrahippocampal injection and depth recording of the hippocampal discharges, referred to as electrohippocampogram (EHG), the guide tubes (0.9 mm in diameter) were placed bilaterally, with coordinates according to Konig and Klippel[8]: anterior–posterior (AP) 3.5 mm, lateral 2 mm, depth 2 mm from the skull surface. All electrodes and guide tubes were anchored to the skull with acrylic cement. Animals were allowed a recovery period of 1–2 weeks. The inner tube for injection was 0.55 mm in diameter at depth 3.2 mm. The bipolar electrodes for recording hippocampal discharges were made with isolated wire (1.5 mm in diameter and depth 3.2 mm). The inner tubes for injection and the recording electrodes were not installed in the guide tubes in advance, but were placed when they were used.

Operated rats with normal electrocorticogram (ECoG) were divided into three groups. The control group received only vehicle (phosphate buffer, pH 7.4, 2 µl/rat, n = 5). A second, GSA-treated, group (n = 5), received 2 µl GSA solution (3.5 nM) through the hippocampal guide tube. Before and after intrahippocampal injection of vehicle or GSA, behaviour, EHG and ECoG activity were continuously observed. General behaviour, time of onset, type and duration of seizures and deaths were recorded.

A third group consisted of drug-treated animals. Three subgroups were considered. Two subgroups were injected intraperitoneally (i.p.) with either sodium phenobarbital (40 mg/kg; n = 5) or phenytoin (20 mg/kg; n = 5), 60 min before intrahippocampal injection of GSA. The third subgroup received ketamine (60 mg/kg, i.p.; n = 5) 15 min before intrahippocampal injection of GSA, and after 1 hour this dose was repeated. After each injection, continuous observation of behavioural changes and recordings of EHG and ECoG discharges were as performed in the GSA and control groups. After 5 days' observation, animals of all groups were killed by transcardial perfusion under deep pentobarbital anaesthesia with 300 ml of 0.9 per cent saline, followed by 300 ml of cold fixative containing 2 per cent paraformaldehyde and 1 per cent glutaraldehyde in 0.1 M phosphate buffer at pH 7.4. Brains were removed after fixation in 3 per cent paraformaldehyde containing 1 per cent $CaCl_2$ overnight and then kept in 0.88 M saccharose solution contained in 1 per cent gum arabic for 12–24 h. Consequently, brains were briefly frozen by dry ice and sectioned by a cryostat (Reichert Histostal) at 10 µm thickness. Sections were mounted on a slide for staining: (1) Nissl stain: parts of sections were stained with 0.1 per cent Cresyl violet; (2) other sections for a modified Kodousek silver impregnation method was used: sections were put in 3 per cent silver nitrate for 1–2 h in a dark room at 37 °C. Then the sections were transferred to reducing fluid for several seconds, washed two or three

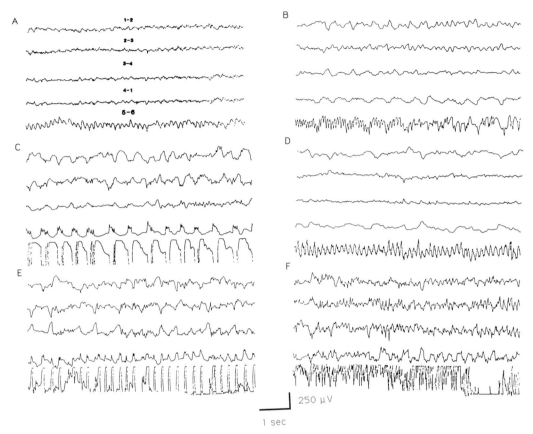

Fig. 1. The characterization of EHG and ECoG produced by intrahippocampal injection of GSA . (A) Normal
ECoG (1–2 right, 2–3 anterior, 3–4 left and 1–4 posterior) and EHG (5–6): the hippocampal discharges are 6–7
Hz rhythmic sinusoidal theta waves. (B) 10 minutes after injection of GSA : discharge frequency is increased in the
EHG (12–13 Hz) and decreased in the ECoG to (2–4 Hz). No behavioural signs were observed. (C) Twenty to
twenty-five minutes after intrahippocampal injection of GSA : periodic bilateral spike and slow-wave complexes
were observed in EHG, and sharp slow high-voltage waves in the ECoG. (D) Partial clonic seizures (PCS) stopped
and EHG and ECoG epileptiform discharges disappeared . PCS repeated every 3–5 min. (E,F) EHG and ECoG
discharges during typical generalized clonic seizures, when the EHG showed bilateral arrythmic bursting spikes.

times with distilled water and then put in 3 per cent NH$_4$OH for 2–3 s. Thereafter, sections were
washed with water and put in 0.5 per cent oxalic acid for 3 min and washed with water. The sections
were dried in air and examined under a light microscope.

## Results

### Behavioural findings

The control group displayed no abnormal behaviour during 6 h of continuous observation. However,
the ECoG of most animals in this group showed slight transient abnormality after injection of
phosphate buffer, returning to baseline in 1–2 h. On the other hand, intrahippocampal injection of
GSA did produce behavioural changes. These changes produced by intrahippocampal injection of

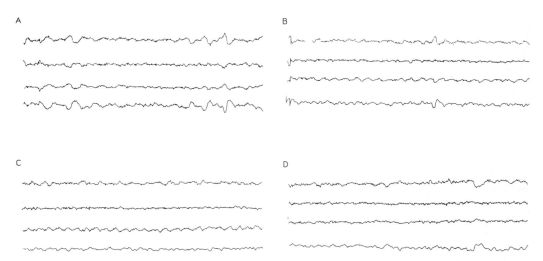

*Fig. 2. ECoG in ketamine-treated rats. (A) ECoG after i.p. injection of ketamine is normal. (B) Thirty minutes after the intrahippocampal injection of GSA, ECoG under influence of ketamine : no epileptiform discharges were observed. Sixty min after intrahippocampal injection of GSA, and after a second dose of ketamine (C): no epileptiform activity was observed. Six hours after GSA (D): ECoG was normal.*

GSA can be divided in several distinct phases. During the first 5–10 min, the spontaneous activity of most animals increased, but in a few cases it decreased. After 15–30 min, the following behavioural effects were observed: twitches in face, eyelids and auris, masticatory movements, and other events, corresponding to partial clonic seizures. These partial clonic seizures repeated every 3–5 min and were followed by numerous wet-dog shakes, contraversive cyclic movements, violent jumping and fast running. Subsequently, all muscles of the forepart of the body underwent clonic twitches, one or both forelimbs were lifted, and the animals thus displayed typical generalized clonic seizures, which occurred about 40–60 min after GSA injection. If the seizure was very severe, loss of postural control occurred. These epileptic fits repeated every 5–10 min and lasted for 6–12 h. In a few animals, the seizures became progressively more complex and prolonged with gradual reduction of the interictal pause. Eventually, these animals displayed a full status epilepticus, i.e. continuous clonic convulsions and epileptiform discharges in the EHG and ECoG. Tonic convulsions and death were not observed with this dose of GSA.

**Electric activity**

Electric activity from the hippocampus and brain cortex were recorded in the three main phases of behavioural abnormalities. Initially (5–10 min after intrahippocampal injection of GSA), the discharges in the EHG and ECoG were nearly normal, without epileptiform discharges. No behavioural change was observed. Characterization of EHG and ECoG discharges during partial and generalized clonic seizures are shown in Fig. 1. The changes of discharges in the EHG and ECoG always appeared prior to the behavioural signs. When the generalized clonic seizures stopped, masticatory movements and twitches in the face and epileptiform discharges (Fig. 1c) in EHG and ECoG persisted for 5 days, until the animals were killed.

Behavioural and electric changes of the animals in the drug-treated groups: results prior to drug treatment were as follows: phenobarbital and phenytoin did not protect against all behavioural effects and epileptiform EHG and ECoG discharges, produced by intrahippocampal injection of GSA. Ketamine however, seemed to be effective. One dose of ketamine given prior to intrahippocampal injection of GSA, did not only prevent behavioural signs of both partial clonic seizures and generalized

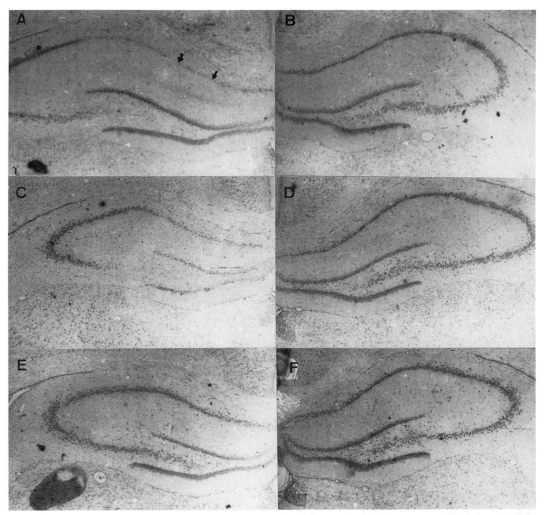

*Fig. 3. Intrahippocampal injection of GSA induced neuronal cell damage in the hippocampus (GSA injected in left side). Hippocampus of a GSA-treated rat (A): in the CA1 region (arrowhead), pyramidal cells showed widespread damage. The right hippocampus (B): no neuronal damage was observed. Under influence of ketamine, only a few pyramidal cells were degenerated in the CA1 region, near the GSA injected point (C). The right hippocampus of the same rat (D): no neuronal cells were damaged. Left and right hippocampus of the srat in control group (E-F): no damaged cells were found.*

clonic seizures but also prevented epileptiform discharges. As the effectiveness of ketamine lasted too short, a second dose was given after 60 min. During 6 h observation neither behavioural signs, nor epileptiform discharges were observed. The results of ECoG are shown in Fig. 2.

The results of light microscopic histological analysis of cresyl violet sections reveal the widespread damage of pyramidal cells in the CA1 region of the left hippocampus in GSA-treated rats only (Fig. 3a). The same region in GSA-treated rats under ketamine protection demonstrated the degeneration of only a few neuronal cells (Fig. 3c). This means that ketamine not only protected against seizures, but also against GSA-induced neuronal damage. The silver-stained sections showed a lot of silver

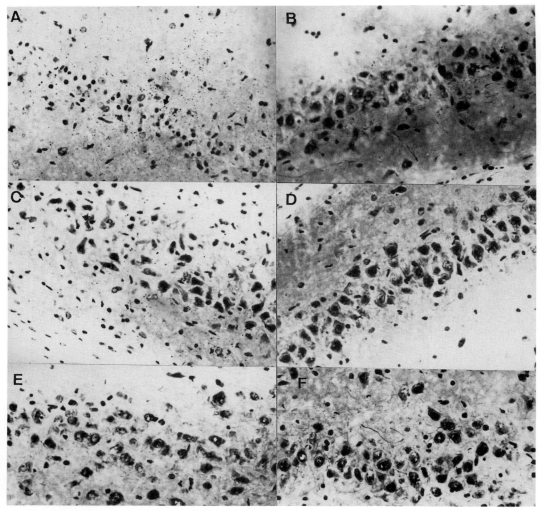

*Fig. 4. Intrahippocampal injection of GSA induced neuronal cells damage in the silver-stained sections (left side-injected GSA). A lot of silver grains were found in the left hippocampus of only GSA-treated rat (B). The right hippocampus of the same rat (B): no silver grains were observed. The left hippocampus of rats priorly treated with ketamine shows only a few silver grains (C). The right hippocampus of the same rat (D): no silver grain was found. Left and right hippocampus in the control group (E-F): no silver grains were found in both sides.*

grains in the left hippocampus of the GSA-treated rats (Fig. 4a). These results show the same effects as the cresyl-violet sections. Results of both procedures are shown in Figs. 3 and 4.

## Discussion

We report here the characterization of seizures, EHG and ECoG pattern and morphological changes, induced by intrahippocampal injection of a microdose of GSA in rats. Our data are of particular interest in view of the endogenous nature of GSA, which has recently been shown to occur in animal and human brain[1]. To our knowledge GSA is the most powerful endogenous convulsant among the guanidino compounds increased in uraemic patients[10]. Intrahippocampal injection of GSA in rats, induces typical generalized clonic seizures, epileptiform EHG and ECoG discharges, and degenerative changes of neuronal cells in the injected-side hippocampus. The behavioural changes correspond with the EHG and ECoG signs. During the first 5–10 min after injection, neither epileptiform dicharges, nor abnormal behaviour were observed. When the EHG showed periodic bilateral rhythmic spike discharges spreading only to part of the cortex (Fig. 1c), only partial clonic seizures appeared. However, when EHG showed arrhythmic bursting spikes, and when discharges spread to the whole brain cortex, generalized clonic seizures appeared (Fig. 1e-f). While generalized clonic seizures stopped, partial clonic seizures persisted for several days. Intrahippocampal injection of GSA not only induced seizures but also neuronal cell damage. The pyramidal cells in CA1 are most vulnerable to GSA, more than, e.g. the granular cells in dentate gyrus. Although phenobarbital and phenytoin are powerful antiepileptics, they did not protect against GSA-induced convulsions, epileptiform discharges in the EHG and ECoG and neurolysis. On the other hand, ketamine, a selective non-competitive NMDA receptor antagonist, not only protected against seizures, but also against neuronal cell damage induced by intrahippocampal injection of GSA.

All these results suggest that convulsions, recurrent epileptiform discharges and subsequently damage of neuronal cells by intrahippocampal injection of GSA are very similar to the effects of the endogenous excitatory amino acid glutamate[11]. Such excitotoxic effects have been shown to be mediated by NMDA-type glutamate receptors, and thus, this type of receptors might also mediate the effects of GSA. This hypothesis is supported by our previous work showing that NMDA receptor antagonists but not antiepileptic drugs selectively antagonize GSA-induced convulsions in mice[8]. This also leads us to suggest that GSA could act as a neurotoxin in some pathophysiological conditions. The suggested link with NMDA receptors therefore constitutes an interesting experimental paradigm for the further elucidation of GSA action on the CNS.

## Acknowledgements

This work was supported by the The Chinese Natural Scientific Foundation (N° 37090801)

# References

1.  Marescau B., Deshmuhk D.R., Cockx M., Possemiers I. & De Deyn P.P. (1992): Guanidino compounds in serum, urine, liver, kidney and brain of man and some ureoletic animals. *Metabolism* **41**, 526–532.

2.  De Deyn, P.P., Marescau, B., Cuykens, J.J., Van Gorp, L., Lowenthal A. & De Potter W.P. (1987): Guanidino compounds in serum and cerebrospinal fluid of non-dialysed patients with renal insufficiency. *Clin. Chim. Acta* **167**, 81–88.

3.  Kishore, B.K. (1983): Some observations on the *in vivo* and *in vitro* effects of guanidinosuccinic acid on the nervous system of laboratory animals. *Acta Med. Biol.* **31**, 79–84.

4.  De Deyn, P.P. & Macdonald, R.L. (1990): Guanidino compounds that are increased in cerebrospinal fluid and brain of uremic patients inhibit GABA and glycine responses on mouse neurons in cell culture. *Ann. Neurol.* **28**, 627–633.

5.  Pei, Y.Q., D'Hooge, R., Marescau, B. & De Deyn, P.P. (1992): Characterization of guanidinosuccinic acid induced seizures. In: *Guanidino compounds in biology and medicine*, eds. P.P. De Deyn *et al.*, pp. 457–459. London: John Libbey.

6.  D'Hooge, R., Pei, Y.Q., Manil, J. & De Deyn, P.P. (1992): The uremic guanidinosuccinic acid induces behavioural convulsions and concomitant epileptiform electrocorticographic discharges in mice. *Brain Res.* **598**, 316–320.

7.  Fleck, M.W., Henze, D.A., Barrionuevo, G. & Palmer, A.M. (1993): Aspartate and glutamate mediate excitatory synaptic transmission in area CA1 of the hippocampus. *J. Neurosci.* **13**, 3944–3955.

8.  D'Hooge, R., Pei, Y.Q. & De Deyn, P.P. (1993): N-methyl-D- aspartate receptors contribute to guanidinosuccinic acid induced convulsions in mice. *Neurosci. Lett.* **157**, 123–126.

9.  Konig, J.F. (1963): *The rat brain: A stereotaxic atlas of the forebrain and lower parts of the brainstem*, eds. Klippel. Baltimore: Williams & Wilkins.

10. D'Hooge, R., Pei, Y.Q., Marescau, B. & De Deyn, P.P. (1992): Convulsive action and toxicity of uremic guanidino compounds: behavioral assessment and relation to brain concentration. *J. Neurol. Sci.* **112,** 96–105.

11. Meldrum, B. (1990): Excitatory amino acid neurotoxicity and neurodegenerative diseases. *Trends Pharmacol. Sci.* **11,** 379–387.

# Section X

## Guanidino compounds in microorganisms, plants and invertebrata

*Guanidino Compounds : 2*, eds. by P.P. De Deyn, B. Marescau, I.A. Qureshi and A. Mori.
©1997 John Libbey & Company Ltd., pp. 389–399.

# Chapter 41

# Arginine catabolism in *Pseudomonas aeruginosa*: key regulatory features of the arginine deiminase pathway

Dieter HAAS[1], Harald WINTELER[1], Van Thanh NGUYEN[2], Catherine TRICOT[2] and Victor STALON[2]

*Laboratoire de Biologie Microbienne, Université de Lausanne, CH-1015 Lausanne, Switzerland, and [2]Laboratoire de Microbiologie, Faculté des Sciences, Université Libre de Bruxelles, B-1070 Bruxelles, Belgium*

## Summary

The ubiquitous bacterium and opportunistic pathogen *Pseudomonas aeruginosa* very effectively utilizes L-arginine as a carbon, nitrogen, and energy source via four catabolic pathways. One of them, the arginine deiminase (ADI) pathway encoded by the *arc* operon, is highly expressed during oxygen limitation and generates ATP by substrate phosphorylation. Here we examine and review some key regulatory features of the ADI pathway. Anaerobic induction of the *arc* operon depends on the ANR protein, a transcriptional activator similar to FNR of *Escherichia coli*. ANR and FNR interact with a sequence (consensus TTGAT....ATCAA) in the –40 region of promoters, but both proteins were found to differ with respect to their recognition specificity in that the sequence TTGAC....ATCAG was recognized well by ANR, but poorly by FNR. High level expression of the first two enzymes, ADI and catabolic ornithine carbamoyltrans-ferase (OTC), correlates with relatively long half-lives of the corresponding *arcA* and *arcB* mRNAs. Insertion of a foreign gene (*lacZ*) of *E. coli* into *arcA* or *arcB* negatively influenced mRNA stability and expression, indicating that the stem-loop structures located at the 3′ ends of the *arcA* and *arcB* transcripts are insufficient to ensure mRNA stability. The reaction catalysed by catabolic OTC is reversible *in vitro*, but not *in vivo* due to a low affinity for carbamoylphos-phate and high cooperativity towards this substrate. Mutant enzymes which lacked this kinetic control and gained anabolic OTC activity were affected either in amino acid residue 105 (Glu → Ala, Gly or Lys) or in residue 321 (Met → Ile). Thus, the expression of the ADI pathway is critically controlled at both the transcript and the protein level.

## Introduction

**P**seudomonas aeruginosa is a Gram-negative bacterium that lives in aquatic and terrestrial habitats and causes infections in compromised human or animal hosts[8,34]. One outstanding characteristic of *P. aeruginosa* and related fluorescent pseudomonads is the ability to degrade a large number of different organic compounds, which can serve as carbon, nitrogen or sulfur sources[34,45]. Moreover, it is not uncommon for pseudomonads to have several catabolic pathways for a single organic substrate. This point is well illustrated by the arginine network of *P. aeruginosa*[17]: L-arginine is degraded via four different pathways (Fig. 1).

In the pathway initiated by arginine deiminase (ADI) in *P. aeruginosa*, L-arginine serves primarily as an energy source. The second reaction of this route, the phosphorolysis of citrulline catalysed by

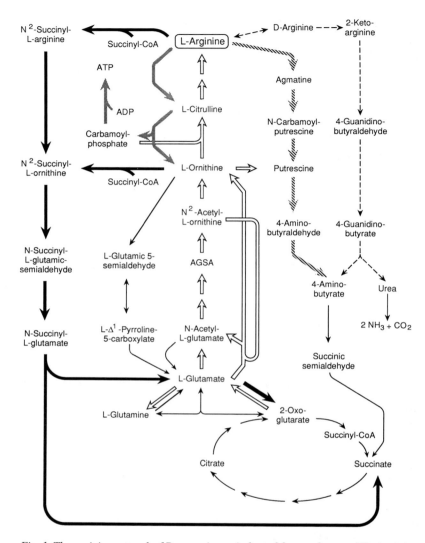

*Fig. 1. The arginine network of P. aeruginosa (adapted from reference 17). Arginine biosynthesis (open arrows), the ADI pathway (grey arrows), the N²-succinylarginine pathway (heavy black arrows), the 2-ketoarginine pathway (dashed arrows) and the arginine decarboxylase pathway (hatched arrows) are described in more detail elsewhere[17].*

catabolic ornithine carbamoyltransferase (OTC), yields carbamoylphosphate, which is used to synthesize ATP from ADP in the carbamate kinase reaction (Figs. 1 and 2)[37,51]. Thus, ATP is produced by substrate phosphorylation and this (albeit meager) supply of energy enables *P. aeruginosa* to activate motility[2] and to grow without respiration (i.e. anaerobically) in rich media containing arginine[51]. The ADI pathway liberates two moles of $NH_3$, which is available to the cells as a nitrogen source[1,27]. By contrast, arginine cannot be used as a carbon source in this route because the end product, ornithine, is excreted quantitatively[51]. One salient feature of the ADI pathway is the mechanism of anaerobic induction; this will be discussed in detail below.

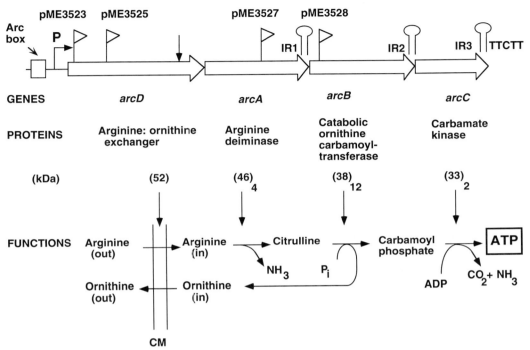

Fig. 2. Structure and functions of the P. aeruginosa arcDABC operon. The arcDABC genes are carried by a 5 kb-fragment, which has been sequenced entirely[3,4,25]. P designates the promoter with the upstream Arc box and CM the cytoplasmic membrane. The inverted repeats IR1 and IR2 specify stem-loop structures in the arc transcript and constitute weak terminators or barriers against 3'-exonuclease attack. IR3 is a rho-independent terminator. The vertical arrow marks the site of mRNA processing[11,12]. Transcriptional lacZ fusions in arcD, arcA and arcB are indicated by flags and plasmid names (see Table 2).

When *P. aeruginosa* grows on L-arginine as the sole carbon and nitrogen source under aerobic conditions, the bulk of arginine is catabolized by the $N^2$-succinylarginine pathway[20]. The first enzyme, arginine $N^2$-succinyltransferase, requires succinyl-CoA[50a]. This compound is later regenerated from 2-oxoglutarate in the citric acid cycle, i.e. during catabolism of N-succinylglutamate via glutamate (Fig. 1). All enzymes of the $N^2$-succinylarginine pathway are inducible by arginine and the corresponding structural genes (*aruABCDE*) are clustered in the *P. aeruginosa* chromosome[20,52]. Mutations causing non-inducibility are closely linked to the *aru* cluster[20]. Interestingly, such mutations also lead to constitutive high expression of anabolic OTC[19], the only arginine biosynthetic enzyme that is repressible by arginine in wild-type *P. aeruginosa*[55]. Thus, it appears that there exists a common regulatory element controlling repression of anabolic OTC and induction of the $N^2$-succinylarginine pathway[19].

The two remaining L-arginine catabolic pathways have a common latter part leading from 4-aminobutyrate to succinate, but differ in their earlier intermediates, agmatine → N-carbamoylputrescine → putrescine and D-arginine → 2-ketoarginine → 4-guanidinobutyrate, respectively (Fig. 1). The principal function of these pathways is to enable *P. aeruginosa* to utilize all of the intermediates mentioned as C and N sources[21,30,47]. An extensive review of the arginine network, with its genes and enzymes in *P. aeruginosa*, has been presented previously[17].

A close relative of *P. aeruginosa*, *P. putida*, possesses essentially the same network of arginine catabolic reactions, with two differences. The first one is characterized by ornithine cyclodeaminase;

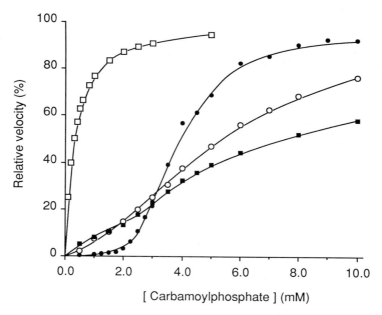

*Fig. 3. CP saturation curves of wildtype catabolic OTC (●) and of its mutant forms E105A (□), E105K (○) and M321I (■). The kinetics of the E105G enzyme (not shown) are similar to those of the E105A enzyme[5]. All enzymes were assayed at pH 6.8. Reproduced with permission from reference 32a.*

this enzyme converts ornithine to proline and contributes to citrulline degradation in *P. putida*, but is missing from *P. aeruginosa*[48]. The second difference concerns the 2-ketoarginine pathway. D-Arginine, which appears to be an intermediate in *P. aeruginosa* [21], does not occur in the oxidative deamination of L-arginine to 2-ketoarginine in *P. putida*[31,48].

The ability of *Pseudomonas* spp. to use arginine and its potential catabolites as a sole source of carbon and/or nitrogen has been investigated intensively by Stalon *et al.*[43,44,47]. Most species display a distinctive utilization pattern. Therefore, these compounds may be useful in the taxonomic classification of pseudomonads[34,44,47].

### Genetic organization and differential expression of the ADI pathway in *P. aeruginosa*

The *arcDABC* operon of *P. aeruginosa* contains the structural genes for the entire ADI pathway[24,25,51]. The first gene, *arcD* (Fig. 2), encodes a 52 kDa transmembrane protein that mediates a stoichiometric, electroneutral exchange between extracellular arginine and intracellular ornithine[25,53]. This reaction is driven by the concentration gradient and does not require additional metabolic energy. In addition, the ArcD protein promotes a slow, proton motive force-driven uptake of arginine and ornithine[53]. The ArcD protein has been localized in the cytoplasmic membrane of *P. aeruginosa* and, like the LysI transport protein of *Corynebacterium glutamicum*[38], has 13 hydrophobic segments, each of which could span the membrane[6]. Mutants defective in *arcD* are unable to utilize arginine as an energy source anaerobically but grow normally on arginine under aerobic conditions[25,51]. These data indicate that the ArcD protein functions specifically in arginine uptake and ornithine excretion during the operation of the ADI pathway.

The cytoplasmic enzymes of the ADI pathway, ADI, catabolic OTC and carbamate kinase, are the products of the *arcABC* genes, respectively (Fig.2)[3,4,25,51]. When the pathway is induced by oxygen limitation, ADI and catabolic OTC are among the most abundant proteins in *P. aeruginosa,* whereas carbamate kinase is expressed somewhat less efficiently. All three enzymes can be visualized by Coomassie Blue staining after SDS-polyacrylamide gel electrophoresis[3,4]. By contrast, the amount of the ArcD integral membrane protein is much lower than that of the ArcABC enzymes, and ArcD has been detected only in maxicells after radioactive labelling[25] or by tagging with PhoA (alkaline phosphatase) or with a colicin A epitope[6].

The *arcDABC* operon is expressed from an anaerobically inducible promoter (Fig. 2). A sequence (TTGAC....ATCAG) in the –40 region of the promoter[10,56a] and a positive control element, ANR[9,59], are both necessary for induction of the operon under oxygen limiting conditions. The ANR protein is also required for nitrate respiration in *P. aeruginosa* and resembles the well-known anaerobic regulator FNR of *Escherichia coli*[23,39,40,58a,59]. In particular, the four essential Cys residues of FNR (which are thought to bind an [4 Fe-4S] cluster [14a,21a,23a]) are all conserved in ANR[13,14,53], and the helix-turn-helix motif of FNR (which interacts with the –40 recognition sequence) has 87 per cent similarity with that of ANR[23,39,40,59]. The consensus FNR binding sequence TTGAT....ATCAA[40] is recognized by ANR[9,56a].

An *E. coli fnr* mutant can be complemented for anaerobic gas production by the cloned *anr* gene of *P. aeruginosa*[9] and, conversely, the FNR protein can partially replace ANR in nitrate respiration of *P. aeruginosa*[56a]. Thus, ANR and FNR perform similar regulatory functions during oxygen limitation but do not have identical recognition specificities. A more comprehensive review of anaerobic control in *P. aeruginosa* has been published elsewhere[18].

Two further aspects of *arcDABC* regulation should be mentioned. Arginine is not essential for induction but enhances maximum enzyme levels about twofold[29]. It is uncertain at present whether this stimulation by arginine involves a particular molecular mechanism or whether it is merely a consequence of arginine being the energy source and a component of protein synthesis. The differential expression of the Arc proteins also deserves a comment. Soon after the onset of induction, *arc* mRNA is processed in the distal part of *arcD*, presumably by an RNase E-like enzyme[11,12]. Moreover, the intergenic stem-loop structures IR1 and IR2 (Fig. 2) determine the 3′ ends of the *arc* mRNAs, by acting as leaky transcription terminators or as barriers against 3′ to 5′ exonucleolytic degradation. As a result, a family of distinct transcripts can be detected: weak bands of *arcDABC*, *arcDAB* and *arcDA* mRNAs and strong, relatively stable *arcABC*, *arcAB* and *arcA* mRNAs. The relative abundance of different *arc* transcripts correlates with the levels of the Arc proteins[11,12].

In conclusion, anaerobic growth of *P. aeruginosa* on arginine requires ANR, each of the *arcDABC* genes and the ANR-dependent *arc* promoter. Whereas these elements are all necessary for anaerobic growth on arginine, they are not sufficient. *P. fluorescens* CHA0 has a functional *anr* gene and an inducible ADI pathway[59], yet cannot grow on arginine without respiration (our unpublished results). The high, inducible expression of the *arc* operon in *P. aeruginosa* has provided the basis for the construction of expression vectors [56b]. In this context, we have clarified the differences in recognition specificity between ANR and FNR. Although it has not been possible to stabilize heterologous, unstable mRNAs by insertion into the more stable *arc* mRNA, we have succeeded in constructing anaerobically controlled expression vectors for *Pseudomonas* spp [56b].

## Recognition specificities of ANR and FNR

Since FNR of *E. coli* can be expressed as a biologically active protein in *P.aeruginosa,* we could directly compare the transcriptional activation abilities of FNR and ANR in *P. aeruginosa*. This was done by constructing a set of analogous plasmids that consisted of a broad-host-range vector (pKT240), an *E. coli lacZ* reporter gene with its own Shine-Dalgarno sequence, the consensus –10 promoter sequence TATAAT, and variable –40 regions including the consensus FNR recognition sequence TTGAT....ATCAA (Table 1). These constructs were introduced into *P. aeruginosa* host strains expressing either ANR or FNR (Table 1). The consensus FNR binding sequence allowed similar anaerobic induction by both FNR and ANR (as determined by the levels of the β-galactosidase reporter enzyme). The ANR recognition sequence originally found upstream of the *arc* operon and termed Arc box (TTGAC....ATCAG; Fig. 2) gave almost maximal induction with ANR but responded poorly to FNR (Table 1). One nucleotide deviating from the FNR consensus (A → G at the 5th position of the second half-site) contributed toward the poor ability of FNR to interact with the Arc box (Table 1). Thus, the DNA recognition domain of ANR, but not of FNR, tolerates a variation in the 5th nucleotide of the second half-site[56a]. Further experiments are needed to determine whether this difference in recognition specificity resides in the helix-turn-helix motifs of ANR and FNR.

**Table 1. Differential recognition of the –40 promoter region by ANR and FNR in *P. aeruginosa***

| Relevant phenotype[a] | Sequence of –40 region[b] | Relative sp. act. of lacZ reporter (%)[c] |
|---|---|---|
| Anr+ | TTGAT....ATCAA | 100 |
| Fnr+ | | 100 |
| Anr+ | TTGAC....ATCAG | 84 |
| Fnr+ | | 33 |
| Anr+ | TTGAC....ATCAA | 87 |
| Fnr+ | | 75 |
| Anr+ | TTGAT....ATCAG | 92 |
| Fnr+ | | 31 |
| Anr+ | CTTCC....GGCCC | < 2 |

[a]The Anr+ host was the wild-type *P. aeruginosa* PAO1. The Fnr+ host was the *anr* mutant PAO6261[18] carrying the *fnr* gene of *E. coli* on pRK2501*fnr* (kindly supplied by S. Busby).
[b]The –40 region shown was combined with a consensus –10 region (TATAAT) in a nucleotide sequence environment similar to that in the *arc* promoter. The artificial promoters created were fused to a translatable *lacZ* reporter gene carried by an IncQ vector plasmid[56a].
[c]The β-galactosidase activity of the *lacZ* reporter gene was monitored in oxygen-limited cells[9].

### Can the *arcDABC* environment stabilize the intrinsically unstable *lacZ* mRNA of *E. coli*?

The stability of bacterial mRNAs is determined largely by endonucleolytic attacks and by 3' to 5' exonucleolytic degradation. Typical half-lives of *E. coli* mRNAs vary from 0.5 to 30 min [7,36]. For instance, wild-type *lacZ* mRNA has a half-life of 1.75 min in *E. coli,* and the rate of mRNA inactivation can be modified by changes at either the 3' or the 5' end of *lacZ* mRNA[32,35]. By contrast, the half-lives of the *arcD* and *arcAB* messages in *P. aeruginosa* are considerably longer; they have been estimated to be about 8 min and 16 min, respectively[11]. Endonucleolytic processing of *arcD* mRNA (Fig. 2) by an RNase E–like enzyme probably accounts for the lower stability of the *arcD* transcripts, compared to that of the *arcAB* mRNAs[12]. If the hairpins in the *arc* mRNA such as IR1 and IR2 (Fig. 2) act as effective barriers against 3' to 5' exonucleolytic degradation, and if these structures are more important for mRNA stability than are endonuclease cleavage sites, then we would expect that *lacZ* mRNA should be stabilized when inserted into an *arcA* or *arcB* environment. We would also expect that an *arcD* environment would be less favourable in terms of *lacZ* stabilization and that, as a consequence, the β-galactosidase levels specified by transcriptional *arcD-lacZ* fusions should be lower than the levels produced by *arcA-lacZ* or *arcB-lacZ* fusions.

This hypothesis, however, was not supported by the experimental data. Transcriptional *lacZ* fusions in *arcD* gave significantly higher β-galactosidase levels than did *lacZ* fusions in *arcA* or *arcB* (Table 2). This suggests that an *arcAB* environment does not enhance the half-life of *lacZ* mRNA. In Northern blots, the strong, discrete *arcA* and *arcAB* transcripts disappeared after *lacZ* insertion into either gene and a smear indicative of unstable mRNA was obtained[56b]. One conclusion is that endonucleolytic attack rather than 3′ to 5′ exonucleolytic degradation is the rate-limiting step in *lacZ* mRNA decay in *P. aeruginosa*. Previous more extensive studies on *lacZ* mRNA degradation in *E. coli* are in agreement with this view[56,58]. Another conclusion is that expression of a heterologous gene does not benefit from transplantation into the *arcAB* genes.

**Table 2. Differential expression of transcriptional *lacZ* fusions in the *arc* operon of *P. aeruginosa*: *lacZ* mRNA does not appear to be stabilized by an *arcAB* environment**

| Plasmid | Fusion | β-Galactosidase sp.act.[a] |
|---|---|---|
| pME3523 | *arcD 253-lacZ* | 20.6 |
| pME3525 | *arcD 557-lacZ* | 20.8 |
| pME3527 | *arcA 2590-lacZ* | 11.8 |
| pME3528 | *arcB ca.3000-lacZ* | 12.7 |
| pME3529 | *lacZ* control | 1.6 |

[a]in $10^3$ Miller units; data are averages of 3 independent experiments. The β-galactosidase activity specified by the *lacZ* reporter gene was monitored in oxygen-limited cells[9].

## Kinetic control of catabolic OTC activity

Anaerobic induction and differential mRNA stabilities are not the only mechanisms regulating the ADI pathway in *P. aeruginosa*. A fine control acts at the level of OTC activity and thereby avoids a futile cycle between ornithine and citrulline (Fig. 1). Anabolic OTC does not function in the catabolic direction because citrulline is a dead-end inhibitor[42], and catabolic OTC is prevented from expressing anabolic activity by a strongly sigmoidal saturation curve for CP and a low affinity for this substrate[15,41,49]. AMP, a signal of energy starvation, allosterically activates catabolic OTC; at 10 mM AMP the enzyme approximates Michaelis-Menten kinetics and exhibits a high affinity for CP[49]. Thus, conditions of low energy supply strongly stimulate catabolic OTC. Another positive effector of this enzyme is phosphate, a CP analogue[49]. In the physiological (catabolic) direction of the OTC reaction, both phosphate and citrulline are cooperative substrates. Allosteric inhibitors are polyamines[49]. However, the physiological significance of these effectors is less evident.

Point mutations in the *arcB* gene can radically alter the homotropic cooperativity of catabolic OTC and endow the enzyme with anabolic activity. Such *arcB* mutations, therefore, will suppress an *argF* deficiency, i.e. a block in anabolic OTC[5,15]. The most dramatic change of kinetic control is brought about by a replacement of the glutamate residue 105 with either alanine or glycine: cooperativity for CP is lost (Fig. 3)[5,16]. Less severe kinetic alterations result from replacements of either glutamate 105 with lysine or methionine 321 with isoleucine (Fig. 3)[32a]. The E105K and M321I mutant forms were obtained in *P. aeruginosa* cells that carried the corresponding *arcB* suppressor alleles on a multi-copy plasmid, thus overproducing the mutant enzymes[32a].

How can these mutational changes be interpreted in terms of the structure of catabolic OTC? The enzyme is a dodecamer composed of identical 38 kDa subunits[3,26,31a]. Recently, the 3D structure of the E105A enzyme has been resolved[54,54a] to 0.3nm allowing the following interpretations. Glutamate 105 may form a salt bridge with the nearby residue arginine 107, which is conserved in OTCs and ATCs (aspartate carbamoyltransferases) and located in the active site[46]. The γ-carboxyl group of glutamate 105 might compete with the carbonyl group of CP for arginine 107; this could in part explain CP cooperativity. In the E105A, E105G and E105K mutants the putative competition for arginine 107 would be eliminated or reduced. Interestingly, anabolic OTC of *E. coli*, an enzyme that normally exhibits hyperbolic substrate saturation curves, manifests substrate cooperativity when arginine 106 (which is homologous to arginine 107 of catabolic OTC) is replaced with glycine[22].

Methionine 321 of catabolic OTC is located in the carboxyterminal helix H12. This helix is in close contact with helix H2, which carries five conserved residues (STRTR, 56 to 60) involved in CP binding[46]. The M321I change might alter the H2-H12 contacts. A small movement of H2 is expected to affect profoundly the active site and hence CP binding.

An alignment of the amino acid sequence of catabolic OTC with the primary sequences of 18 anabolic OTCs[54] would not have predicted an important role of glutamate 105 and methionine 321 in catabolic OTC; both amino acids are conserved in anabolic OTC of *Neisseria gonorrhoeae*[28]. This also means that further residues contribute to the kinetic control of catabolic OTC. They may reside in the N- and C-termini of the enzyme. In ATC of *E. coli*, the N- and C-terminal residues of the catalytic chain are located in close proximity to each other at the protein surface[57]. In the same vein, the 3D model of catabolic OTC shows that the N- and C-termini are close to each other within the CP domain[54,54a]. Mutational modifications at the C-terminus (deletion of one or two amino acids or extension) strongly alter the homotropic cooperativity and allosteric behaviour of catabolic OTC[5,50]. Similar changes occur when certain N-terminal fragments of catabolic OTC are exchanged for homologous fragments of anabolic OTC from *E. coli (argF)*[33]. Modifications of the N- and C-termini probably do not affect the effector binding sites directly, but rather exert long-range effects on intra- and inter-subunit interactions[31a,33,50].

## Concluding remarks

The tools of molecular genetics have allowed us to identify three checkpoints in the regulation of the ADI pathway: initiation of transcription by the anaerobic regulator ANR, modulation of translational efficiencies via processing of *arc* mRNA, and kinetic control of the allosteric catabolic OTC.

## Acknowledgements

V.T. Nguyen was supported by a fellowship of the 'Internationale Brachet Stiftung'. V.S. is a Research Associate of the National Fund for Scientific Research (Belgium). This work was supported by the Schweizerische Nationalfonds (project 31-28570. 90), the Fund for Joint Basic Research (Belgium; grant 2.4507.91) and a grant from ETH-Zürich. We thank M. Monnin for secretarial assistance.

## References

1. Abdelal, A.T., Bibb, W. F. & Nainan, O. (1982): Carbamate kinase from *Pseudomonas aeruginosa:* purification, characterization, physiological role, and regulation. *J. Bacteriol.* **151**, 1411–1419.

2. Armitage, J.P. & Evans, M.C.W. (1988): The motile and tactic behaviour of *Pseudomonas aeruginosa* in anaerobic environments. *FEBS Lett.* **156**, 113–118.

3. Baur, H., Stalon V., Falmagne, P., Lüthi, E. & Haas, D. (1987): Primary and quaternary structure of the catabolic ornithine carbamoyltransferase from *Pseudomonas aeruginosa*. *Eur. J. Biochem.* **166**, 111–117.

4. Baur, H., Lüthi, E., Stalon, V., Mercenier, A. & Haas, D. (1989): Sequence analysis and expression of the arginine-deiminase and carbamate-kinase genes of *Pseudomonas aeruginosa*. *Eur. J. Biochem.* **179**, 53–60.

5. Baur, H., Tricot, C., Stalon, V. & Haas, D. (1990): Converting catabolic ornithine carbamoyltransferase to an anabolic enzyme. *J. Biol. Chem.* **265**, 14728–14731.

6. Bourdineaud, J.-P., Heierli D., Gamper M., Verhoogt, H. J.C., Driessen, A. J.M., Konings, W.N., Lazdunski, C. & Haas, D. (1993): Characterization of the *arcD* arginine:ornithine exchanger of *Pseudomonas aeruginosa*. *J. Biol. Chem.* **268**, 5417–5424.

7. Bouvet, P. & Belasco, J.G. (1992): Control of RNase E-mediated RNA degradation by 5'-terminal base pairing in *E. coli. Nature* **360**, 488–491.

8. Campa, M., Bendinelli, M. & Friedman, H. (1993): Pseudomonas aeruginosa *as an opportunistic pathogen. New York: Plenum Press.*

9. Galimand, M., Gamper, M., Zimmermann, A. & Haas D. (1991): Positive FNR-like control of anaerobic arginine degradation and nitrate respiration in *Pseudomonas aeruginosa*. *J. Bacteriol.* **173**, 1598–1606.

10. Gamper, M., Zimmermann, A. & Haas, D. (1991): Anaerobic regulation of transcription initiation in the *arcDABC* operon of *Pseudomonas aeruginosa. J. Bacteriol.* **173,** 4742–4750.

11. Gamper, M., Ganter, B., Polito, M.R. & Haas, D. (1992): RNA processing modulates the expression of the *arcDABC* operon in *Pseudomonas aeruginosa. J. Mol. Biol.* **226,** 943–957.

12. Gamper, M. & Haas, D. (1993): Processing of the *Pseudomonas arcDABC* mRNA requires functional RNase E in *Escherichia coli. Gene* **129,** 119–122.

13. Green, J. & Guest, J.R. (1993): A role for iron in transcriptional activation by FNR. *FEBS Lett.* **329,** 55–58.

14. Green, J., Sharrocks, A.D., Green, B., Geisow, M. & Guest, J.R. (1993): Properties of FNR proteins substituted at each of the five cysteine residues. *Mol. Microbiol.* **8,** 61–68.

14a. Green, J., Bennett, B., Jordan, P., Ralph, E.T., Thomson, A.J. & Guest, J. (1996): Reconstruction of the [4Fe-4S] cluster in FNR and demonstration of the aerobic-anaerobic transcription switch in vitro. *Biochem J.* **316,** 887-892.

15. Haas, D., Evans, R., Mercenier, A., Simon, J.-P. & Stalon, V. (1979): Genetic and physiological characterization of *Pseudomonas aeruginosa* mutants affected in the catabolic ornithine carbamoyltransferase. *J. Bacteriol.* **139,** 713–720.

16. Haas, D., Baur, H., Tricot, C., Galimand, M. & Stalon, V. (1989): Catabolic ornithine carbamoyltransferase of *Pseudomonas*: expression in stationary phase and experimental evolution. In: *Recent advances in microbial ecology*, eds. T. Hattori, Y. Ishida, Y. Maruyama, R.Y. Morita & A. Uchida, pp. 617–621. Tokyo: Japan Scientific Societies Press.

17. Haas, D., Galimand, M., Gamper, M. & Zimmermann, A. (1990): Arginine network of *Pseudomonas aeruginosa*: specific and global controls. In: *Pseudomonas: Biotransformations, pathogenesis, and evolving biotechnology*, eds. S. Silver, A.M. Chakrabarty, B. Iglewski, & S. Kaplan, pp. 303–316. Washington, D.C.: American Society for Microbiology.

18. Haas, D., Gamper, M. & Zimmermann, A. (1992): Anaerobic control in *Pseudomonas aeruginosa*. In: *Pseudomonas: Molecular biology and biotechnology*, eds. S. Silver, E. Galli & B. Witholt, pp. 177–187. Washington, D.C.: American Society for Microbiology.

19. Itoh, Y. & Matsumoto, H. (1992): Mutations affecting regulation of the anabolic *argF* and the catabolic *aru* genes in *Pseudomonas aeruginosa* PAO. *Mol. Gen. Genet.* **231,** 417–425.

20. Jann, A., Stalon, V., Vander Wauven, C., Leisinger, T. & Haas, D. (1986): $N^2$-Succinylated intermediates in an arginine catabolic pathway of *Pseudomonas aeruginosa. Proc. Natl. Acad. Sci. USA* **83,** 4937–4941.

21. Jann, A., Matsumoto, H. & Haas, D. (1988): The fourth arginine catabolic pathway of *Pseudomonas aeruginosa. J. Gen. Microbiol.* **134,** 1043–1053.

21a. Khoroshilova, N., Beinert, H. & Kiley, P.J. (1995): Association of a polynuclear iron-sulfur center with a mutant FNR protein enhances DNA binding. *Proc. Natl. Acad. Sci. USA* **92,** 2499-2503.

22. Kuo, L.C., Zambidis, I. & Caron, C. (1989): Triggering of allostery in an enzyme by a point mutation: ornithine transcarbamoylase. *Science* **245,** 522–524.

23. Lazazzera, B.A., Bates, D.M. & Kiley, P.J. (1993): The activity of the *Escherichia coli* transcription factor FNR is regulated by a change in oligomeric state. *Genes Develop.* **7,** 1993–2005.

23a. Lazazzera, B.A., Beinert, H., Khoroshilova, N., Kennedy, M.C. & Kiley, P.J. (1996): DNA binding and dimerization of the Fe-S containing FNR protein from *Escherichia coli* are regulated by oxygen. *J Biol. Chem.* **271,** 2762-2768.

24. Lüthi, E., Mercenier, A. & Haas, D. (1986): The *arcABC* operon required for fermentative growth of *Pseudomonas aeruginosa* on arginine: Tn*5-751*-assisted cloning and localization of structural genes. *J. Gen. Microbiol.* **132,** 2667–2675.

25. Lüthi, E., Baur, H., Gamper, M., Brunner, F., Villeval, D., Mercenier, A. & Haas, D. (1990): The *arc* operon for anaerobic arginine catabolism in *Pseudomonas aeruginosa* contains an additional gene, *arcD*, encoding a membrane protein. *Gene* **87,** 37–43.

26. Marcq, S., Diaz-Ruano, A., Charlier, P., Dideberg, O., Tricot, C., Piérard, A. & Stalon, V. (1991): Molecular size and symmetry of *Pseudomonas aeruginosa* catabolic ornithine carbamoyltransferase: an X-ray crystallography analysis. *J. Mol. Biol.* **220,** 9–12.

27. Marquis, R.E., Bender, G.R., Murray, D.R. & Wong, A. (1987): Arginine deiminase system and bacterial adaptation to acid environments. *Appl. Environ. Microbiol.* **53,** 198–200.

28. Martin, P.R., Cooperider, J.W. & Mulks, M.H. (1990): Sequence of the *argF* gene encoding ornithine transcarbamoylase from *Neisseria gonorrhoeae. Gene* **94,** 139–140.

29.  Mercenier, A., Simon, J-P., Vander Wauven, C., Haas, D. & Stalon, V. (1980): Regulation of enzyme synthesis in the arginine deiminase pathway of *Pseudomonas aeruginosa. J. Bacteriol.* **144,** 159–163.

30.  Mercenier, A., Simon, J.-P., Haas, D. & Stalon, V. (1980): Catabolism of L-arginine by *Pseudomonas aeruginosa. J. Gen. Microbiol.* **116,** 381–389.

31.  Miller, D.L. & Rodwell, V.W. (1971): Metabolism of basic amino acids in *Pseudomonas putida. J. Biol. Chem.* **246,** 5053–5058.

31a. Mouz, N., Tricot, C., Ebel, C., Petillot, Y., Stalon, V. & Dideberg, O. (1996): Use of a designed fusion protein dissociates allosteric properties from the dodecameric state of *Pseudomonas aeruginosa* catabolic ornithine carbamoyltransferase. *Proc. Natl. Acad. Sci. USA* **93,** 9414-9419.

32.  Murakawa, G.J., Kwan, C., Yamashita, J. & Nierlich, D.P. (1991): Transcription and decay of the *lac* messenger: role of an intergenic terminator. *J. Bacteriol.* **173,** 28–36.

32a. Nguyen, V.T., Tricot, C., Stalon, V., Dideberg, O., Villeret, V. & Haas, D. (1994): Methionine-321 in the C-terminal α-helix of catabolic ornithine carbamoyltransferase from *Pseudomonas aeruginosa* is important for homotropic cooperativity. *FEMS Microbiol. Lett* **124,** 411-418.

33.  Nguyen, V.T, Baker, D., Tricot, C., Baur, H., Villeret, V., Dideberg, O., Gigot, D., Stalon, V. & Haas, D. (1996): Catabolic ornithine carbamoyltransferase of *Pseudomonas aeruginosa*. Importance of the N-terminal region for dodecameric structure and homotropic carbamoylphosphate cooperativity. *Europ. J. Biochem.* **236,** 283-293.

34.  Palleroni, N.J. (1993): *Pseudomonas* classification. *Antonie van Leeuwenhoek* **64,** 231–251.

35.  Petersen, C. (1991): Multiple determinants of functional mRNA stability: sequence alterations at either end of the *lacZ* gene affect the rate of mRNA inactivation. *J. Bacteriol.* **173,** 2167–2172.

36.  Petersen, C. (1992): Control of functional mRNA stability in bacteria: multiple mechanisms of nucleolytic and non-nucleolytic inactivation. *Mol. Microbiol.* **6,** 277–282.

37.  Ramos, F., Stalon, V., Piérard, A. & Wiame, J.M. (1967): The specialization of the two ornithine carbamoyltransferases of *Pseudomonas. Biochim. Biophys. Acta* **139,** 98–106.

38.  Seep-Feldhaus, A.H., Kalinowski, J. & Pühler, A. (1991): Molecular analysis of the *Corynebacterium glutamicum lysI* gene involved in lysine uptake. *Mol. Microbiol.* **5,** 2995–3005.

39.  Spiro, S., Gaston, K.L., Bell, A.I., Roberts, R.E., Busby, S.J.W. & Guest, J.R. (1990): Interconversion of the DNA-binding specificities of two related transcription regulators, CRP and FNR. *Mol. Microbiol.* **4,** 1831–1838.

40.  Spiro, S. & Guest, J.R. (1991): Adaptive responses to oxygen limitation in *Escherichia coli. Trends Biochem Sci.* **16,** 310–314.

41.  Stalon, V., Ramos, F., Piérard, A. & Wiame, J.-M. (1972): Regulation of the catabolic ornithine carbamoyltransferase of *Pseudomonas fluorescens. Eur. J. Biochem.* **29,** 25–35.

42.  Stalon, V., Legrain, C. & Wiame J.-M. (1977): Anabolic ornithine carbamoyltransferase of *Pseudomonas. Eur. J. Biochem.* **74,** 319–327.

43.  Stalon, V. & Mercenier, A. (1984): L-Arginine utilization by *Pseudomonas* species. *J. Gen. Microbiol.* **130,** 69–76.

44.  Stalon, V., Vander Wauven, C., Momin, P. & Legrain, C. (1987): Catabolism of arginine, citrulline and ornithine by *Pseudomonas* and related bacteria. *J. Gen. Microbiol.* **133,** 2487–2495.

45.  Stanier, R.Y., Palleroni, N.J. & Doudoroff, M. (1966): The aerobic pseudomonads: a taxonomic study. *J. Gen Microbiol.* **43,** 159–271.

46.  Stevens, R.C., Chook, Y.M., Cho, C.Y., Lipscomb, W.N. & Kantrowitz, E.R. (1991): *Escherichia coli* aspartate carbamoyltransferase: the probing of crystal structure analysis via site-specific mutagenesis. *Protein Engineering* **4,** 391–408.

47.  Tricot, C., Piérard, A. & Stalon, V. (1990): Comparative studies on the degradation of guanidino and ureido compounds by *Pseudomonas J. Gen. Microbiol.* **136,** 2307–2317.

48.  Tricot, C., Stalon, V. & Legrain, C. (1991): Isolation and characterization of *Pseudomonas putida* mutants affected in arginine, ornithine and citrulline catabolism: function of the arginine oxidase and arginine succinyltransferase pathways. *J. Gen. Microbiol.* **137,** 2911–2918.

49.  Tricot, C., Nguyen, V.T. & Stalon, V. (1993): Steady-state kinetics and analysis of pH dependence on wild-type and a modified allosteric *Pseudomonas aeruginosa* ornithine carbamoyltranferase containing the replacement of glutamate 105 by alanine. *Eur. J. Biochem.* **215,** 833–839.

50. Tricot, C., Schmid, S., Baur, H., Villeret, V., Dideberg, O., Haas, D. & Stalon, V. (1994): Catabolic ornithine carbamoyltransferase of *Pseudomonas aeruginosa*: changes of allosteric properties resulting from modifications at the C-terminus. *Eur. J. Biochem.* **221,** 555–561.

50a. Tricot, C., Vander Wauven, C., Wattez R., Falmagne, P. & Stalon, V. (1994): Purification and properties of a succinyltransferase from *Pseudomonas aeruginosa* specific for both arginine and ornithine. *Eur. J. Biochem.* **224,** 853-861.

51. Vander Wauven, C., Piérard, A., Kley-Raymann, M. & Haas, D. (1984): *Pseudomonas aeruginosa* mutants affected in anaerobic growth on arginine: evidence for a four-gene cluster encoding the arginine deiminase pathway. *J. Bacteriol.* **160,** 928–934.

52. Vander Wauven, C., Jann, A., Haas, D., Leisinger, T. & Stalon, V. (1988): $N^2$-Succinylornithine in ornithine catabolism of *Pseudomonas aeruginosa. Arch. Microbiol.* **150,** 400–404.

53. Verhoogt, H.J.C., Smit, H., Abee, T., Gamper, M., Driessen, A.J.M., Haas, D. & Konings, W.N. (1992): *arcD*, the first gene of the *arc* operon for anaerobic arginine catabolism in *Pseudomonas aeruginosa*, encodes an arginine-ornithine exchanger. *J. Bacteriol.* **174,** 1568–1573.

54. Villeret, V. (1994): Etudes structurales de l'ornithine carbamoyltransferase catabolique de *Pseudomonas aeruginosa*. Ph. D. Thesis, Université de Liège.

54a. Villeret, V., Tricot, C., Stalon, V. & Dideberg, O. (1995): Crystal structure of *Pseudomonas aeruginosa* ornithine transcarbamoylase at 3.0 Å resolution: a different oligomeric organization in the transcarbamoylase family. *Proc. Natl. Acad. Sci. USA* **92,** 10762-10766.

55. Voellmy, R. & Leisinger, T. (1978): Regulation of enzyme synthesis in the arginine biosynthetic pathway of *Pseudomonas aeruginosa. J. Gen. Microbiol.* **109,** 25–35.

56. Wagner, L.A., Gesteland, R.F. Dayhuff, T.J. & Weiss, R.B. (1994): An efficient Shine-Dalgarno sequence but not translation is necessary for *lacZ* mRNA stability in *Escherichia coli. J. Bacteriol.* **176,** 1683–1688.

56a. Winteler, H.V. & Haas, D. (1996): The homologous regulators ANR of *Pseudomonas aeruginosa* and FNR of *Escherichia coli* have overlapping but distinct specificities for anaerobically inducible promoters. *Microbiology* **142,** 685-693.

56b. Winteler, H.V., Schneidinger, B., Jaeger, K.-E. & Haas, D. (1996): Anaerobically controlled expression system derived form the *arcDABC* operon of *Pseudomonas aeruginosa*: application to lipase production. *Appl. Environ. Microbiol.* **62,** 3391-3398.

57. Yang, Y.R. & Schachman, H.K. (1993): Aspartate transcarbamoylase containing circularly permuted catalytic polypeptide chains. *Proc. Natl. Acad. Sci. USA* **90,** 11980–11984.

58. Yarchuk, O., Jacques, N., Guillerez, J. & Dreyfus, M. (1992): Interdependence of translation, transcription and mRNA degradation in the *lacZ* gene. *J. Mol. Biol.* **226,** 581–596.

58a. Ye, R.W., Haas, D., Ka, J.-O., Krishnapilai, V., Zimmermann, A., Baird, C. & Tiedje, J. M. (1995): Anaerobic activation of the entire denitrification pathway in *Pseudomonas aeruginosa* requires Anr, an analog of Fnr. *J. Bacteriol.* **177,** 3606-3609.

59. Zimmermann, A., Reimmann, C., Galimand, M. & Haas, D. (1991): Anaerobic growth and cyanide synthesis of *Pseudomonas aeruginosa* depend on *anr*, a regulatory gene homologous with *fnr* of *Escherichia coli. Mol. Microbiol.* **5,** 1483–1490.

*Guanidino Compounds : 2*, eds. by P.P. De Deyn, B. Marescau, I.A. Qureshi and A. Mori.
©1997 John Libbey & Company Ltd., pp. 401–405.

Chapter 42

# Ecological significance of canavanine and biological activity of some canavanine derivatives

J. MIERSCH[1], K. GRANCHAROV[2], T. PAJPANOVA[2], R. BAUMBACH[1],
S. TABAKOVA[2],S. STOEV[2], G.-J. KRAUSS[1], and E. GOLOVINSKY[2]

[1]*Martin-Luther-University Halle-Wittenberg, Institute of Biochemistry, Kurt-Mothes Str.3, 06120 Halle (S.),Germany;*
[2]*Bulgarian Academy of Sciences, Institute of Molecular Biology, G. Bonchev Street Bl. 21, 1113 Sofia, Bulgaria*

## Summary

The non-protein amino acid L-canavanine (C), a natural arginine analogue, occurs in many leguminous plants and plays an important ecological role. During the development of seedlings of alfalfa canavanine is synthesized in addition to arginine and asparagine, and is released in the environment of roots[8]. We have synthesized a series of canavanine derivatives, some of which were biologically active[9]. In this study we report the influence of a novel compound, canavanine bis (2-chloroethyl)hydrazide (CBCH), on the growth of bacteria and plants, using C as a positive control. CBCH inhibited bacterial growth in concentrations ranging from 0.4 to 2 mM, and most susceptible to its action were Bacillus subtilis and Proteus vulgaris. C was generally more active than CBCH, but its effect was reversed by equimolar amounts of arginine, while 10-fold higher molar arginine concentrations were required to abolish the effect of CBCH. The novel compound had no influence on the growth of cultured tomato cells, but reduced significantly the radicle growth of cress and amaranth. It was less active against both species in comparison to C, but it was more inhibitory than the related derivative, canavanine hydrazide[9].

## Introduction

Plants in the biocoenoses interact among each another and with their environment. These interactions guarantee a coordinating development of organisms in their living communities. In the investigation of plant communities were also found, e.g. the influence of natural occurring chemicals. Such plant metabolites, known as allelochemicals, originated from various plant organs and, released in the environment, caused in many cases distinct physiological and biochemical influences after their uptake by an other species. These influences of allelochemicals are well known as so-called allelopathic effects[5,7,11,14,15]. There is a special interest in studying of natural compounds influencing crop rotation in agriculture and growth of weeds[12].

In former time we investigated the toxic amino acid canavanine (2-amino-4-guanidinooxy-butyric acid), an analog of arginine[1,2]. This compound occurs in many leguminous plants and can be released in the environment by roots[3,4,8]. Canavanine is also toxic against insects[13] and causes growth redartation of radicle of cress, amaranth and cabbage[6,8]. In this study we report the influence of a novel compound, canavanine bis (2-chloroethyl)hydrazide on the growth of bacteria and plants using canavanine as a positive control and compare its activity with other guanidino compounds.

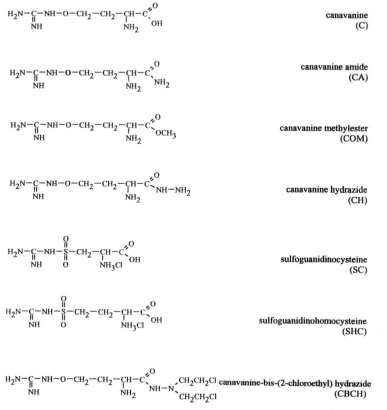

*Fig. 1. Chemical structure of L-canavanine derivatives.*

## Material and Methods

Plant material. Seeds of alfalfa (*Medicago sativa* L. cv Verko) were obtained from VEG Pflanzenproduktion Langenstein, Germany. The details of cultivation were given by Miersch and co-workers[6].

Biotests. The influence of canavanine and its derivatives on cell growth of tomato suspension culture (*Lycopersicon esculentum* Mill.) was measured as described by Grancharov *et al.*[1]. The inhibition studies of radicle growth of cabbage (*Brassica oleracea* L.), cress (*Lepidium sativum* L.), and amaranth (*Amaranthus* paniculatus L) were carried out as reported previously[3]. The antimicrobial screening (Table 2) was performed according to Grancharov *et al.*[1]. Synthesis of canavanine derivatives. The synthesis of canavanine derivatives (Fig. 1) was principally carried out as described elsewhere[9,10]. We have obtained intermediates, new antimetabolites and precursors of biologiocally active peptides. CBCH was prepared by stirring of N-protected canavanine, bis-(-chloroethyl)hydrazide and TBTU, dissolved inDMF, for 3 h at room temperature in the presence of the organic base DIPEA. Canavanine was protected at $N_1$ and $N_g$-guanidino groups by a t-butyloxycarbonyl group (Boc) which was after synthesis readily removed by HCl-saturated organic solvent. The structure of CBCH was confirmed by ¹H-NMR and mass spectrometry.

## Results and Discussion

The metabolism and possible function of canavanine during germination of alfalfa seeds as well as development of seedlings were described earlier[3,8]. Based on the structure and the properties of canavanine it could be assumed that derivatives of this compound could also exert interesting biological effects[2].

## Effects on the growth of cultured tomato cells.

The inhibitory effects of canavanine derivatives (at 1 mM) on the growth of tomato cells up to 7 days of cultivation are presented in Table 1. The most active compound proved to be canavanine, reaching on the 3rd day 57 per cent, and on the 7th day 80 per cent growth inhibition. Some canavanine

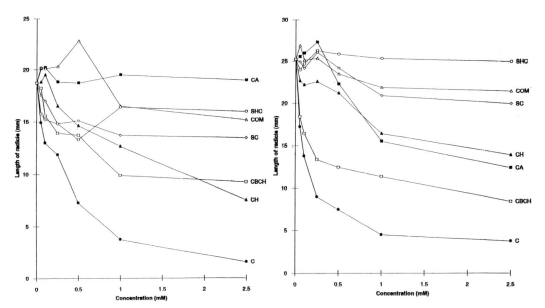

Fig. 2. Growth inhibition of radicles of cress (Lepideum sativum L.) caused by L-canavaine (C), canavanine hydrazide (CH), canavanine methylester (COM), canavanine amide (CA), canavanine-bis-(-2-chloroethyl) hydrazide (CBCH), sulfoguanidino homocysteine (SHC).The points are means of 33 to 136 measurements and are statistically significamt by F- and Chi²-test (C, CH, CBCH).

Fig. 3. growth inhibition of radicles of amaranth (Amaranthuspaniculatus Λ.) caused by L-cana-vanine (C), canavanine (C), canavnine hydra-zide (CBCH), sulfoguanidinocyteine (SC) and sulfoguanidinohomocysteine (SHC). The points are means of 42 to 125 measurements and are statistically significant by F- and Chi²-test (C, CH, CBCH, CA).

**Table 1. Influence of canavanine-derivatives on the growth of cultured tomato cells (*Lycopersicon esculentum*)**

| | | | | Time of Cultivation | | | | |
|---|---|---|---|---|---|---|---|---|
| | 2 days | | 3 days | | 4 days | | 7 days | |
| Compound | DM mg/ml | Inh. % | DM mg/ml | Inh. % | DM mg/ml | Inh. % | DM mg/ml | Inh. % |
| Control | 9.1 | 0 | 13 | 0 | 26.8 | 0 | 38.3 | 0 |
| C | 6.4 | 30 | 5.8 | 57 | 9.3 | 69 | 7.7 | 80 |
| CA | 8.9 | 2 | 10.4 | 23 | 12.4 | 54 | 23.3 | 39 |
| CH | 8.8 | 4 | 12.2 | 9 | 23.8 | 11 | 45 | −17* |
| COM | 7.6 | 17 | 9.3 | 31 | 14.4 | 46 | 33.8 | 12 |
| CBCH | 7.3 | 20 | 11.6 | 11 | 17.7 | 34 | 33.9 | 11 |
| SC** | 9.2 | 0 | 11.3 | 15 | 25.3 | 6 | 37.9 | 2 |
| SHC** | 8.7 | 4 | 11.4 | 5 | 24.4 | 9 | 44.1 | −15* |

*negative values = stimulation of growth. **0.5 mM; all others = 1 mM. DM = dry mass; Inh. = inhibition; C = canavanine, CA = canavanineamide; CH = canavanine hydrazide; COM = canavanine methylester; CBCH = canavanine-bis-(2-chloroethyl) hydrazide; SC = sulfoguanindinocyteine; SHC = sulfoguanidinohomocysteine.

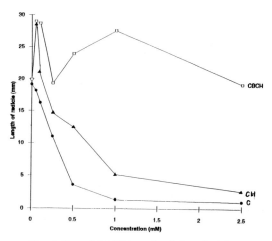

*Fig. 4. Growth inhibition of radicles of cabbage (Brassica oleracea L.) caused by L-canavanine (C), ca-navanine-bis- (2-chloroethyl) hydrazide (CBCH) and canavanine hydrazide (CH). The points are means of 30 to 54 measurements and are statistically significant by F-teset (C, CH).*

derivatives ( CA, COM and CBCH ) exerted also an inhibitory activity (54 per cent, 46 per cent and 34 per cent growth inhibition, respectively), the effect being, unlike canavanine, non-lethal. Upon prolonged treatment with these compounds (7 days) the cells recovered from the inhibition reaching dry mass value near to that of controls. Resuming the above results, the following order of decreasing activity could be given: C CA COM CBCH CH SHC SC.

Plant seedling inhibitory effects. The influence of canavanine derivatives on the growth of plants was further studied using plant seedlings of cress, amaranth and cabbage as model systems. As seen from Figures 2, 3 and 4, in all cases, again, canavanine was most inhibitory, reducing (at 1mM) the length of radicles significantly by 80–94 per cent. Of the remaining compounds, the novel substance CBCH exhibited the highest activity against cress and amaranth (47 per cent and 54 per cent, respectively). CH, a relative derivative (Fig. 1), was less effective. The following sequence of inhibitory activity towards these species were found: C CBCH CH SC COM SHC CA (cress) and C CBCH CH = CA COM SHC (amaranth) However, against cabbage CH was found to be more inhibitory than CBCH (Fig .4), indicating the role of other factors for the specific action of these compounds.

Antimicrobial activity. The antimicrobial effects of various canavanine derivatives as expressed by the minimum inhibitory concentration (MIC), is illustrated in Table 2. C, CBCH, and to lower extent CH, showed activity against *Bacillus subtilis, Bac. cereus*, and *Proteus vulgaris*. C was generally more active than CBCH, but its effect was reversed by equimolar amount of arginine, while 10-fold higher molar concentrations of arginine were required to abolish the effect of CBCH. This interesting property of CBCH could be related to the presence of alkylating function in its molecule.

**Table 2. Antibacterial activity of some canavanine derivatives**

| Microorganisms | Minimal inhibitory concentration (mM) | | | | |
| --- | --- | --- | --- | --- | --- |
| | C | CH | CA | COM | CBCH |
| *Klebsiella pneum. 450* | >1.6 | n.b. | n.b. | n.b. | >1.6 |
| *Kl. pneumoniae 52940* | >1.6 | n.b. | n.b. | n.b. | >1.6 |
| *Escherichia coli DI* | >1.6 | n.b. | n.b. | n.b. | >1.6 |
| *E. coli DH-Amp* | >0.4 | n.b. | n.b. | n.b. | >1.6 |
| *E. coli 387* | >1.6 | >2.5 | >5.0 | >5.0 | >1.6 |
| *E. coli pR 55* | >1.6 | >5.0 | >5.0 | >5.0 | >1.6 |
| *Proteus vulgaris* | 0.07–0.03 | 0.25–0.13 | >5.0 | >5.0 | 0.1 |
| *Pseudomonas aeruginosa* | >1.6 | n.b. | n.b. | n.b. | 0.4 |
| *Bacillus cereus* | 0.63–0.31 | 1.8–0.63 | >5.0 | >5.0 | 0.4 |
| *B. subtilis* | 0.02–0.01 | 0.14–0.07 | n.b. | n.b. | 0.02 |
| *Sarcina lutea* | >2.5 | >2.5 | >5.0 | >5.0 | n.b. |
| *Candida albicans* | >5.0 | >5.0 | >5.0 | >5.0 | n.b. |

n.b. = not determined

## Acknowledgement

We are grateful to Mrs. Marianne Dübler for her skilful technical help.

# References

1.  Grancharov, K., Krauss, G.-J., Spassovska, N., Miersch, J., Maneva, L., Mladenova, J., & E. Golovinsky: Inhibitory effects of pyruvic acid semi- and thiosemicarbazones on the growth of bacteria, yeasts, experimental tumours, and plant cells. *Pharmazie* **40,** 574–575.

2.  Krauss, G.-J., Miersch, J., & E. Golovinsky (1994): Synthetische und natürliche Meta-bolit-Analoga als Stoffwechselinhibitoren. *Scientia halensis* **2,** 19–20.

3.  Miersch, J. (1992): Possible function of canavanine during germination of alfalfa seeds (*Medicago sativa* L.). In: *Guanidino Compounds in Biology and Medicine* eds. De Deyn, P.P. *et al.*) pp. 39–45, London: John Libbey.

4.  Miersch, J., Pajpanova, T., Grancharov, K., Krau, G.-J., Tintemann, H., & E. Golovinsky (1993): Some new canavanine derivatives: synthesis and biological activity. *Amino Acids* **5,** 126.

5.  Miersch, J., Schlee, D., and Rosche, I. (1994): Ecological significance of allelochemicals: Occurance of canavanine in Medicago sativa L. In: Allelopathy in Agriculture and Forestry (Eds. Narval, S.S. and Tauro), Chapter 9, p.153-166, Scient. Publ. Jodhpur.

6.  Miersch, J., Jühlke, C. & D. Schlee (1988): Zum Canavaninmetabolismus während der Sämlingsentwicklung von Luzerne (Medicago sativa L.). Wiss. Beiträge der Martin-Luther-Universität **33** (Sonderheft 65), 62–71.

7.  Miersch, J., Krauss, G.-J. & D. Schlee (1989): Allelochemische Wechselwirkungen zwischen höheren Pflanzen – eine kritische Wertung. Wiss. Zeitscvhr. Univ. Halle **38,** 59–74.

8.  Miersch, J., Jühlke, C., Sternkopf, G., & G.-J. Krauss (1992): Metabolism and exudation of canavanine during development of alfalfa (Medicago sativa). *J. Chem. Ecol.* **18,** 2117–2129.

9.  Pajpanova, T., Grancharov, K., Miersch, J., Tintemann, H., Krauss, G.-J., Stoev, S., & E. Golovinsky (1992): The action of some l-canavanine derivatives as metabolic inhibitors. *Compt. Rend. Bulg. Acad. Sci.* **45,** 49–52.

10. Pajpanova, T., Stoev, S., Golovinsky, E., Krauss, G.-J., and J. Miersch (1996): Canavanine derivatives useful in peptide synthesis. *Amino acids* (in press).

11. Rice, E.L. (1994): ed. In: *Allelopathy*, 2nd Ed.  New York: Academic Press 1984.

12. Rizvi, S.J.H. & V. Rizvi (1992): eds. In: *Allelopathy*: basic and aplied aspects. London: Chapman and Hall.

13. Rosenthal, G. (1988): The protective action of a higher plant toxic product. *Bioscience* **38,** 104–108.

14. Schlee, D., Miersch, J., Krauss, G.-J.& Ch. Müller-Uri (1989): Allelopathie-Chemische Wechselwirkungen zwischen höheren Pflanzen. Bibliographie, Universitäts-und Landesbibliothek Halle, 255 S.

15. Waller, G.R. (1987): In: *Allelochemicals*: role in agriculture and forestry. ACS Symp. Series 330, eds. Washington: American Chemical Society.